ORCHID MONOGRAPHS
VOLUME 2

Orchid Monographs

VOLUME 2

A TAXONOMIC REVISION OF THE CONTINENTAL AFRICAN BULBOPHYLLINAE

J. J. VERMEULEN

Drawings by the author

E. J. BRILL
LEIDEN • NEW YORK • KØBENHAVN • KÖLN
1987

Published and distributed for the Rijksherbarium, State University Leiden,
by E. J. Brill, Leiden.

Library of Congress Cataloging-in Publication Data

Vermeulen, J. J., 1955-
A taxonomic revision of the continental African
Bulbophyllinae.

(Orchid monographs, ISSN 0920-1998 ; v. 2)
Bibliography: p.
Includes index.
1. Bulbophyllum—Africa—Classification.
2. Orchids—Africa—Classification. 3. Botany—Africa—
Classification. I. Title. II. Title: Bulbophyllinae.
III. Series
QK495.064V395 1987 584'.15 87-10372
ISBN 90-04-08458-4 (pbk.)

ISSN 0920-1998
ISBN 90 04 08458 4

PRINTED IN THE NETHERLANDS

This revision is dedicated to

MR. P.S.J. GROOT

Borculo, The Netherlands
because he convinced me
now some ten years ago
that *Bulbophyllum* is at least as interesting
a genus to study as *Pleurothallis*

CONTENTS

A TAXONOMIC REVISION OF THE CONTINENTAL AFRICAN BULBOPHYLLINAE

J.J. VERMEULEN

Rijksherbarium, Leiden, The Netherlands
Drawings by the author

SUMMARY

This paper contains a taxonomic revision of the *Bulbophyllinae*, with the genera *Bulbophyllum* (including *Cirrhopetalum* and *Megaclinium*) and *Chaseella*, from continental Africa (including Bioko (Fernando Poo), Sao Tomé, Principe, Annobon and Zanzibar). Keys are given to the genera and species. For each species full synonymy, descriptions, notes on distribution, habitat etc. and a line drawing are presented. Many species are also illustrated by colour photographs. Three new taxa are described: *Bulbophyllum subligaculiferum*, *B. bidenticulatum* ssp. *joyceae* and *B. scaberulum* var. *crotalicaudatum*. Some new combinations are made. The botanical terminology used is clarified.

INTRODUCTION

The largest genus of the pantropical subtribe *Bulbophyllinae* (sensu Dressler, 1981) is the genus *Bulbophyllum* with an estimated number of 1000–1200 species (c. 2200 described). The status of most of the other genera of the subtribe (some 10, totaling less than 100 species), is a matter upon which there is little agreement among botanists: some prefer to recognize them whereas others merely see them as infrageneric taxa within *Bulbophyllum*.

In this paper a revision is presented of the continental African species of the genus. Continental Africa here includes Bioko (Fernando Poo), Sao Tomé, Principe and Pagulu (Annobon), as well as Zanzibar. The orchid flora of these islands is essentially the same as that of the African continent. Excluded, however, are the Seychelles, the Comores, Madagascar, Réunion, Mauritius and the smaller islands in this area. The orchid flora of these islands is endemic and only shows a remote similarity with continental Africa.

In addition to *Bulbophyllum*, representatives of three other genera of the *Bulbophyllinae* occur in continental Africa: *Chaseella*, *Cirrhopetalum* and *Megaclinium*. In the present revision only *Chaseella* is maintained as a genus (1 species). *Cirrhopetalum* (1 species in Africa) and *Megaclinium* (18 species) have been included in *Bulbophyllum* (68 species) as infrageneric groups.

For the convenience of the user three figures and a glossary have been added explaining the botanical terminology used.

THE GENUS BULBOPHYLLUM

The oldest name for the genus now known as *Bulbophyllum* is *Phyllorkis* from Du Petit Thouars (1809). However, the later name *Bulbophyllum*, also from Du Petit Thouars (1822), was conserved because it became widely accepted. In his table to the genera and species of orchids, Du Petit Thouars mentioned it in the same column as other now accepted names such as *Cymbidium* and *Epidendrum*. The name *Phyllorkis* is mentioned in another column, apparently reserved for artificially made French names.

Until the beginning of the 20th century the number of species described from Africa remained rather limited: Rolfe (1897), in the Flora of Tropical Africa mentioned 61 species (under *Bulbophyllum* and *Megaclinium*). In the first decades of this century, however, the genus became the happy hunting ground of authors such as Kränzlin, Schlechter and De Wildeman, each describing large numbers of new species.

De Wildeman (1921), enumerated 165 species. After that the first local revision was executed by Summerhayes (1936), for the first edition of the Flora of West Tropical Africa. He was the first person who rightfully reduced a number of species to synonyms. This tendency towards a somewhat more practical and useful way of grouping was continued in his revision for the second edition of the same flora (1968), and by the recently published revision for the flora of Tropical East Africa by Cribb (1985). Other floristic work on African *Bulbophyllum* has hardly been done; the volumes of other large African flora projects (Flore du Cameroun, Flore du Gabon, Flore d'Afrique Centrale, Flora Zambeziaca) covering the genus have not yet been published. A revision of *Bulbophyllum* for the Flore d'Afrique Centrale has been prepared by the present author and will be published soon.

Altogether, about 300 *Bulbophyllum* species have been described from Continental Africa. In the present revision 92 taxa have been recognized, 68 as species, 24 as subspecies and varieties. Thirteen species and 3 infrageneric taxa proved to be new. Most of these taxa have already been described in precursory papers (Vermeulen 1984a and b, 1986a and b), only one species and two infrageneric taxa are described new in the present paper.

THE STATUS OF MEGACLINIUM AND CIRRHOPETALUM

The genus *Megaclinium* was established by Lindley (1824). Although Reichenbach f., as early as 1861, considered it merely a section of *Bulbophyllum*, most 19th century authors followed Lindley. As in *Bulbophyllum*, the number of species described increased rapidly in the first decades of the 20th century and the discussions as to the status of *Megaclinium* gained impetus. Authors as Pfitzer and Kränzlin (in his later publications) still stuck to Lindley's point of view (Kränzlin even published a division into 3 sections in 1923). Others, however, became progressively convinced of the close similarity between *Megaclinium* and *Bulbophyllum* in spite of the singular shape of the rhachis in many species of the former genus. Schlechter (1915) gave short comments. De Wildeman (1921) kept *Megaclinium* only as a subgenus but never expressed himself on this matter (he made combinations in both genera for the new species he described). Summerhayes (1935) considered *Megaclinium* as a section of *Bulbophyllum*.

In my opinion *Megaclinium* cannot be kept as a genus but it is not yet possible to say in which infrageneric rank it will end up; this depends on the infrageneric division of *Bulbophyllum* as a whole.

First of all it must be stated that *Megaclinium* is probably a natural group (see the description of Group 5, below). However, giving a definition of it, and of most of the other infrageneric groups ever recognized in *Bulbophyllum*, is very difficult. Often not a single character can be found which is shared by all species of a group; a group can only be defined by a combination of characters, each of which occurs in most, but not in all its species. Moreover, each character may separately occur in other species elsewhere.

As a consequence, a species can already be included in a group as soon as it shows most, but not all characters of that group.

When keeping *Megaclinium* as a genus, other groups defined by comparable combinations of characters, but often much less apparently natural should be raised to genus level as well. This would result in an endless row of small and ill defined genera. The recently performed studies on *Dendrobiinae* by Brieger (1981) are deterrent examples of this procedure.

During its life span of some 160 years the generic name *Cirrhopetalum* has undergone a fate similar to that of *Megaclinium*. It was created by Lindley in 1824, and formally merged into *Bulbophyllum* in 1861 by Reichenbach f. The discussions about its status have lasted until the present day.

Arguments to keep *Cirrhopetalum* within *Bulbophyllum* are given by J.J. Smith (1912) and Seidenfaden (1973). In the present paper *Cirrhopetalum* is included in *Bulbophyllum* for reasons which come down to the same as those for *Megaclinium*.

Since only one species of *Cirrhopetalum* occurs among the numerous *Bulbophyllum* species from Africa, I feel that the discussion should not be continued here.

THE PREVIOUS INFRAGENERIC DIVISIONS OF BULBOPHYLLUM

With the ever increasing number of species that became known to science, division of *Bulbophyllum* into subgenera and sections became necessary and much work has been done on this matter.

For the Madagascar species a framework was made by Schlechter (1925). It was later refined by Perrier (1939) in his account for the Flore de Madagascar (1939). Later authors invariably used Perrier's division to fit in their new species.

For Asia sections were described by several authors, often based on names of genera now included in *Bulbophyllum*. The only comprehensive survey was again by Schlechter (1912-13) for the numerous species of New Guinea. In spite of its many shortcomings it appears that it can be successfully applied as a frame for Asiatic species outside New Guinea as well.

For South America a reasonable division into sections was made by Cogniaux (1902).

Viewed in this light it is surprising that, apart from the segregation of *Megaclinium*, the division of continental African *Bulbophyllum* has never got much beyond De Wildeman's (1921) establishment of the subgenus *Eubulbophyllum* for everything that is not *Megaclinium*. Only very few other names at section level have been proposed, most of them only with the mentioning of a type species and without any remarks about the distinction between sections.

INFRAGENERIC DIVISION OF CONTINENTAL AFRICAN BULBOPHYLLUM

I am of the opinion that an infrageneric division which will survive inclusion of new information without the need for fundamental changes, will have to embrace the genus as a whole. No local classification, however useful it may be for identifying species, meets or can possibly meet this requirement. The *Bulbophyllum* species occurring in any restricted area are never so much isolated that morphological similarities with species outside that area can be ignored, as has been done in the past.

Considering that the present state of knowledge is insufficient for a worldwide approach I prefer to abstain from introducing a number of official section names for the groups of African *Bulbophyllum* distinguished below. This division largely follows my earlier division of continental African *Bulbophyllum* (1982) except for the fact that many groups recognized there are lumped here.

The groups have been assembled as follows: small numbers of similar looking species have served as cores. Less similar species were added to these cores, often blurring the distinctness between the cores and causing them to melt together.

This was continued until the groups obtained reached a level comparable to the level of *Megaclinium* as it was circumscribed by previous authors. Thus an infrageneric division was obtained which does not deviate more than necessary from widely accepted concepts.

In this revision the species within one group were arranged so that those included in one core (which show a distinct overall similarity) were placed together. The sequence in which the cores are placed was also defined by similarity, and by species or series of species which constitute links between cores because they are intermediate.

Doing so a sequence was developed which, as far as possible, places the most similar species closest together. I feel this is the most convenient for the user: it facilitates comparison of similar looking (often 'difficult') species.

The numbered groups are arranged arbitrarily. The group which includes the type species of the genus (from Madagascar) is placed first.

It is emphasized that by no means phylogenetic relationships are suggested by the order of either the species or the groups.

On specific level the continental African *Bulbophyllum* species are all endemic except for four: *B. elliotii* (Continental E. Africa and Madagascar); *B. humblotii* (Continental E. Africa, Seychelles, Madagascar); *B. intertextum* (Continental Africa, Seychelles) and *B. longiflorum* (Continental E. Africa, Reunion, Mauritius and SE Asia).

This high rate of endemism justifies revision of the continental African *Bulbophyllum* species as a whole.

In the continental African *Bulbophyllum* species the following groups are distinguished in this revision:

Group 1 (spp. 1–30). - This group is placed first because, throughout the range of the genus *Bulbophyllum*, numerous species occur which are very similar to species of this group. The continental African species can only be defined as a rest group after definition of all other groups. As a consequence, the discussion of this group has been inserted after group 11.

Group 2 (sp. 31). - Monotypic. This group is defined by the combination of characters: pseudobulbs 2-leafed; flowers scattered; lip with glabrous margins; protrusion of the anther not dorsoventrally flattened. Species showing the combination of the first two characters occur on Madagascar in the section *Ploiarum* Schlechter (1916), *Alcistachys* Schlechter (1925) and *Lemuraea* Schlechter (1925, = *Loxosepalum-diphylli* Perrier, 1951). Otherwise the groups mentioned show little similarity.

Group 3 (sp. 32). - Monotypic. This group is defined by the combination of characters: rhachis spindle shaped, sepals with long hairs abaxially. The only other species with a similarly swollen rhachis is *B. bisetum* Lindl. from continental SE Asia, which has very different flowers.

Group 4 (spp. 33–35). - The species of this group are defined by the combination of characters: scales on rhizome and inflorescence only tubular at their very base; lateral sepals reflexed at their very base and appressed against the rhachis; anther without protrusion overtopping the front margin. The first character occurs also in *B. scariosum* and *B. pandanetorum*, included in group 1. The combination of characters, however, is unique among African *Bulbophyllum*. Elsewhere no comparable species occur. On account of a distinct overall similarity of the species this group is considered natural.

Group 5 (spp. 36–53; '*Megaclinium*'). - The species of this group show at least one of the following characters: rhachis flattened with the flowers inserted along the midnerve; lip very soft and easy to compress. Many species of this group show both characters, some, however, only one. The character of the rhachis is unique among *Bulbophyllum*, but does not occur in all species of this group. Species with equally fleshy and very soft lips (but with a much more complicated structure) occur in New Guinea, in the section *Hapalochilus* Schlechter (1912-13). No species comparable to those of group 5 occur elsewhere.

In spite of its unsatisfactory definition the group is considered natural on account of the character of the rhachis, which is unique to it, and because of an overall similarity in the structure of the flowers of all species.

Group 6 (sp. 54). - Monotypic. This group is defined by the combination of characters: stelidia sharply acute to acuminate; anther slightly dorsoventrally flattened. The second character connects this group to group 7, 8 and 9, but it cannot be included in anyone of these groups due to a lack of overall similarity, especially in the shape of the lip and the column. Elsewhere no comparable species occur.

Group 7 (spp. 55–56). - The species of this group are defined by the following combination of characters: flowers not much open; sepals fleshy; petals glabrous; lip very fleshy, margins of the lip with seamlike, thin lateral lobes with long, multicellular hairs; column with rather long, sharply acute stelidia; anther with a dorsoventrally flattened or with a bifid protrusion in front. Similar species occur on Madagascar (*B. bicoloratum* Schltr, *B. cirrhoglossum* Perr., *B. occultum* Thou., *B. pentasticha* Pfitz., *B. rubrum* Jum. & Perr., *B. sambiranense* Jum. & Perr. and *B. trifarium* Rolfe). Together they undoubtedly constitute a natural group; the species show a distinct overall similarity in the structure of the flowers.

Group 8 (spp. 57–59). - The species of this group are defined by the combination of characters: floral bracts 1.5 times as long as the flowers or longer (pedicel and ovary included); margins of the lip with seamlike, thin lateral lobes with denticulate margins; column with short, shortly acute stelidia. It differs mainly from group 7 in the second and the third character. No comparable species occur elsewhere. The group is very homogeneous and undoubtedly natural.

Group 9 (spp. 60–66). - The species of this group are defined by the combination of characters: floral bracts as long as the flowers or shorter (pedicel and ovary included); lip without seamlike lateral lobes; column with short, obtuse to shortly acute stelidia. No comparable species occur elsewhere. On account of a distinct overall similarity of the constituent species this group is considered natural. The most obvious differences between the species are mainly in the structure of the lip.

Group 10 (sp. 67). - This group comprises an oddity among continental African *Bulbophyllum*, defined by the combination of characters: leaves colliculate adaxially, jugae ending in distinct teeth, column extremely elongated. Although the characters mentioned occur separately in other species of *Bulbophyllum*, the combination is unique.

Group 11 (sp. 68; '*Cirrhopetalum*'). - The only African species of this group is defined by the combination of characters: inflorescence subumbellate; lateral sepals connate along the upper margins. Numerous species comparable to the African one occur in SE. Asia.

As stated above, group 1 cannot be defined in a positive way. It is also very difficult to split it up into smaller groups which are easier to define. The cores constituting this group show a wide range of morphological alliances with species outside continental Africa.

B. schinzianum, *B. subligaculiferum*, *B. saltatorium*, *B. barbigerum*, *B. vulcanicum*, *B. cochleatum*, all 1- or 2-leafed species with much specialized flowers are connected to group 1 through species as *B. burttii* and *B. kivuense* and should therefore be included. *B. penicillium* Par. & Rchb.f. from Bhutan and Burma is very similar to *B.*

schinzianum and *B. saltatorium* but differs in having almost orbicular, ciliate petals.

B. inornatum, *B. bavonis*, *B. stolzii* and *B. prorepens* are mainly distinguished by having 2-leafed pseudobulbs. Their flowers are relatively undifferentiated and similar to those of many other species of group 1. Comparable species occur on Madagascar in the sections *Lemuraea* Schlechter (1925, = *Loxosepalum-diphylli* Perrier, 1951), e.g. the lectotype species of the genus: *B. nutans* Thou.

Species comparable to *B. intertextum*, *B. vanum*, *B. capituliflorum*, *B. pipio*, *B. pumilum*, *B. coriscense* and *B. cocoinum* occur on Madagascar in the sections *Trichopus* Schlechter (1925) and *Pantoblepharon* l.c. Remarkably similar species occur also in SE. Asia, e.g. *B. limbatum* Lindl., *B. nigrescens* Rolfe, *B. parviflorum* Par. & Rchb.f., *B. secundum* Hook.f. and *B. thomsoni* Hook.f.

B. josephii, *B. calvum*, *B. expallidum*, *B. mediocre* and *B. schimperanum* are species with 1-leafed pseudobulbs and rather undifferentiated flowers. Species with strikingly similar flowers can be found throughout the range of the genus: sect. *Loxosepalum* Schlechter (1925, e.g. *B. baronii* Ridl., *B. leptostachyum* Schltr) in Madagascar; species as *B. reptans* (Lindl.) Lindl. in India; sect. *Racemobulbum* J.J. Smith (1914, e.g. *B. compressum* T. & B.) in Indonesia; species of sect. *Ischnopus* Schlechter (1912-13, e.g. *B. gracilliscapum* Schltr) in New Guinea; and finally sect. *Micrantha* Cogniaux (1902, e.g. *B. micranthum* Barb. Rodr.) in S. America. In Africa as well as elsewhere numerous species can be found which are similar but which have developed one or more unique characters, especially in the shape of the flowers (In continental Africa: *B. ivorense*, *B. oxychilum* and *B. nigritianum*).

B. scariosum and *B. pandanetorum* are distinguished by their rhizome and peduncle scales which are only tubular at their very base, 2-leafed pseudobulbs and distichous, large flowers. Species with similar rhizome scales are found in group 4; species with 2-leafed pseudobulbs and equally large flowers of very similar shape (but not distichous, however) occur on Madagascar in the section *Pachychlamys* Schlechter (1925, e.g. *B. molossus* Rchb.f.).

Inside continental Africa as well as elsewhere most of the species of group 1 mentioned above are morphologically connected by intermediate species or a series of species. Due to its heterogeneity it is impossible to say whether group 1 is natural or not.

THE GENUS CHASEELLA

The monotypic genus *Chaseella* was established by Summerhayes (1961) to accommodate a species from E. Africa which has a single unique character compared to *Bulbophyllum*: it has 6–12-leafed pseudobulbs with very narrow leaves. Another character ascribed to it by Summerhayes (only two pollinia present instead of 4) does not hold; many *Bulbophyllum* species also have 2 pollinia. Except for the character of the leaf number the species could be included in *Bulbophyllum*.

Nevertheless, it has been kept in a separate genus because the character '6–12-pseudobulbs' is alien to *Bulbophyllum*. Inclusion in that genus would rob *Bulbophyllum* of one of its most outstanding features: '1- or 2-leafed pseudobulbs'.

A comparable situation occurs in the S. American *Maxillariinae*: whereas most genera of this subtribe have 1 to few leafed pseudobulbs, the oligotypical genus *Pityphyllum*

Schlechter (1920, p. 126) has equally many-leafed pseudobulbs with very narrow leaves.

THE USE OF INFRASPECIFIC TAXONOMICAL RANKS

In this revision the subspecies concept has been applied next to the variety concept. Entities within a species are ranked as subspecies if a geographical isolation between them could be proved on account of the (herbarium) material available to the author. In its main lines this use corresponds with v. Steenis (1957, p. CXC and CCXXV, sub 16).

Sympatric infraspecific entities may well show e.g. ecological isolation, or isolation caused by having different species of pollinators. Because this is impossible to prove without detailed observation in the field, the variety concept has provisionally been applied in these cases.

No formal status has been given in case that infraspecific entities merely constitute a 'trend' in the variability of a species: a number of characteristical specimens linked to other entities by a series of intermediates. These entities have only been indicated with an a., b., etc.

NOTES TO THE USERS OF THIS REVISION

Botanical terminology is often ambiguous, which may seriously impede correct use of keys and descriptions in a taxonomic revision. Moreover, users may not be entirely familiar with the terminology used. To overcome this problem figs. A, B and C are added for explanation, as well as a short glossary.

In general the terminology used in this paper is based on Daydon Jackson's 'Glossary of Botanic terms', ed. 1971.

abaxially: The side or face of an organ turned away from the axis of the compound the organ is part of. See also: *adaxially*.

adaxially: The side or face of an organ turned to the axis of the compound the organ is part of. In contrast with: *abaxially*. In this context an organ is e.g. a petal; that petal is part of the compound: flower. The definition breaks down with *B. longiflorum*: the lateral sepals of that species are twisted so that the abaxial side is turned towards the axis of the flower. In this case the lateral sepals are stretched before indicating the adaxial and the abaxial side.

back margin: Margin of lateral lobe facing towards the base of the organ it is part of (e.g. the lip). See fig. C.

central: Relating to the central part of a body. See fig. C.

distichous: Arranged in two rows (e.g. the flowers along the rhachis).

entire: without incisions in the margin, with even margin. In contrast with: *dentate* (beset with teeth), etc.

front margin: margin of lateral lobes facing towards the top of the organ it is part of (e.g. the lip). See fig. C.

glabrous: Margin or surface without pubescence or papillae, smooth. In contrast with: *papillose* (with papillae) etc.

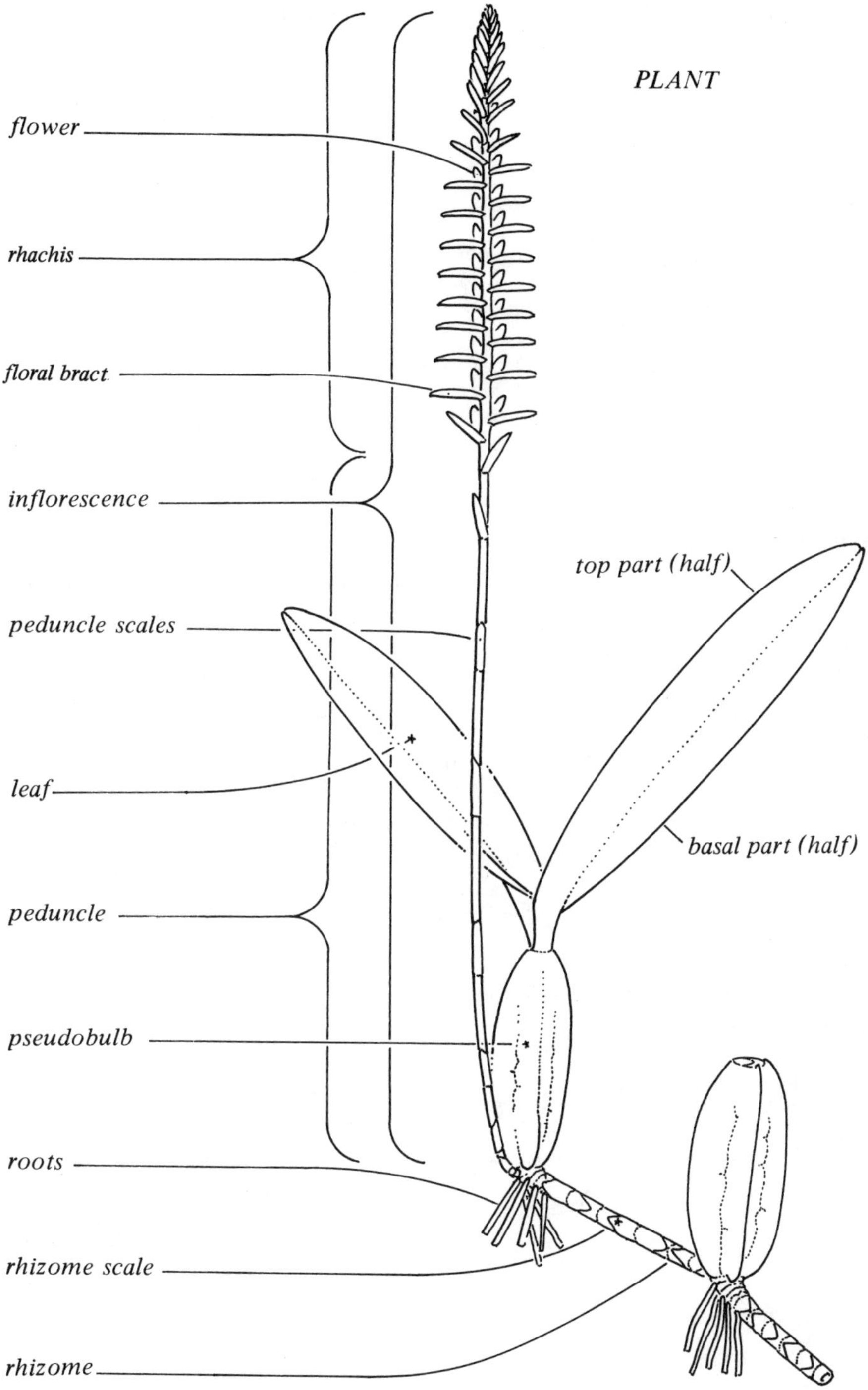

Fig. A. Terminology.

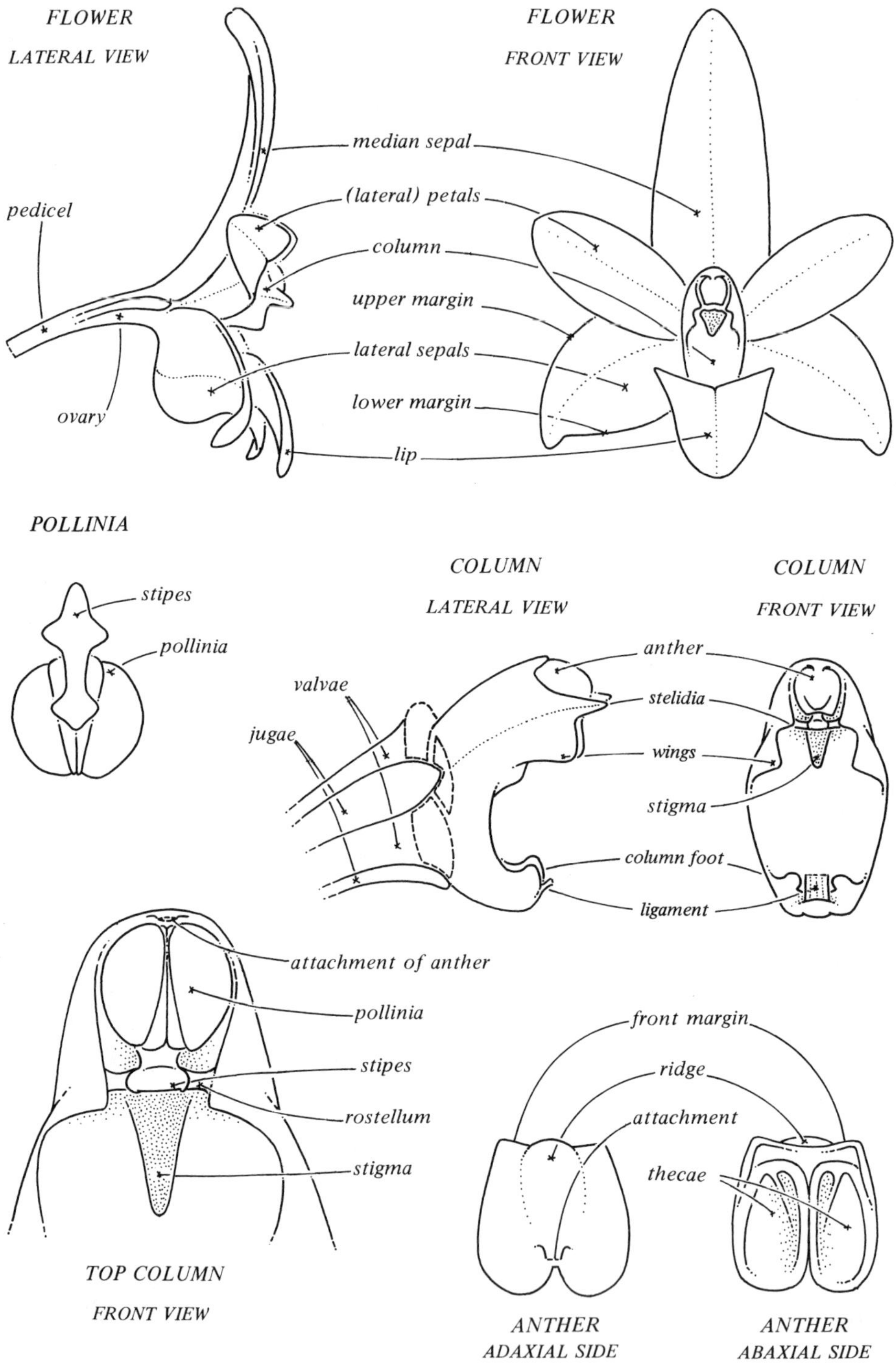

Fig. B. Terminology.

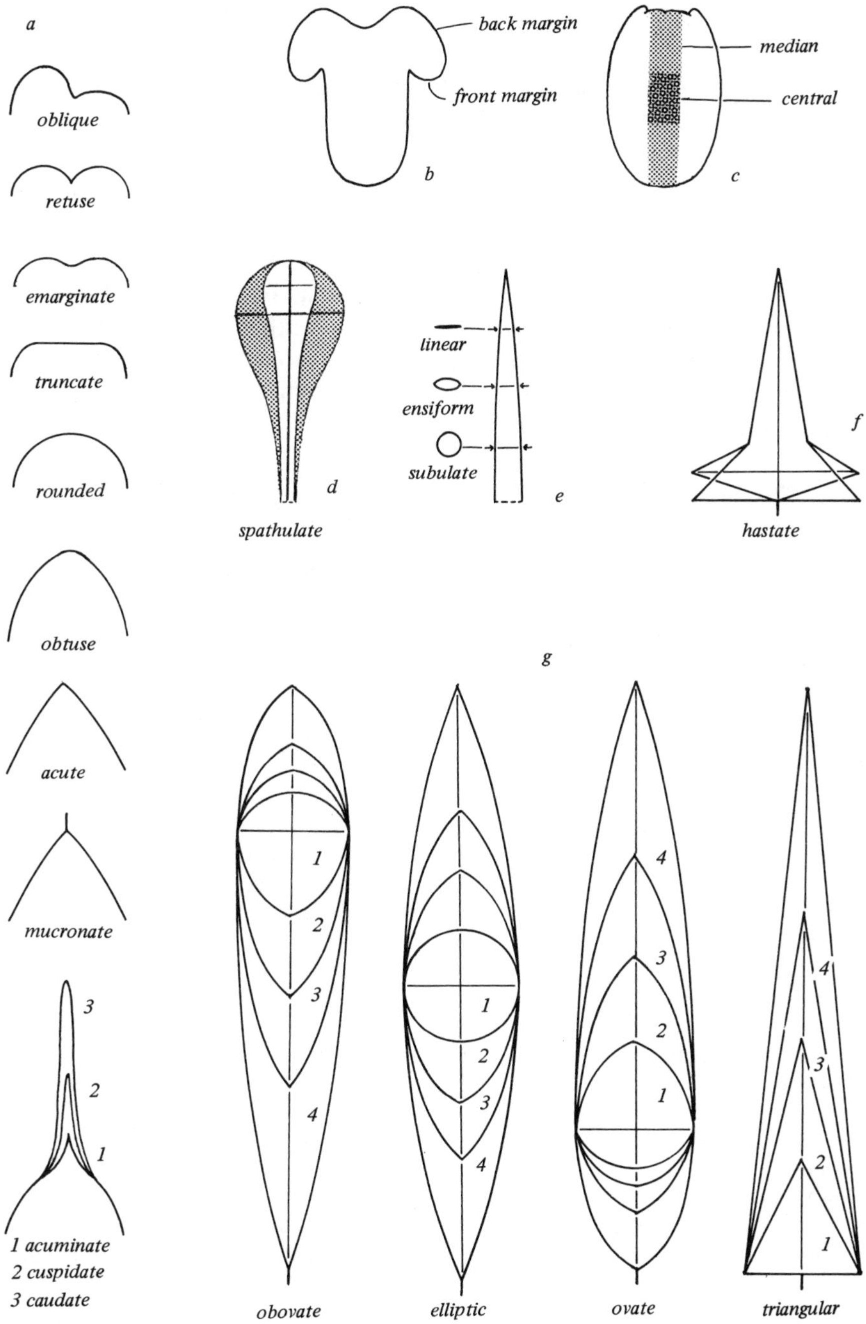

Fig. C. Terminology. a - Tips of leaves and floral parts. b - Front margin and back margin of the lateral lobes of the lip. c - Central/median part of a floral part. d - Spathulate. e - Cross sections through leaf or floral part: linear, ensiform, subulate. f - Hastate. g - Shape of leaf and floral part: *Obovate series*: 1 - broadly obovate (index length/width c. 1), 2 - obovate (index 1–2), 3 - obovate-oblong (index 2–3), 4 obovate-lanceolate (index 3–5); *elliptic series*: 1 - orbicular (index c. 1), 2 - elliptic (index 1–2), 3 - oblong (index 2–3), 4 - lanceolate (index 3–5); *ovate series*: 1 - broadly ovate (index c. 1), 2 - ovate (index 1–2), 3 - ovate-oblong (index 2–3), 4 - ovate lanceolate (index 3–5); *triangular series*: 1 - deltoid (index c. 1), 2 - triangular (index 1–2), 3 narrowly triangular (index 2–3), 4 - very narrowly triangular (index 3–5).

median: Relating to the imaginary line that divides a body into two (often symmetric) parts. See fig. C.
scattered: Arranged without apparent order (e.g. the flowers along the rhachis).
secund: Turned to one side (e.g. the flowers along the rhachis).

To indicate the range of the species the following geographical delimitations have been applied: For the geographical division within countries the large floras have been followed (the Flore d'Afrique Centrale for Zaire, Rwanda and Burundi; the Flora of Tropical East Africa for Uganda, Kenya and Tanzania and the Flora Zambeziaca for Zambia, Malawi, Zimbabwe and Mozambique). For other countries the official national subdivision has been followed in most cases; occasionally unofficial terms as 'Coastal part' or 'Northern part' have been applied. The countries are indicated with the following abbreviations:

AETH.: Aethiopia.
ANG.: Angola.
BUR.: Burundi.
CAM.: Cameroun.
C.A.R.: Central African Republic.
CONGO: Congo (French Congo).
EQ. GUIN.: Equatorial Guinea.
GAB.: Gabon.
GHANA: Ghana.
GUIN.: Guinea (French Guinea).
IV.C.: Cote d'Ivoire.
KEN.: Kenya.
  K1: Northern Frontier Province.
  K2: Turkana Province.
  K3: Rift Valley Province.
  K4: Central Province.
  K5: Nyanza Province.
  K6: Masai Province.
  K7: Coast Province.
LIB.: Liberia.
MAL.: Malawi (Nyasaland).
MOZ.: Mozambique.
NIG.: Nigeria.
R.S.A.: South African Republic.
RW.: Rwanda.
S.L.: Sierra Leone.
S.T. & PR.: Sao Tomé & Principe.
SUDAN: Sudan.
TANZ.: Tanzania (Tanganyika Terr.).
  T1: Lake Province.
  T2: Northern Province.
  T3: Tanga Province.
  T4: Western Province.
  T5: Central Province.
  T6: Eastern Province.
  T7: Southern Highlands Province.
  T8: Southern Province.
TOGO: Togo.
UG.: Uganda.
  U1: Northern Province.
  U2: Western Province.
  U3: Eastern Province.
  U4: Buganda Province.
ZAI.: Zaire (Belgian Congo).
  Z1: Côtier.
  Z2: Mayombe.
  Z3: Bas-Congo.
  Z4: Kasai.
  Z5: Bas Katanga.
  Z6: Forestier Central.
  Z7: Ubangi-Uele.
  Z8: Lac Albert.
  Z9: Lacs Edouard et Kivu.
  Z10: Rwanda-Burundi Z11: Haut-Katanga.
ZAM.: Zambia (Northern Rhodesia).
ZIM.: Zimbabwe (Southern Rhodesia).

The number of specimens seen from each geographical unit mentioned is given between brackets.

THE RELEVANCE OF LIVING ORCHIDS IN BOTANICAL GARDENS AND PRIVATE ORCHID COLLECTIONS

Several botanical gardens and private orchid growers have played an important role in the realization of this revision. Especially in *Orchidaceae* living material is indispensable for the taxonomist since it allows him to check morphological interpretations based on herbarium specimens, which often make up the bulk of the material at his disposal. Living plants are also very useful when drawings are prepared; dried flowers often need extensive treatment (boiling in water or diluted ammonia, reconstruction) before they are suitable for this purpose, not to mention the numerous species which can never be forced back into their natural shape due to the diaphanous structure of their flowers.

The botanical gardens of Leiden and Wageningen (The Netherlands) have contributed to our knowledge of *Bulbophyllum* in another, very important way. It is common knowledge that less than 10 per cent of the orchid species present in tropical forests are in flower simultaneously. It is therefore impossible for a botanist, who generally visits an area only for a short period, to gain an impression of the orchid flora actually present.

A collecting method amongst others practiced by Seidenfaden in Thailand overcomes this problem: portions of not flowering plants of apparently different species are collected, sent to a botanical garden, grown until flowering and then partly conserved or dried. In this way a much more complete picture is achieved of the orchid flora of an area.

This strategy has been successfully applied during the collecting trips to Gabon, organized by the Laboratory of Plant Systematics in Wageningen in 1983 and 1985. The collected orchids (among which many *Bulbophyllum* for the present revision) have been distributed among the botanical gardens of Wageningen and Leiden. In this way an extensive collection has been built up from an area which was until then only covered by the collecting work of very few botanists. Among the specimens, otherwise undoubtedly passed by unnoticed as just another lump of leathery leaves on a branch of a tree, were some new species and several new records for the area. Gabon can now be counted among the African countries rather well explored with this respect.

ACKNOWLEDGMENTS

I wish to thank all staff of the Rijksherbarium in Leiden for their hospitality during the past years, especially my colleague Dr. Ed de Vogel, for his always pleasant, stimulating and critical cooperation. Dr. C. Kalkman and Dr. G.J.C.M. van Vliet critically read the manuscript.

I am much obliged to the keepers of the following herbaria for putting their material at my disposal: AMES, BM*, BR*, C, E, G, H, K*, L*, NY, P*, S, U, UPS, W, WAG, Z. The herbaria marked with an * have been visited during my investigations. Dr. P.J. Cribb should be especially mentioned here for his kind criticism of my work.

The role that the botanical gardens of Leiden and Wageningen, and private orchid growers (Mr. P.S.J. Groot, Borculo; Mr. P. Jongejan, Amersfoort; Mr. J. Lodewijk, Haarlem) have played has been explained above. Mr. F.M. van der Laan, keeper of the orchid collection in Wageningen should be mentioned here.

Last but not least I wish to mention Mr. P.T. Allersma for scanning the text for slips of the pen and Miss Jolanda Huisman for general support in many ways.

REFERENCES

Bentham, G. & Hooker, J.D. 1880. *Bulbophyllum & Megaclinium*. Genera Plantarum 3: 501–505. London.

Brieger, F.G. 1981. *Dendrobiinae*. In: Schlechter, Die Orchideen, ed. 2: 636–752. Berlin.

Cogniaux, A. 1902. *Bulbophyllum*. In: Martius, Flora Brasiliensis 3, pars 5: 589–634. Monaco.

Cribb, P.J. 1985. *Bulbophyllum*. In: Polhill, R.M., Flora of Tropical East Africa, Orchidaceae, pars 2: 237–305. Rotterdam. (F.T.E.A.)

Daydon Jackson, B. 1971. A glossary of botanic terms, ed. 4. Hafner, N.Y.

Dressler, R.L. 1981. The Orchids: 231. Harv. Un. Press.

Kränzlin, F. 1923. *Orchidaceae*. In: Schinz, H. Beiträge zur Kenntnis der Afrikanischen Flora 31. Vierteljahrresschr. Naturf. Gesellsch. Zürich, 68: 420–425.

Lindley, J. 1824. *Brassia caudata* Lindl. Bot. Reg. 10: pl. 832.

Perrier de la Bathie, H. 1939. *Cirrhopetalum* and *Bulbophyllum*. In: Flore de Madagascar, 49e fam., 1: 308–477.

Perrier de la Bathie, H. 1951. Orchidées de Madagascar et des Comores. Nouvelles observations. Not. Syst. 14, 2: 138–165.

Petit Thouars, Du, A. 1809. Nouv. Bull. Sci. Soc. Philom. 1: 319.

Petit Thouars, Du, A. 1822. Histoire particulière des Plantes... etc. Paris (In this revision mentioned as: Orch. Iles Austr. Afr.).

Reichenbach, H.C. 1861. *Bulbophyllaria* to *Bulbophyllum*. Walpers Annales 6: 241–265.

Rolfe, R.A. 1897. *Bulbophyllum* and *Megaclinium*. In: Thiselton-Dyer, W.T., Flora of Tropical Africa 7: 22–44. London. (F.T.A.)

Schlechter, R. 1912-13. Die Orchideen von Deutsch Neu Guinea. Fedde, Rep., Beih. 1: 693–888.

Schlechter, R. 1915. Die Orchideen. Berlin.

Schlechter, R. 1916. Orchidaceae Perrierianae. Beih. Bot. Centralbl. 34, 2: 294–341.

Schlechter, R. 1920. Die Orchideenfloren der Süd Amerikanischen Kordillerenstaaten. Fedde, Rep., Beih. 7: 1–301.

Schlechter, R. 1925. Orchidaceae Perrierianae. Fedde, Rep., Beih. 33: 1–391.

Seidenfaden, G. 1973. Notes on *Cirrhopetalum* Lindl. Dansk Bot. Ark. 29: 1–260.

Seidenfaden, G. 1979. Orchid genera in Thailand VIII: *Bulbophyllum* Thou. Dansk Bot. Ark. 33, 3: 9–228.

Smith, J.J. 1912. *Bulbophyllum* Thou. sect. *Cirrhopetalum*. Bull. J. Bot. Buit. ser. 2, 8: 19–29.

Smith, J.J. 1914. Neue Orchideen des Malayischen Archipels 7. Bull. J. Bot. Buit. ser. 2, 13: 1–52.

Smith, J.J. 1927. *Orchidaceae*. Mitt. Bot. Inst. Hamb. 7, 1: 9–75.

Steenis, C.G.G.J. van., 1957. Specific and infraspecific delimitation. Flora Malesiana, ser. 1, vol. 5, 3: CLXVII–CCXXXIV.

Summerhayes, V.S. 1935. African Orchids 7. Kew Bull. 1935: 195–204.

Summerhayes, V.S. 1936. *Bulbophyllum*. In: Hutchinson, J. & Dalziel, J.M., Flora of West Tropical Africa, ed. 1, 2: 432–440. London. (F.W.T.A.).

Summerhayes, V.S. 1961. A new genus of Orchids from Southern Rhodesia. Kirkia 1, 1960-61: 88–89.

Summerhayes, V.S. 1968. *Bulbophyllum*. In: Hutchinson, J. & Dalziel, J.M., Flora of West Tropical Africa, ed. 2, 3: 226–242. London. (F.W.T.A.).

Vermeulen, J.J. 1982. A provisional checklist of the continental African taxa of the genus *Bulbophyllum* Thou. (Orchidaceae). Selbyana 7: 10–19.

Vermeulen, J.J. 1984a. Five new species and one new variety of continental African *Bulbophyllum* (Orchidaceae). Bull. J. Bot. Nat. Belg. 54: 137–150.

Vermeulen, J.J. 1984b. A new species of *Bulbophyllum* (Orchidaceae) from E. Africa. Blumea 29: 589–592.

Vermeulen, J.J. 1986a: New species and new combinations in central African *Bulbophyllum* (Orchidaceae). Bull. J. Bot. Nat. Belg. 56: 229–240.

Vermeulen, J.J. 1986b: Two new species of *Bulbophyllum* (Orchidaceae) in Africa. Kew Bull. 42: 263–266.

Wildeman, De, E. 1921. *Bulbophyllum*. Plantae Bequaertianae 1: 53–122.

BULBOPHYLLINAE Pfitzer

(All descriptions below include continental African material only).

Subtribus *Bulbophyllinae* Pfitz. in Engl. & Prantl, Nat. Pflanzenfam., Teil 2, Abt. 6 (1889) 176; Dressler, The Orchids (1981) 231.–Type genus: *Bulbophyllum* Thou.

Herbs sympodial, heteranthous. *Roots* glabrous to villous, usually sprouting from below the pseudobulb (in *B. longiflorum* everywhere along the rhizome). *Rhizome* creeping, consisting of the entire stem of each shoot except for the last internode, with few to many nodes, with a scale arising from each node. *Young shoots* lateral, usually sprouting from an internode close to the base of the pseudobulb. *Rhizome scales* distichous, inarticulate, usually tubular (not in *B. scariosum*, *B. pandanetorum*, *B. bidenticulatum*, *B. nigericum* and *B. bifarium*), with a triangular top part, usually more or less glabrous (papillose in *B. acutebracteatum* var. *rubrobrunneopapillosum*). *Pseudobulbs* 1–2 (– 3)-leafed (6–12-leafed in *Chaseella*), consisting of the last internode of each shoot, ovoid, ellipsoid or cylindrical, often with distinct edges, often partly or entirely covered by the rhizome scales when young. *Leaves* duplicate, inarticulate, often with a petiole, orbicular to linear, tip acuminate to emarginate, in the latter case often with an extended mid-nerve, often oblique; thickly herbaceous to coriaceous, usually glabrous (in *B. gilgianum* slightly papillose adaxially). *Inflorescence* usually sprouting from the node below the pseudobulb, a raceme. *Peduncle* usually terete, with few to several nodes bearing scales. *Peduncle scales* almost similar to the rhizome scales. *Rhachis* terete, swollen into a spindle-shaped body, or widened and bilaterally flattened; glabrous, papillose or pubescent. *Floral bracts* appressed to reflexed. *Flowers* either distichous (when the rhachis is flattened the flowers are either situated along the midnerve of the flattened side of the rhachis or along the margins) or scattered (spirally arranged in the young inflorescence), either (partly) resupinate or not, appressed against the rhachis to reflexed, either flowering simultaneously or in succession, almost closed to wide open. *Pedicel and ovary* inarticulate; the jugae of the ovary may end distally in a tooth extending over the base of the petals and the lip. *Sepals*: the median and the lateral ones more or less equal except for the shape (in *B. longiflorum* the median differing very much from the lateral), usually free (adnate along the edges in *B. curvimentatum*; the lateral ones connate along their upper margins in *B. longiflorum*); the lateral ones often distinctly oblique, joined with the column foot along the basal part of their lower margin but hardly forming a distinct mentum (most conspicuous in *B. pumilum*, *B. schimperanum* and *B. cocoinum*). *Petals* (lateral petals) either as the sepals (except for the shape) or not. *Lip* (median petal) differentiated from the other parts of the perianth, hinged to the column foot by a very thin, flat strip of tissue allowing the lip to move freely parallel to the bilateral plane of symmetry of the flower; usually entire or inconspicuously three-lobed (in *B. intertextum*, *B. nigritianum* and *B. fuscum* var. *fuscum* distinctly three-lobed), margins either entire and/or glabrous or not, adaxially either with ridges or not, abaxially usually with a median ridge. *Column* ending in two more or less developed stelidia; two additional wings of various shape may be present along the adaxial margins of the column; clinandrium (that part of the column in which the anther is concealed) terminal, separated from the stigma by a seamlike, thin or fleshy rostellum; stigma concave, orbicular to longitudinally slitlike, either with a basal tooth or not. *Anther* connected to the top of the column by a very thin strip of tissue, usually easily to be detached (firmly attached in *B. bidenticulatum* ssp. *joyceae*), more or less cordate, flattened to distinctly concave, with two thecae which each may partly be split in two; abaxially often with a distinct ridge of various shape. *Pollinia* 2 or 4 (in *B. bidenticulatum* ssp. *joyceae* 2 pollinia are present which are partly split), when 4 either of about equal shape and size or with a much smaller and/or differently shaped inner pair; a stipes may be present, which is either firmly attached to the pollinia or easily falling off. Fruit either stalked or not, opening with 3 valves.

Ecology. Mainly low and high altitude primary forest, either wet (rain forest and montane forest) or with a drier period (savanna woodlands). Less common in low and high altitude shrub vegetations and mangrove forest. Some species may occur in secondary forest and plantations. Epiphytic herbs, less frequently occurring lithophytic, rarely terrestrial.

Distribution. Tropical part of Africa, northwards to the S. part of GUIN., IV.C., GHANA, TOGO, NIG, CAM., C.A.R., S. part of SUDAN, S. part of AETH., W. part of KEN; southwards to ANG., ZAM., ZIM., in R.S.A. along the E. coast S. to Alexandria.
Distribution outside Africa: Comores, Madagascar, Seychelles, Reunion and Mauritius (ca. 200 species described). Asia: Northwards up to the E. part of India, Nepal, Bhutan, SE China and S. Japan; Southwards to N. Australia; Eastwards to Tahiti (ca. 1500 species described). S. America: Northwards up to Mexico and Florida; Southwards to Brazil, Paraguay, Bolivia and Peru (ca. 100 species described).

Notes. 1 - All continental African species have a creeping rhizome. Elsewhere species occur which have an erect, patent or pendulous rhizome, only attached to the substratum at the very base. This is present in only one Madagascar species (*B. protectum* Perr.) and in many SE Asiatic species (especially from New Guinea). All these species otherwise show little similarity.

2 - Species with 2-leafed pseudobulbs are only common in Africa (Madagascar etc. included). Elsewhere this character is only present in a very restricted number of *Bulbophyllum* species: in SE Asia in *B. blepharistes* Rchb. f. only, and in S. America in some 6 species similar to *B. bracteolatum* Lindl. and *B. pachyrrhachis* (A. Rich.) Griseb. In combination with very thin and soft, deciduous leaves this character also occurs in S.E. Asia in the section *Pleiophyllus* J.J. Smith (1914, p. 34; 14 species, similar species with 1-leafed pseudobulbs and deciduous leaves also occur) and the section *Tripudianthes* Seidenfaden (1979, p. 188; 13 species). Otherwise these groups show little similarity.

3 - Knobs along the edges of the pseudobulbs may occur in the continental African species *B. cochleatum* var. *gravidum*, *B. maximum*, *B. sandersonii* ssp. *sandersonii* and *B. scaberulum* var. *scaberulum*. They are almost exclusively found on specimens growing in drier conditions such as savanna woodland; they may be the result of adaptation to these conditions. Comparable knobs occur as a specific character in two species of sect. *Macrouris* Schlechter (1912-13): *B. elisae* (F. Muell.) Benth. (Australia) and *B. corythium* Hallé (New Caledonia). Both species are mainly humid forest dwellers.

4 - The rhachis may be flattened in two directions perpendicular to each other. As a result the flowers are either arranged along the midnerve or along the margins of the blade-like rhachis. Species showing the first character (flowers arranged along the midnerve) occur exclusively in continental Africa (in group 5, '*Megaclinium*'). Species showing the second character (flowers arranged along the margins) occur in continental Africa in the groups 1, 4, 6, 8, 9. Outside this area this character occurs in a small number of SE Asiatic species, most conspicuously in sect. *Hymenobractea* Schlechter (1912-13, p. 752, and sect. *Diadochanthe* J.J. Smith, 1927, p. 66).

5 - Distichous flowers are found in many continental African species of the groups 1, 4, 5, 6, 8, 9. This character occurs only sparsely outside this area: in SE Asia in the sections *Dialeipanthe* Schlechter (1912-13), *Harpobrachium* l.c., *Hymenobractea* l.c., *Diadochanthe* J.J. Smith, *Ischnopus* Schlechter (1912-13) and *Macrouris* l.c.; in S America in the sections *Xiphizuza* (Reichenb. f.) Cogniaux (1902) (including *B. fallax* Rolfe, which has been erroneously described as originating from Assam), *Malachadenia* (Lindl.) Bentham & Hooker (1880) and *Napelli* Reichenb. f., 1861). Most of these sections show otherwise very little similarity.

6 - The flowers may either be resupinate or not. In case the flowers are not resupinate the lip is turned upwards (towards the top of the inflorescence). This situation occurs in e.g. *B. schimperanum*. Resupination may amount to 90 degrees (the lip is turned sidewards as in *B. cochleatum*), or 180 degrees (the lip is turned downwards, e.g. in *B. schinzianum*). Some species have flowers which always assume a position with their axis of symmetry vertical, e.g. *B. falcatum*.

7 - In the descriptions distinction is made between the stelidia and the various ornaments which may occur along the adaxial margins of the column. The stelidia, generally considered as the rudiments of two stamens, usually consist of two teeth projecting from the top of the column at each side of the anther. They always have a vascular bundle which, however, is often inconspicuous. Wings or teeth along the adaxial margin of the column do not have a vascular bundle. When the stelidia, however small they may be, are the most frontwards projecting features at the top of the column distinction between the two is easy, but in species such as *B. renkinianum* (fig. 53) or *B. purpureorhachis* (fig. 66) confusion may occur because the wings are as long as or even projecting beyond the stelidia. In species as *B. fuscum* and *B. acutebracteatum* (fig. 89–92) each stelidium may be fused to the wing or tooth along an adaxial margin of the column forming one winglike structure.

8 - A stipes (appendage on the pollinia which consist of the upper layers of tissue from the rostellum) has been observed in various species. It may be of different shape. In some specimens of one species they may be present, whereas in other specimens of the same species they are not. In case they are easily detached from the pollinia one may assume that they got lost occasionally, but even in species with a firmly attached stipes this has been observed. In *B. falcatum* a much thicker, frayed lump of tissue of unknown origin is present in some specimens.

KEY TO THE GENERA OF BULBOPHYLLINAE OCCURRING IN CONTINENTAL AFRICA

1 - Pseudobulbs 1–2(– 3)-leafed.. **Bulbophyllum** (1)
- Pseudobulbs 6–12-leafed.. **Chaseella** (2)

1 - **BULBOPHYLLUM** Thouars

Bulbophyllum Thou., Orch. Iles Austr. Afr. (1822) t. esp. 3, sub u, nomen conservandum; Lindl., Gen. & Sp. Orch. Pl. (1830) 47 ('*Bolbophyllum*'); Rchb. f., Walp. Ann. 6 (1861) 243; Benth. & Hook., Gen. Pl. 3 (1880) 501; Pfitz. in Engl. & Prantl, Nat. Pflanzenfam., Teil 2, Abt. 6 (1889) 178; Rolfe in F.T.A. 7 (1897) 22; De Wild., Pl. Beq. 1 (1921) 53; Summerh. in F.W.T.A. ed. 1, 2 (1936) 432; id. ed. 2, 3 (1968) 226; Cribb in F.T.E.A., Orch. 2 (1984) 237.-*Phyllorkis* Thou., Nouv. Bull. Sc. Soc. Philom. 1 (1809) 319; Kuntze, Rev. Gen. Pl. 2 (1891) 675 ('*Phyllorchis*').-Type species: *Bulbophyllum nutans* Thou., typus conservandum.

Cirrhopetalum Lindl., Bot. Reg. 10 (1824) sub t. 832; id., Gen. & Sp. Orch. Pl. (1830) 58; Benth. & Hook., Gen. Pl. 3 (1880) 504; Pfitz. in Engl. & Prantl, Nat. Pflanzenfam., Teil 2, Abt. 6 (1889) 178; Schltr, Fedde, Rep., Beih. 33 (1925) 252.-*Bulbophyllum* sect. *Cirrhopetalum* (Lindl.) Rchb. f., Walp. Ann. 6 (1861) 259; J.J. Smith, Bull. J. Bot. Buit. ser. 2, 8 (1912) 19; Seidenf., Dansk Bot. Ark. 29 (1973) 7.-Type species: *Cirrhopetalum thouarsii* Lindl., = *Bulbophyllum longiflorum* Thou.

Tribrachia Lindl., Bot. Reg. 10 (1824) sub t. 832, nomen; id., Coll. Bot. (1821) sub t. 41a.-Type species: *Tribrachia reptans* Lindl., = *Bulbophyllum reptans* (Lindl.) Lindl.

Megaclinium Lindl., Bot. Reg. 10 (1824) sub t. 832, nomen; id., Bot. Reg. 12 (1826) t. 989; id., Gen. & Sp. Orch. Pl. (1830) 47; Benth. & Hook., Gen. Pl. 3 (1880) 505; Benth., J. Linn. Soc. Bot. 18 (1881) 300; Pfitz. in Engl. & Prantl, Nat. Pflanzenfam., Teil 2, Abt. 6 (1889) 180; Rolfe in F.T.A. 7 (1897) 36; Krzl., Vierteljahresschr. Naturf. Gesellsch. Zurich 68 (1923) 424.-*Bulbophyllum* subg. *Megaclinium* De Wild., Pl. Beq. 1 (1921) 71.-*Bulbophyllum* sect. *Megaclinium* Summerh., Kew Bull. 1935 (1935) 200.-Type species: *Megaclinium falcatum* Lindl., = *Bulbophyllum falcatum* (Lindl.) Rchb. f. var. *falcatum*.

Taurostalix Rchb. f., Bot. Zeit. (1852) 933.-Type species: *Taurostalix herminiostachys* Rchb. f., = *Bulbophyllum pumilum* (Sw.) Lindl.

Bulbophyllum sect. *Ptiloglossum* Lindl., J. Linn. Soc. Bot. 6 (1862) 125; Pfitz. in Engl. & Prantl, Nat. Pflanzenfam., Teil 2, Abt. 6 (1889) 179. Lectotype species: *Bulbophyllum barbigerum* Lindl.

Bulbophyllum sect. *Cocoina* Pfitz. in Engl. & Prantl, Nat. Pflanzenfam., Teil 2, Abt. 6 (1889) 179.-Type species: *Bulbophyllum cocoinum* Lindl.

Bulbophyllum sect. *Calamaria* Krzl., Engl., Bot. Jahrb. 28 (1900) 163, nomen (sub *B. urbanianum*); Schltr, Fedde, Rep., Beih. 33 (1925) 216.-Type species: ? *B. calamarium* Lindl., = *B. saltatorium* Lindl. var. *calamarium* (Lindl.) J.J. Vermeulen.

Bulbophyllum sect. *Pendula* Pfitz. in Engl. & Prantl, Nat. Pflanzenfam., Teil 2, Abt. 6 (1889) 179.-Type species: &Tribrachia pendula Lindl., = *Bulbophyllum pumilum* (Sw.) Lindl.

Bulbophyllum sect. *Lupulina* Pfitz. in Engl. & Prantl, Nat. Pflanzenfam., Teil 2, Abt. 6 (1889) 179.-Type species: *Bulbophyllum occultum* Thou. (Madagascar).

Bulbophyllum subg. *Eubulbophyllum* De Wild., Pl. Beq. 1 (1921) 101.-Type species: not indicated.

Megaclinium sect. *Eumegaclinium* Krzl., Vierteljahresschr. Naturf. Gesellsch. Zürich 68 (1923) 424.-Type species: not mentioned, probably *Bulbophyllum falcatum* (Lindl.) Rchb. f. var. *falcatum*.

Megaclinium sect. *Psilorhachis* Krzl., Vierteljahresschr. Naturf. Gesellsch. Zürich 68 (1923) 424.-Type species: *Megaclinium buchenavianum* Krzl., = *Bulbophyllum calyptratum* Krzl. var. *calyptratum*.

Megaclinium sect. *Pseudomegaclinium* Krzl., Vierteljahresschr. Naturf. Gesellsch. Zürich 68 (1923) 424.-Type species: *Megaclinium hebetatum* Krzl., = *Bulbophyllum imbricatum* Lindl.

Description of the genus: Pseudobulbs 1-2(-3)-leafed. Pollinia 2 or 4.

Ecology. See description of the subtribe.

Distribution. See description of the subtribe.

KEY TO THE INFRAGENERIC GROUPS OF BULBOPHYLLUM OCCURRING IN CONTINENTAL AFRICA

1 - Lateral sepals connate along their upper margins................**Group 11** (sp. 68)
 - Lateral sepals not connate along their upper margins................................2

2 - Abaxial side of the sepals with long hairs............................**Group 3** (sp. 32)
- Abaxial side of the sepals without long hairs..3

3 - Lateral sepals reflexed at the very base and appressed against the rhachis........
....**Group 4** (spp. 33–35)
- Lateral sepals either reflexed or not, but not appressed against the rhachis.....4

4 - Anther with a dorsoventrally flattened, rounded, truncate to bifid protrusion in front which overtops the front margin..5
- Anther either without a protrusion in front, or with a protrusion which may overtop the front margin but which is not dorsoventrally flattened................9

5 - Flowers scattered, or flowers 1–2..6
- Flowers distichous, 3 or more...7

6 - Flowers 3 or less...**Group 10** (sp. 67)
- Flowers 8 or more..**Group 7** (spp. 55–56)

7 - Stelidia 1–1.2 mm long, about as long as the body of the column, narrowly acute to acuminate..**Group 6** (sp. 54)
- Stelidia 0.5 mm or shorter, much shorter than the body of the column, at most shortly acute...8

8 - Floral bracts 1.5 times as long as the flower (pedicel and ovary included) or longer; lip with seamlike, thin lateral lobes which are distinctly separated from the median lobe and which have denticulate margins.......................**Group 8** (spp. 57–59)
- Floral bracts as long as the flower (pedicel and ovary included) or shorter; either lip without lateral lobes; or lip with semi-orbicular to obovate lateral lobes......
....**Group 9** (spp. 60–66)

9 - Rhachis flattened with the flowers inserted along the midnerve; or lip fleshy but soft and very easy to compress (very thin when dry); or both characters present
....**Group 5** (spp 36–53)
- Rhachis not flattened with the flowers inserted along the midnerve, and lip either not fleshy or fleshy but not soft and easy to compress.............................10

10 - Pseudobulbs 1-leafed...**Group 1** (spp. 1–30)
- Pseudobulbs 2-leafed...11

11 - Flowers scattered...**Group 2** (sp. 31)
- Flowers distichous...**Group 1** (spp. 1–30)

KEY TO THE SPECIES OF BULBOPHYLLUM OCCURRING IN CONTINENTAL AFRICA

Note: An (*) indicates that an explanatory sketch can be found on figure D (see p. 20). These sketches are marked with the number of the couplet in the key they refer to. An a or b is added to the number on the figure and to the asterisk in the key when there are two sketches referring to one couplet.

Good MAGNIFICATION EQUIPMENT (up to 30 times) is needed when using this key.

1 - Pseudobulbs 1-leafed (incidentally a single 2-leafed pseudobulb may be present) (*a)..2

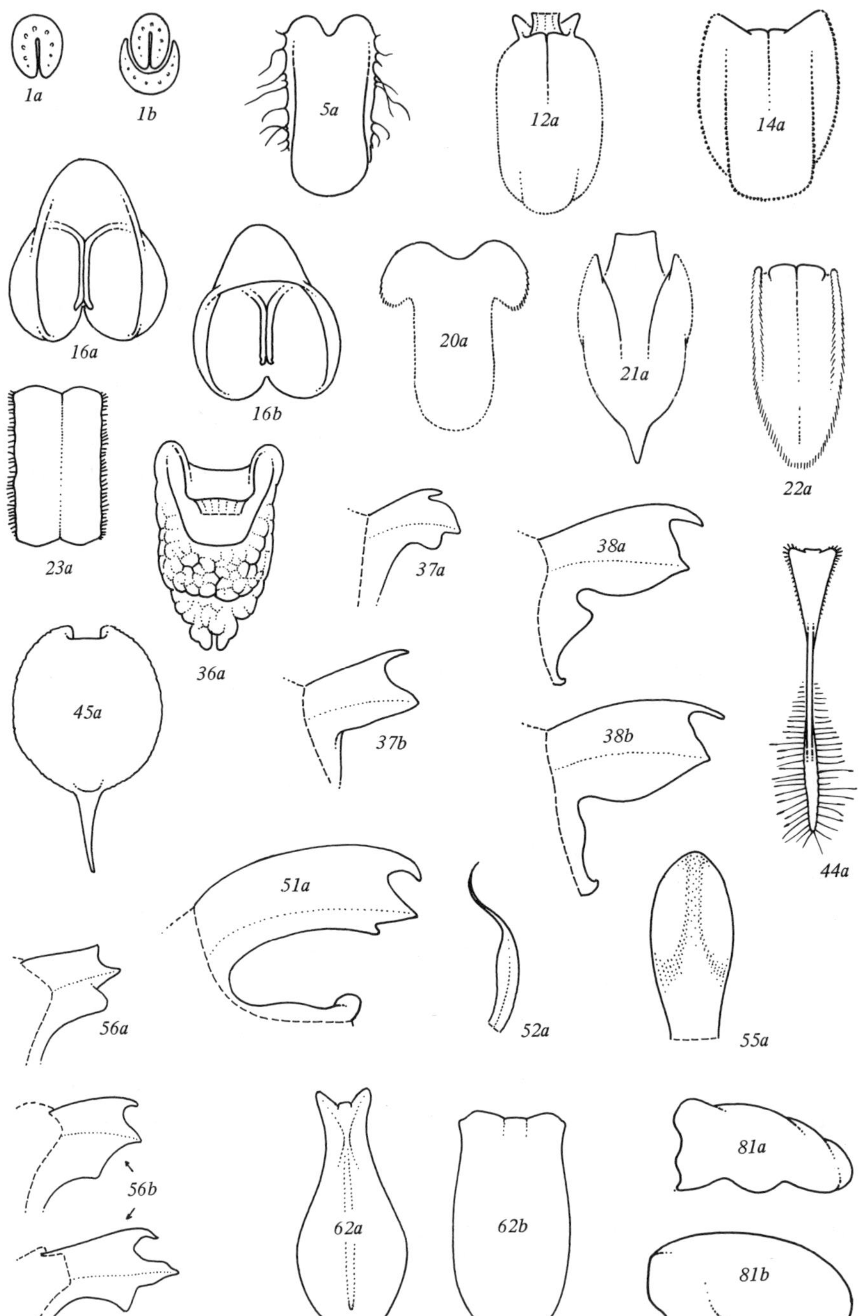

Fig. D. Explanatory sketches with the key. 1a, b - Leaf marks on pseudobulb; 5a - lip *B. ballii*, adaxially; 12a - lip *B. capituliflorum*, adaxially; 14a - lip *B. pumilum*, adaxially; 16a - anther *B. pipio*, adaxially; 16b - anther *B. pumilum*, adaxially; 20a - lip *B. nigritianum*, adaxially; 21a - lip *B. oxychilum*, adaxially; 22a - lip *B. cocoinum*, adaxially; 23a - lip *B. schimperanum*, adaxially; 36a - lip *B. encephalodes*, abaxially; 37a - top column *B. unifoliatum*; 37b - top column *B. concatenatum*; 38a - column *B. tetragonum*; 38b - column *B. lupulinum*; 44a - lip *B. vulcanicum*, adaxially; 45a - lip *B. renkinianum*, adaxially; 51a - column *B. imbricatum*; 52a - petal *B. purpureorhachis*; 55a - median sepal *B. falcatum*; 56a - top column *B. falcatum*; 56b - top column *B. calyptratum* (above) and *B. carnosilabium* (below); 62a - lip *B. inornatum*, adaxially; 62b - lip *B. stolzii*, adaxially; 81a - lip *B. acutebracteatunm*, lateral view; 81b - lip *B. fuscum* (var. *melinostachyum*) lateral view.

- Pseudobulbs 2(–3)-leafed (incidentally a single 1-leafed pseudobulb may be present) (*b)..39

2 - Inflorescence subumbellate; lateral sepals more than twice as long as the median sepal, connate along their upper margins........................**B. longiflorum** (68)
- Inflorescence a raceme or flowers solitary; lateral sepals less than twice as long as the median sepal, either free or connate along their lower margins...............3

3 - Sepals abaxially with long hairs.......................................**B. comatum** (32)
- Sepals abaxially glabrous, papillose or shortly tomentose............................4

4 - Margins of petals entire and margins of lip with long hairs........................5
- Margins of petals either entire or not; margins of lip entire, papillose, ciliate, denticulate or lacerate, if lip with long hairs the petals have finely and irregularly denticulate margins towards the tip..9

5 - Median sepal 3.2 mm or shorter; margins of lip with only few, widely spaced hairs (*)..**B. ballii** (55)
- Median sepal 5.6 mm or longer; margins of lip densely beset with hairs.........6

6 - Lowermost floral bracts 16 mm or longer..7
- Lowermost floral bracts 13 mm or shorter...8

7 - Lip adaxially at the very base with a forwards projecting, dorsoventrally flattened, deltoid, about 0.8 mm long median tooth with an emarginate tip, which covers a distinctly concave basal part...................................**B. subligaculiferum** (2)
- Lip adaxially without a basal tooth.................................**B. schinzianum** (1)

8 - Petals 2 mm or longer..**B. saltatorium** (3)
- Petals 1 mm or shorter...**B. barbigerum** (4)

9 - (4) Median sepal about 19.5 mm long (one specimen studied)...................... ....**B. dolabriforme** (37)
- Median sepal 14 mm or shorter..10

10 - Flowers not distichous or rhachis with 2–4 flowers................................11
- Flowers distichous and rhachis with 5 flowers or more............................27

11 - Petals about 1/4 the length of the column (stelidia included)....**B. ivorense** (23)
- Petals about 2/3 the length of the column or longer (stelidia included).........12

12 - Lip with 2 distinct, backwards directed knobs abaxially of the attachment to the column foot (*)..**B. capituliflorum** (15)
- Lip without such knobs...13

13 - Lateral sepals abaxially with distinct keels which extend to the ovary............. ....**B. carnosisepalum** (54)
- Lateral sepals abaxially without keels or with inconspicuous keels...............14

14 - Lip adaxially with a distinct furrow between the thick median part and the much thinner, seamlike lateral lobes (one or two median furrows may also be present) (*)..15
- Lip adaxially either divided into a thick median part and thinner lateral lobes or not, but without furrows as above (one or two median furrows may be present)17

15 - Stelidia much shorter than the body of the column.............**B. coriscense** (18)
- Stelidia as long as or longer than the body of the column..........................16

16 - Front margin of the anther extended into an obtuse beak with a concave adaxial side (*a)..**B. pipio** (16)
- Front margin of the anther not extended, overtopped by a solid conical crest at

the abaxial side of the anther (*b)......................................**B. pumilum** (17)

17 - (14) Median sepal distinctly wider than the lateral ones.........**B. humblotii** (28)
- Median sepal equaling the lateral ones in width or narrower.......................18

18 - Inflorescence with 2–4 flowers..19
- Inflorescence with 5 or more flowers...20

19 - Ciliae along margins of lip about equally long everywhere; flowers usually 1 mm apart or less...**B. vanum** (14)
- Ciliae along margins of lip distinctly shorter towards the tip of the lip or even absent, or lip (almost) entirely glabrous; flowers 3 mm apart or more...**B. intertextum** (13)

20 - Lip with lateral lobes which are distinct, forwards directed, obliquely semi-elliptic, rounded to obtuse, thin, attached near the base and separated from the median lobe by a more or less acute sinus (*)..............................**B. nigritianum** (27)
- Lip either without lateral lobes or with seamlike or backwards directed lateral lobes..21

21 - Lip with fleshy, more or less backwards directed lateral lobes which give the lip a more or less cordate or hastate appearance when flattened (dried specimens) (*)..**B. oxychilum** (26)
- Lip either without sidelobes or with seamlike sidelobes..............................22

22 - Lip adaxially with two knobs near the base, which more or less form a transverse ridge (often shriveled, sometimes difficult to see in dried specimens) (*)...**B. cocoinum** (19)
- Lip adaxially without such knobs...23

23 - Lip adaxially (almost) glabrous, margins finely but distinctly stiffly hirsute (if the lip is slightly and very finely papillose towards the margins these papillae are much finer than the stiff hairs along the margins) (*).............**B. schimperanum** (25)
- Lip adaxially either entirely glabrous, or more or less papillose with ditto margins (papillae on surface lip and along the margins about equally fine)...............24

24 - Median sepal about 4 times as long as the petals or longer...**B. expallidum** (22)
- Median sepal about 3 times as long as the petals or shorter.......................25

25 - Lip widest in the top half...**B. mediocre** (24)
- Lip widest in the basal half..26

26 - Lateral sepals less than 3 times as long as wide (one specimen studied)..**B. calvum** (21)
- Lateral sepals more than 3 times as long as wide...................**B. josephii** (20)

27 - (10) Margins of lip ciliate..28
- Margins of lip glabrous, papillose, denticulate or lacerate.........................29

28 - Margins of petals finely crenulate towards the tip; lip adaxially with a distinct median furrow...**B. intertextum** (13)
- Margins of petals glabrous; lip adaxially without a median furrow..**B. kivuense** (8)

29 - Lip more or less fleshy but soft and easy to compress when fresh or preserved in liquid, membranous and very thin when dry...30
- Lip fleshy and solid when fresh, remaining solid after drying (lip may have thin lateral lobes)...33

30 - Lateral sepals obtuse, acute or very shortly acuminate..............................31

- Lateral sepals cuspidate or caudate..32

31 - Petals ovate-oblong to ovate-lanceolate; margins of lip entire........................ ....**B. magnibracteatum** (43)

- Petals linear-lanceolate; margins of lip usually lacerate near the base............. ....**B. imbricatum** (42)

32 - Leaves 11 cm or longer..**B. colubrinum** (39)

- Leaves c. 3.5 cm long (one specimen studied)..............**B. curvimentatum** (38)

33 - (29) Floral bracts shorter than or more or less equaling the flowers in length (pedicel and ovary included)..34

- Floral bracts 1.5 times as long as the flowers..38

34 - Rhachis distinctly wiry..**B. intertextum** (13)

- Rhachis sturdy and/or flattened...35

35 - Stelidia 1–1.2 mm long, about as long as the body of the column................. ....**B. carnosisepalum** (54)

- Stelidia 0.5 mm or shorter, much shorter than the body of the column........36

36 - Lip abaxially with brain-like surface due to large, irregularly shaped warts (*) ....**B. encephalodes** (64)

- Lip without such warts...37

37 - Column with thin, more or less deltoid, rounded to obtuse wings along the adaxial margins near the tip, just below the stelidia (*a)...............**B. unifoliatum** (66)

- Column without wings along the adaxial margins (*b).....**B. concatenatum** (65)

38 - (33) Column foot with a distinct, transverse ridge (*a)........**B. tetragonum** (59)

- Column foot without a transverse ridge but usually with two weak lateral knobs (*b)...**B. lupulinum** (58)

39 - (1) Margins of lip ciliate or with long hairs...40

- Margins of lip glabrous, papillose, denticulate or lacerate..........................45

40 - Flowers 2–3...**B. gilgianum** (67)

- Flowers 8 or more...41

41 - Flowers scattered or in 3–5 rows..42

- Flowers distichous..43

42 - Leaves about orbicular to oblong; margins of lip densely ciliate..**B. elliotii** (56)

- Leaves lanceolate; margins of lip with some hairs only...............**B. ballii** (55)

43 - Median sepal 4 times as long as the lip or longer......................**B. burttii** (7)

- Median sepal 3 times as long as the lip or shorter.................................44

44 - Basal part of the lip triangular, much wider than the linear, ensiform top part (*) ....**B. vulcanicum** (5)

- Basal part of the lip as wide as the top part or narrower, or lip gradually tapering from a wide base, top part not narrow as above................**B. cochleatum** (6)

45 - (39) Lip orbicular or broadly elliptic in outline, abruptly narrowing into a long cuspidate to caudate, slender top part with acute tip (*)....**B. renkinianum** (36)

- Lip otherwise, top part not slender, tip retuse to slightly acuminate............46

46 - Margins of lip (lateral lobes included) partly or entirely serrulate, denticulate or lacerate 47

- Margins of lip (lateral lobes included) entire...54

47 - Floral bracts 1.5 times as long as the flowers or longer (pedicel and ovary included) 48

- Floral bracts as long as the flowers or shorter (pedicel and ovary included)...50

48 - Rhachis and abaxial surface of the floral bracts glabrous...**B. porphyrostachys** (57)

- Rhachis and abaxial surface of the floral bracts with fine, dark hairs..........49

49 - Column foot with a distinct, transverse ridge (* see fig. 37a)..**B. tetragonum** (59)

- Column foot without a transverse ridge, usually with two weak lateral teeth (* see fig. 37b)..**B. lupulinum** (58)

50 - (47) Lip with a distinctly thickened, solid top part...................**B. fuscum** (62)

- Lip with a thickened basal part, gradually thinning towards the tip.............51

51 - Adaxial margins of the column wingless or with a narrow, seamlike wing which ends in a very small, deltoid tooth at the base of the stelidia (*)..**B. imbricatum** (42)

- Adaxial margins of the column with a distinct wing..................................52

52 - Petals caudate (*)..**B. purpureorhachis** (46)

- Petals at most shortly acuminate...53

53 - Lateral sepals cuspidate to caudate...................................**B. maximum** (40)

- Lateral sepals acute or slightly acuminate...........................**B. injoloense** (41)

54 - (46) Flowers scattered...**B. denticulatum** (31)

- Flowers distichous...55

55 - Top half of the median sepal with a fleshy callus along both margins (*).....56

- Top half of the median sepal without callus along both margins or entirely swollen 58

56 - Adaxial margin of the stelidia passing with a sharp angle (90 degrees or less) into the front margin of the wing below the stelidia, the latter often folded inwards, semi-elliptic or obtusely triangular and forwards projecting when folded out, in the direction of the petals (*a)..**B. falcatum** (51)

- Adaxial margin of the stelidia passing gradually or with an obtuse angle into the anterior margin of the wing below the stelidia, the latter not folded inwards, inconspicuous to about triangular, not or hardly forwards projecting (*b).....57

57 - Tip of median sepal acuminate...................................**B. carnosilabium** (53)

- Tip of median sepal emarginate to acute.........................**B. calyptratum** (52)

58 - (55) Stelidia distinct and slender, with an acuminate to subulate tip............59

- Stelidia minute, with a bifid, truncate, rounded, obtuse, acute or shortly acuminate tip...63

59 - Top of petals truncate, with a cuspidate to caudate tip.........**B. nigericum** (35)

- Top of petals slightly emarginate to acute...60

60 - Lowermost floral bracts 5 mm or longer...........................**B. cochleatum** (6)

- Lowermost floral bracts 4.5 mm or shorter...61

61 - Sepals adaxially very finely papillose towards the margins; lip almost entirely and distinctly papillose adaxially..**B. bavonis** (10)

- Sepals glabrous; lip glabrous or papillose only near the base.....................62

62 - Lip distinctly constricted in the basal half, slightly widened again at the base (*) ....**B. inornatum** (9)

- Lip not constricted in the basal half, gradually narrowing towards the base (*). ....**B. stolzii** (11)

63 - (58) Lateral sepals abaxially with a winglike keel with a deeply lacerate crest.... .... **B. horridulum** (60)
- Lateral sepals abaxially without a keel or with an inconspicuous keel...........64

64 - Top part of the petals slightly swollen, ensiform or subulate in cross-section, somewhat thicker than the median part (make sure to have complete petals, in some cases the tip may easily break off)..65
- Top part of petals not swollen...67

65 - Median sepal c. 28 mm long (one specimen studied)...................**B. lizae** (45)
- Median sepals 15 mm or shorter...66

66 - Lateral sepals spreading from their very base when flower fully open............. .... **B. sandersonii** (44)
- Lateral sepals straight or recurved, at most with a reflexed top part.............. .... **B. scaberulum** (48)

67 - (64) Petals up to 2 times as long as wide...68
- Petals 3 times as long as wide or longer..72

68 - Lateral sepals slightly recurved to spreading...69
- Lateral sepals entirely reflexed and pressed against the ovary and the rhachis... 71

69 - Margins of lip with two distinct undulations in the basal half....................... .... **B. prorepens** (12)
- Margins of lip folded upwards in the basal half but without waves.............70

70 - Floral bracts c. 14 mm long, overtopping the flowers.....**B. pandanetorum** (30)
- Floral bracts 5–8 mm long, about as long as the flowers (pedicel and ovary included).. **B. scariosum** (29)

71 - (68) Lip recurved about half-way.......................................**B. bifarium** (33)
- Lip not recurved or only with a slightly recurved extreme tip (make sure to have flowers in good condition).....................................**B. bidenticulatum** (34)

72 - (67) Lip more or less fleshy but soft and very easy to compress when fresh or preserved in liquid, membranous and very thin when dry.............................73
- Lip fleshy and solid when fresh, remaining solid after drying (lip may have thin lateral lobes)...78

73 - Petals caudate (*)..**B. purpureorhachis** (46)
- Petals at most shortly acuminate...74

74 - Floral bracts half-way the fully developed part of the rhachis 7–13 mm wide; petals 6–7 mm long...**B. scaberulum** (48)
- Floral bracts half-way the fully developed part of the rhachis 1.2–4 mm wide, petals 1.4–5.5 mm long..75

75 - Rhachis more or less widened and distinctly flattened (in case the rhachis is only slightly widened it has at least one sharp edge..76
- Rhachis not widened and not or hardly flattened (but may be slightly flattened just above the place where each flower is inserted), without sharp edges............77

76 - Leaves linear-lanceolate, 2.1–4.5 cm long.......................**B. falcipetalum** (47)
- Leaves generally longer, up to 28 cm, when less than 4.5 cm long not linear-lanceolate but wider..**B. scaberulum** (48)

77 - Leaves 0.6–0.9 cm long; inflorescence 1.2–1.6 cm long; petals linear-lanceolate, widest slightly below the middle (one specimen studied)................**B. parvum** (50)

- Leaves 1.3–8.5 cm long; inflorescence 1.8–40 cm long; petals narrowly lanceolate to linear-lanceolate; maximum width not below the middle..**B. resupinatum** (49)

78 - (72) Top part of the lip abaxially distinctly convex, neither with an indenture nor a (minute) downwards bent apiculus.................................**B. oreonastes** (63)

- Top part of the lip abaxially with a (small) indenture and/or with a (minute) downwards bent apiculus...79

79 - Rhachis and the abaxial surface of the floral bracts and the sepals partly or entirely papillose...**B. acutebracteatum** (61)

- Rhachis, floral bracts and sepals glabrous...80

80 - Lip with lateral lobes near the base which arise from the main body of the lip with a distinct sinus at their anterior end.....................................**B. fuscum** (62)

- Lip without such lateral lobes...81

81 - Lip abaxially with a large lump about half-way which gives the lip a hunch backed appearance when observed from aside; lip adaxially with an obtuse median ridge towards the tip (*a)..**B. acutebracteatum** (61)

- Lip abaxially without such a lump; lip adaxially without a ridge towards the tip (dried specimens may occasionally develop a ridge-like structure due to shriveling) (*b)...**B. fuscum** (62)

GROUP 1 (spp. 1–30)

Pseudobulbs 1–2-leafed. Inflorescence a raceme, many-flowered. Rhachis either terete (not thickened) or flattened, with the flowers inserted along the margins. Flowers scattered or distichous. Sepals glabrous abaxially, rarely with a few dark hairs; the lateral ones not connate along their upper margins, straight to reflexed, but not appressed against the rhachis. Lip thin to thick, not soft and easy to compress; lateral lobes present or absent; margins glabrous to long hairy. Stelidia very long and acuminate to short and truncate. Anther with or without a not dorsoventrally flattened protrusion overtopping the front margin.

1 - **Bulbophyllum schinzianum** Krzl. - Figs. 1–3; Plate 1a, b.

Literature: see under the varieties.

Roots 0.8–3.5 mm diam. *Rhizome* 3.5–10 mm diam., nodes not or hardly swollen. *Rhizome scales* tubular, tip acute, the lower rounded, either mucronate or not; membranous, with or without nerves; keeled abaxially. *Pseudobulbs* 1-leafed, orbicular to narrowly ovoid, 0.6–7 cm apart, 1–6 by 0.8–3 cm, slightly to moderately flattened, obtusely to about sharply 2–4-angled. *Petiole* 8–40 mm long. *Leaf blade* oblong to linear lanceolate, maximum width occasionally somewhat above the middle, 7.5–32 by 1.4–6,5 cm, tip rounded to subacute, either emarginate and/or slightly oblique or not, coriaceous, thick; midrib slightly prominent abaxially. *Inflorescence* 28–102 cm long, 11–102-flowered. *Peduncle* erect to patent, 25–77 cm by 1.8–6 mm, orbicular in section, glabrous, with 7–12 scales. *Peduncle scales* tubular, the longest 18–28 mm long; tip acute to acuminate, the lower obtuse; membranous, with some nerves, either keeled towards the tip or not and glabrous abaxially. *Rhachis* erect to slightly arching, terete, about orbicular in section, either slightly zigzag bent or not, 4–47 cm long, glabrous. *Floral bracts* recurved, the lower often reflexed, slightly concave, broadly elliptic to broadly lanceolate, 16–30 by 5–13 mm, tip obtuse to acute; margins slightly rolled inwards; base narrowly attached; membranous, with some nerves; not or hardly keeled and glabrous abaxially. *Flowers* scattered, in a dense to rather lax raceme, recurved to spreading, few open simultaneously, not fully open. *Pedicel and ovary* 8–24 mm long, glabrous. *Median sepal* either slightly recurved or not, elliptic or ovate to (ovate)-linear-lanceolate, occasionally somewhat constricted about half-way, 7–14 by 1.6–4.5 mm, tip subacute to acute; margins entire, glabrous; base broadly to rather narrowly attached; rather thin to rather thick; glabrous; not keeled abaxially. *Lateral sepals* free, recurved, hardly falcate, ovate-lanceolate, oblique, 6–13 by 1.8–3 mm, base broadly attached; otherwise as the median sepal. *Petals* not recurved, falcate, linear-lanceolate to linear, oblique, 2.5–8 by 0.2–1 mm, tip acute to acuminate; margins entire, glabrous; base broadly to rather narrowly attached; thin; glabrous; not keeled abaxially. *Lip* slightly recurved near the base, oblong to narrowly linear-lanceolate in outline (not spread), either constricted in the basal half or not, 5–11 by 1–3.5 mm, tip rounded to acute; margins entire, beset with short and long hairs (see varieties); rather thick, with a thick base; adaxially with two very weak to rather distinct ridges in the basal half, either with a third median ridge in front of them or not, with a slightly concave basal part, with a median slit starting at about the base and continuing up to 1/3–1/2 of the lip, surface glabrous to finely papillose or pubescent, or coarsely papillose in the top half; abaxially with a weak, obtuse ridge near the base, surface glabrous to finely papillose. *Column* from the base of the free part up to the tip of the stelidia 3–6 mm long, with distinct, gradually tapering, slightly falcate, acute to acuminate stelidia of 1–3 mm long; with weak, rounded wings along its adaxial margins just below the base of the stelidia; stigma without a basal callus. *Anther* 1.2–1.7 by 0.7–0.8 mm; front drawn out, thin, not concave; front margin retuse, entire, about glabrous; basally either emarginate or not; abaxially with a distinct, rounded, colliculate ridge over about the entire length of the anther which overtops the front margin, surface glabrous, either colliculate in the median part or not. *Pollinia* 4, very narrowly ellipsoid, distinctly flattened on either side, all of about equal length; a thin stipes has been observed. *Fruit* obovoid, up to 2.8–3.5 by 0.8–1.5 cm, stalk 4–5 mm.

KEY TO THE VARIETIES

1 - Petals 6.5 mm or longer .... var. **phaeopogon** (c)
- Petals 6 mm or shorter 2

2 - Lip oblong to broadly linear-lanceolate, top part about as wide as the basal part or much wider .... var. **schinzianum** (a)
- Lip narrowly linear-lanceolate, top part much narrower than the basal part .... var. **irigaleae** (b)

1a. var. **schinzianum** - Fig. 1; Plate 1a.

Bulbophyllum schinzianum Krzl., Bull. Soc. R. Bot. Belg. 38 (1899) 57 ('*Bolbophyllum*'); id. in Durand & De Wild., Mat. Fl. Congo 3 (1899) 49 ('*Bolbophyllum*'); De Wild., Not. Pl. Ut. et Int. Fl. Congo 1 (1903) 123; id., Et. Fl. B. et Moy. Congo 1 (1904) 116; id., 2 (1908) 237; Durand & Durand, Syll. Fl. Congol. (1909) 516; De Wild., Et. Fl. B. et Moy. Congo 3 (1910) 182; De Wild., Pl. Beq. 1 (1921) 119; Summerh., Kew Bull. 12 (1957) 115; Ake Assi, Et. Fl. C. Iv. (1964) 245; Summerh. in F.W.T.A. ed. 2, 3 (1968) 236; Johansson, Acta Phyt. Suec. 59 (1974) 33; Geerinck, Bull. Soc. R. Bot. Belg. 109 (1976) 176; Adam, Fl. Descr. Mt. Nimba 5 (1981) 1684, p.p.–Type: Laurent s.n. (BR).

Non *Bulbophyllum schinzianum* auct.: Segerbaeck, Orch. Nig. (1983) 59 (= var. *phaeopogon*).

Bulbophyllum gentilii Rolfe, Gard. Chron., ser. 3, 35, no. 929 (1904) 266, p.p. (see below); De Wild., Et. Fl. B. et Moy. Congo 3 (1910) 181, pl. 35; id., Pl. Beq. 1 (1921) 111, p.p.-Syntypes: Kew cult. Hort. (1903) = Gentil (1898) s.n. (Lecto K, here proposed); Zenker 832 ('823' in Gard. Chron.) (BM, E, G. L) (= var. *phaeopogon*); Preuss 1225 (BM, S) (= var. *phaeopogon*).

Non *Megaclinium gentilii* De Wild., Bel. Col. 8 (1902) 425, n.v. (= *B. falcatum* var. *bufo*).

Bulbophyllum calamarium auct. non Lindl. (1843): Lindl., J. Linn. Soc. Bot. 6 (1862) 125 ('*Bolbophyllum*'); Durand & Schinz, Consp. F. Afr. 5 (1895) 11; Rolfe in F.T.A. 7 (1897) 33, p.p.; Summerh. in F.W.T.A. ed. 1, 2 (1936) 437, p.p.

Median sepal narrowly ovate or elliptic to (ovate) linear-lanceolate, not constricted about half-way, 7–14 by 1.6–4.5 mm. Lateral sepals 6–13 by 1.8–3 mm. Petals linear-lanceolate to linear, 2.5–6 by 0.2–0.7 mm. Lip oblong to linear-lanceolate in outline, the epichilium wider than the hypochilium; 5–11 by 1.3–3.5 mm, margins beset with rather thick hairs, those that arise along the margins of the adaxial side of the lip short, placed in a very thin rim, occasionally hardly constituting a rim at all along the epichilium or over the entire length of the lip, those that arise along the margins of the abaxial side of the lip up to 4–8 mm long; adaxially with very weak ridges near the base, surface glabrous or very finely papillose towards the tip. Column 3.2–5 mm long. Stelidia 1–2 mm long.

Colours. Pseudobulbs yellowish. Leaves occasionally slightly glaucous. Floral bracts yellowish, dry when the flowers are open. Sepals yellowish or greenish, occasionally brownish adaxially, usually spotted with red-purple or entirely red-purple; (slightly glaucous) purple or brownish abaxially. Petals red-purple, often with white markings. Lip yellowish to glossy green, flushed red-purple; hairs brown.

Ecology. (Dry) primary forest, 0–800 m alt. Epiphyte in the crown of trees, on bark or in minor humus deposits (Johansson, 1974). Fragrant (1 rec.). Flowering mainly X in LIB. to GAB., I–XII in ZAI.

Distribution. LIB. (5 coll.). IV.C. (6). NIG.: Niger D. (1). CAM.: S. part (7). GAB. (4). ZAI.: Kasai Z4 (3); Forestier Central Z6 (17); Ubangi-Uele Z7 (1). Material seen from AMES, BM, BR, E, K, L, P, UPS, WAG.

Note. *B. schinzianum* var. *schinzianum* is variable in several characters. In some specimens the raceme is very dense and semi-globose in outline due to the large, spreading, very pale bracts. In others the raceme is lax, with only slightly recurved, rather stiff bracts which are of dark colour.

The lip is also subject to extreme variability, especially in the length/width index and in the way the hairs are arranged along the margins.

Var. *schinzianum* and var. *irigaleae* do not differ very much. Only the general shape of the lip provides a more or less reliable character to distinguish between the two varieties. Specimens of var. *schinzianum* closest to the type of *B. schinzianum* have a broad lip, whereas others have much narrower lips which in shape approach var. *irigaleae*. Specimens of var. *schinzianum* with a narrow lip mainly originate from Central Africa where var. *irigaleae* does not occur. In West Africa (LIB., S.L. and IV.C.) where the two varieties are sympatric only specimens of var. *schinzianum* with a broad lip have been found.

The rim of hairs along the adaxial surface of the lip varies in thickness. In some specimens this rim is very thick and almost entirely covers the adaxial surface of the lip. In others it is much thinner or even entirely absent along the top part of the lip. The length of the hairs along the margins of the abaxial side of the lip is also variable.

Some of the characters mentioned above usually occur together in one specimen, others do not: A relatively narrow lip often occurs together with an elongate raceme but not with a thick rim of hairs on the adaxial side of the lip, and a relatively broad lip with a well developed rim of hairs along its adaxial side rarely occurs together with a very dense raceme.

1b. var. **irigaleae** (Cribb & Perez-Vera) J.J. Vermeulen, stat. nov. - Fig. 2.

Bulbophyllum irigaleae Cribb & Perez-Vera, Adansonia, ser. 2, 15 (1975) 200, 206.-Type: Perez-Vera 571 (K).

Bulbophyllum phaeopogon auct. non Schltr: Johansson, Acta Phyt. Suec. 59 (1974) 32, phot. p. 46; Adam, Fl. Descr. Mt. Nimba 5 (1981) 1682, pl. 913.

Bulbophyllum schinzianum auct. non Krzl.: Adam, Fl. Descr. Mt. Nimba 5 (1981) 1684, p.p.

Median sepal ovate-lanceolate, constricted about half-way, 9.5–12 by 3–3.6 mm. Lateral sepals 9.2–12 by 1.8–3 mm. Petals linear, 2.6–4.8 by 0.2–0.3 mm. Lip linear-lanceolate to linear in outline, the epichilium narrower than the hypochilium; 8.8–10.5 by 0.8–1 mm, margins beset with rather thick hairs, those that arise along the margins of the adaxial side of the lip short, very few, hardly constituting a rim at all, those that arise along the margins of the abaxial side of the lip up to 6 mm long; adaxially with very weak ridges near the base, surface glabrous. Column 3.8–4.6 mm long. Stelidia 2–2.2 mm long.

Colours. Pseudobulbs and leaves yellowish green. Sepals and petals yellowish, spotted purple adaxially, brownish purple abaxially. Lip brownish purple, basal half yellowish brown, at the base green; hairs purple.

Ecology. Primary forest up to 800 m alt. Epiphyte in the crown of trees, on bark or in humus deposits (Johansson, 1974, sub *B. phaeopogon*). Flowering observed in V, VI, XI.

Distribution. LIB. (5 coll.). IV.C.: W. part (2). Material seen from K, UPS.

Note. See under var. *schinzianum*.

1c. var. **phaeopogon** (Schltr) J.J. Vermeulen - Fig. 3; Plate 1b.

Bulbophyllum schinzianum var. *phaeopogon* (Schltr) J.J. Vermeulen, Bull. J. Bot. Nat. Belg. 56 (1986) 229.-*Bulbophyllum phaeopogon* Schltr, Engl., Bot. Jahrb. 38 (1906) 157; Rendle, Cat. Talb. S. Nig. Pl. (1913) 146; De Wild., Pl. Beq. 1 (1921) 116; Schltr, Fedde, Rep. Beih. 68 (1932) pl. 68; Summerh. in F.W.T.A. ed. 1, 2 (1936) 437; id., ed. 2, 3 (1968) 236; Geerinck, Bull. Soc. R. Bot. Belg. 109 (1976) 176.-Syntypes: Winkler 14a (B, lost; isolecto Z, here proposed); Schlechter 15791 (B, lost, sketch of Summerhayes in K).

Non *Bulbophyllum phaeopogon* auct.: Johansson, Acta Phyt. Suec. 59 (1974) 32; Adam, Fl. Descr. Mt. Nimba 5 (1981) 1682 (= both var. *irigaleae*).

Bulbophyllum calamarium auct. non Lindl.: Rolfe in F.T.A. 7 (1897) 33.

Bulbophyllum gentilii Rolfe, Gard. Chron., ser. 3, 35, no. 929 (1904) 266, p. sp. Zenker 832 ('823') & Preuss 1225 (Syntype specimens of *B. gentilii*), lectotype specimen of *B. gentilii* (Kew cult. Hort. [1903] = Gentil [1898] s.n.) excluded (= var. *schinzianum*); De Wild., Pl. Beq. 1 (1921) 111, p.p.

Bulbophyllum schinzianum auct. non Krzl.: Segerbaeck, Orch. Nig. (1983) 59, col. phot.

Median sepal elliptic to oblong, not constricted about half-way, 9.5–11 by 1.4–4 mm. Lateral sepals 9–11 by 2–2.9 mm. Petals linear, 6.5–8 by 0.3–1 mm. Lip oblong to lanceolate in outline, the epichilium wider than the hypochilium; 6–7.5 by 2–2.5 mm, margins beset with thick hairs, those that arise along the margins of the adaxial side of the lip short, placed in a dense but narrow rim, those that arise along the margins of the abaxial side of the lip up to 4 mm long; adaxially with two distinct, rounded, fleshy ridges near the base, occasionally with a third, much thinner median ridge starting between them and continuing up to about half-way the lip; adaxial surface coarsely papillose in the top half. Column 3–6 mm long. Stelidia 2–3 mm long.

Colours. Pseudobulbs and leaves pale green. Floral bracts yellow, dry when the flowers are open. Sepals yellowish, cream coloured or greenish adaxially, with purple-brown spots, or entirely purple-brown or dirty purple. Petals yellowish, with purple tips or margins. Lip glossy yellowish or green, with dark purple and/or brown hairs. Column white.

Ecology. Primary forest up to 500 m alt., relict forest in savanna (Sanford, 1974). Epiphyte. Flowering II–VI.

Distribution. IV.C. (1 coll.). GHANA (2). NIG.: E part, Cross P. & Rivers P. (3). CAM.: S. part (17). GAB. (3). ZAI.: Bas Congo Z3 (1). Material seen from AMES, BM, BR, C, E, G, K, L, P, S, U, WAG, Z.

Note. The ridges on the adaxial side of the lip may be either distinct and reach over half the length of the lip, or inconspicuous and much shorter.

2 - **Bulbophyllum subligaculiferum** J.J. Vermeulen, spec. nov. – Fig. 4; Plate 1c.

Bulbophyllum subligaculiferum J.J. Vermeulen, spec. nov. Bulbophyllo schinziano var. irigaleae similis, in denticulo adaxialiter a basi labelli prorsum versus triangulari complanato differt.–Type: De Wilde, Arends, Louis, Bouman & Karper 183 (WAG).

Roots 0.5–1.2 mm diam. *Rhizome* c. 2 mm diam., nodes not or hardly swollen. *Rhizome scales* not seen. *Pseudobulbs* 1-leafed, orbicular to ovoid, c. 1 cm apart, c. 3.2 by 3 cm, distinctly flattened, about sharply 2-angled. *Petiole* 15 mm long. *Leaf blade* lanceolate, c. 9.5 by 2,5 cm, tip obtuse, hardly emarginate, slightly oblique, coriaceous, thick; midrib slightly prominent abaxially. *Inflorescence* c. 95 cm long, c. 45-flowered. *Peduncle* patent, with a tortuose base, c. 80 cm by 2.5 mm, orbicular in section, glabrous, with 11 scales. *Peduncle scales* tubular, the longest c. 28 mm long; tip acute, the lower obtuse; membranous, with some nerves; slightly keeled towards the tip and glabrous abaxially. *Rhachis* c. erect, terete, about orbicular in section, c. 15 cm long, glabrous. *Floral bracts* recurved, hardly concave, (slightly obovate)-lanceolate, c. 22 by 7 mm, tip acute to slightly acuminate; base narrowly attached; membranous, with some nerves; slightly keeled towards the tip and glabrous abaxially.

Flowers scattered, in a rather dense raceme, recurved to spreading, few open simultaneously, fully open. *Pedicel and ovary* c. 8 mm long, glabrous. *Median sepal* slightly recurved, ovate-lanceolate, c. 12 by 3.6 mm, tip acute; margins entire, glabrous; base rather broadly attached; rather thin; glabrous; not keeled abaxially. *Lateral sepals* free, recurved, not falcate, slightly oblique, c. 12 by 3 mm, otherwise as the median sepal. *Petals* not recurved, falcate, linear, not oblique, c. 4.8 by 0.3 mm, tip acute; margins entire, glabrous; base broadly attached; thin; glabrous; not keeled abaxially. *Lip* slightly recurved near the base, linear-lanceolate in outline (not spread), slightly constricted in the basal half, c. 8.5 by 1.2 mm, tip subacute; margins entire, beset with rather thick hairs of various length, those that arise along the margins of the adaxial side of the lip short, placed in a thick rim in the basal half which continues as a single row of hairs towards the tip, only at the very tip some longer hairs, those that arise along the margins of the abaxial side of the lip short in the basal half of the lip but up to 6 mm long in the top half; lip rather thick, with a thick base; adaxially with two rather distinct ridges in the basal half, with a distinctly concave basal part which continues as a median slit over c. 1/3 of the lip and which is covered by a forwards projecting, dorsoventrally flattened, deltoid, about 0.8 mm long median tooth with an emarginate tip which is inserted at the very base of the lip, surface glabrous, pubescent near the base; abaxially with a weak obtuse ridge near the base, surface glabrous. *Column* from the base of the free part up to the tip of the stelidia c. 3 mm long, with distinct, gradually tapering, falcate, acute stelidia of c. 1.5 mm long; with weak rounded wings along its lower margins at the base of the stelidia; stigma without a basal callus. *Anther* c. 1.2 by 0.8 mm; front not drawn out, thin, not concave; front margin straight, entire, about glabrous; basally hardly emarginate; abaxially with a distinct, rounded, colliculate ridge over about the entire length of the anther which distinctly overtops the front margin as an acute beak, surface colliculate. *Pollinia* 4, about ellipsoid; the small ones more than half as long as the large ones, distinctly flattened, the large ones flattened; a thin stipes has been observed. *Fruit* not seen.

Colours. Floral bracts yellowish. Sepals and petals yellowish. Lip with purple hairs.

Ecology. Primary forest at 570 m alt. Epiphyte. Flowering I (once observed).

Distribution. GAB. (1 coll.).

Note. *B. subligaculiferum* is very similar to *B. schinzianum* var. *irigaleae*. Close scrutiny, however, reveals differences which are considered important enough to justify treatment as a new species: the median tooth at the base of the adaxial surface of the lip covering a concavity, the front margin of the anther which is not drawn out and the pollinia which are obtuse at both ends.

A markedly sinuose peduncle as in *B. subligaculiferum* is not observed in *B. schinzianum*

3 - Bulbophyllum saltatorium Lindl. – Fig. 5–9; Plate 1d, 2a-c.

Literature: see under the varieties.

Roots 0.4–5 mm diam. *Rhizome* 1.1–1.5 mm diam., nodes not or hardly swollen. *Rhizome scales* tubular, tip subacute, the lower occasionally obtuse; membranous, with nerves; slightly keeled abaxially. *Pseudobulbs* 1-leafed, orbicular to narrowly ovoid, 0.8–7 cm apart, 0.7–5 by 0.6–3.3 cm, more or less flattened, obtusely to sharply 2–4(–5)-angled. *Petiole* 7–35 mm long. *Leaf blade* narrowly elliptic to linear-lanceolate, maximum width occasionally slightly above the middle, 1.3–21 by 0.8–5.5 cm, tip rounded to subacute, slightly emarginate, occasionally oblique; coriaceous, thick; midrib not or hardly prominent abaxially. *Inflorescence* 3.5–84 cm long, 2–105-flowered. *Peduncle* wiry to sturdy, erect to patent, 1.5–58 cm by 0.5–2.5 mm, orbicular in section, glabrous, with 6–11 scales. *Peduncle scales* tubular, the longest 4.5–13 mm long; tip acute, the lower obtuse to acute; membranous, with or without some nerves; slightly keeled and glabrous abaxially. *Rhachis* erect to slightly arching, terete, about orbicular in section, either slightly zigzag bent or not, 0.7–40 cm long, glabrous. *Floral bracts* recurved, narrowly elliptic to oblong, 2–9.5 by 1–3.5 mm, tip obtuse to acute; base rather narrowly attached; membranous, with some nerves; not or hardly keeled and glabrous abaxially. *Flowers* distichous or scattered, in a dense to rather lax raceme, spreading, either secund or not, few to many open simultaneously, not fully to wide open. *Pedicel and ovary* 3.2–10 mm long, glabrous. *Median sepal* either recurved or not, narrowly ovate-oblong to ovate-lanceolate, 5.6–15 by 1.8–3.8 mm, tip acute to subacute; margins entire, finely ciliate; base broadly attached; rather thick; glabrous; not keeled abaxially. *Lateral sepals* free, spreading to reflexed, hardly falcate, oblique, 5.4–15 by 1.7–3.6 mm, otherwise as the median sepal. *Petals* not recurved, slightly falcate, linear, oblique, (2 –) 2.5–6 (– 7) by 0.15–0.6 mm, tip acute to acuminate; margins entire, glabrous; base broadly attached; thin to rather thick; glabrous; not keeled abaxially. *Lip* recurved near the base, ovate-oblong to linear in outline (not spread), 4–9 (– 12) by 0.7–1.8 mm, tip obtuse to acute; margins entire, beset with short and long hairs (see varieties), just above the base with or without obliquely semi-elliptic lateral lobes of less than 1 mm long which have an obtuse tip, finely papillose margins, a broadly attached base and which are thin and finely papillose; rather thick, with a thick base; surface glabrous; adaxially with two rather weak ridges in the basal half, with a slightly concave basal part, with a median slit starting at about the base and continuing over 1/3 to 1/2 of the lip; abaxially with a weak obtuse ridge near the base. *Column* from the base of the free part up to the tip of the stelidia 1.5–3.5 mm long, with distinct, gradually tapering, more or less falcate, acute to acuminate stelidia of 0.6–1.8 mm long; with weak, rounded wings along its adaxial margins just below the base of the stelidia; stigma with a weak to distinct basal tooth. *Anther* 0.6–0.9 by 0.5–0.9 mm; front hardly drawn out; front margin rounded, entire, glabrous to papillose; basally hardly emarginate; abaxially with a distinct, slightly conical, obtuse, colliculate ridge near the top of the anther which distinctly overtops the front margin, surface colliculate, glabrous towards the margins. *Pollinia* 4; the small ones more than half as long as the large ones, narrowly ellipsoid to obovoid, distinctly flattened on either side; the large ones ellipsoid to obovoid, flattened on one side; a rather thick stipes has been observed. *Fruit* obovoid, up to 2.2 by 0.6 cm, stalk 7 mm.

KEY TO THE VARIETIES

1 - The very fine and velvety or wooly hairs that arise along the margins of the adaxial side of the lip constitute a thick rim up to the extreme tip of the lip; in the top part

of the lip they are never coarser and longer......................var. **calamarium** (c)

- The very fine and velvety or wooly hairs that arise along the margins of the adaxial side of the lip constitute a thick rim only to about half-way the lip; in the top half of the lip they either gradually disappear or become much coarser and longer..2

2 - Inflorescence 8 cm or shorter, not or only slightly overtopping the leaves; 15–30 flowers in a dense raceme..var. **saltatorium** (a)

- Inflorescence 6.5 cm or longer (if less than 8 cm long with only very few, widely spaced flowers), usually largely overtopping the leaves; raceme either dense or lax .... var. **albociliatum** (b)

3a. var. **saltatorium** - Fig. 5; Plate 1d.

Bulbophyllum saltatorium Lindl., Bot. Reg. 23 (1837) pl. 1970 ('*Bolbophyllum*'); Rchb. f., Walp. Ann. 6 (1861) 252 ('*Bolbophyllum*'); Durand & Schinz, Consp. Fl. Afr. 5 (1895) 14; Rolfe in F.T.A. 7 (1897) 34; Schltr, Die Orch. (1914) 329; De Wild., Pl. Beq. 1 (1921) 119; Summerh. in F.W.T.A. ed. 1, 2 (1936) 437; id., ed. 2, 3 (1968) 236; Johansson, Acta Phyt. Suec. 59 (1974) 32; Cribb & Perez-Vera, Adansonia, ser. 2, 15 (1975) 206; Adam, Fl. Descr. Mt. Nimba 5 (1981) 1683.–*Phyllorchis saltatoria* (Lindl.) Kuntze, Rev. Gen. Pl. 2 (1891) 678, incorr. name.–Type: Loddiges s.n. (K).

Pseudobulbs 1.4–3 cm apart, 1.2–2.7 cm long, distinctly flattened. Leaves 3–11.5 by 1–2.3 cm. Inflorescence 3.5–8 cm long. Peduncle 1.5–4 cm long. Rhachis 1.5–4 cm long, 15–30-flowered. Floral bracts 4.5–6.5 by 1.8–2.7 mm. Flowers opening about simultaneously. Pedicel and ovary 4–6 mm long. Median sepal 6–7.5 by 2–2.2 mm. Lateral sepals narrowly ovate-oblong to ovate-lanceolate, 5.8–7.5 by 2–2.2 mm. Petals 3.4–5 by 0.15–0.4 mm, more than half as long as the median sepal. Lip 5–6.5 by 1–1.8 mm, those hairs that arise along the margins of the adaxial side of the lip are very fine and velvety or wooly and constitute a thick rim only to about half-way the lip; in the top half of the lip they either gradually disappear or become much coarser and longer. Column 1.5–2.2 mm. Stelidia 0.6 mm.

Colours. Pseudobulbs and leaves green. Sepals dark purple adaxially; brownish purple abaxially. Petals purple. Lip dark purple-red with purple or brownish purple hairs.

Ecology. Primary forest up to 600 m alt. Epiphyte on tree trunks and the central part of the crown, on bark or in minor humus deposits (Johansson, 1974).

Distribution. S.L. (3 coll.). LIB. (1). IV.C. (5). GHANA (1). NIG.: W. part (1). EQ.GUIN.: Mbini (fide Sanford, 1974).

Note. Unlike var. *albociliatum* and, to a lesser extent, var. *calamarium* this variety shows very little variability.

3b. var. **albociliatum** (Finet) J.J. Vermeulen. - Fig. 6–8; Plate 2a, b.

Bulbophyllum saltatorium Lindl. var. *albociliatum* (Finet) J.J. Vermeulen, Bull. J. Bot. Nat. Belg. 56 (1986) 230.–*Bulbophyllum calamarium* Lindl. var. *albociliatum* Finet, Not. Syst. 1 (1911) 383.–Type: Lousteau s.n. (P).

Bulbophyllum distans Lindl., J. Linn. Soc. Bot. 6 (1862) 125 ('*Bolbophyllum*'); Durand & Schinz, Consp. Fl. Afr. 5 (1895) 11; Rolfe in F.T.A. 7 (1897) 34; De Wild., Pl. Beq. 1 (1921) 108; Summerh. in F.W.T.A. ed. 1, 2 (1936) 437; Ake Assi, Et. Fl. C. Iv. (1964) 243; Summerh., Kew Bull. 12 (1957) 113; id. in

F.W.T.A. ed. 2, 3 (1968) 236; Piers, Orch. E. Afr. (1968) 99; Johansson, Acta Phyt. Suec. 59 (1974) 31; Geerinck, Bull. Soc. R. Bot. Belg. 109 (1976) 176; Adam, Fl. Descr. Mt. Nimba 5 (1981) 1675; Segerbaeck, Orch. Nig. (1983) 59, phot.; Cribb, in F.T.E.A., Orch. 2 (1984) 305.–*Phyllorchis distans* (Lindl.) Kuntze, Rev. Gen. Pl. 2 (1891) 677, incorr. name.–Type: Mann 525 (K).

Non *Bulbophyllum distans* J.J. Smith, Bull. J. Bot. Buit. ser. 2, 26 (1918) 57.

Bulbophyllum miniatum Anon., Orch. Rev. 12 (1904) 118, nom. nud.; Moore, J. R. Hort. Soc. 33 (1908) 380, nom. nud.; Summerh., Kew Bull. 1931 (1931) 389.–Neotype: Kew cult. Hort. (1931) (K, here proposed; most probably this is the specimen Summerhayes used for his description because a printed copy of it has been attached to the sheet).

Bulbophyllum kindtianum De Wild., Not. Pl. Ut. et Int. Fl. Congo 2 (1904) 309; id., Bel. Col. 10 (1904) 449, n.v.; id., Miss. Laur. (1905) 53; Durand & Durand, Syll. Fl. Congol. (1909) 516; De Wild., Et. Fl. B. et Moy. Congo 3 (1910) 182; id., Pl. Beq. 1 (1921) 113; Summerh. in F.W.T.A. ed. 1, 2 (1936) 437.– Type: 'Rivière Loliva, juillet 1903 (Marcel Laurent)' (BR?, see note below).

Bulbophyllum mildbraedii Krzl., Engl., Bot. Jahrb. 43 (1909) 340; Krzl. in Mildbraed, Deutsch. Zentr. Afr. Exp. 1907-08, 2 (1914) 84; De Wild., Pl. Beq. 1 (1921) 114.–Type: Mildbraed 3312 (B, lost).

Bulbophyllum nudiscapum Rolfe, Kew Bull. 1909 (1909) 365; Rendle, Cat. Talb. S. Nig. Pl. (1913) 146; De Wild., Pl. Beq. 1 (1921) 114. Syntypes: Mantin s.n. (K), V. Imschoot s.n. (K), Afzelius s.n., n.v.

Bulbophyllum gracilliscapum Summerh., Kew Bull. 8 (1954) 579.–*Bulbophyllum flexiliscapum* Summerh., Kew Bull. 11 (1956) 232. Type: Le Testu 5787 (Holo K; iso P).

Non *Bulbophyllum graciliscapum* Schltr, Schum. & Laut., Nachtr. (1905) 203; nec *Bulbophyllum graciliscapum* Ames & Rolfe in Ames, Orch. 5 (1915) 175; nec *Bulbophyllum graciliscapum* Perrier, Not. Syst. 6 (1937) 107.

Pseudobulbs 0.8–7 cm apart, 0.7–5 cm long, either distinctly flattened or not. Leaves 1.3–20.5 by 0.8–4.6 cm. Inflorescence 6.5–84 cm long. Peduncle 5–58 cm long. Rhachis 1.5–40 cm long, 4–105-flowered. Floral bracts 2–9,5 by 1–3.5 mm. Flowers opening either about simultaneously or few at a time. Pedicel and ovary 3.2–10 mm long. Median sepal 5.6–12 (- 15) by 1.8–3.5 mm. Lateral sepals ovate-lanceolate, 5.4–12 (- 15) by 1.7–3 mm. Petals 2–6 by 0.2–0.35 mm, either more than half as long as the median sepal or less. Lip 4–8.5 (- 12) by 0.7–1.8 mm, those hairs that arise along the margins of the adaxial side of the lip are very fine and velvety or wooly and constitute a thick rim only to about half-way the lip; in the top half of the lip they either gradually disappear or become much coarser and longer. Column 1.5–3 mm. Stelidia 0.8–1.5 mm.

Colours. Pseudobulbs and leaves pale to dark green. Sepals yellowish or greenish adaxially, usually suffused with red-purple or brown-purple, or entirely red-purple. Petals red-purple. Lip greenish, usually suffused with dark red-purple, or entirely dark red-purple, occasionally white towards the tip. Hairs purple, pink or white.

Ecology. Primary forest at 0–750 m alt; low coastal forest (2 rec.). Epiphyte in the crown of trees, on bark or in minor humus deposits (Johansson. 1974). Occasionally terrestrial or lithophytic. Flowering I–XII.

Distribution. LIB. (7 coll.). IV.C. (3). GHANA (1). NIG.: Cross P. (2). CAM.: S. part (7). C.A.R. (1). EQ.GUIN.: Bioko (1). GAB. (26). CONGO (1). ZAI.: Kasai Z4 (1); Forestier Central Z6 (8); Lac Albert Z8 (1); Lacs Edouard & Kivu Z9 (2). UG.: West U2 (2). ANG. (2). Material seen from AMES, BM, BR, K, L, P, UPS, WAG.

Notes. 1 - *B. saltatorium* var. *albociliatum* includes a series of specimens which are of a rather different appearance. Two entities can be distinguished which have not been given a taxonomical rank because numerous intermediate specimens exist:

a. - The entity which includes the type specimen of *B. distans*, *B. kindtianum* and *B. nudiscapum*. Inflorescence up to 84 cm long, comparatively sturdy (up to 2.5 mm diam.), with up to 105 flowers. Peduncle up to 58 cm long. Rhachis up to 40 cm long, not zigzag bent. Flowers scattered, one or few open at the time. Floral bracts up to 9.5 mm long. Lip with purple hairs.

b. - The entity which includes the type specimen of *B. calamarium* var. *albociliatum* and *B. miniatum*. The type of *B. mildbraedii* may also belong here but has not been studied. Inflorescence up to 40 cm long, comparatively wiry (up to 0.9 mm diam.), with up to 27 flowers. Peduncle up to 24.5 cm long. Rhachis up to 15.5 cm long, zigzag bent. Flowers distichous, about all open simultaneously. Floral bracts up to 4.4 mm long. Lip with white hairs (the shorter, upwards facing hairs may be purple).

In most parts of its range entity a. shows little variability. Only where entity b. occurs (CAM., EQ.GUIN., GAB. and ZAI.) intermediates between entity a. and b. have been found. More or less in that same restricted area specimens occur which fit well within var. *albociliatum* except for a single character:

- The specimen Bos 6423 from CAM. has the longer hairs on its lip slightly widened and flattened.

- The specimens Dekker 434 (= Wageningen cult. Hort. 79-083) and v.d. Laan 844 (= Wageningen cult. Hort. 83-236 = Leiden cult. Hort. 23686), both from GAB. have a short inflorescence hardly overtopping the leaves. This is usual in var. *saltatorium*; however, the specimens are otherwise identical with var. *albociliatum*.

- The specimens v.d. Meer 1264 from CAM. and Talbot 3309 from NIG. have a very densely flowered raceme.

- The specimen Tisserant in Le Testu 2638 from C.A.R. is remarkable for its very large flowers (sepals 13 mm long) and especially for its long lip, which is 12 mm long but only 0.7 mm wide. The petals are slightly shorter than usual (only 2 mm).

- The specimens Le Testu 5787 (type of *B. gracilliscapum*) and v.d. Laan 746 (= Wageningen cult. Hort. 83-376) both from GAB. have a lip with much thicker and stiffer hairs than usual. The plants are extremely small.

The variability which var. *albociliatum* shows in a small area of its extensive range, and the existence of specimens as those mentioned above in that same area may possibly be explained in terms of beginning speciation.

2 - The type specimen of *B. kindtianum* has been cited by De Wildeman in 1904 as 'Rivière Loliva, juillet 1903 (Marcel Laurent)'. The specimen was cultivated in the Bot. Garden in Laeken. In 1921 he mentioned the same specimen as 'Loliva, 1903 (M. Laurent 819)'. A photograph of what is probably the Laeken specimen, published with the original description of *B. kindtianum* in 1904, depicts a specimen which clearly belongs to var. *albociliatum*. The only specimen 'Laurent 819' which is kept in BR is not marked as a type specimen and originates from the Bot. Garden in Eala. The latter specimen might have been a duplicate from the Laeken specimen if only it did not belong to var. *calamarium*. Other specimens of var. *albociliatum* in BR, collected by others and as well cultivated in Eala, have also been marked '819'. Most likely this is a number used in the Eala Garden, not a collector's number.

The real type specimen of *B. kindtianum* is probably lost or has never existed.

3c. var. **calamarium** (Lindl.) J.J. Vermeulen. – Fig. 9; Plate 2c.

Bulbophyllum saltatorium Lindl. var. *calamarium* (Lindl.) J.J. Vermeulen, Bull. J. Bot. Nat. Belg. 56 (1986) 230.–*B. calamarium* Lindl., Bot. Reg. 29 (1843) misc. 70 ('*Bolbophyllum*'); Hook., Bot. Mag. 70 (1844) pl. 4088 ('*Bolbophyllum*'); Hook., Cent. Orch. Pl. (1849) 17, pl. 4 ('*Bolbophyllum*'); Rchb. f., Walp. Ann. 6 (1861) 252 ('*Bolbophyllum*'); Rolfe in F.T.A. 7 (1897) 33; Schltr, Westafr. Kautchuk Exp. (1901) 281; De Wild., Not. Pl. Ut. et Int. Fl. Congo 1 (1903) 123; id., Et. Fl. B. et Moy. Congo 1 (1904) 115; id., Miss. Laur. (1905) 53; id., Et. Fl. B. et Moy. Congo 2 (1908) 237; Durand & Durand, Syll. Fl. Congol. (1909) 516; Schltr, Die Orch. (1914) 323; Krzl. in Mildbraed, Deutsch. Zentr. Afr. Exp. 1907-08, 2 (1914) 84; De Wild., Pl. Beq. 1 (1921) 107; Schltr in Mildbraed, 2e Deutsch. Zentr. Afr. Exp. 1910-11, 2 (1922) 30; Summerh. in F.W.T.A. ed. 1, 2 (1936) 437; id., Kew Bull. 12 (1957) 113; id. in F.W.T.A. ed. 2, 3 (1968) 236; Johansson, Acta Phyt. Suec. 59 (1974) 29; Cribb & Perez-Vera, Adansonia, ser. 2, 15 (1975) 206; Adam, Fl. Descr. Mt. Nimba 5 (1981) 1673.–*Phyllorchis calamaria* (Lindl.) Kuntze, Rev. Gen. Pl. 2 (1891) 677, incorr. name.–Type: Fielding s.n. (K).

Non *Bulbophyllum calamarium* auct.: Lindl., J. Linn. Soc. Bot. 6 (1862) 125 ('*Bolbophyllum*'); Durand & Schinz, Consp. Fl. Afr. 5 (1895) 11 (= both *B. schinzianum* var. *schinzianum*).

Bulbophyllum rupincola Rchb. f., Flora 48 (1865) 184 ('*Bolbophyllum*'); Durand & Schinz, Consp. Fl. Afr. 5 (1895) 14; Rolfe in F.T.A. 7 (1897) 33; Rendle, Cat. Welw. Afr. Pl. 2 (1899) 3; De Wild., Pl. Beq. 1 (1921) 119.–Type: Welwitsch 697 (Holo BM; iso G, K, P).

Pseudobulbs 1–4.5 cm apart, 1.5–4.2 cm long, either distinctly flattened or not. Leaves 6.5–21 by 1.6–5.5 cm. Inflorescence 4–60 cm long. Peduncle 3.3–50 cm long. Rhachis 0.7–20 cm long, 2–80-flowered. Floral bracts 6.5–13 by 2–5 mm. Flowers opening few at the time. Pedicel and ovary 5.8–9 mm long. Median sepal 9–12.5 by 2.8–3.8 mm. Lateral sepals ovate-lanceolate to very narrowly triangular, 8.5–12.5 by 2.2–3.5 mm. Petals 3.5–4.6 (– 7) by 0.3–0.6 mm, less than half as long as the median sepal. Lip 7–9 by 1–1.5 mm, those hairs that arise along the margins of the adaxial side of the lip are very fine and velvety or wooly and constitute a thick rim up to the extreme tip of the lip; in the top half of the lip they are never coarser or longer. Column 2–3.5 mm. Stelidia 0.9–1.8 mm.

Colours. Pseudobulbs and leaves green. Sepals yellowish or greenish adaxially, stained or spotted red-purple or brown-purple; abaxially slightly stained purple. Petals red-purple. Lip red-purple or brown-purple, often glossy green near the base; short hairs yellowish, whitish or purple, long hairs purple.

Ecology. Primary forest at 0–900 m alt. Epiphyte in the crown of trees, on bark or in minor humus deposits (Johansson, 1974). Occasionally lithophytic.

Distribution. S.L. (2 coll.). LIB. (7). IV.C. (7). GHANA (1). CAM. (2). EQ.GUIN.: Mbini (1). GAB. (1). CONGO (1). ZAI.: Forestier Central Z6 (4). ANG. (1). Material seen from AMES, BM, BR, G, HBG, K, L, P, UPS, WAG.

Notes. 1 - The bracts along the rhizome and the inflorescence tend to have acute tips, unlike the other varieties of *B. saltatorium* which usually have bracts with slightly more obtuse tips.

2 - The specimen Laurent 819: see note 2 under var. *albociliatum*.

4 - **Bulbophyllum barbigerum** Lindl. - Fig. 10; Plate 2d.

Bulbophyllum barbigerum Lindl., Bot. Reg. 23 (1837) pl. 1942 ('*Bolbophyllum*'); Rchb. f., Walp. Ann. 6 (1861) 252 ('Bolbophyllum); Veitch, Man. Orch. Pl. 3 (1888) 93, pl.; Durand & Schinz, Consp. Fl. Afr. 5 (1895) 10; Rolfe in F.T.A. 7 (1897) 34; Schltr, Westafr. Kautchuk Exp. (1901) 280; De Wild., Not. Pl. Ut. et Int. Fl. Congo 2 (1904) 309; id., Et. Fl. B. et Moy. Congo 1 (1904) 115; id., Miss. Laur. (1905) 55; id., Et. Fl. B. et Moy. Congo 2 (1908) 237; Durand & Durand, Syll. Fl. Congol. (1909) 516; De Wild., Et. Fl. B. et Moy. Congo 3 (1910) 181; Schltr, Die Orch. (1914) 321, pl.; De Wild., Pl. Beq. 1 (1921) 103; Schltr in Mildbraed, 2e Deutsch. Zentr. Afr. Exp. 1910-11, 2 (1922) 30; Summerh. in F.W.T.A. ed.

1, 2 (1936) 437, pl.; id., ed. 2, 3 (1968) 236, pl.; Geerinck, Bull. Soc. R. Bot. Belg. 109 (1976) 175; Adam, Fl. Descr. Mt. Nimba 5 (1981) 1671, pl. 904; Stewart & Hennesy, Orch. Afr., sel. rev. (1981) 86, pl. 20; Lückel, Die Orch. 35 (1984) 125, col. phot.-*Phyllorchis barbigera* (Lindl.) Kuntze, Rev. Gen. Pl. 2 (1891) 677, incorr. name.-Type: Loddiges s.n. (K).

Roots 0.5-1.2 mm diam. *Rhizome* 2-5 mm diam., nodes not or hardly swollen. *Rhizome scales* tubular, tip acute, membranous with well-spaced nerves, keeled abaxially towards the tip. *Pseudobulbs* 1-leafed, orbicular to narrowly ellipsoid or slightly ovoid, 1-2.5 cm apart, 1.5-3.5 by 1.5-3 cm, distinctly flattened, obtusely 2-4-angled. *Petiole* 1-4 mm long. *Leaf blade* narrowly elliptic to broadly linear-lanceolate, 3-13 by 1.7-3 cm, tip obtuse, slightly emarginate, either slightly oblique or not; coriaceous, thick; midrib hardly prominent abaxially. *Inflorescence* 7-18 cm long, 12-32-flowered. *Peduncle* erect to patent, 3-10 cm by 0.9-1.5 mm, orbicular in section, glabrous, with 4-7 scales. *Peduncle scales* tubular, the longest 5-8 mm long; tip acute to acuminate; membranous, without nerves; keeled and glabrous abaxially. *Rhachis* about erect, terete, about orbicular in section, either slightly zigzag bent or not, 6-14 cm long, glabrous. *Floral bracts* about spreading, ovate-oblong, 6-10 by 2.5-4 mm, tip acute to acuminate; base rather broadly attached; membranous, with some nerves; not or hardly keeled and glabrous abaxially. *Flowers* distichous, in a rather lax raceme, spreading, occasionally slightly secund, some open simultaneously, fully open. *Pedicel and ovary* 0.6-1.5 mm long, glabrous. *Median sepal* either slightly recurved or not, ovate-lanceolate to broadly linear-lanceolate, 11-14 by 2-4 mm, tip acute; margins entire, glabrous; base rather broadly attached; rather thin; glabrous; not keeled abaxially. *Lateral sepals* free, reflexed at their base, hardly falcate, ovate-lanceolate, oblique, 9-13.5 by 2-3 mm, base broadly attached; otherwise as the median sepal. *Petals* not recurved, not falcate, ovate to ovate-oblong, oblique, 0.3-0.8 by 0.2-0.35 mm, tip subacute; margins entire, glabrous; base broadly attached; thick; glabrous; not keeled abaxially. *Lip* recurved near the base, narrowly linear-lanceolate in outline (not spread), 8-11 by 1-1.5 mm, tip about acute; margins entire, beset with short and long hairs, the short hairs constituting a rim along the whole margin except for the base, thin, wooly, the long hairs arising from the adaxial side of the lip constituting a dense tuft near the tip of the lip, the long hairs arising from the abaxial side of the lip constituting a rim starting about half-way the basal half of the lip and continuing, less dense, up to the tip, some or many of those near the tip ending in small paleae; lip rather thick, with a thick base; surface glabrous; adaxially without ridges, with a slightly concave basal part, with a median slit near the base; abaxially with a weak obtuse ridge near the base. *Column* from the base of the free part up to the tip of the stelidia 3-3.5 mm long, with distinct, gradually tapering, bent upwards, acute stelidia of 1.5-2 mm long; column with weak, rounded wings along the adaxial margins, just below the stelidia; stigma with a weak basal tooth. *Anther* c. 1 by 0.8 mm; front hardly drawn out; front margin rounded, about entire, glabrous; basally hardly emarginate; abaxially with a distinct , more or less conical, obtuse, coliculate to papillose ridge near the top of the anther which distinctly the front margin; surface colliculate, glabrous towards the margins. *Pollinia* 4, all about elliptic; the small ones more than half as long as the large ones; distinctly flattened on either side; the large ones flattened on either side; a rather thick stipes has been observed. *Fruit* not seen.

C o l o u r s. Pseudobulbs and leaves pale (glaucous) green. Sepals yellowish or greenish adaxially, suffused or spotted red-purple; abaxially yellowish brown. Petals very pale greenish. Lip yellowish or greenish, marked red-purple; hairs at the base of the lip white, at the top of the lip purple. Column very pale greenish, stelidia red-purple.

E c o l o g y. Evergreen and semi-deciduous primary forest up to 900 m alt. Epiphyte on the trunks and in the crown of trees, on bark or in minor humus deposits (Johansson, 1974). Flowering I–XII.

D i s t r i b u t i o n. S.L. (6 coll.). LIB. (7). IV.C. (4). NIG.: S. part (2). CAM.: S. part (2). GAB. (1). C.A.R. (1). CONGO (3). ZAI.: Forestier Central Z6 (5). Material seen from AMES, BM, BR, G, K, L, P, UPS, WAG.

5 - **Bulbophyllum vulcanicum** Krzl. - Fig. 11; Plate 3a.

Bulbophyllum vulcanicum Krzl., Engl., Bot. Jahrb. 51 (1914) 383; De Wild., Pl. Beq. 1 (1921) 121; Robyns, Fl. Sp. P. Alb. 3 (1955) 494; Cribb, in F.T.E.A., Orch. 2 (1984) 315.–T y p e: Kassner 2340 (Holo B, lost; iso BM, BR, E, K, P, Z).

Roots 0.4–1 mm diam. *Rhizome* 1.5–4.5 mm diam., nodes not or hardly swollen. *Rhizome scales* tubular, tip acute to subacute, the lower obtuse, membranous, the upper with nerves; slightly keeled abaxially towards the tip. *Pseudobulbs* 2-leafed, about cylindrical, slightly thickened towards the base, 3.5–11 cm apart, 3.5–11 by 0.4–1.3 cm, either slightly flattened or not, inconspicuously and obtusely 2–4-angled. *Petiole* 5–30 mm long. *Leaf blade* lanceolate to linear, 10.5–17 by 0.6–3.1 cm, tip obtuse, either slightly emarginate and/or oblique or not; coriaceous, rather thick; midrib slightly prominent abaxially. *Inflorescence* 18–40 cm long, 28–72-flowered. *Peduncle* erect to patent, occasionally slightly swollen towards the tip, 11.5–24 cm by 1–2.5 mm, about orbicular in section, flattened towards the base, glabrous, with 3–5 scales. *Peduncle scales* tubular, the longest 7–11 mm long; tip about acute; membranous, with some nerves; slightly keeled and glabrous abaxially. *Rhachis* about erect, terete to slightly swollen, about orbicular in section, either slightly zigzag bent or not, 4.5–18.5 by at most 3.2 mm, glabrous. *Floral bracts* recurved, ovate to broadly ovate-lanceolate, 8–9.5 by 2.5–5 mm, tip acute; base rather broadly attached; membranous, with some nerves; keeled and glabrous abaxially. *Flowers* distichous, in a dense raceme, spreading, secund, many open simultaneously, not fully open. *Pedicel and ovary* 1.7–4.5 mm long, glabrous. *Median sepal* slightly recurved or not, narrowly ovate-oblong to ovate-lanceolate, 5.4–7.5 by 1.7–2.5 mm, tip about acute; margins entire, glabrous; base broadly attached; rather thick; glabrous; not keeled abaxially. *Lateral sepals* free, reflexed at the base, hardly falcate, very narrowly triangular, oblique, 4.8–7.2 by 1.2–2.1 mm, otherwise as the median sepal. *Petals* not recurved, hardly falcate, very narrowly triangular, oblique, 1.5–2 by 0.2–0.5 mm, tip about acute; margins entire, glabrous; base broadly attached; rather thin; glabrous; not keeled abaxially. *Lip* recurved near the base, consisting of a narrowly triangular hypochilium and a linear epichilium which is separated by an inconspicuous constriction (lip not spread); 3.8–5.5 by 0.4–0.9 mm; tip obtuse to acute; margins entire, hypochilium finely ciliate towards the base, epichilium thinly beset with much longer hairs which are slightly club-shaped at their tip; rather thick, with a very thick base; adaxially without ridges, with a hardly concave basal part with a small, cuplike structure near the base, without a median slit, surface finely papillose; abaxially with a weak obtuse ridge near the base, surface glabrous. *Column* from the base of the free part up to the tip of the stelidia 1.3–2 mm long, with distinct, gradually tapering, acute stelidia of 0.6–0.9 mm long; column with weak, deltoid, obtuse teeth along its lower margins near the base of the stelidia; stigma without a basal callus. *Anther* c. 0.3 by 0.25 mm; front not drawn out; front margin rounded, entire, about glabrous; basally distinctly emarginate; abaxially with a distinct, conical, obtuse, colliculate ridge near the top of the anther which distinctly overtops the front margin, surface colliculate, glabrous towards the margins. *Pollinia* 4, about ellipsoid, flattened; the small ones about half as long as the large ones; no stipes has been observed. *Fruit* ellipsoidal, up to 1 by 0.3 cm, stalk 2 mm.

Colours. Pseudobulbs and leaves green. Sepals green, suffused with (pale) purple or entirely purple-red, or pink (1 rec.). Petals purple-red. Lip dark purple-red.

Ecology. Montane (moss) forest, 1900–2400 m alt. Epiphytic, rarely lithophytic. Flowering I–III and VI–X.

Distribution. ZAI.: Lac Edouard & Kivu Z9 (16 coll.). UG.: West U2 (1). RW.: W. part (12). BU. (1). KEN.: Nyanza K5 (fide Stuart). Material seen from AMES, BM, BR, C, E, K, L, NY, P, UPS, Z.

6 - Bulbophyllum cochleatum Lindl. – Fig. 12–17; Plate 3b, c.

Literature: see under the varieties.

Roots 0.4–3 mm diam. *Rhizome* 1–7 mm diam., nodes not or hardly swollen. *Rhizome scales* tubular, tip acute, the lower obtuse to acute; membranous, with some well-spaced nerves; either slightly keeled abaxially or not. *Pseudobulbs* 2-leafed, either about orbicular or broadly to narrowly ovoid or ellipsoid, or about cylindrical, slightly thickened towards the base, 0.6–10 cm apart, 0.7–14 by 0.4–1.4 cm, either flattened or not, inconspicuously to distinctly, obtusely to about sharply 2–7-angled, edges either with large knobs or not. *Petiole* 1–25 mm long. *Leaf blade* narrowly oblong to linear, 1.8–24 by 0.3–2 cm, tip obtuse, emarginate, usually slightly oblique; coriaceous, rather thick; midrib slightly prominent abaxially. *Inflorescence* 6.5–60 cm long, 8–180-flowered. *Peduncle* erect to patent, occasionally slightly swollen towards the tip, 3–43 cm by 0.5–4.5 mm, about orbicular in section, flattened towards the base, glabrous, with 3–13 scales. *Peduncle scales* tubular, the longest 4.5–22 mm long; tip acute to acuminate; membranous, with some nerves; keeled and glabrous abaxially. *Rhachis* erect to slightly nodding or arching, terete to slightly swollen and flattened, about orbicular to slightly elliptic in section, either slightly zigzag bent or not, 1.5–39 cm by at most 4 mm, glabrous or with fine dark hairs. *Floral bracts* recurved to about spreading, elliptic or ovate to (ovate-)oblong, 3.7–12 by 2.4–8 mm, tip acute to slightly acuminate; base rather narrowly attached; membranous, with some nerves; keeled and glabrous abaxially. *Flowers* distichous, in a dense to rather lax raceme, recurved to spreading, secund, many open simultaneously, not fully open. *Pedicel and ovary* 1.5–4 mm long, glabrous or with fine, dark hairs. *Median sepal* either slightly recurved or not, ovate to ovate-lanceolate, 2.5–7.5 by 1-4 mm, tip rounded to acute; margins entire, glabrous; base (rather) broadly attached; very thin to very thick; glabrous adaxially; not keeled and glabrous or with fine dark hairs abaxially. *Lateral sepals* free, recurved to reflexed at the base, hardly falcate, ovate or triangular to ovate-lanceolate or very narrowly triangular, oblique, 2.3–7.8 by 0.8–3.2 mm, base broadly attached; otherwise as the median sepal. *Petals* not recurved, hardly falcate, ovate-oblong to broadly linear-lanceolate, oblique, 0.8–2.5 by 0.2–0.8 mm, tip obtuse to acute; margins entire, glabrous, the lower either with a distinct tooth near its base or not; base broadly attached; very thin to rather thick; glabrous; not keeled abaxially. *Lip* recurved near the base, (ovate)-oblong to linear in outline (not spread), 1.2–6.3 by 0.3–1.8 mm, tip rounded to acute; margins entire, glabrous, very finely papillose or partly or entirely short or long ciliate; thin to thick, with a very thick base; adaxially either with two weak, obtuse ridges near the base and/or a weak to distinct, obtuse median ridge towards the tip or not, with a hardly concave basal part, with a small, cuplike structure near the base, either with a median slit in the basal part of the lip or not, surface partly or entirely, finely to rather coarsely papillose, towards the base either shortly pubescent or not; abaxially with a weak to distinct, obtuse ridge near the base, surface glabrous or nearly so. *Column* from the base of the free part up to the tip of the stelidia 1–2.5 mm long, with distinct, gradually tapering, either slightly falcate or not, acute to acuminate stelidia of 0.5–1.4 mm long which may have a rather weak (rarely distinct), obliquely forward directed, obtuse to subacute tooth along their upper margins near the base; column with weak to distinct, deltoid, obtuse to subacute teeth along its adaxial margins near the base of the stelidia; stigma without a basal callus. *Anther* 0.4–0.7 by 0.4–0.6 mm; front not drawn out; front margin rounded, entire, about glabrous; basally distinctly emarginate; abaxially with a distinct, conical, obtuse, colliculate ridge near the top of the anther which distinctly overtops the front margin, surface colliculate, glabrous towards the margins. *Pollinia* 4, ellipsoid to ovoid; the small ones about half as long as the large ones, distinctly flattened; the large ones more or less flattened; no stipes has been observed. *Fruit* ellipsoid, up to 1.2 by 0.7 mm, stalk 1 mm.

KEY TO THE VARIETIES

1 - Margins of lip entirely glabrous or papillose........................var. **bequaertii** (c)
- Margins of lip ciliate..2

2 - Ciliae along the top part of the lip distinctly longer than those along the basal part; lip usually 2 times as long as the petals or longer, when only 1.4–2 times as long as the petals the lip has a top part which is wider than the basal part............3
- *Either* ciliae along the lip everywhere about equally long or slightly longer in the basal part; *or* lip 0.8–1.6 (– 1.7) times as long as the petals, its top part as wide as the basal part or narrower; *or* both possibilities mentioned above present........4

3 - Mature pseudobulbs cylindrical, at most slightly swollen near the base, with smooth edges..var. **cochleatum** (a)
- Mature pseudobulbs ovoid, ellipsoid or about orbicular, often (not always) with very irregular edges with large knobs.................................var. **gravidum** (b)

4 - Top part of lip very thin..var. **bequaertii** (c)
- Top part of lip moderately thin to rather thick..5
5 - Sepals abaxially with fine, dark hairs; top part of lip usually distinctly recurved; sepals rather thick to very thick..................................var. **brachyanthum** (e)
- Sepals abaxially glabrous; top part of lip slightly recurved or not; sepals very thin to rather thick..var. **tenuicaule** (d)

6a. var. **cochleatum** – Fig. 12–13; Plate 3b.

Bulbophyllum cochleatum Lindl., J. Linn. Soc. Bot. 6 (1862) 125 ('*Bolbophyllum*'); Durand & Schinz, Consp. Fl. Afr. 5 (1895) 11; Rolfe in F.T.A. 7 (1897) 28; De Wild., Pl. Beq. 1 (1921) 107; Schltr in Mildbraed, 2e Deutsch. Zentr. Afr. Exp. 1910-11, 2 (1922) 180; Summerh. in F.W.T.A. ed. 1, 2 (1936) 438; Jaeger & Summerh., Kew Bull. 3 (1949) 482; Petersen, Bot. Tidsskr. 49 (1952) 159; Summerh. in Andrews, Fl. Pl. Sudan 3 (1956) 313; id., Bull. Inst. Fr. d'Afr. Noire, ser. A, 21 (1959) 470; id. in F.W.T.A. ed. 2, 3 (1968) 236; Piers, Orch. E. Afr. (1968) 98; Johansson, Acta Phyt. Suec. 59 (1974) 31; Stewart in Agnew, Upl. Ken. W. Fl. (1974) 763; Williamson, Orch. S. Centr. Afr. (1977) 136; Adam, Fl. Descr. Mt. Nimba 5 (1981) 1673, pl. 907; La Croix et al., Orch. Mal. 1 (1983) 50; Cribb, in F.T.E.A., Orch. 2 (1984) 315.–*Phyllorchis cochleata* (Lindl.) Kuntze, Rev. Gen. Pl. 2 (1891) 677, incorr. name.– Type: Mann 643 (K).

Bulbophyllum mannii Hook. f., J. Linn. Soc. Bot. 7 (1864) 219 ('*Bolbophyllum*'); Durand & Schinz, Consp. Fl. Afr. 5 (1895) 12; Rolfe in F.T.A. 7 (1897) 28; Schltr, Westafr. Kautchuk Exp. (1901) 282; De Wild., Pl. Beq. 1 (1921) 113; Summerh. in F.W.T.A. ed. 1, 2 (1936) 438; id, ed. 2, 3 (1968) 236.–*Phyllorchis mannii* (Lindl.) Kuntze, Rev. Gen. Pl. 2 (1891) 677, incorr. name.–Syntypes: Mann 1337, 2111 (K).

Non *Bulbophyllum mannii* Rchb. f., Flora 55 (1872) 275.

Bulbophyllum talbotii Rendle, Cat. Talb. S. Nig. Pl. (1913) 100; De Wild., Pl. Beq. 1 (1921) 120.–Type: Talbot 931 (Holo BM; iso K).

Bulbophyllum pholidotoides Krzl., Engl., Bot. Jahrb. 51 (1914) 383; De Wild., Pl. Beq. 1 (1921) 116.– Type: Ledermann 6168 (B, lost, note from Summerhayes in K).

Bulbophyllum jungwirthianum Schltr, Notizbl. Berl.-Dahl. 8 (1922) 122; Schltr, Fedde, Rep. Beih. 68 (1932) pl. 66.–Type: Berlin cult. Hort. (Deistel s.n.) (B, lost, conspecific according to Summerhayes in F.W.T.A. ed. 2).

? *Bulbophyllum* sp. Williamson, Orch. S. Centr. Afr. (1977) 137.–Specimen: Williamson 18, n.v.

Rhizome 1–4 mm diam. Pseudobulbs cylindrical, at most slightly thickened near the base, 0.8–7 cm apart, 1.5–11 by 0.4–1.3 cm, with at most inconspicuous edges (when shriveled with weak edges which do not bear knobs). Leaves linear-lanceolate to linear, 2.8–23 by 0.3–1.8 cm. Inflorescence 8–55 cm. Peduncle 5.2–43 cm long, not thickened towards the rhachis, with 3–12 scales. Peduncle scales up to 7–22 mm long. Rhachis not thickened, 2.8–12 cm long, 14–64 (- 84)-flowered. Floral bracts 4.5–12 by 2.5–6.2 mm. Pedicel and ovary glabrous. Median sepal ovate to broadly ovate-lanceolate, 3.2–7.5 by 1–4 mm, tip acute; rather thin; abaxially glabrous. Lateral sepals reflexed at their base, (very) narrowly triangular, 3–7.8 by 1–3.2 mm, otherwise as the median sepal. Petals 1.1–2.5 by 0.2–0.8 mm, tip acute to subacute, usually no tooth along the lower margin near the base. Lip about equally wide over most of its length or with a wider top part, 2.1–6.3 by 0.3–1.6 mm, top not bent backwards; margins ciliate, ciliae longest along the top part; (rather) thin; adaxially glabrous to very finely papillose or pubescent, near the base occasionally slightly coarser pubescent towards the margins, no median ridge towards the tip. Column 1–2.5 mm long. Stelidia 0.6–1.4 mm long.

Colours. Pseudobulbs yellowish or green, often heavily marked with purple. Leaves green. Floral bracts very pale green, dry when the flowers are open, then greyish. Sepals and petals green, often stained purple-red or entirely purple-red with a green base. Lip dark purple-red, occasionally with a yellow centre. Column usually white.

Ecology. Hill and montane forest at 500–2400 m alt. Epiphyte on tree trunks and in the central part of the crown, on bark or in minor humus deposits (Johansson, 1974). Occasionally lithophytic. Flowering mainly VIII–III.

Distribution. GUIN.: Dalaba-Labe Plateaux (6 coll.), Mt. Nimba (2). S.L. (17). LIB. (5). IV.C. (6). NIG.: S. and SE part (5). CAM.: N. to Adamawa Mts. (20). EQ.GUIN.: Bioko (5). S.T. & PR.: Sao Tomé (1). GAB. (6). SUDAN: S. Region, E. Equatoria, Imatong Mts. (1). UG.: North U1, Imatong Mts. (1); East U3, Mt. Elgon (2). RW.: W. part (1). KEN.: Rift Valley K3/Nyanza K5 (3). TANZ.: Tanga T3 (1). ZAM.: West, Mwinilunga (4). MAL.: North (1). R.S.A.: Eastern Cape (1?, Rudatis 1549, possibly wrongly labeled). Material seen from AMES, BM, BR, C, G, K, L, P, S, UPS, WAG, Z.

Notes. Within var. *cochleatum* two entities can be distinguished:

a. - The entity which includes the type specimens of *Bulbophyllum cochleatum*, *Bulbophyllum talbotii*, *Bulbophyllum pholidotoides*, *Bulbophyllum jungwirthianum* and *Bulbophyllum* sp.

Pseudobulbs 1.5–10.5 cm long. Leaves 2.8–17 cm long. Inflorescence 8–27 (- 30) cm long, 14–40 (- 84) flowered. Rhachis 2.8–11 cm long. Median sepal 3.2–5.2 (- 7) by 1–2.5 (- 3) mm. Lateral sepals 3–5 (- 7.5) by 1–1.8 (- 2.4) mm. Petals 1.2–1.8 (- 2.4) by 0.2–0.8 mm. Lip 2.2–4.2 (- 5.5) by 0.3–1.1 mm.

This entity occurs throughout the range of var. *cochleatum*.

The specimen Talbot 931 from NIG., type specimen of *Bulbophyllum talbotii*, has very dense racemes with up to 84 flowers and a remarkably narrow, linear lip. The latter character it shares with the specimen Rudatis 1549 from (?) R.S.A.

The specimens Jongejan 63 (of unknown origin), Sanford 5457 (CAM.) and Johansson 631 (LIB., Mt. Nimba) are slightly larger in most parts and therefore intermediate between this and the next entity. They account for most of the measurements between brackets in the description of entity a.

b. - The entity which includes the syntype specimens of *Bulbophyllum mannii*.

Pseudobulbs 4.5–11 cm long. Leaves 10–23 cm long. Inflorescence 34–55 cm long, 20–46 flowered. Rhachis 6–12 cm long. Median sepal 6.5–7.5 by 2.6–4 mm. Lateral sepals 7–7.8 by 2–3.2 mm. Petals 2–2.5 by 0.3–0.8 mm. Lip 5–6.3 by 1.2–1.8 mm.

The specimens included differ from entity a. in being larger. The entity only occurs in the mountainous areas of CAM. and Bioko together with entity a. as well as intermediate specimens.

6b. var. **gravidum** (Lindl.) J.J. Vermeulen. - Fig. 14.

Bulbophyllum cochleatum Lindl. var. *gravidum* (Lindl.) J.J. Vermeulen, Bull. J. Bot. Nat. Belg. 56 (1986) 230.-*Bulbophyllum gravidum* Lindl., J. Linn. Soc. Bot. 6 (1862) 126 ('*Bolbophyllum*'); Durand & Schinz, Consp. Fl. Afr. 5 (1895) 12; Rolfe in F.T.A. 7 (1897) 27; De Wild., Pl. Beq. 1 (1921) 111; Schltr in Mildbraed, 2e Deutsch. Zentr. Afr. Exp. 1910-11, 2 (1922) 180; Summerh. in F.W.T.A. ed. 1, 2 (1936) 438; id., ed. 2, 3 (1968) 236; La Croix et al., Orch. Mal. 1 (1983) 50.-*Phyllorchis gravida* (Lindl.) Kuntze, Rev. Gen. Pl. 2 (1891) 677, incorr. name.-Type: Mann 650 (K).

Bulbophyllum monticolum Hook. f., J. Linn. Soc. Bot. 7 (1864) 219 ('*Bolbophyllum*'); Durand & Schinz, Consp. Fl. Afr. 5 (1895) 12.-*Phyllorchis monticola* (Hook. f.) Kuntze, Rev. Gen. Pl. 2 (1891) 677, incorr. name.-Type: Mann 2126 (K).

Bulbophyllum rugosibulbum Summerh., Kew Bull. 14 (1960) 138; id., Hook. Ic. Pl. 37 (1969) pl. 3670; Williamson, Orch. S. Centr. Afr. (1977) 138, col. phot.; La Croix et al., Orch. Mal. 1 (1983) 55; Cribb, in F.T.E.A., Orch. 2 (1984) 315.-Type: Holmes 05 (K).

Bulbophyllum near *gravidum* Morris, Proc. Linn. Soc. London 179 (1968) 63; Morris, Ep. Orch. Mal. (1970) 51.-Specimen: Morris 98 (K).

Rhizome 1–4 mm diam. Pseudobulbs ovoid or ellipsoid to about orbicular, 0.6–10 cm apart, 0.7–4.5 by 0.4–1.3 cm, with distinct edges when fully grown, which may or may not bear a row of large, conical, radially striated knobs. Leaves narrowly oblong to linear, 2.5–12 by 0.3–1.8 cm. Inflorescence 7–23 cm. Peduncle 6.2–19 cm long, not thickened towards the rhachis, with 3–12 scales. Peduncle scales up to 7–22 mm long. Rhachis not thickened, 1.5–9.5 cm long, 10–40-flowered. Floral bracts 4.5–12 by 2.5–6.2 mm. Pedicel and ovary glabrous. Median sepal ovate to broadly ovate-lanceolate, 4.4–7 by 2–3.3 mm, tip (sub)acute; rather thin; abaxially glabrous. Lateral sepals reflexed at their base, (very) narrowly triangular, 4.2–7.2 by 1.5–2.4 mm, otherwise as the median sepal. Petals 1.8–2.5 by 0.25–0.7 mm, tip obtuse to acute, usually no tooth along the lower margin near the base. Lip about equally wide over most of its length or with a wider top part, 3.5–5.8 by 0.7–1.5 mm, top not bent backwards; margins ciliate, ciliae longest along the top part; (rather) thin; adaxially glabrous to very finely papillose or pubescent, near the base occasionally slightly coarser pubescent towards the margins, no median ridge towards the tip. Column 1–2.5 mm long. Stelidia 0.6–1.4 mm long.

Colours. Pseudobulbs green, often suffused with red. Leaves green. Floral bracts very pale greenish, dry when the flowers are open and then greyish. Sepals and petals yellow or greenish, often suffused with purple-red or entirely dark purple-red. Lip dark purple-red. Column white.

Ecology. Epiphytic in (lower) montane forest, riverine forest (large plants with widely spaced pseudobulbs; also growing lithophytic with lichens, fide La Croix, pers. comm.), and Brachystegia-Uapaca woodland (very small plants with close pseudobulbs; on tree trunks and lower branches, fide La Croix, pers. comm.), lithophytic in open grassland; at 900–2400 m alt. Flowering XI–II.

Distribution. CAM.: W. part (2 coll). EQ.GUIN.: Bioko (4). ZAI.: Haut Katanga Z11 (1). TANZ.: Southern Highlands T7 (1); South T8, Songea (1). ZAM.: West, Mwinilunga (3), Solwezi (1), Copperbelt (1); East, Niyka Foothills (1). MAL.: North (6); South (2). Material seen from BM, BR, K, WAG.

Notes. 1 - Two entities can be distinguished within var. *gravidum* (entity c. and d.):

c. - The entity which includes the type specimens of *Bulbophyllum gravidum* and *Bulbophyllum monticolum*.

Pseudobulbs 2.1–4 cm long, (narrowly) ellipsoid to (narrowly) ovoid, 4-angled, edges without knobs. Leaves 4.8–12 cm long, narrowly linear-lanceolate to linear. Inflorescence 12 - 23 cm long, 10–18 flowered. Rhachis 1.5–4 cm long. Median sepal 6–7 by 2.7–3.3 mm. Lateral sepals 6–7.2 by 1.5–2.4 mm. Petals 2–2.5 by 0.3–0.4 mm. Lip 5–5.8 by 1.2–1.5 mm.

This entity has about the same distribution as entity b. of var. *cochleatum*: it is restricted to CAM. (Cameroun Mts.) and Bioko. The specimens included in this entity usually have a longer peduncle but a shorter rhachis than specimens of entity d. The petals tend to be slightly more veined (observed in dried specimens) than in entity d.

d. - The entity which includes the type specimen of *Bulbophyllum rugosibulbum*, and the specimens identified as *B. gravidum* auct. La Croix (1983) and *B.* near *gravidum* Morris (1970).

Pseudobulbs 0.7–4.5 cm long, orbicular to narrowly ellipsoid or ovoid, 4–7-angled, edges with rather inconspicuous to very distinct conical knobs when the pseudobulb is mature. Leaves 2.5–9 cm long, narrowly oblong to linear. Inflorescence 7–21 cm long, 10–40 flowered. Rhachis 1.8–9.5 cm long. Median sepal 4.4–6 by 2–3.2 mm. Lateral sepals 4.2–6.3 by 1.5–2 mm. Petals 1.8–2 by 0.25–0.7 mm. Lip 3.5–5 by 0.7–1 mm.

This entity is widely distributed throughout the range of var. *gravidum*, except for those areas were entity c. occurs.

2 - Var. *cochleatum* and var. *gravidum* mainly differ in the shape of the pseudobulbs. Summerhayes has observed that young pseudobulbs of entity d. have the same shape as mature pseudobulbs of entity a. Otherwise the two entities are identical.

6c. var. **bequaertii** (De Wild.) J.J. Vermeulen – Fig. 15; Plate 3c.

Bulbophyllum cochleatum Lindl. var. *bequaertii* (De Wild.) J.J. Vermeulen, Bull. J. Bot. Nat. Belg. 56 (1986) 230.-*Bulbophyllum bequaertii* De Wild., Rev. Zool. Afr. 9, suppl. bot. (1921) 28; De Wild., Pl. Beq. 1 (1921) 104; Summerh., Kew Bull. 8 (1953) 144; id., in F.W.T.A. ed. 2, 3 (1968) 236; Piers, Orch. E. Afr. (1968) 97; Stewart in Agnew, Upl. Ken. W. Fl. (1974) 763; Cribb, in F.T.E.A., Orch. 2 (1984) 314, p.p.-Type: Bequaert 6413 (BR).

Rhizome 2.5–7 mm diam. Pseudobulbs cylindrical, at most slightly thickened near the base, 1–6 cm apart, 3–14 by 0.7–1.3 cm, with at most inconspicuous edges (when shriveled with weak edges which do not bear knobs). Leaves linear-lanceolate, 6.5–24 by 0.6–2 cm. Inflorescence 13–45 cm. Peduncle 5–20 cm long, slightly thickened towards the rhachis and below the nodes, with 8–13 scales. Peduncle scales up to 11–16 mm long. Rhachis usually slightly thickened and flattened, 5–27 cm long, 36–160-flowered. Floral bracts 5–10 by 3–8 mm. Pedicel and ovary glabrous or with fine, dark hairs. Median sepal ovate to broadly ovate-oblong, 3.2–5.7 by 1.8–3.4 mm, tip obtuse to subacute; rather thin; abaxially glabrous or with fine, dark hairs. Lateral sepals more or less recurved towards the tip, (narrowly) triangular, 3.2–5.7 by 1.2–2 mm, otherwise as the median sepal. Petals 1–1.5 by 0.2–0.5 mm, tip acute to subacute, usually with a small tooth along the lower margin near the base. Lip about equally wide over most of its length or with a wider top part, occasionally gradually tapering towards the tip, 2.5–4.5 by 0.4–0.9 mm, top not bent backwards; margins glabrous, papillose or very shortly ciliate, if ciliate with longest ciliae along the basal part; (very) thin; adaxially glabrous to very finely papillose, no median ridge towards the tip. Column 1.2–1.8 mm long. Stelidia 0.6–1 mm long.

Colours. Pseudobulbs green, suffused with red. Leaves green. Floral bracts pale green, dry and dark greyish brown when the flowers are open. Sepals and petals yellowish, suffused with brown-red, or entirely brown-red with a yellow base. Lip entirely dark brown-red. Column pale yellow.

Ecology. Epiphyte in (lower) montane forest at 900–2400 m alt. Flowering mainly VIII–IV.

Distribution. CAM.: W. part (3 coll.). ZAI.: Bas Katanga Z5 (1); Lacs Edouard & Kivu Z9 (2). UG.: West U2, S. part (8). RW.: W. part (2). TANZ.: Lake T1 (1); Tanga T3 (5). Material seen from BR, K, L, NY.

Note. Typical specimens of var. *bequaertii* can be recognized by the lip which has a very thin top part and more or less glabrous margins. Some specimens, however, tend to show a single or a few characters which are more typical for the other varieties of *Bulbophyllum cochleatum*, e.g. Eggeling 3191 and Lankester 28, both from UG., which have shortly but distinctly ciliate lips. They can be separated from var. *cochleatum* a.o. by the ciliae which are longest along the basal part of the lip, and not along the top part as in var. *cochleatum*.

Specimens as the above mentioned occasionally make distinction difficult between var. *bequaertii* and the other varieties within *Bulbophyllum cochleatum*.

6d. var. **tenuicaule** (Lindl.) J.J. Vermeulen - Fig. 16.

Bulbophyllum cochleatum Lindl. var. *tenuicaule* (Lindl.) J.J. Vermeulen, Bull. J. Bot. Nat. Belg. 56 (1986) 230.-*Bulbophyllum tenuicaule* Lindl., J. Linn. Soc. Bot. 6 (1862) 126 ('*Bolbophyllum*'); Hook. f., J. Linn. Soc. Bot. 7 (1864) 218 ('*Bolbophyllum*'); Durand & Schinz, Consp. Fl. Afr. 5 (1895) 14; Rolfe in F.T.A. 7 (1897) 27; De Wild., Pl. Beq. 1 (1921) 120; Schltr in Mildbraed, 2e Deutsch. Zentr. Afr. Exp. 1910-11, 2 (1922) 180; Summerh. in F.W.T.A. ed. 1, 2 (1936) 438; Exell, Cat. Vasc. Pl. S. Tomé (1944) 328; Summerh. in F.W.T.A. ed. 2, 3 (1968) 236; Adam, Fl. Descr. Mt. Nimba 5 (1981) 1684.-Type: Mann 648 (K).

Bulbophyllum thomense Summerh., Kew Bull. 1937 (1937) 464.-Type: Exell 390 (Holo BM, iso K).

Bulbophyllum bequaertii De Wild. var. *brachyanthum* Summerh., Kew Bull. 8 (1953) 144, p.p., type specimen excluded.

Bulbophyllum bequaertii auct.: Cribb, in F.T.E.A., Orch. 2 (1984) 314, p.p.

Rhizome 1.2–6 mm diam. Pseudobulbs usually cylindrical, at most slightly thickened near the base, occasionally narrowly ovoid, 1–9 cm apart, 1–10 by 0.4–1.2 cm, if cylindrical with at most inconspicuous edges (when shriveled with weak edges without knobs), if ovoid with distinct edges without knobs. Leaves linear-lanceolate to linear, 1.8–16 by 0.4–1.5 cm. Inflorescence 6.5–22 cm. Peduncle 3–14.5 cm long, either thickened towards the rhachis and below the nodes or not, with 3–9 scales. Peduncle scales up to 4.5–10 (- 14) mm long. Rhachis either thickened and slightly flattened or not, 2–12 cm long, 8–60-flowered. Floral bracts 3.7–11 by 2.4–5 mm. Pedicel and ovary glabrous or with fine, dark hairs. Median sepal ovate to broadly ovate-lanceolate, 2.5–4 (- 6.5) by 1.3–2.9 mm, tip acute to subacute; very thin to rather thick; abaxially glabrous, rarely with some fine, dark hairs. Lateral sepals more or less recurved towards the tip, ovate to ovate-lanceolate, 2.3–3.8 (- 5.8) by 0.8–2.1 mm, otherwise as the median sepal. Petals 1.2–2 by 0.2–0.5 mm, tip acute to subacute, usually with a tooth along the lower margin near the base, occasionally without. Lip about equally wide at the base and at the top or tapering towards the tip, 1.4–2.5 by 0.3–0.9 mm, top either slightly bent backwards or not; margins ciliate, ciliae either longest along the top part or not; rather thin to rather thick; adaxially finely to coarsely papillose or pubescent, usually no median ridge towards the tip. Column 1–2.1 mm long. Stelidia 0.5–1 mm long.

Colours. Sepals and petals greenish, suffused with brown-red or purple-red, or entirely brown-red or purple-red. Lip dark brown-red or purple-red.

Ecology. Epiphyte in montane forest at 1000–2400 m alt. Flowering mainly II–III and VI–IX.

Distribution. NIG.: SW. part (1 coll.). CAM.: W. part (2). EQ.GUIN.: Bioko (1). S.T. & PR.: Sao Tomé (2). ZAI.: Lacs Edouard & Kivu Z9 (6). UG.: West U2 (3). RW.: W. part (6). KEN.: Nyanza K5 (2). Material seen from BM, BR, K, L, P, WAG.

Notes. 1 - Together var. *tenuicaule* and var. *brachyanthum* cover a series of specimens showing considerable variability in numerous characters. Some characters tend to occur together, others do not or only to a lesser extent. Those occurring together have been used in the key to distinguish between the two varieties.

Only few specimens (Cribb & Grey Wilson 10171 from TANZ., Leonard 5109 from ZAI. and Brunt 1252 from CAM.) do not show the combination of key characters.

2 - The material ascribed to var. *tenuicaule* is slightly heterogeneous mainly due to some specimens originating from CAM. and adjacent areas:

- The specimens Exell 390 (type of *Bulbophyllum thomense*) and Chevalier 13657, both from Sao Tomé differ in having ovoid or ellipsoid, 4 angled, not cylindrical pseudobulbs. They are very small and slender, with unusual wiry, few-flowered inflorescences. The floral bracts are short (3.5–4 mm long, usually 5–11 mm long). The flowers are large (median sepal 5.5–6.5 mm long, usually 2.5–4 mm long); the sepals are relatively narrow. The petals lack a tooth along their lower margin near the base (present in most other specimens). The stelidia of the column have a very distinct winglike tooth along their upper margins.

- The specimen Mann 648 from Bioko (type of *Bulbophyllum tenuicaule*) is similar to these two specimens but has cylindrical pseudobulbs, flowers of normal size, sepals of normal shape and lacks the winglike tooth along the upper margins of the stelidia.

- The specimen Wimbush & King 150 from NIG. is also similar to these two specimens but is of normal size with a sturdy, many flowered inflorescence. The flowers are larger than usual (median sepal 5.1 mm long) but the sepals are of normal shape. A winglike tooth along the upper margins of the stelidia is absent.

Just as var. *gravidum* has been separated from var. *cochleatum* on account of the shape of the pseudobulbs, Exell 390, Chevalier 13657 and Wimbush & King 150 could be separated from var. *tenuicaule*. However, doing so, the strikingly slender appearance shared by Exell 390, Chevalier 13657 and Mann 648 is ignored by accomodating them in different entities. Besides, the variety thus obtained lacks homogeneity due to the above mentioned differences between the three specimens included. Therefore the present author prefers to keep them within var. *tenuicaule*

3 - Within var. *tenuicaule* the ratio: length lip/length petals usually amounts to 1.1–1.6, rarely 1.7 (Bamps 3225 from RW.). Most specimens of var. *cochleatum* and var. *gravidum* show a ratio of 2 or more. In two specimens of the latter variety, however, a ratio of 1.4 and 1.7 has been found, which in fact makes this character unreliable and unfit for use in the key to the varieties. However, most specimens can be identified when using some additional characters, cfr. the key. I prefer to accept the existence of a few specimens which are intermediate, rather than joining together var. *cochleatum* and var. *tenuicaule*, which, as a logical consequence, would imply inclusion of var. *gravidum* and var. *brachyanthum* as well. The entity thus achieved would be heterogeneous beyond acceptance.

6e. var. **brachyanthum** (Summerh.) J.J. Vermeulen. - Fig. 17.

Bulbophyllum cochleatum Lindl. var. *brachyanthum* (Summerh.) J.J. Vermeulen, Bull. J. Bot. Nat. Belg. 56 (1986) 231.-*Bulbophyllum bequaertii* De Wild. var. *brachyanthum* Summerh., Kew Bull. 8 (1953) 144, p.p., type specimen included; Piers, Orch. E. Afr. (1968) 98.-Type: Gardner 2548 (K).
Bulbophyllum bequaertii auct.: Cribb, in F.T.E.A., Orch. 2 (1984) 314, p.p.

Rhizome 1.2–6 mm diam. Pseudobulbs cylindrical, at most slightly thickened near the base, 1–9 cm apart, 5–13 by 0.6–1.4 cm, with at most inconspicuous edges (when shriveled with weak edges which do not bear knobs). Leaves linear-lanceolate to linear, 10–23 by 0.8–2.4 cm. Inflorescence 15–60 cm. Peduncle 5–21 cm long, usually thickened towards the rhachis and below the nodes, with 3–9 scales. Peduncle scales up to 4.5–10 (- 14) mm long. Rhachis usually thickened and slightly flattened, 4.5–39 cm long, 20–180-flowered. Floral bracts 5.5–12 by 3–8 mm. Pedicel and ovary with fine, dark hairs. Median sepal ovate, 2.7–3.8 by 1.5–2.8 mm, tip rounded to subacute, or apiculate; rather thick to very thick; abaxially with fine, dark hairs. Lateral sepals slightly recurved towards the tip, ovate to ovate-oblong, 2.5–3.8 by 1.2–1.7 mm, otherwise as the median sepal. Petals 0.8–1.8 by 0.2–0.4 (- 0.6) mm, tip acute to subacute, with a tooth along the lower margin near the base. Lip about equally wide over most of its length or tapering towards the tip, 1.2–1.6 by 0.4–0.7 mm, top distinctly bent backwards; margins ciliate, ciliae either longest along the top part or not; (very) thick; adaxially glabrous to very finely papillose, often with a distinct median ridge towards the tip. Column 1–2.1 mm long. Stelidia 0.5–1 mm long.

Colours. Sepals brown-red, occasionally with a greenish base. Lip dark brown-red.

Ecology. Montane forest at 1000–2800 m alt. Epiphytic. Flowering mainly I–V.

Distribution. ZAI.: Lacs Edouard & Kivu Z9 (8 coll.). UG.: North U1, E. part (2); West U2 (2). RW.: W. part (7). BUR. (1). KEN.: Rift Valley K3/Nyanza K5 (7). TANZ.: Tanga T3 (1). Material seen from BR, K, L, UPS, WAG.

Notes. 1 - See note 1 below var. *tenuicaule*.

2 - A few specimens are slightly aberrant:

The specimen Deuse 1153 from RW. has a very long, lax inflorescence with narrow floral bracts.

The specimen Troupin 15646 from RW. has very wide fleshy petals with fine, dark hairs on either side.

3 - Vegetatively var. *brachyanthum* is identical with var. *bequaertii*.

7 - Bulbophyllum burttii Summerh. - Fig. 18.

Bulbophyllum burttii Summerh., Kew Bull. 8 (1953) 144; Robyns, Fl. Sp. P. Alb. 3 (1955) 496.-Type: Burtt 3164 (Holo K; iso BR).

Roots 0.4–1 mm diam. *Rhizome* 2.5–6 mm diam., nodes not or hardly swollen. *Rhizome scales* tubular, tip obtuse to about acute; membranous, the upper with nerves; not or hardly keeled abaxially. *Pseudobulbs* 2-leafed, very narrowly ovoid, 1.8–9 cm apart, 2.8–6 by 0.8–1.8 cm, slightly to moderately flattened, 4-angled. *Petiole* 0.5–1.5 mm long. *Leaf blade* linear-lanceolate to linear, 5.5–17 by 0.7–1.6 cm, tip obtuse, emarginate, usually oblique; coriaceous, rather thick; midrib slightly prominent abaxially. *Inflorescence* 13-28 cm long, 24-48-flowered. *Peduncle* erect to patent, occasionally slightly swollen towards the tip, 9-15 cm by 1.2-2 mm, about orbicular in section, flattened towards the base, glabrous, with 3-5 scales. *Peduncle scales* tubular, the longest 6-8 mm long; tip acute to acuminate; membranous, without nerves; slightly keeled and glabrous abaxially. *Rhachis* about erect, terete or slightly swollen, slightly zigzag bent, 6.5-14 cm long, glabrous. *Floral bracts* recurved, narrowly triangular, 5.5-7 by 3-3.5 mm, tip acute; base broadly attached; membranous, without nerves; keeled and glabrous abaxially. *Flowers* distichous, in a dense raceme, recurved to spreading, secund, many open simultaneously, not fully open. *Pedicel and ovary* 1.8-3.5 mm long, glabrous. *Median sepal* either slightly recurved or not, narrowly triangular, 6-8 by 2-3 mm, tip slightly obtuse to acute; margins entire, glabrous; base (rather) broadly attached; rather thin; glabrous; not keeled abaxially. *Lateral sepals* free, reflexed at the base, hardly falcate, very narrowly triangular, oblique, 5.2-8 by 1.5-2 mm, base broadly attached; otherwise as the median sepal. *Petals* not recurved, hardly falcate, narrowly lanceolate to linear-lanceolate, oblique, 1.4-1.8 by 0.2-0.4 mm, tip about obtuse; margins entire, glabrous, the lower with or without a distinct tooth near the base; base broadly attached; rather thin; glabrous; not keeled abaxially. *Lip* recurved in the basal half, ovate-oblong in outline (not spread), 1.4-2 by 0.5-0.9 mm, tip obtuse; margins entire, with fine, short ciliae near the base which turn into fine papillae towards the tip; rather thick, with a thick base; adaxially without ridges; with a hardly concave basal part with a small, cuplike structure near the base, without a median slit, surface glabrous to very finely papillose; abaxially with a weak obtuse ridge towards the base, surface about glabrous. *Column* from the base of the free part up to the tip of the stelidia 1.5-1.8 mm long, with distinct, gradually tapering, either slightly falcate or not, acute stelidia of c. 1 mm long; column with distinct, deltoid, obtuse, teeth along its adaxial margins, near the base of the stelidia; stigma without a basal callus. *Anther* c. 0.4 by 0.4 mm; front not drawn out; front margin rounded, entire, glabrous; basally distinctly emarginate; abaxially with a distinct, more or less conical, obtuse, colliculate ridge near the top of the anther which distinctly overtops the front margin, surface colliculate, glabrous towards the margins. *Pollinia* 4; the small ones about half as long as the large ones, ovoid, distinctly flattened; the large ones obovoid, flattened; a stipes has not been observed. *Fruit* not seen.

Colours. Flowers entirely yellow or greenish.

Ecology. Montane forest at 1900-2100 m alt. Epiphytic. Flowering observed in XII-II and IV.

Distribution. ZAI.: Lacs Edouard & Kivu Z9 (3 coll.). RW.: W. part (7). Material seen from BR, K.

8 - Bulbophyllum kivuense J.J. Vermeulen. - Fig. 19.

Bulbophyllum kivuense J.J. Vermeulen, Bull. J. Bot. Nat. Belg. 56 (1986) 231.-Type: Troupin 15652 (BR).

Roots 0.5-1.2 mm diam. *Rhizome* 1.8-2.8 mm diam., nodes not or hardly swollen. *Rhizome scales* tubular, tip obtuse; membranous, with some nerves; keeled abaxially towards the tip. *Pseudobulbs* 1-leafed, about cylindrical, slightly thickened towards the base, 5–5.5 cm apart, c. 7.5 by 0.8 cm, slightly flattened, inconspicuously and obtusely c. 5-angled. *Petiole* c. 5 mm long. *Leaf blade* linear, 9–15 by 0.6–1 cm, tip obtuse, emarginate, either oblique or not; coriaceous, rather thick; midrib slightly prominent abaxially. *Inflorescence* c. 20 cm long, 15–26-flowered. *Peduncle* about erect, slightly swollen towards the tip, c. 14 cm by 1.5–2.5 mm, slightly flattened, glabrous, with c. 2 scales. *Peduncle scales* tubular, the longest c. 12 mm long; tip acute to subacute; membranous, without nerves; keeled towards the tip and glabrous abaxially. *Rhachis* about erect, terete or slightly swollen, slightly zigzag bent, 6–12 cm by 1.5–2.5 mm, glabrous. *Floral bracts* recurved, oblong, 7.5–9 by 3.6–4 mm, tip acute; base broadly attached; membranous, with some nerves; slightly keeled towards and glabrous abaxially. *Flowers* distichous, in a rather dense raceme, recurved to spreading, secund, many open simultaneously, not fully open. *Pedicel and ovary* 5–5.5 mm long, with some fine dark hairs. *Median sepal* not recurved, ovate to about triangular, c. 4.2 by 2.7–3 mm, tip acute to acuminate; margins entire, glabrous; base rather narrowly attached; thick except for two thin patches along the margins near the base; glabrous adaxially; not keeled and with fine dark hairs abaxially. *Lateral sepals* free, slightly recurved, hardly falcate, triangular, oblique, c. 4 by 2.1–2.2 mm, tip acuminate; thick except for the thin patch along the upper margin near the base; otherwise as the median sepal. *Petals* not recurved, hardly falcate, oblong, oblique, c. 2 by 1.2–1.3 mm, tip subacute; margins entire, glabrous, the upper with a weak edge and the lower with a distinct tooth near the base; base broadly attached; rather thick except for the thin top part; glabrous; not keeled abaxially. *Lip* slightly recurved in the basal half, ovate in outline (not spread), 2–2.2 by c. 1.3 mm, tip rounded; margins entire, ciliate, ciliae shorter towards the tip and towards the base; thick; adaxially with a very weak, obtuse median swelling towards the tip, with a slightly convex basal part, with a small, conical median tooth just above the base, without a median slit but slightly channeled half way the lip, surface about glabrous; abaxially with a weak, about truncate ridge towards the base, surface glabrous, locally slightly papillose. *Column* from the base of the free part up to the tip of the stelidia 2–2.2 mm long, with distinct, gradually tapering, slightly falcate, acute stelidia of 1.1–1.3 mm long; column with distinct, deltoid, obtuse teeth along its lower margins near the base of the stelidia; stigma without a basal callus. *Anther* c. 0.8 by 0.6 mm; front not drawn out; front margin rounded, slightly irregular, glabrous; basally distinctly emarginate; abaxially with a distinct, conical, obtuse, colliculate ridge near the top of the anther which distinctly overtops the front margin, surface colliculate, glabrous towards the margins. *Pollinia* 4, somewhat ovoid; the small ones about half as long as the large ones, distinctly flattened; the large ones flattened; a stipes has not been observed. *Fruit* not seen.

Colours. Probably greenish, suffused with purple.

Ecology. Montane forest at 1900 m alt. Epiphytic. Flowering observed in XII.

Distribution. RW.: W. part (1 coll.).

9 - Bulbophyllum inornatum J.J. Vermeulen – Fig. 20.

Bulbophyllum inornatum J.J. Vermeulen, Kew Bull. 42 (1987) 266.–Type: Stolze & Scharff s.n. (K).

Roots 0.5–0.8 mm diam. *Rhizome* c. 1 mm diam., nodes not or hardly swollen. *Rhizome scales* tubular, tip subacute; membranous, with some nerves; not keeled abaxially. *Pseudobulbs* 2-leafed, about orbicular to ovoid, c. 2 cm apart, 0.8–1 by 0.5–0.7 cm, slightly flattened, obtusely 3–4-angled. *Petiole* c. 0.5 mm long. *Leaf blade* narrowly elliptic to broadly lanceolate, 0.9–1.5 by c. 0.5 cm, tip rounded, emarginate, not oblique; coriaceous, thick; midrib slightly prominent abaxially. *Inflorescence* 2.5–4.8 cm long, 7–12-flowered. *Peduncle* erect to patent, wiry, 1.5–3.2 cm by c. 0.4 mm, orbicular in section, glabrous, with c. 6 scales. *Peduncle scales* tubular, the longest 2–3 mm long; tip acute, the lower obtuse; membranous, without nerves; not or hardly keeled and glabrous abaxially. *Rhachis* about erect, terete, slightly zigzag bent, 1–1.5 cm long, glabrous. *Floral bracts* recurved, ovate, c. 1.8 by 1.4 mm, tip obtuse, base broadly attached; membranous, without nerves; slightly keeled and glabrous abaxially. *Flowers* distichous, in a rather dense raceme, about recurved, many open simultaneously, wide open. *Pedicel and ovary* c. 1.7 mm long, with a few brown appendages. *Median sepal* recurved about half-way, ovate, c. 2.8 by 1.6 mm, tip rounded; margins entire, glabrous; base broadly attached; rather thin; glabrous; obtusely keeled abaxially. *Lateral sepals* free, slightly falcate, ovate oblong, oblique, c. 2.5 by 1.2 mm, tip obtuse to subacute; otherwise as the median sepal. *Petals* not recurved, hardly falcate, narrowly obovate-oblong, slightly oblique, c. 1.5 by 0.6 mm, tip rounded; margins entire, hardly irregular towards the tip, glabrous; base broadly attached; very thin, slightly thickened towards the base; glabrous or with some papillae adaxially; not keeled and glabrous abaxially. *Lip* recurved in the basal half, obovate oblong in outline (not spread), constricted in the basal half, c. 2.5 by 1.2 mm, tip rounded; margins entire, almost glabrous; rather thick, with a thick base; adaxially with a rather conspicuous, obtuse median ridge which starts at the base as a small, flat callus, continues in a shallow furrow over a short distance and which ends at about 4/5 of the length of the lip, with a slightly concave basal part, surface about glabrous, papillose near the base; abaxially with a weak, rounded ridge near the base which turns into a shallow furrow near the tip, surface glabrous. *Column* from the base of the free part up to the tip of the stelidia c. 1.5 mm long, with distinct, gradually tapering, slightly falcate, acute stelidia of c. 0.9 mm long; column with weak, deltoid, obtuse teeth along its adaxial margins, near the base of the stelidia; stigma without a basal callus. *Anther* c. 0.5 by 0.4 mm; front not drawn out; front margin rounded, entire, glabrous; basally emarginate; abaxially with a distinct, more or less conical, obtuse, c. glabrous ridge near the top of the anther which distinctly overtops the front margin, surface colliculate. *Pollinia* 4, ellipsoid; the small ones more than half as long as the large ones, distinctly flattened; the large ones flattened; a stipes has not been observed. *Fruit* not seen.

Colours. Unknown.

Ecology. Unknown, at 1100 m alt. Flowering observed in VIII.

Distribution. TANZ.: Southern Highlands T7, Uzungwa Mts. near Sanje (1 coll.).

10 - Bulbophyllum bavonis J.J. Vermeulen – Fig. 21.

Bulbophyllum bavonis J.J. Vermeulen, Blumea 29 (1984) 589.–Type: Dowsett-Lemaire 674 (K).

Roots 0.5–0.8 mm diam. *Rhizome* 0.8–1.2 mm diam., nodes not or hardly swollen. *Rhizome scales* tubular, tip acute or subacute, the lower obtuse; membranous, with some nerves; keeled abaxially towards the tip. *Pseudobulbs* 2-leafed, ovoid to narrowly ovoid, 2–9 cm apart, 1.2–3.5 by 0.3–1.3 cm, slightly flattened, 4-angled. *Petiole* 1–2 mm long. *Leaf blade* oblong to lanceolate, 1.7–5.5 by 0.4–1.3 cm, tip rounded to obtuse, usually emarginate, either oblique or not; coriaceous, thick; midrib slightly prominent abaxially. *Inflorescence* 3.5–11 cm long, 4–18-flowered. *Peduncle* erect to patent, rather wiry, 1.2–3.6 cm by 0.4–0.5 mm, orbicular in section, glabrous, with 5–6 scales. *Peduncle scales* tubular, the longest 3–4 mm long; tip acute, the lower acuminate; membranous, without nerves; keeled towards the tip and glabrous abaxially. *Rhachis* slightly arching or nodding, terete, either slightly zigzag bent or not, 2.3–7.4 cm long, glabrous. *Floral bracts* recurved, deltoid to ovate, 3.2–4.5 by 3–3.2 mm, tip acute; base broadly attached; membranous, without nerves; keeled and glabrous abaxially. *Flowers* distichous, in a rather lax raceme, recurved, many open simultaneously, not fully open. *Pedicel and ovary* 2–3 mm long, glabrous. *Median sepal* not recurved, narrowly elliptic to oblong, 5.2–6 by 2–3 mm, tip obtuse to subacute, or slightly apiculate; margins entire, very finely papillose, especially in the top half; base broadly attached; rather thin; very finely papillose adaxially towards the margins; not keeled and glabrous abaxially. *Lateral sepals* free, slightly recurved, hardly falcate, oblong to ovate-oblong, oblique, 5–5.5 by 2–2.6 mm, tip acute; otherwise as the median sepal. *Petals* not recurved, hardly falcate, obovate-oblong to obovate-lanceolate, oblique, 2–2.8 by 0.8–0.9 mm, tip slightly emarginate to subacute; margins slightly irregular and very finely papillose towards the tip; base broadly attached; thin; glabrous; not keeled and abaxially. *Lip* recurved near the base, oblong to broadly lanceolate in outline (not spread), 3.6–4 by 1.2–1.8 mm, tip obtuse; margins entire, papillose, thick, adaxially with two very weak rounded ridges in the lower half, with a hardly concave basal part, without a median slit but slightly channeled in the median part, surface papillose except for a median strip which is glabrous; abaxially with a very weak retuse ridge near the base, surface about glabrous. *Column* from the base of the free part up to the tip of the stelidia 2.3–3 mm long, with distinct, gradually tapering, slightly falcate, about subulate stelidia of 1.2–1.6 mm long, occasionally with a distinct, obliquely forwards directed, triangular, acute tooth along their upper margin, just above the base of the stelidia; column with distinct, obliquely backwards directed, deltoid, obtuse to subacute teeth along its adaxial margins, near the base of the stelidia; stigma without a basal callus. *Anther* c. 0.6 by 0.5 mm; front not drawn out; front margin about truncate, entire or slightly irregular, glabrous; basally distinctly emarginate; abaxially with a rather distinct, slightly conical, rounded, colliculate ridge near the top of the anther which distinctly overtops the front margin, surface colliculate. *Pollinia* 4, ovoid, flattened; the small ones about half as long as the large ones; a stipes has not been observed. *Fruit* not seen.

Colours. Pseudobulbs suffused with red. Sepals and petals pale green, tinged purplish, or entirely brownish. Lip yellow.

Ecology. Montane forest, epiphyte on trunks and branches of trees, among moss and ferns (La Croix, written comm.); at 1500–1700 m alt. Flowering III–IV.

Distribution. TANZ.: Southern Highlands T7 (3 coll.). MAL.: North (1). Material seen from K.

Note. The type specimen (see fig. 9) has a slightly wider lip than some of the other specimens seen.

11 - Bulbophyllum stolzii Schltr - Fig. 22.

Bulbophyllum stolzii Schltr, Engl., Bot. Jahrb. 53 (1915) 571; id., Fedde, Rep. Beih. 68 (1932) pl. 69; Morris, Proc. Linn. Soc. London 179 (1968) 63; id., Ep. Orch. Mal. (1970) 51; La Croix et all., Orch. Mal. 1 (1983) 58; Cribb, in F.T.E.A., Orch. 2 (1984) 313.–Type: Stolz 767 (Holo B, lost; iso G, L, S, U, UPS, WAG, Z).

Roots 0.4–0.8 mm diam. *Rhizome* 0.5–0.8 mm diam., nodes not or hardly swollen. *Rhizome scales* tubular, tip about acute; membranous, with some nerves; not keeled abaxially. *Pseudobulbs* 2-leafed, about orbicular to narrowly ovoid, 3–11 cm apart, 0.6–3.7 by 0.3–1.2 cm, not or hardly flattened, obtusely 4–5-angled. *Petiole* c. 1 mm long. *Leaf blade* narrowly elliptic to linear, 0.5–6.8 by 0.2–0.9 cm, tip rounded to obtuse, not or hardly emarginate, slightly oblique; coriaceous, thick; midrib slightly prominent abaxially. *Inflorescence* 3–12.5 cm long, 4–16(– 18)-flowered. *Peduncle* erect to patent, wiry, 2–5.8 cm by 0.4–0.5 mm, orbicular in section, glabrous, with 4–7 scales. *Peduncle scales* tubular, the longest 3–5 mm long; tip acute; membranous, without nerves; not keeled and glabrous abaxially. *Rhachis* arching or nodding, terete, slightly to distinctly zigzag bent, 1–6 cm long, glabrous. *Floral bracts* about spreading, ovate, 2.8–3.5 by 1.6–2.5 mm, tip acuminate; base rather broadly attached; membranous, without nerves; slightly keeled and glabrous abaxially. *Flowers* distichous, in a lax raceme, about recurved, many open simultaneously, not fully open. *Pedicel and ovary* 1–2.2 mm long, glabrous. *Median sepal* not recurved, narrowly ovate to ovate-oblong, 3.7–6 by 1.8–4 mm, tip obtuse to acute; margins entire, glabrous; base rather narrowly attached; thin; glabrous; slightly keeled abaxially. *Lateral sepals* free, either slightly recurved or not, hardly falcate, ovate oblong, oblique, 3.7–5.7 by 1.4–2.4 mm, base broadly attached; otherwise as the median sepal. *Petals* not recurved, hardly falcate, narrowly obovate to broadly obovate-oblong, oblique, 1.3–2.3 by 0.6–0.9 mm, tip emarginate to rounded; margins slightly irregular towards the tip, glabrous; base broadly attached; very thin; glabrous or with some papillae adaxially; not keeled and glabrous abaxially. *Lip* recurved about half-way, broadly lanceolate in outline (not spread), 1.8–3.3 by 0.8–1.5 mm, tip obtuse; margins entire, glabrous; rather thin to rather thick; adaxially without ridges, with a slightly concave basal part, without a median slit, surface about glabrous; abaxially with a weak, rounded ridge near the base, surface glabrous. *Column* from the base of the free part up to the tip of the stelidia 1.2–2 mm long, with distinct, gradually tapering, hardly falcate, acuminate stelidia of 0.4–0.8 mm long; column with weak, obliquely backwards directed, rounded to obtuse teeth along its adaxial margins, near the base of the stelidia; stigma without a basal callus. *Anther* c. 0.5 by 0.4 mm; front not drawn out; front margin rounded, entire, glabrous; basally distinctly emarginate; abaxially with a distinct, more or less conical, obtuse, colliculate ridge near the top of the anther which distinctly overtops the front margin, surface glabrous or nearly so. *Pollinia* 4, obovoid; the small ones about half as long as the large ones, distinctly flattened; the large ones flattened; a stipes has not been observed. *Fruit* not seen.

Colours. Pseudobulbs suffused with red. Leaves green. Sepals and petals whitish, greenish or pink, usually with purple nerves. Lip yellowish, occasionally with a purple base.

Ecology. Montane forest, growing epiphytic with moss and hygrophylous ferns (La Croix, written comm.); 1300–2500 m alt. Flowering IV–VI and VIII–I.

Distribution. TANZ.: North T2, Kilimanjaro (2 coll.); Tanga T3 (fide F.T.E.A); East T6, Uluguru Mts (1); Southern Highlands T7 (6). MAL.: North (6); South (7). Material seen from BM, C, E, G, K, L, S, U, UPS, WAG, Z.

Note. The specimen Moreau 477 from TANZ. has flowers with relatively short, obtuse sepals of a slightly thicker texture.

12 - **Bulbophyllum prorepens** Summerh. - Fig. 23.

Bulbophyllum prorepens Summerh., Kew Bull. 8 (1954) 582; Robyns, Fl. Sp. P. Alb. 3 (1955) 496.–Type: Burtt 3152 (Holo K; iso AMES).

Roots 0.6–1 mm diam. *Rhizome* 1–1.5 mm diam., nodes not or hardly swollen. *Rhizome scales* tubular, tip obtuse, the upper occasionally apiculate, membranous, with some nerves; the upper keeled abaxially towards the tip. *Pseudobulbs* 2-leafed, narrowly ovoid, 2.5–12 cm apart, 1.5–3(–4) by 0.4–1 cm, either slightly flattened or not, 4–6-angled. *Petiole* 2–6 mm long. *Leaf blade* narrowly lanceolate to linear-lanceolate, 2.5–6 by 0.4–0.9 cm, tip obtuse, not or hardly emarginate and/or oblique; coriaceous, thick; midrib slightly prominent abaxially. *Inflorescence* 7.5–12 cm long, 10–16-flowered. *Peduncle* erect to patent, rather wiry, 5–6 cm by c. 1 mm, orbicular in section, glabrous, with c. 5 scales. *Peduncle scales* tubular, the longest 5–8 mm long; tip acuminate; membranous, without nerves; keeled and glabrous abaxially. *Rhachis* slightly arching or nodding, terete, slightly zigzag bent, 3.5–7 cm long, glabrous. *Floral bracts* about spreading, ovate to ovate-oblong, 3.5–6 by 1.6–3 mm, tip subacute; base rather broadly attached; membranous, without nerves; keeled and glabrous abaxially. *Flowers* distichous, in a lax raceme, recurved, many open simultaneously, not fully open. *Pedicel and ovary* 1.3–2 mm long, glabrous. *Median sepal* not recurved, elliptic to oblong, 4.6–5 by 2.2–3 mm, tip about acute; margins entire, glabrous; base rather broadly attached; thin; glabrous; not keeled abaxially. *Lateral sepals* free, either slightly recurved or not, hardly falcate, ovate to ovate-oblong, oblique, 4–5 by 2–2.3 mm, base broadly attached; slightly keeled abaxially; otherwise as the median sepal. *Petals* not recurved, not falcate, narrowly elliptic, oblique, c. 2 by 0.8–1 mm, tip subacute; margins entire, glabrous; base rather narrowly attached; very thin; glabrous; not keeled abaxially. *Lip* recurved, elliptic to narrowly oblong in outline (not spread), 2.8–3 by 1–1.7 mm, tip acute to slightly acuminate; each margin with two distinct undulations which start at the base and reach up to 1/3–2/3 of the lip, margins entire, glabrous; thick; adaxially with a weak to distinct, obtuse median ridge in the basal half, with a slightly concave basal part, without a median slit, surface glabrous to very finely papillose; abaxially with a weak, obtuse ridge near the base, surface about glabrous. *Column* from the base of the free part up to the tip of the stelidia 1.5–1.7 mm long, with distinct, triangular, not falcate, emarginate, truncate or acute stelidia which may have a weak, acute tooth along their upper margin near the base; column without teeth along its adaxial margins; stigma without a basal callus. *Anther* c. 0.5 by 0.5 mm; front not drawn out; front margin about truncate, entire, about glabrous; basally distinctly emarginate; abaxially with a weak, slightly conical, rounded, colliculate ridge about half-way the anther which does not overtop the front margin; surface colliculate, glabrous towards the margins. *Pollinia* 4, ovoid; the inner pair almost as long as the outer pair, distinctly flattened on either side; the outer flattened on one side; a stipes has not been observed. *Fruit* not seen.

Colours. Sepals and petals pale yellow or pale green. Lip yellow.

Ecology. Montane forest, 1900–2400 m alt. Epiphytic. Flowering XII (1 rec.).

Distribution. ZAI.: Lacs Edouard & Kivu Z9 (2 coll.). RW.: W. part (2). Material seen from AMES, BM, BR, K.

13 - Bulbophyllum intertextum Lindl. - Fig. 24.

Bulbophyllum intertextum Lindl., J. Linn. Soc. Bot. 6 (1862) 127 ('*Bolbophyllum*'); Durand & Schinz, Consp. Fl. Afr. 5 (1895) 12; Rolfe in F.T.A. 7 (1897) 29; De Wild., Pl. Beq. 1 (1921) 113; Summerh. in F.W.T.A. ed. 1, 2 (1936) 438; Robyns, Fl. Sp. P. Alb. 3 (1955) 493; Morris, Proc. Linn. Soc. London 179 (1968) 63; Piers, Orch. E. Afr. (1968) 104; Summerh. in F.W.T.A. ed. 2, 3 (1968) 234; Hunt, Kew Bull. 22 (1968) 491; Morris, Ep. Orch. Mal. (1970) 49; Stewart in Agnew, Upl. Ken. W. Fl. (1974) 760; Johansson, Acta Phyt. Suec. 59 (1974) 31; Geerinck, Bull. Soc. R. Bot. Belg. 109 (1976) 176; Grosvenor, Excelsa 6 (1976) 79; Williamson, Orch. S. Centr. Afr. (1977) 138; Ball, S. Afr. Ep. Orch. (1978) 76, pl.; Adam, Fl. Descr. Mt. Nimba 5 (1981) 1677, pl. 907; La Croix et al., Orch. Mal. 1 (1983) 46; Cribb, in F.T.E.A., Orch. 2 (1984) 309.-*Phyllorchis intertexta* (Lindl.) Kuntze, Rev. Gen. Pl. 2 (1891) 677, incorr. name-Type: Mann 527 (K).

Bulbophyllum seychellarum Rchb.f., Linnaea 41 (1877) 93.-*Phyllorchis seychellarum* (Rchb.f.) Kuntze, Rev. Gen. Pl. 2 (1891) 678, incorr. name.-Type: Horne 1874 (K); Seychelles.

Bulbophyllum quintasii Rolfe, Bol. Soc. Brot. 9 (1891) 138; Durand & Schinz, Consp. Fl. Afr. 5 (1895) 14; Rolfe in F.T.A. 7 (1897) 29; De Wild., Pl. Beq. 1 (1921) 116; Exell, Cat. Vasc. Pl. S. Tomé (1944) 326.-Type: Quintas 10 (holo K; iso COI, n.v.)

Bulbophyllum viride Rolfe, Kew Bull. 1893 (1893) 170; Rolfe in F.T.A. 7 (1897) 29; Rendle, Cat. Talb. S. Nig. Pl. (1913) 146; De Wild., Pl. Beq. 1 (1921) 121.-Type: Crowley cult. s.n. (K).

Bulbophyllum usambarae Krzl., Engl., Bot. Jahrb. 34 (1904) 58; De Wild., Pl. Beq. 1 (1921) 121.-Type: Engler 1025a (B, lost).

Bulbophyllum triaristellum Krzl. & Schltr, Orchis 2 (1908) 98; De Wild., Pl. Beq. 1 (1921) 120.-Type: Schlechter 15756 (Holo B, lost; iso BM, BR, G, K, L, P, WAG, Z).

Bulbophyllum amauryae Rendle, Cat. Talb. S. Nig. Pl. (1913) 101, 146; De Wild., Pl. Beq. 1 (1921) 103.-Type: Talbot 952 (Holo BM; iso K).

Bulbophyllum pertenue Krzl., Engl., Bot. Jahrb. 51 (1914) 385; De Wild., Pl. Beq. 1 (1921) 116.-Type: Ledermann 6201 (B, lost).

Bulbophyllum intertextum Lindl. var. *parvilabium* Williamson, Pl. Syst. Evol. 134 (1980) 62.-Type: Williamson & Simon 442 (Holo K; iso SRGH, n.v.).

Non *Bulbophyllum parvilabium* Schltr, Engl., Bot. Jahrb. 45, Beibl. 104 (1911) 52; nec *Bulbophyllum parvilabium* Schltr, Fedde, Rep. 16 (1919) 128.

Roots 0.3–0.6 mm diam. *Rhizome* 0.7–1.2 mm diam., nodes not or hardly swollen. *Rhizome scales* tubular, tip acute or subacute; membranous, with some nerves; not keeled abaxially. *Pseudobulbs* 1-leafed, ovoid to about orbicular, 0.2–2.5 (- 4) cm apart, 0.4–1 by 0.3–0.7 cm, either flattened or not, obtusely 4 (- 5)-angled. *Petiole* 0.2–2 mm long. *Leaf blade* elliptic to linear-lanceolate, 0.7–10 by 0.3–1.1 cm, tip rounded to acute, either emarginate or not, not oblique; coriaceous, thick; midrib hardly prominent abaxially. *Inflorescence* 2–30 cm long, 2–14(- 20)-flowered. *Peduncle* erect to patent, usually wiry, 1.8–11 cm by 0.3–0.6 mm, orbicular in section, glabrous, with 3-5 scales. *Peduncle scales* tubular, the longest 3–5 mm long; tip acute; membranous, without nerves; not keeled and glabrous abaxially. *Rhachis* arching or nodding, terete, usually zigzag bent, 0.2–19 cm long, glabrous. *Floral bracts* recurved, elliptic to oblong, 1.5–3 (– 4) by 1–2 mm, tip acuminate; base rather broadly attached; membranous, without nerves; not keeled and glabrous abaxially. *Flowers* distichous, in a lax raceme, recurved, many open simultaneously, not fully open. *Pedicel and ovary* 1–2 mm long, glabrous. *Median sepal* not or hardly recurved, ovate-oblong to ovate-lanceolate, 3–6 by 0.9–2 mm, tip cuspidate; margins entire, glabrous to finely papillose; base broadly attached; thin; locally very finely papillose adaxially; not keeled abaxially. *Lateral sepals* free, slightly recurved, hardly falcate, (very narrowly) triangular, oblique, 2.8–6 by 0.8–3 mm, otherwise as the median sepal. *Petals* not or hardly recurved, hardly falcate, broadly elliptic to oblong, oblique, 1–3 by 0.5–1.6 mm, tip rounded to acute; margins slightly irregular towards the tip, glabrous; base rather broadly to rather narrowly attached; very thin; glabrous or with a few coarse papillae; not keeled abaxially. *Lip* recurved, triangular, ovate or elliptic in outline (not spread), 1.2–4 by 0.7–2 mm, tip rounded to subacute; margins entire, with short or long hairs, or about glabrous, top part glabrous to finely puberulous; thick; adaxially with a more or less thickened median part, with a slightly concave basal part, with a median slit starting at about the base and continuing over 1/2–2/3 of the lip, surface glabrous, papillose towards the margins; abaxially with a weak rounded ridge in the basal half, surface glabrous, locally slightly papillose. *Column* from the base of the free part up to the tip of the stelidia 0.8–2.2 mm long, with distinct, (very narrowly) triangular, not or hardly falcate, rounded to acuminate or about truncate stelidia of 0.3–1 mm long; column occasionally with either folded inwards or not, weak to distinct, deltoid, rounded to subacute wings or teeth along its adaxial margins just below the base of the stelidia; stigma with or without a weak to distinct, obtuse basal tooth or callus. *Anther* c. 0.5 by 0.6 mm; front not drawn out; front margin rounded, entire, papillose; basally distinctly emarginate; abaxially with a very weak, rounded, colliculate ridge near the top of the anther which hardly overtops the front margin, surface colliculate. *Pollinia* 2; ellipsoid or obovoid, flattened on one side; a stipes has not been observed. *Fruit* obovoid to ellipsoid, up to 0.6 by 0.4 cm, stalk 1 mm.

Colours. Pseudobulbs often stained with red. Leaves green. Flowers very pale yellowish or greenish, often partly suffused with purple or purple-red, or entirely purple or purple-red.

Ecology. Lowland forest, montane forest, also recorded from 'semi-montane grasslands' (Sanford, 1974). Epiphytic in the crown of trees, on bark, minor humus deposits or moss cushions (Johansson, 1974; La Croix, written comm.). Rare in Brachystegia woodland (La Croix, written comm.). Occasionally lithophytic. Flowering mainly VIII–IV.

Distribution. GUIN. (1 coll). S.L. (2). LIB. (8). IV.C.: W. part (1). NIG.: W. part, Cross P. & Rivers P. (5). CAM.: S. part (10). EQ.GUIN.: Mbini (2); Bioko (2). S.T. & PR.: Sao Tomé (1); Principe (1). GAB. (15). ZAI.: Bas Katanga Z5 (1); Forestier Central Z6 (2); Lacs Edouard & Kivu Z9 (1). AETH.: SE part, Ilubabor P. (1). KEN.: Rift Valley K3/Nyanza K5 (2); Coast K7 (2). TANZ.: Tanga T3, Usambara Mts (5); East T6, Uluguru Mts (1); Southern Highlands T7, Iringa (5). ANG.: NE part, Lunda P. (2). ZAM.: West, Mwinilunga (3). MAL.: North (2); South (2). ZIM.: East, N. part (2). Material seen from AMES, BM, BR, C, G, HBG, K, L, P, S, UPS, WAG, Z.

Distribution outside Africa: Seychelles (Horne 1874, type of *Bulbophyllum seychellarum* Rchb.f.).

Notes. 1 - *Bulbophyllum intertextum* is very variable in all its parts. In spite of this it is impossible to distinguish varieties because all extremes are linked by series of intermediates.

The variability in the vegetative parts may partly be due to differences in habitat: as far as can be concluded from herbarium labels humid forest specimens tend to be much more elongate in the vegetative parts than specimens from drier environments.

The lip especially is variable in general outline, and the length as well as the number of the ciliae along the margins.

The ciliae along the lip may be long, short or even almost absent. The latter character occurs in the specimens Louis, Breteler & De Bruin 1242 from GAB. and, according to the description, in the holotype specimen of var. *parvilabium* from ZAM. (not studied, an isotype specimen in K certainly has a ciliate lip). Other characters ascribed to var. *parvilabium* do not hold because of the existence of series of intermediate specimens. Therefore var. *parvilabium* can not be maintained as a separate variety.

The column is also variable. The projections next to the stelidia, described above, may either be present or absent. Besides, in some specimens (e.g. Le Testu 5707 from GAB.) the column is much more elongated than in others (e.g. Linder 334a from LIB.). Here, too, series of intermediates exist.

All the characters mentioned in the notes are distributed in a reticulate pattern over the material observed. Only two characters occur more or less together: specimens with wide leaves tend to have an elliptic to ovate lip, not a triangular one.

2 - The specimen Le Testu 5720 (from GAB.) may well be a natural hybrid with *B. pumilum*. It is intermediate between the two species.

14 - Bulbophyllum vanum J.J. Vermeulen – Fig. 25.

Bulbophyllum vanum J.J. Vermeulen, Bull. Soc. R. Bot. Belg. 54 (1984) 148.–Type: Breteler & De Wilde 642 (WAG).

Roots 0.15–0.3 mm diam. *Rhizome* 0.4–0.6 mm diam., nodes not or hardly swollen. *Rhizome scales* tubular, tip acute or subacute; membranous, with some nerves; not keeled abaxially. *Pseudobulbs* 1-leafed, ovoid to about orbicular, 0.2–0.8 cm apart, 0.2–0.5 by 0.2–0.4 cm, slightly flattened, inconspicuously and obtusely 3 (– 5)-angled. *Petiole* 0.3–1 mm long. *Leaf blade* oblong to broadly linear-lanceolate, 0.4–1.2 (– 1.8) by 0.4–0.6 cm, tip rounded to subacute, emarginate, not oblique; coriaceous, thick; midrib slightly prominent abaxially. *Inflorescence* 1.8–2.8 cm long, 1–3-flowered. *Peduncle* about erect, wiry, slightly swollen towards the tip, 1.7–2.6 cm by c. 0.3 mm, orbicular in section, glabrous, with 3 scales at the base and a fourth about half-way. *Peduncle scales* tubular, the longest c. 2 mm long; tip about acute; membranous, without nerves; not keeled and glabrous abaxially. *Rhachis* erect, slightly swollen, about orbicular in section, zigzag bent, 0.08–0.2 cm by c. 0.5 mm, glabrous. *Floral bracts* recurved, oblong, c. 2.8 by 1.2 mm, tip acute; base broadly attached; membranous, without nerves; not keeled and glabrous abaxially. *Flowers* (?) scattered, in a dense raceme, recurved, about all open simultaneously, not much open. *Pedicel and ovary*

c. 1 mm long, glabrous. *Median sepal* not recurved, ovate, c. 2.5 by 1.2 mm, tip obtusely cuspidate; margins entire, very finely papillose towards the tip; base broadly attached; rather thin; very finely papillose adaxially towards the tip; not keeled and slightly colliculate abaxially. *Lateral sepals* free, hardly falcate, oblique, c. 2.7 by 1.3 mm, otherwise as the median sepal. *Petals* not recurved, hardly falcate, obovate, oblique, c. 1.3 by 0.8 mm, tip acuminate; margins slightly irregular towards the tip, glabrous; base rather broadly attached; very thin; glabrous; not keeled abaxially. *Lip* slightly recurved about half-way, elliptic to ovate in outline (not spread), 1–1.5 by 1.0 mm, tip rounded; margins entire, ciliate; thick; surface glabrous, papillose towards the margins; adaxially slightly thickened in the median part, with a hardly concave basal part, with a median slit starting at about the base and continuing over about the entire lip; abaxially with a weak rounded ridge near the base. *Column* from the base of the free part up to the tip of the stelidia c. 1 mm long, with distinct, not falcate, narrowly triangular, subacute stelidia of 0.5 mm long which have a distinct, deltoid, obliquely forward directed, obtuse tooth along their upper margin; column with very weak, rounded wings along its lower margins, near the base of the stelidia; stigma without a basal callus. *Anther* c. 0.4 by 0.4 mm; front not drawn out; front margin rounded, entire, glabrous; basally distinctly emarginate; abaxially with a rather distinct, obtuse, colliculate ridge towards the top of the anther which overtops the front margin, surface colliculate. *Pollinia* 4; the small ones about half as long as the large ones, ellipsoid, distinctly flattened on either side; the large ellipsoid to obovoid, flattened on one side; a stipes has not been observed. *Fruit* 0.4 by 0.2 cm, ellipsoid, without a stalk.

Colours. Pseudobulbs pale green. Flowers pale brownish red, hairs along the lip white.

Ecology. Hill forest at 500 m alt. (1 rec.). Epiphytic in the crown of trees, in moss cushions or humus deposits.

Distribution. GAB. (2 coll.). ZAI.: Forestier Central Z6 (2, both sterile but probably belonging here). Material seen from BR, P, WAG.

15 - **Bulbophyllum capituliflorum** Rolfe – Fig. 26.

Bulbophyllum capituliflorum Rolfe, Kew Bull. 1906 (1906) 84; De Wild., Pl. Beq. 1 (1921) 107.–Type: Glasnevin cult. s.n. (K).

Roots c. 0.5 mm diam. *Rhizome* 1.5–1.8 mm diam., nodes not or hardly swollen. *Rhizome scales* tubular, tip acute to acuminate; membranous, with some nerves; either keeled abaxially or not. *Pseudobulbs* 1-leafed, ovoid to about orbicular, 0.7–1.2 cm apart, 0.4–0.8 by 0.5–0.8 cm, slightly flattened, obtusely 4-angled. *Petiole* 0.5–1.5 mm long. *Leaf blade* broadly elliptic to lanceolate, 0.9–4 by 0.7–1.3 cm, tip acute to shortly acuminate, either emarginate or not, not oblique; coriaceous, thick; midrib hardly prominent abaxially. *Inflorescence* 1.5–6 cm long, 3–8-flowered. *Peduncle* about erect, either wiry or not, slightly swollen towards the tip and below the nodes, 1.2–5.8 cm by 0.4–1 mm, orbicular in section, glabrous, with 4–5 scales. *Peduncle scales* tubular, the longest 3.5–4.5 mm long; tip about acute to acuminate; membranous, without nerves; keeled and glabrous abaxially. *Rhachis* erect or nodding, terete or slightly swollen, 0.2–0.5 cm long, glabrous. *Floral bracts* recurved, elliptic to ovate, 2.8–3 by

1.8–2 mm, tip acute to acuminate; base rather broadly attached; membranous, without nerves; not keeled and glabrous abaxially. *Flowers* scattered, in a (rather) dense raceme, recurved, many open simultaneously, not fully open. *Pedicel and ovary* 1.5–2 mm long, glabrous. *Median sepal* recurved, ovate to ovate-oblong, 3.2–5.2 by 1.5–2 mm, tip about acuminate; margins entire, either finely papillose towards the tip or not; base broadly attached; rather thin; finely papillose adaxially towards the tip; not keeled and almost entirely finely papillose abaxially. *Lateral sepals* free, hardly falcate, ovate to narrowly triangular, oblique, 3–5 by 2–2.2 mm, otherwise as the median sepal. *Petals* not recurved, hardly falcate, obovate-oblong, oblique, 1.5–2 by 0.6–0.7 mm, tip acute to acuminate; margins slightly irregular towards the tip, glabrous; base rather broadly attached; very thin; glabrous; not keeled abaxially. *Lip* with a distinctly recurved top part, elliptic in outline (not spread), 2–2.2 by 1–1.2 mm, tip rounded; margins entire, very finely papillose; thick; adaxially with a thickened median part, with a hardly concave basal part, with a median slit starting at about the base and continuing up to about half-way the lip, surface glabrous, very finely papillose towards the margins; abaxially with a weak rounded ridge in the basal half, with two distinct, triangular, slightly concave, obtuse, thick auricles abaxially of the attachment to the column foot, surface glabrous, very finely papillose towards the margins, slightly irregularly colliculate near the base. *Column* from the base of the free part up to the tip of the stelidia 1.3–1.8 mm long, with distinct, narrowly triangular, distinctly falcate, acute stelidia of 0.5–0.7 mm long which have a rather distinct, obtuse to acute edge along their upper margin near the base; column with either folded inwards or not, distinct, obliquely backwards directed, deltoid, obtuse teeth along its lower margins just below the base of the stelidia; stigma with a distinct, obtuse basal tooth; top of column foot with two distinct, about orbicular lateral swellings. *Anther* c. 0.8 by 0.5 mm; front not drawn out; front margin rounded, entire, papillose; basally distinctly emarginate; abaxially with a distinct, slightly conical, rounded, colliculate ridge near the top of the anther which distinctly overtops the front margin, surface glabrous, colliculate to papillose towards the top. *Pollinia* 4; the small ones about half as long as the large ones, ovoid, distinctly flattened on either side; the large ones ellipsoid, flattened on one side; a stipes has not been observed. *Fruit* ellipsoid, up to 0.6 by 0.4 cm, without a stalk.

Colours. Sepals and petals cream coloured or greenish, either or not suffused with brown-purple. Lip yellow or red-purple.

Ecology. Primary lowland forest. Epiphytic. Flowering time unknown.

Distribution. GAB. (3 coll.). CONGO (1). ZAI.: Forestier Central Z6 (1). Material seen from BR, K, P, WAG.

16 - Bulbophyllum pipio Rchb. f. – Fig. 27; Plate 3d.

Bulbophyllum pipio Rchb. f., Linnaea 41 (1877) 92; Rchb. f., Xenia Orch. 3 (1881) 45, pl. 219; Krzl., Mitth. Deutsch. Schutzgeb. 2 (1889) 155; Durand & Schinz, Consp. Fl. Afr. 5 (1895) 13; Rolfe in F.T.A. 7 (1897) 32; De Wild., Pl. Beq. 1 (1921) 116; Summerh., Kew Bull. 6 (1952) 471; id. in F.W.T.A. ed. 2, 3 (1968) 234; Cribb & Perez-Vera, Adansonia, ser. 2, 15 (1975) 206; Vermeulen, Tijdschr. Ned. Orch. Ver. 44 (1982) 45.–*Phyllorchis pipio* (Rchb. f.) Kuntze, Rev. Gen. Pl. 2 (1891) 677, incorr. name.–Type: 'Cult. Hort. Hammoniae, no. 2302 & 2307' (W, small portion in K).

Bulbophyllum milesii Summerh., Kew Bull. 1935 (1935) 200; Summerh. in F.W.T.A. ed. 1, 2 (1936) 438.–Type: Miles 19 (K).

Roots 0.3–0.6 mm diam. *Rhizome* 0.9–2 mm diam., nodes not or hardly swollen. *Rhizome scales* tubular, tip acute or shortly acuminate; membranous, slightly woolly when old, without nerves; slightly keeled abaxially. *Pseudobulbs* 1-leafed, (broadly) ovoid, 0.5–1.2 cm apart, 0.5–1.5 by 0.4–0.8 cm, slightly flattened, obtusely 3–4-angled. *Petiole* 1.5–4 mm long. *Leaf blade* narrowly lanceolate to linear, 1.7–7 by 0.2–0.5 cm, tip acute, not emarginate, not or hardly oblique; coriaceous, thick; midrib prominent abaxially. *Inflorescence* 0.8–6.5 cm long, 4–25-flowered. *Peduncle* usually erect, 0.6–2.5 cm by 0.7–1.5 mm, orbicular in section, glabrous, with 4–7 scales. *Peduncle scales* tubular, the longest 3.5–7 mm long; tip acuminate; membranous, without nerves; keeled towards the tip and glabrous abaxially. *Rhachis* erect, terete, 0.2–4.5 cm long, glabrous. *Floral bracts* recurved, occasionally slightly concave, ovate-oblong to ovate-lanceolate, 5–7 by 1.5–2.8 mm, tip acute to acuminate; base rather broadly attached; membranous, without nerves; not or hardly keeled and glabrous abaxially. *Flowers* scattered, in a dense raceme, recurved, either secund or not, many open simultaneously, not fully open. *Pedicel and ovary* 1.8–2 mm long, glabrous. *Median sepal* recurved, top part spreading, narrowly ovate-lanceolate to broadly linear-lanceolate, 4–5 by 0.8–1.2 mm, tip (sub)acute; margins entire, glabrous; base broadly attached; rather thin; glabrous; not keeled abaxially. *Lateral sepals* free, hardly falcate, ovate-lanceolate to broadly linear-lanceolate, oblique, 5–6 by 1–1.5 mm, otherwise as the median sepal. *Petals* not recurved, hardly falcate, lanceolate to broadly linear-lanceolate, oblique, 2–2.4 by 0.3–0.6 mm, tip obtuse to about acute; margins entire, glabrous to very finely papillose; base rather broadly attached; thin; glabrous; not keeled abaxially. *Lip* recurved, narrowly elliptic to oblong in outline (not spread), maximum width either slightly below the middle or not, 1.3–1.9 by 0.5–0.9 mm, tip rounded; over about the entire length (or at least over c. 2/3 of the entire length) with seamlike lateral lobes which have entire to slightly irregular and slightly irregularly papillose margins and which are decurrent towards the tip of the lip, thin, separated from the median part of the lip by a distinct furrow, irregularly papillose adaxially and about glabrous abaxially; lip rather thick; adaxially without ridges, with a slightly concave basal part, with a median slit starting at about the base and continuing up to half-way of the lip, surface glabrous, irregularly papillose towards the tip; abaxially with a weak, rounded to truncate ridge near the base, surface about glabrous. *Column* from the base of the free part up to the tip of the stelidia 1.2–1.6 mm long, with distinct, gradually tapering, not falcate, acute stelidia of 0.6–0.9 mm long; column with weak, rounded wings along its lower margins just below the base of the stelidia; stigma without a basal callus. *Anther* c. 0.5 by 0.5 mm; front distinctly drawn out, thin, concave; front margin rounded, entire, papillose; basally distinctly emarginate; abaxially with a rather weak, rounded, colliculate ridge near the top of the anther which does not overtop the front margin, surface glabrous but colliculate towards the top. *Pollinia* 4, obovoid; the small ones about half as long as the large ones, distinctly flattened on either side; the large ones flattened on one side; a stipes has not been observed. *Fruit* ellipsoid, up to 0.4 by 0.25 cm, without a stalk.

Colours. Pseudobulbs olive green, often suffused with red. Leaves green. Sepals cream-coloured, often with a bright yellow top part. Petals cream-coloured. Lip cream-coloured or green, usually with a purple base. Column very pale greenish.

Ecology. Mangrove forest (1 rec.). Lowland forest, savanna woodland (Sanford, 1974). Epiphytic. Flowering (II, VI) X–XII.

Distribution. S.L. (3 coll.). IV.C. (5). GHANA (3). NIG. (fide Sanford, 1974). CAM. (1). Material seen from BR, K, L, P.

17 - **Bulbophyllum pumilum** (Sw.) Lindl. - Fig. 28–30; Plate 4a-d, 5a.

Bulbophyllum pumilum (Sw.) Lindl., Gen. and Sp. Orch. Pl. (1830) 54 ('? *Bolbophyllum*'); Durand & Schinz, Consp. Fl. Afr. 5 (1895) 14.–*Dendrobium pumilum* Sw., Schrad. N. J. Bot. 1, 1 (1805) 97; Persoon, Syn. Pl. 2 (1807) 523.–*Genyorchis pumila* (Sw.) Schltr, Westafr. Kautchuk Exp. (1901) 280, pro comb.; Schltr, Engl., Bot. Jahrb. 38 (1905) 12, pro comb.–Type: Afzelius s.n. in Herb. Swartz (S).

Non *Genyorchis pumila* auct.: Schltr, Westafr. Kautchuk Exp. (1901) 280; Schltr, Engl., Bot. Jahrb. 38 (1905) 12 (= both *Genyorchis apetala*, see under excluded species no. 1: *Bulbophyllum apetalum*).

Bulbophyllum recurvum Lindl., Gen. and Sp. Orch. Pl. (1830) 53 ('*Bolbophyllum*'); Lindl., Bot. Reg. 30 (1844) misc. 72 ('*Bolbophyllum*') ('Mrs. Loddiges received it from Guiana 1839, the specimens that have flowered having proved to be identical in all aspects with the African plant.' Possibly a specimen wrongly labeled as of American origin); Rchb. f., Walp. Ann. 6 (1861) 254 ('*Bolbophyllum*'); Durand & Schinz, Consp. Fl. Afr. 5 (1895) 14; Rolfe in F.T.A. 7 (1897) 31; De Wild., Pl. Beq. 1 (1921) 117; Summerh. in F.W.T.A. ed. 1, 2 (1936) 438; id., ed. 2, 3 (1968) 234; Johansson, Acta Phyt. Suec. 59 (1974) 32; Adam, Fl. Descr. Mt. Nimba 5 (1981) 1682, pl. 914; Vermeulen, Tijdschr. Ned. Orch. Ver. 44 (1982) 186.–*Phyllorchis recurva* (Lindl.) Kuntze, Rev. Gen. Pl. 2 (1891) 677, incorr. name.–*Tribrachia pendula* Lindl., Coll. Bot. (1825) sub pl. 41a; Lindl., Bot. Reg. 12 (1826) pl. 963. Type: Don s.n. (? K, n.v.).

Non *Bulbophyllum recurvum* auct.: Exell, Cat. Vasc. Pl. S. Tomé (1944) 327 (= *Bulbophyllum mediocre*).

Non *Bulbophyllum pendulum* Thou, Orch. Isles Austr. Afr. (1822) pl. 103.

Bulbophyllum flavidum Lindl., Bot. Reg. 26 (1840) misc. 83 ('*Bolbophyllum*'); Rchb. f., Walp. Ann. 6 (1861) 257 ('*Bolbophyllum*'); Durand & Schinz, Consp. Fl. Afr. 5 (1895) 12; Rolfe in F.T.A. 7 (1897) 30; Durand & Durand, Syll. Fl. Congol. (1909) 516; De Wild., Pl. Beq. 1 (1921) 110; Summerh. in F.W.T.A. ed. 1, 2 (1936) 438; Summerh., Kew Bull. 6 (1952) 469; Ake Assi, Et. Fl. C. Iv. (1964) 243; Summerh. in F.W.T.A. ed. 2, 3 (1968) 234; Johansson, Acta Phyt. Suec. 59 (1974) 31; Adam, Fl. Descr. Mt. Nimba 5 (1981) 1675; Vermeulen, Tijdschr. Ned. Orch. Ver. 44 (1982) 168, 186; Segerbaeck, Orch. Nig. (1983) 59, phot.–*Phyllorchis flavida* (Lindl.) Kuntze, Rev. Gen. Pl. 2 (1891) 677, incorr. name.–Type: Loddiges s.n. (K).

Non *Bulbophyllum flavidum* auct.: Exell, Cat. Vasc. Pl. S. Tomé (1944) 326 (= *Bulbophyllum mediocre*).

Bulbophyllum herminiostachys (Rchb. f.) Rchb. f., Walp. Ann. 6 (1861) 254 ('*Bolbophyllum*'); Durand & Schinz, Consp. Fl. Afr. 5 (1895) 12; Rolfe in F.T.A. 7 (1897) 35; De Wild., Pl. Beq. 1 (1921) 111.–*Taurostalix herminiostachys* Rchb. f., Bot. Zeit. (1852) 933.–*Phyllorchis herminiostachys* (Rchb. f.) Kuntze, Rev. Gen. Pl. 2 (1891) 677, incorr. name.–Type: Hb. Reichenbach (W).

Bulbophyllum pavimentatum Lindl., J. Linn. Soc. Bot. 6 (1862) 128 ('*Bolbophyllum*'); Hook., Bot. Mag. 88 (1862) pl. 5329; Krzl., Mitth. Deutsch. Schutzgeb. 2 (1889) 155; Durand & Schinz, Consp. Fl. Afr. 5 (1895) 13; Rolfe in F.T.A. 7 (1897) 32; Malmquist, Orchis 8 (1914) 69; De Wild., Pl. Beq. 1 (1921) 115; Summerh. in F.W.T.A. ed. 1, 2 (1936) 438; id., ed. 2, 3 (1968) 234; Johansson, Acta Phyt. Suec. 59 (1974) 32; Geerinck, Bull. Soc. R. Bot. Belg. 109 (1976) 176; Adam, Fl. Descr. Mt. Nimba 5 (1981) 1682; Vermeulen, Tijdschr. Ned. Orch. Ver. 44 (1982) 168, 187.–*Phyllorchis pavimentata* (Lindl.) Kuntze, Rev. Gen. Pl. 2 (1891) 677, incorr. name.–Type: Mann 519 (K).

Bulbophyllum gabonis Lindl. & Rchb. f., Flora 48 (1865) 185 ('*Bolbophyllum*'); Durand & Schinz, Consp. Fl. Afr. 5 (1895) 12; Rolfe in F.T.A. 7 (1897) 35; De Wild., Pl. Beq. 1 (1921) 110.–Type: Hb. Reichenbach (W).

Bulbophyllum drallei Rchb. f., Gartenflora 35 (1885) 20.–Type: Dralle s.n. (W).

Bulbophyllum porphyroglossum Krzl., Engl., Bot. Jahrb. 22 (1895) 24 ('*Bolbophyllum*'); Rolfe in F.T.A. 7 (1897) 34; De Wild., Pl. Beq. 1 (1921) 116; Summerh. in F.W.T.A. ed. 1, 2 (1936) 438; id., Kew Bull. 12 (1957) 115; id. in F.W.T.A. ed. 2, 3 (1968) 234; Cribb & Perez-Vera, Adansonia, ser. 2, 15 (1975) 206; Vermeulen, Tijdschr. Ned. Orch. Ver. 44 (1982) 185.–Type: Preuss 1299 ('1279') (Holo B, lost; iso K).

Bulbophyllum nanum De Wild., Not. Pl. Ut. et Int. Fl. Congo 1 (1903) 122; id., Et. Fl. B. et Moy. Congo

1 (1904) 116; Durand & Durand, Syll. Fl. Congol. (1909) 516; De Wild., Pl. Beq. 1 (1921) 114.–Type: Laurent 178 (BR).

Bulbophyllum papillosum Finet, Bull. Mus. Hist. Nat. Paris 9 (1903) 303 ('*Bolbophyllum*'); De Wild., Pl. Beq. 1 (1921) 115.–Syntypes: Vergnes s.n. (P), Lousteaux s.n. (P, n.v.).

Non *Bulbophyllum papillosum* J.J.S., Fl. Buitenz. 6 (1905) 464; nec *Cirrhopetalum papillosum* Rolfe, Kew Bull. 1908 (1908) 70.–*Bulbophyllum papillosum* (Rolfe) Seidenf. & Smitin., Orch. Thail. 3 (1961) 364.

Bulbophyllum calabaricum Rolfe, Kew Bull. 1906 (1906) 114; De Wild., Pl. Beq. 1 (1921) 106.–Type: Holland s.n. (K).

Bulbophyllum moliwense Schltr, Engl., Bot. Jahrb. 38 (1906) 157; De Wild., Pl. Beq. 1 (1921) 114; Schltr, Fedde, Rep. Beih. 68 (1932) pl. 67.–Type: Schlechter 15757 (B, lost).

Bulbophyllum winklerii Schltr, Engl., Bot. Jahrb. 38 (1906) 158; De Wild., Pl. Beq. 1 (1921) 121; Schltr, Fedde, Rep. Beih. 68 (1932) pl. 69; Summerh. in F.W.T.A. ed. 1, 2 (1936) 438; id., ed. 2, 3 (1968) 234; Johansson, Acta Phyt. Suec. 59 (1974) 33; Cribb & Perez-Vera, Adansonia, ser. 2, 15 (1975) 208; Geerinck, Bull. Soc. R. Bot. Belg. 109 (1976) 176; Adam, Fl. Descr. Mt. Nimba 5 (1981) 1685; Vermeulen, Tijdschr. Ned. Orch. Ver. 44 (1982) 188.–Type: Winkler 157 (B, lost).

Non *Bulbophyllum winklerii* Schltr, Orchis 8 (1914) 132 (= *Bulbophyllum josephii* var. *josephii*).

Bulbophyllum leucopogon Krzl., Engl., Bot. Jahrb. 48 (1912) 391; De Wild., Pl. Beq. 1 (1921) 113.–Type: Zenker 3903 (Holo B, lost; iso BM, E, G, K, L, P, S, Z).

Bulbophyllum dorotheae Rendle, Cat. Talb. S. Nig. Pl. (1913) 100, 146; De Wild., Pl. Beq. 1 (1921) 108.–Type: Talbot 934 (BM).

Bulbophyllum imogeniae Hamilton, Trans. Bot. Soc. Edin. 27 (1917) 228.–Type: Edinburg cult. s.n. (E).

Bulbophyllum verecundum Summerh., Kew Bull. 8 (1954) 582.–Type: Le Testu s.n. = Le Testu 6370b (Holo K; iso P).

Bulbophyllum yangambiense Louis & Mullenders ex Geerinck, Bull. Soc. R. Bot. Belg. 109 (1976) 176.–Type: Louis 9090 (BR).

Bulbophyllum elachon J.J. Vermeulen, Selbyana 7 (1982) 11.–*Bulbophyllum* sp. J.J. Vermeulen, Tijdschr. Ned. Orch. Ver. 44 (1982) 167, 185.–Type: Jongejan 51 (L).

Roots 0.2–1 mm diam. *Rhizome* 1–3.6 mm diam., nodes not or hardly swollen. *Rhizome scales* tubular, tip acute to shortly acuminate; membranous, with nerves; the upper keeled abaxially towards the tip. *Pseudobulbs* 1-leafed, orbicular to narrowly ovoid, 0.2–2.8 cm apart, 0.3–4 by 0.3–1.8 cm, slightly to moderately flattened, obtusely 2–4-angled. *Petiole* 1–20 mm long. *Leaf blade* about orbicular to broadly linear, maximum width occasionally either slightly below or above the middle, 1.1–23 by 0.6–4 cm, tip obtuse, acute or shortly acuminate, not or hardly emarginate or oblique; coriaceous, (rather) thick; midrib hardly prominent abaxially. *Inflorescence* 2–31 cm long, 3–65-flowered. *Peduncle* erect to pendulous, occasionally wiry, 1.4–20 cm by 0.6–1.5 mm, orbicular in section, glabrous, with 3–8 scales. *Peduncle scales* tubular, the longest 4.5–13 mm long; tip acute to mucronate, occasionally obtuse; membranous, with or without some nerves; either or not slightly keeled and glabrous abaxially. *Rhachis* erect to pendulous, terete, either slightly zigzag bent or not, 0.6–20 cm long, glabrous. *Floral bracts* recurved, elliptic to ovate or (ovate)-oblong, 2.5–6 by 1–3 mm, tip rounded to acute; base rather broadly to rather narrowly attached; membranous, without nerves; not or hardly keeled and glabrous abaxially. *Flowers* scattered, in a dense to lax raceme, recurved to spreading, either secund or not, many open simultaneously, not fully open. *Pedicel and ovary* 0.6–3.3 mm long, glabrous. *Median sepal* hardly recurved to entirely reflexed, triangular to very narrowly triangular, ovate to ovate-lanceolate or elliptic to oblong, or about spathulate, 2.3–7.5 by 0.9–2.8 mm, tip rounded to acuminate, or apiculate; margins entire, glabrous to finely ciliate; base (rather) broadly attached; rather thick, top part slightly to strongly thickened; ciliate or very finely to very coarsely papillose adaxially towards the tip; not keeled and glabrous abaxially. *Lateral sepals* free, slightly falcate, oblique, 2.7–8 by 0.9–3 mm, base broadly attached, otherwise as the median sepal. *Petals* not or hardly recurved, hardly falcate, elliptic to broadly lanceolate, maximum width occasionally either below or above the middle, oblique, 1.2–2.5 by 0.5–1.2 mm, tip obtuse to acuminate; margins entire to slightly irregular and glabrous to finely papillose towards the tip; base broadly to rather narrowly attached; thin; about glabrous; not keeled abaxially. *Lip* recurved, about ovate in outline (not spread), 0.7–2.3 by 0.5–1.2 mm, tip rounded to about acuminate; margins over about the entire length with seamlike lateral lobes which have finely to coarsely papillose or ciliate margins and which are either decurrent towards the tip of the lip or not and which are thin, set apart from the median part of the lip by a distinct furrow, about glabrous to finely pubescent or finely to coarsely papillose towards the tip of the lip and towards the margins adaxially (texture may be different from the median part of the lip) and about glabrous abaxially; lip rather thick; adaxially without ridges, with a slightly concave basal part, with a median slit starting at about the base and continuing up to 1/4–3/4 of the lip, surface glabrous to finely pubescent or finely to coarsely papillose; abaxially with a weak, broadly rounded to truncate ridge towards the base, surface about glabrous. *Column* from the base of the free part up to the tip of the stelidia 0.9–2 mm long, with distinct, gradually tapering, more or less falcate, subacute to acute stelidia of 0.7–1.4 mm long; column with rather weak to rather distinct, deltoid, obtuse teeth along its adaxial margins just below the base of the stelidia; stigma without a basal callus. *Anther* 0.3–0.5 by 0.3–0.4 mm; front not drawn out; front margin rounded, entire, papillose; basally distinctly emarginate; abaxially with a distinct, conical, obtuse, colliculate ridge near the top of the anther which distinctly overtops the front margin, surface glabrous but colliculate towards the top. *Pollinia* 4, ellipsoid; the small ones about half as long as the large ones, distinctly flattened on either side; the large ones flattened on one side; a stipes has not been observed. *Fruit* obovoid, up to 0.8 by 0.35 cm, stalk c. 1 mm.

Colours. Pseudobulbs and leaves occasionally heavily suffused with purple. Sepals white, cream-coloured or greenish, often suffused with purple-red or purple, or entirely purple-red or purple. Petals usually white. Lip white, yellowish, greenish, red or purple.

Ecology. Lowland forest, relict forest in savanna, montane forest up to 1900 m alt. Also in littoral forest (fide Sanford, 1974). Epiphyte on the trunks and in the central part of the crown of trees, on bark or in minor humus deposits. Occasionally lithophytic. Flowering mainly III–XI in GUIN. to NIG., mainly II–VI in CAM. and GAB., II - IV in ZAI.

Distribution. GUIN.: SE part (2 coll.). S.L. (20). LIB. (16). IV.C. (13). GHANA (2). NIG.: S. part (17). CAM.: S. part (15). EQ.GUIN.: Bioko (2). GAB. (12). CONGO (1). ZAI.: Kasai Z4 (1); Bas Katanga Z5 (1); Forestier Central Z6 (6); Lacs Edouard & Kivu Z9 (2). Material seen from AMES, BM, BR, C, E, G, HBG, K, L, P, S, UPS, W, WAG, Z.

Notes. 1 - Assigned to *Bulbophyllum pumilum* are specimens which show an evident similarity in general concept but which are varying in some aspects. The most striking variability exists in the shape and the texture of the sepals which are the most eye catching elements of the flowers of this species.

However, using the shape and the texture of the sepals to define a number of infraspecific entities is impossible due to the existence of numerous intermediate specimens. Moreover, it appears the entities obtained are not convincingly supported by any other character.

Therefore the entities described below have not been given any formal status.

a. - This entity comprises the types of *Dendrobium pumilum* (see note below), *Bulbophyllum flavidum*, *Taurostalyx herminiostachys*, *Bulbophyllum gabonis*, *Bulbophyllum drallei*, *Bulbophyllum porphyroglossum*, *Bulbophyllum calabaricum*, *Bulbophyllum moliwense*, *Bulbophyllum leucopogon* and *Bulbophyllum nanum*.

Sepals with glabrous margins, adaxially finely papillose towards the tip. Median sepal triangular, rarely slightly ovate, 4–7.5 by 0.9–2 mm, tip acute to subacute. Lateral sepals 4–8 by 0.9–2.4 mm, index 2.1–4.4; tip acute to subacute; hardly to moderately, but very gradually widened towards the base. Petals with an obtuse to finely acuminate tip. Lip with finely to coarsely papillose margins and adaxial surface.

Specimens with a finely papillose lip tend to have slightly more robust vegetative parts, longer sepals (the median 5–7.5 mm long) and a lip with a more acuminate tip than specimens with a coarsely papillose lip (sepals 4–6 mm long, lip with a rounded tip).

Specimens with a finely papillose lip often have yellowish green to creamy white flowers, incidentally tinged with red-purple or entirely red-purple (the petals excepted), with a lip of the same colour. Specimens with a coarsely papillose lip tend to have creamy white flowers with a red-purple lip.

b. - This entity comprises the types of *Bulbophyllum verecundum* and *Bulbophyllum elachon*.

Sepals with glabrous margins, adaxially finely papillose towards the tip. Median sepal triangular to ovate, 3–4 by 1.2–1.5 mm, tip acute. Lateral sepals 2.8–3.8 by 1.4–1.6 mm, index 1.8–2.7; tip acute; not or hardly widened towards the base. Petals with an acute tip. Lip with finely to coarsely papillose margins and adaxial surface.

The sepals are green or yellowish, partly or entirely suffused with red-purple. The petals are white. The lip is yellowish, with some red-purple along the margins or entirely red-purple.

Vegetatively this entity is characterized by orbicular to elliptic leaves of 2 cm long or shorter. It resembles most entity c., but differs a.o. in having sepals with glabrous margins.

c. - This entity comprises the type of *Tribrachia pendula*.

Sepals with more or less ciliate margins, adaxially ciliate towards the tip. Median sepal triangular, 2.3–4.5 by 1.4–2 mm, tip acute. Lateral sepals 2.9–4.5 by 1.5–2.3 mm, index 1.7–2.3; tip acute; either slightly but rather abruptly widened towards the base or not. Petals with an acute to acuminate, rarely obtuse tip. Lip with very finely papillose margins and adaxial surface, occasionally with coarsely papillose or ciliate margins and a coarsely papillose adaxial surface.

The sepals are green or yellowish. The petals are white. The lip is red-purple.

The specimens Bunting 74 from S.L. and Louis 3570 from ZAI. have the flowers of entity c. but the general appearance of resp. entity b. and e.

The specimens Adames 46 and Barter s.n. both from S.L. are intermediate between entity c. and a.

d. - This entity comprises the types of *Bulbophyllum pavimentatum*, *Bulbophyllum papillosum*, *Bulbophyllum dorotheae* and *Bulbophyllum yangambiense*.

Sepals with glabrous margins, adaxially very finely to very coarsely papillose towards the tip. Median sepal ovate to elliptic or about spathulate, rarely almost triangular, 3–4.8 by 1.2–2.8 mm, tip rounded to acute, occasionally apiculate. Lateral sepals 3–5.2 by 1.2–3 mm, index 1.4–2.3; tip rounded to acute, occasionally apiculate; usually more or less widened towards the base, either gradually or abruptly. Petals with an obtuse, rarely with an acute tip. Lip with papillose to coarsely papillose, occasionally distinctly ciliate margins, adaxial surface finely to coarsely papillose.

The sepals are dark red-purple, occasionally greenish towards the base, rarely entirely greenish. The petals are white. The lip is red-purple, rarely greenish.

Specimens with an almost triangular median sepal (e.g. Brenan FHI 25420 from NIG.) may approach entity c. but do not have sharply acute sepals with ciliate margins.

Specimens with elliptic or spathulate sepals tend to have sepals with distinctly thickened tips (e.g. the types of *Bulbophyllum pavimentatum* from NIG. and *Bulbophyllum yangambiense* from ZAI.).

The specimen Herbigneaux 1 from ZAI. is reminiscent of entity b. in its vegetative parts. It is remarkable for its very wide, short sepals (the median 3.2 by 2.2 mm, the lateral 3.3 by 2.4 mm) which have an obtuse tip.

The specimen Laurent 2045 from ZAI. has extremely narrow leaves. The sepals are as in entity b. but have slightly obtuse tips.

e. - This entity comprises the types of *Bulbophyllum winklerii* and *Bulbophyllum imogeniae*.

Sepals with glabrous margins, adaxially glabrous to very finely papillose towards the tip. Median sepal triangular to ovate or elliptic, 2.8–4.4 by 1.2–2 mm, tip acute to acuminate. Lateral sepals 2.7–5 by 1.5–2.5 mm, index 1.2–2.1; tip subacute to acuminate; distinctly widened towards the base, either or not in an abrupt way. Petals with acuminate, rarely with obtuse tip. Lip with finely papillose or ciliate margins, adaxial surface finely papillose.

A cultivated specimen in Leiden Hort. has its vegetative parts entirely suffused with purple. The sepals are dark red-purple or purple. The petals are white. The lip is orange suffused with purple, or entirely purple.

Vegetatively this entity is characterized by the slightly ovate or obovate leaves which are acuminate. The inflorescences are short, with a dense raceme. The peduncle scales tend to be inflated at their tips.

The entities a., c. and e. mainly occur in West Africa (S.L. to NIG.), entity b. in GAB. and ZAI., and entity d. in NIG, GAB. and ZAI.

In the Western part of the range of entity a. only specimens with a finely papillose lip are found; in the Eastern part of its range (E of NIG.) they are entirely replaced by specimens with a coarsely papillose lip, except for an incidental specimen with a finely papillose lip from Bioko. In ZAI. most specimens which are intermediate between two entities have been found.

2 - A probable natural hybrid with *Bulbophyllum intertextum* has been found, see p. 56, note 2.

18 - **Bulbophyllum coriscense** Rchb. f. - Fig. 31.

Bulbophyllum coriscense Rchb. f., Otia Hamb. 2 (1881) 118; Durand & Schinz, Consp. Fl. Afr. 5 (1895) 11; Rolfe in F.T.A. 7 (1897) 30; De Wild., Pl. Beq. 1 (1921) 107.-*Phyllorchis coriscensis* (Rchb. f.) Kuntze, Rev. Gen. Pl. 2 (1891) 677, incorr. name.-Type: Mann 1883 (K).

Non *Bulbophyllum coriscense* auct.: Engler, Pflanzenw. O-Afr. 2, c (1895) 157 (= *Bulbophyllum josephii* var. *josephii*).

Roots 0.5–0.8 mm diam. *Rhizome* 0.8–1 mm diam., nodes not or hardly swollen. *Rhizome scales* tubular, tip acute, the lower obtuse; membranous, with some nerves; not keeled abaxially. *Pseudobulbs* 1-leafed, ovoid to about orbicular, 0.2–0.6 cm apart, 0.5–0.8 by 0.3–0.6 cm, more or less flattened, (2 -) 4-angled. *Petiole* 0.5–1.5 mm long. *Leaf blade* elliptic to broadly oblong, 1.3–1.9 by 0.6–0.9 cm, tip obtuse, emarginate, either slightly oblique or not; coriaceous, thick; midrib hardly prominent abaxially. *Inflorescence* 3.5–4.5 cm long, 4–5-flowered. *Peduncle* erect, 2.5–3 cm by 0.5–1 mm, orbicular in section, glabrous, with 4–5 scales. *Peduncle scales* tubular, the longest 4–5 mm long; tip acuminate, the lower obtuse to acute; membranous, without nerves; the upper keeled and all glabrous abaxially. *Rhachis* erect, terete, either slightly zigzag bent or not, 1–1.5 cm long, glabrous. *Floral bracts* recurved to spreading, elliptic, c. 3 by 1.8 mm, tip acute; base rather broadly attached; membranous, without nerves; slightly keeled towards the tip and glabrous abaxially. *Flowers* scattered, in a rather lax raceme, recurved to spreading, (?) secund, many open simultaneously, not fully open. *Pedicel and ovary* c. 1.3 mm long, glabrous. *Median sepal* hardly recurved, narrowly ovate-lanceolate, c. 6.2 by 1.4 mm, tip subacute; margins entire, glabrous; base broadly attached; rather thin; very finely papillose adaxially towards the tip; not keeled and glabrous abaxially. *Lateral sepals* free, recurved, hardly falcate, narrowly ovate-oblong, oblique, c. 6.2 by 2.4 mm, otherwise as the median sepal. *Petals* not recurved, hardly falcate, about elliptic, oblique, c. 2.1 by 1.1 mm, tip about emarginate; margins slightly irregular towards the tip, glabrous; base rather narrowly attached; thin; glabrous; not keeled abaxially. *Lip* recurved, about elliptic in outline (not spread), c. 1.5 by 0.9 mm, tip rounded; along the basal half with seamlike lateral lobes which have entire, coarsely papillose margins and which are decurrent towards the tip of the lip, thin, separated from the median part of the lip by a distinct furrow, glabrous but papillose towards the tip of the lip and towards the margins adaxially, about glabrous abaxially; lip rather thin; adaxially without ridges, with a concave basal part, with a median slit over about the entire length of the lip, surface glabrous, coarsely papillose towards the tip; abaxially with a weak, retuse ridge in the basal half, surface about glabrous. *Column* from the base of the free part up to the tip of the stelidia c. 1 mm long, with rudimentary, triangular, not falcate, acute stelidia, column without teeth along its adaxial margins; stigma without a basal callus. *Anther* c. 0.5 by 0.5 mm; front slightly drawn out, thin, not concave; front margin rounded, entire, papillose; basally weakly emarginate; abaxially with a rather distinct, obtuse, colliculate ridge near the top of the anther which hardly overtops the front margin, surface glabrous but colliculate towards the top. *Pollinia* 2, ellipsoid, flattened on one side, a stipes has not been observed. *Fruit* not seen.

Colours. Unknown.

Ecology. Lowland vegetation.

Distribution. GAB.: Estuaire P., Corisco Bay (1 coll., found in 1862, now extinct?).

19 - **Bulbophyllum cocoinum** Lindl. - Fig. 32.

Bulbophyllum cocoinum Lindl., Bot. Reg. 23 (1837) pl. 1964 ('*Bolbophyllum*'); Rchb. f., Walp. Ann. 6 (1861) 255 ('*Bolbophyllum*'); Durand & Schinz, Consp. Fl. Afr. 5 (1895) 11; Rolfe in F.T.A. 7 (1897) 31; Schltr, Westafr. Kautchuk Exp. (1901) 281; Schltr, Die Orch. (1914) 324, pl.; De Wild., Pl. Beq. 1 (1921) 107; Summerh. in F.W.T.A. ed. 1, 2 (1936) 438; Mildbraed, Notizbl. Berl.-Dahl. 13 (1937) 698; Exell, Cat. Vasc. Pl. S. Tomé (1944) 325; Petersen, Bot. Tidsskr. 49 (1952) 162; Summerh., Kew Bull. 20 (1966) 187; id. in F.W.T.A. ed. 2, 3 (1968) 234; Johansson, Acta Phyt. Suec. 59 (1974) 31; Cribb & Perez-Vera, Adansonia, ser. 2, 15 (1975) 206; Geerinck, Bull. Soc. R. Bot. Belg. 109 (1976) 176; Adam, Fl. Descr. Mt. Nimba 5 (1981) 1674, pl. 908, 909; Vermeulen, Tijdschr. Ned. Orch. Ver. 44 (1982) 42, pl.; Cribb, in F.T.E.A., Orch. 2 (1984) 310.-Type: Loddiges s.n. (K).

Bulbophyllum andongense Rchb. f., Flora 48 (1865) 184 ('*Bolbophyllum*'); Durand & Schinz, Consp. Fl. Afr. 5 (1895) 10; Rolfe in F.T.A. 7 (1897) 31; Rendle, Cat. Welw. Afr. Pl. 2 (1899) 3; De Wild., Et. Fl. B. et Moy. Congo 1 (1904) 115; Durand & Durand, Syll. Fl. Congol. (1909) 515; De Wild., Pl. Beq. 1 (1921) 103.-Type: Welwitsch 689 (Holo BM, iso K, G).

Bulbophyllum vitiense Rolfe, Kew Bull. 1893 (1893) 5.-Type: Yeoward s.n. (K), wrongly labeled 'Fiji, Botanical Station'; undoubtedly the specimen is of African origin.

Bulbophyllum brevidenticulatum De Wild., Bull. J. Bot. Brux. 5 (1916) 179; id., Pl. Beq. 1 (1921) 105; Robyns, Fl. Sp. P. Alb. 3 (1955) 494.-Syntypes: Gentil s.n. (1903) (BR), Boone 105 (BR).

Roots 0.7–1.2 mm diam. *Rhizome* 2.5–7 mm diam., nodes not or hardly swollen. *Rhizome scales* tubular, tip acute, the lower obtuse; membranous, slightly woolly when old, with nerves; keeled abaxially towards the tip. *Pseudobulbs* 1-leafed, ovoid to narrowly ovoid, 1–3 cm apart, 2–5 by 1–2.7 cm, either slightly flattened or not, sharply 3–4-angled. *Petiole* 7–25 mm long. *Leaf blade* lanceolate, occasionally broadly linear, maximum width usually slightly above the middle, 9–27 by 1.2–3.5 cm, tip acuminate, occasionally acute, neither emarginate nor oblique; coriaceous, thick; midrib slightly prominent abaxially. *Inflorescence* 9–38 cm long, 15–150-flowered. *Peduncle* usually erect, 5–10 cm by 1.2–2.3 mm, orbicular in section, glabrous, with 3–5 scales. *Peduncle scales* tubular, the longest 9–19 mm long; tip acute to acuminate; membranous, with some nerves; keeled and glabrous abaxially. *Rhachis* erect to pendulous, terete, 4–29 cm long, glabrous. *Floral bracts* recurved, narrowly elliptic to broadly lanceolate, 3.5–5.5 by 1–1.9 mm, tip acute; base rather broadly attached; membranous, with some nerves; not or hardly keeled and glabrous abaxially. *Flowers* scattered, in a dense to lax raceme, recurved to about spreading, many open simultaneously, not much open. *Pedicel and ovary* 1.5–3.5 mm long, glabrous. *Median sepal* erect or slightly recurved towards the tip, narrowly ovate-lanceolate to broadly linear-lanceolate, 6.7–12 by 0.8–1.8 mm, tip (sub)acute; margins entire, glabrous; base broadly attached; thin; glabrous; not keeled abaxially. *Lateral sepals* free, slightly falcate, oblique, 9–14 by 1.5–2.5 mm, otherwise as the median sepal. *Petals* not recurved, hardly falcate, narrowly oblong to lanceolate, oblique, 2–2.8 by 0.4–0.8 mm, tip obtuse to acute; margins irregular and finely papillose towards the tip; base rather broadly attached; thin; finely papillose towards the tip; not keeled abaxially. *Lip* recurved, ovate to ovate-oblong in outline (not spread), 1.5–3 by 0.8–1.2 mm, tip obtuse; margins irregularly denticulate towards the base, puberulous towards the tip; rather thin, with a rather thick base; adaxially without ridges, with a slightly concave basal part with two slightly protruding, obtuse knobs just above the base, with a median slit starting at about the base and continuing over about the entire length of the lip, surface finely puberulous towards the margins; abaxially with a weak truncate ridge near the base, surface about glabrous, slightly papillose towards the margins. *Column* from the base of the free part up to the tip of the stelidia 1.6–2.5 mm long, with distinct, gradually tapering, hardly falcate, about acute stelidia of 0.7–1.2 mm long; column with very weak, rounded wings along its lower margins; stigma with a distinct, obtuse basal callus. *Anther* c. 0.7 by 0.7 mm; front drawn out, thin, slightly concave; front margin rounded, entire, papillose; basally slightly emarginate; abaxially without a ridge, surface glabrous, papillose along the median line and near the top. *Pollinia* 2, ellipsoid, flattened on one side; a thin stipes has been observed. *Fruit* obovoid, up to 0.9 by 0.5 cm; stalk 0.5–1 mm long.

Colours. Sepals white, often pink towards the tip. Petals white. Lip white or cream coloured. Column white. Very fragrant.

Ecology. Lowland to lower montane forest, up to 1200 m alt., in UG. found in montane forest at 2000 m alt. Epiphyte in the crown of trees, mainly in humus deposits (Johansson, 1974). Flowering V–VIII and XII.

Distribution. S.L. (2 coll.). LIB. (5). IV.C. (2). GHANA (6). GAB. (1). ZAI.: Kasai Z4 (1); Forestier Central Z6 (5). UG.: West U2, S part (1). ANG. (1). Material seen from BM, BR, G, K, L, P, UPS.

Note. The specimen Hall & Enti GC 38111 from GHANA has branched inflorescences: branches of 6.5–9 cm long arise from the upper peduncle scales.

20 - **Bulbophyllum josephii** (Kuntze) Summerh. – Fig. 33 - 35; Plate 5b.

Literature: see under the varieties.

Roots 0.2–1.3 mm diam. *Rhizome* 1.5–6 mm diam., nodes not or hardly swollen. *Rhizome scales* tubular, tip about acute, the lower about obtuse; membranous, slightly woolly when old, with nerves; not keeled abaxially. *Pseudobulbs* 1-leafed, usually ovoid, occasionally about orbicular or narrowly ovoid, 0.7–3 cm apart, 1.5–4 by 0.6–2.4 cm, slightly flattened, obtusely 2–4-angled. *Petiole* 2–35 mm long. *Leaf blade* (linear-)lanceolate, occasionally linear, 4.5–28 by 0.9–3.2 cm, tip obtuse to acute, occasionally finely mucronate, not or hardly emarginate, not oblique; coriaceous, thick; midrib hardly prominent abaxially. *Inflorescence* 8.5–40 cm long, 7–80-flowered. *Peduncle* erect to patent, 3–20 cm by 0.9–3.2 mm, orbicular in section, glabrous, with 3–7 scales. *Peduncle scales* tubular, the longest 7–22 mm long; tip acute to mucronate; membranous, usually with some nerves; keeled towards the tip and glabrous abaxially. *Rhachis* arching to pendulous, terete, 3–22 cm long, glabrous. *Floral bracts* recurved, oblong to broadly lanceolate, 2.5–7.5 by 1–3 mm, tip obtuse to acute; base rather broadly attached; membranous, without nerves; not or hardly keeled and glabrous abaxially. *Flowers* scattered, in a rather dense to rather lax raceme, recurved to spreading, either secund or not, many open simultaneously, not much open. *Pedicel and ovary* 1.2–6.5 mm long, glabrous. *Median sepal* either slightly recurved or not, ovate-lanceolate to very narrowly triangular, 4–11.7 by 1–2.2 mm, tip obtuse to acute; margins entire, glabrous; base broadly attached; rather thin; glabrous; not keeled abaxially. *Lateral sepals* free, hardly falcate, oblique, 4–12 by 1.4–3.2 mm, otherwise as the median sepal. *Petals* not recurved, slightly falcate, (obovate-)oblong to (obovate-)lanceolate, oblique, 1.8–5 by 0.6–1.7 mm, tip obtuse to acute; margins entire to irregular, glabrous to coarsely papillose; base broadly to rather narrowly attached; thin; either glabrous or papillose towards the tip adaxially; not keeled and glabrous abaxially. *Lip* recurved in the basal half, ovate in outline (not spread), 1.5–2.7 by 0.8–1.6 mm, tip rounded to subacute; margins entire, glabrous to papillose; rather thick; adaxially without ridges, with a slightly concave basal part, with or without a median slit starting at about the base and continuing up to 1/2–3/4 of the lip, surface glabrous, either papillose towards the tip or not; abaxially with a weak rounded ridge in the basal half, surface glabrous, either finely papillose towards the margins or not. *Column* from the base of the free part up to the tip of the stelidia 1.3–2.4 mm long, with distinct, more or less gradually tapering, either slightly falcate or not, acute stelidia of 0.5–0.9 mm long; column with or without rather weak, deltoid, obtuse teeth along its adaxial margins, just below the base of the stelidia; stigma without a basal callus. *Anther* c. 0.4 by 0.4 mm; front not drawn out; front margin rounded, entire, papillose; basally distinctly emarginate; abaxially with a weak, rounded, colliculate ridge near the top of the anther which overtops the front margin, surface colliculate. *Pollinia* 4, ellipsoid; the small ones about half as long as the large ones, distinctly flattened on either side; the large ones flattened on one side; a stipes has not been observed. *Fruit* ellipsoid, up to 0.7 by 0.4 cm; stalk 1–2 mm long.

Notes. 1 - Living specimens always have arching or pendulous inflorescences (La Croix, written comm., obs. in MAL.; compare also note under *Bulbophyllum expallidum*). In herbarium specimens this is often not visible.

2 - The specimen Breteler & De Wilde 83 from GAB. is aberrant in having a more elongate lip with a flattened, slightly irregularly folded top part and only very finely papillose margins. It may well be a hybrid between *Bulbophyllum schimperanum* and *Bulbophyllum josephii*, because it is intermediate in some aspects.

KEY TO THE VARIETIES

1 - Column with rather weak, deltoid, obtuse teeth along the adaxial margins, just below the base of the stelidia...var. **mahonii** (b)
- Column without teeth along the adaxial margins.....................var. **josephii** (a)

20a. var. **josephii** – Fig. 33; Plate 5b.

Bulbophyllum josephii (Kuntze) Summerh., Bot. Mus. Leafl. Harv. Un. 11 (1945) 250; Ake Assi, Et. Fl. C. Iv. (1964) 244; Summerh. in F.W.T.A. ed. 2, 3 (1968) 234, p.p.; Adam, Fl. Descr. Mt. Nimba 5 (1981) 1678; Vermeulen, Tijdschr. Ned. Orch. Ver. 44 (1982) 42.–*Phyllorchis josephii* Kuntze, Rev. Gen. Pl. 2 (1891) 676, nom. incorr.–*Bulbophyllum aurantiacum* Hook. f., J. Linn. Soc. Bot. 7 (1864) 219 ('*Bolbophyllum*'); Krzl., Mitth. Deutsch. Schutzgeb. 2 (1889) 154; Durand & Schinz, Consp. Fl. Afr. 5 (1895) 10; Rolfe in F.T.A. 7 (1897) 30; Schltr, Westafr. Kautchuk Exp. (1901) 280; De Wild., Pl. Beq. 1 (1921) 103; Schltr in Mildbraed, 2e Deutsch. Zentr. Afr. Exp. 1910-11, 2 (1922) 160.–*Bulbophyllum gustavi* Schltr, Fedde, Rep. 9 (1911) 165; Summerh. in F.W.T.A. ed. 1, 2 (1936) 438.–Type: Mann 2124 (K).

Non *Bulbophyllum josephii* auct.: Johansson, Acta Phyt. Suec. 59 (1974) 31 (= var. *mahonii*).

Non *Bulbophyllum aurantiacum* F. Muell., Fragm. Phyt. Aust. 3 (1862) 39.

?*Bulbophyllum amanicum* Krzl., Engl., Bot. Jahrb. 51 (1914) 382; De Wild., Pl. Beq. 1 (1921) 103; Vermeulen, Tijdschr. Ned. Orch. Ver. 44 (1982) 40.-Type: Braun ('in Herb. Amani n. 1879', n.v.).

Bulbophyllum winklerii Schltr, Orchis 8 (1914) 132; Schltr, Engl., Bot. Jahrb. 53 (1915) 572.-*Bulbophyllum schlechterii* De Wild., Pl. Beq. 1 (1921) 120; Robyns, Fl. Sp. P. Alb. 3 (1955) 493; Piers, Orch. E. Afr. (1968) 107; Stewart in Agnew, Upl. Ken. W. Fl. (1974) 760; Vermeulen, Tijdschr. Ned. Orch. Ver. 44 (1982) 45.-Type: Breslau cult. (Winkler s.n.) (B, lost).

Non *Bulbophyllum winklerii* Schltr, Engl., Bot. Jahrb. 38 (1906) 158 (= *Bulbophyllum pumilum*).

Non *Bulbophyllum schlechterii* Krzl., Fedde, Rep. 25 (1928) 56 = *Bulbophyllum schlechterii* Perrier, Cat. Pl. Madag. Orch. (1930) 39.

Bulbophyllum sennii Chiovenda, Atti R. Accad. d'Italia 11 (1940) 57; Cufodontis, En. Pl. Aeth. Sperm. (1972) 1619.-Type: Senni 2434 (FT, n.v., phot. in K).

Bulbophyllum coriscense auct. non Rchb. f.: Engler, Pflanzenw. O-Afr. 2, c (1895) 157.

Bulbophyllum mahonii auct. non Rolfe: Grosvenor, Excelsa 6 (1976) 79, p.p.; Ball, S. Afr. Ep. Orch. (1978) 72, pl., p.p.; Morris, Ep. Orch. Mal. (1970) 47, p.p.; La Croix et al., Orch. Mal. 1 (1983) 47, p.p.; Cribb, in F.T.E.A., Orch. 2 (1984) 310 p.p.

Bulbophyllum sp. 1 La Croix, Orch. Mal. 1 (1983) 50.-Specimen: La Croix 138 (K, n.v.).

Column without teeth along the adaxial margins; base of stigma usually slightly protruding.

Colours. Pseudobulbs and leaves often suffused with red. Sepals white, cream coloured or greenish, often tipped pink, occasionally entirely brownish pink. Petals white or very pale greenish. Lip white, pale greenish or yellowish. Flowers rarely entirely bright yellow (CAM. specimens). Often distinctly fragrant.

Ecology. Lowland forest (rare) to montane forest, Brachystegia-Uapaca woodland; at 0–2000 m alt. in W. Africa, at 1000–2400 m alt. in E. Africa. Epiphytic, occasionally lithophytic. Flowering I–XII.

Distribution. CAM.: S. part (4 coll.). ZAI.: Lacs Edouard & Kivu Z9 (4). UG.: West U2, S. part (2); North U1/East U3, Mt. Elgon (1). RW.: W. part (5). BUR. (1). AETH.: SE. part, Kefa P., Sidamo P. (2). KEN.: Rift Valley K3/Nyanza K5 (2). TANZ.: North T2, S. part (1); Tanga T3, Usambara Mts (5); East T6, Uluguru Mts (1); Southern Highlands T7 (7). MAL.: North (1); South (6). ZIM.: East (1). MOZ.: Zambezia (1); Manica e Sofala, Chimanimani Mts (1). Material seen from AMES, BM, BR, C, G, K, L, NY, S, U, UPS, WAG, Z.

Notes. 1 - The petals and the lip of East African specimens have an obtuse tip and are coarsely papillose adaxially towards the tip and along the margins. Most specimens from CAM., however, differ in having petals and a lip with a more or less acute tip and an almost glabrous adaxial surface. Some East African specimens tend to be intermediate between these two extremes.

2 - The specimen Schlieben 3578 (BM) from TANZ. has one 2-leafed pseudobulb.

20b. var. **mahonii** (Rolfe) J.J. Vermeulen - Fig. 34–35.

Bulbophyllum josephii (Kuntze) Summerh. var. *mahonii* (Rolfe) J.J. Vermeulen, Bull. J. Bot. Nat. Belg. 56 (1986) 232.-*Bulbophyllum mahonii* Rolfe, Kew Bull. 1906 (1906) 32; De Wild., Pl. Beq. 1 (1921) 113; Morris, Proc. Linn. Soc. London 179 (1968) 63; id., Ep. Orch. Mal. (1970) 47, p.p.; Grosvenor, Excelsa 6 (1976) 79, p.p.; Williamson, Orch. S. Centr. Afr. (1977) 138; Ball, S. Afr. Ep. Orch. (1978) 72, p.p; Cribb, in F.T.E.A., Orch. 2 (1984) 310, p.p.-Type: Mahon s.n. (K).

Non *Bulbophyllum mahonii* auct.: Geerinck, Bull. Soc. R. Bot. Belg. 109 (1976) 176 (= *Bulbophyllum expallidum*); La Croix et al., Orch. Mal. 1 (1983) 48 (= *B. josephii* var. *josephii*).

Bulbophyllum modicum Summerh., Kew Bull. 12 (1957) 114; id., in F.W.T.A. ed. 2, 3 (1968) 234; Adam, Fl. Descr. Mt. Nimba 5 (1981) 1681; Vermeulen, Tijdschr. Ned. Orch. Ver. 44 (1982) 44.–Type: Gregory 193 (K).

Bulbophyllum josephii auct. non (Kuntze) Summerh.: Johansson, Acta Phyt. Suec. 59 (1974) 31.

Column with short, deltoid, obtuse teeth along the adaxial margins, near the base of the stelidia; base of stigma not or hardly protruding.

Colours. Specimens of entity a. and the CAM. specimens of entity b. (see below): Pseudobulbs often suffused with red. Sepals white to very pale greenish, suffused with pink or very pale purplish. Petals white or pink. Lip suffused with purple. Specimens of entity b., except the CAM. specimens: Pseudobulbs suffused with red. Flowers almost entirely yellow to bright orange red.

Ecology. W. and Central African specimens: Lowland forest and montane forest, also in 'semi-montane grassland' (Sanford, 1974), at 0–2300 m alt. Epiphyte on trunks and in the central part of the crown of trees, in humus deposits (Johansson, 1974, sub *Bulbophyllum josephii*). E. African specimens: Montane forest and Brachystegia woodland at 1300–2000 m alt. Specimens of entity a. and the CAM. specimens of entity b. (see below): flowering II–III and X–XI. Specimens of entity b., except the CAM. specimens flowering III–V.

Distribution. GUIN.: SE part, Nimba Mts (1 coll). LIB. (3). IV.C. (2). NIG.: SE part, Gongola P. (1). CAM.: W. part, Mt. Cameroun (4). EQ.GUIN.: Bioko (1). ZAI.: Haut Katanga Z11 (4). ZAM.: West, Mwinilunga (2). MAL.: North (2); South (1). Very locally. Material seen from BR, K, UPS, Z.

Notes. Within var. *mahonii* two entities can be distinguished, either of which is confined to a part of the range of the variety.

a. - The entity which occurs in West Africa (West of CAM.), and on Bioko.

Leaves 11–28 cm long. Inflorescence 17–40 cm long. Sepals 9.5–12 mm long. Petals 3.5–5 mm long. Lip 1.8–2.7 mm long. Column 2–2.4 mm long. Stelidia 0.5–0.7 mm long.

b. - The entity which occurs in central and East Africa (from CAM. eastwards). This entity includes the types of *Bulbophyllum modicum* and *Bulbophyllum mahonii.*

Leaves 4.5–15 cm long. Inflorescence 9–21 cm long. Sepals 4.6–7 (– 8) mm long. Petals 1.8–3 mm long. Lip 1.7–2.1 mm long. Column 1.4–1.8 mm long. Stelidia 0.7–0.8 mm long.

The CAM. specimens (including the type of *Bulbophyllum modicum*) differ from the East African specimens (including the type of *Bulbophyllum mahonii*) in the slightly narrower leaves (resp. 5.5–15 by 0.9–2 cm and 4.5–9.5 by 1.2–2 cm), the shorter floral bracts (resp. 2.5–3.5 mm long and 3.5–5.5 mm long), and the slightly wider, more obtuse lip.

The specimen Schayes 1878 from ZAI. has a median sepal of 8 mm long. It should be considered a monstrosity: next to some normal flowers it has flowers with a very short, truncate column with a third stelidium along one of its adaxial margins.

21 - Bulbophyllum calvum Summerh. - Fig. 36.

Bulbophyllum calvum Summerh., Kew Bull. 20 (1966) 185, pl.; id. in F.W.T.A. ed. 2, 3 (1968) 234, pl.; Vermeulen, Tijdschr. Ned. Orch. Ver. 44 (1982) 40, pl.–Type: King 178 (K).

Roots c. 1.2 mm diam. *Rhizome* c. 4.5 mm diam. *Rhizome scales* not seen. *Pseudobulbs* 1-leafed, broadly ovoid, close together, c. 3.5 by 3.2 cm, hardly flattened, obtusely c. 6-angled. *Petiole* c. 23 mm long. *Leaf blade* lanceolate, c. 19 by 3.8 cm, tip obtuse, emarginate, hardly oblique; coriaceous, thick; midrib prominent abaxially. *Inflorescence* (?) cm long. *Peduncle* erect, sturdy, c. 10 cm by 2 mm, orbicular in section, glabrous, with c. 4 scales. *Peduncle scales* tubular, the longest c. 12 mm long; tip acute to acuminate; membranous, without nerves; slightly keeled towards the tip and glabrous abaxially. *Rhachis* erect, terete, glabrous. *Floral bracts* recurved, oblong, c. 7 by 3 mm, tip subacute; base broadly attached; membranous, without nerves; keeled and glabrous abaxially. *Flowers* scattered, in a rather dense raceme, recurved, secund, many open simultaneously, not fully open. *Pedicel and ovary* c. 4 mm long, glabrous. *Median sepal* slightly recurved, ovate lanceolate, c. 10 by 2.5 mm, tip acute; margins entire, glabrous; base broadly attached; thick; glabrous; not keeled abaxially. *Lateral sepals* free, hardly falcate, narrowly triangular, oblique, c. 10.2 by 4 mm, otherwise as the median sepal. *Petals* not recurved, hardly falcate, ovate-oblong, oblique, c. 5.5 by 2 mm, tip acute; margins entire, very finely papillose towards the tip; base rather narrowly attached; rather thin; glabrous; not keeled abaxially. *Lip* recurved in the basal half, ovate-oblong in outline (not spread), c. 4.8 by 2 mm, tip subacute; margins entire, glabrous, with a small wave about half-way; rather thick; adaxially without ridges, with a hardly concave basal part, with a median slit starting at about the base and continuing up c. 1/3 of the lip, surface glabrous; abaxially with a weak, truncate ridge in the basal half, surface glabrous. *Column* from the base of the free part up to the tip of the stelidia c. 2.8 mm long, with distinct, gradually tapering, not falcate, acute stelidia of c. 0.8 mm long; column with very weak, rounded wings along its lower margins, near the base of the stelidia; stigma without a basal callus. *Anther* 0.5 by 0.7 mm, front not drawn out; front margin rounded, entire, papillose; basally slightly emarginate; abaxially with a distinct, rounded, papillose ridge over almost the entire length of the anther, surface entirely papillose. *Pollinia* 4; the small ones c. half as long as the large ones, ellipsoid, distinctly flattened; the large ones ovoid, flattened on one side. *Fruit* not seen.

Colours. Sepals yellowish green. Petals white, yellowish towards the tip. Lip red, abaxially with a bright yellow base, purple along the margins.

Ecology. Montane forest at 1800 m alt. Flowering X (once recorded).

Distribution. NIG.: Gongola P., Mambila Mts (1 coll.).

Note. The type specimen is not complete. Some data have been copied from Summerhayes' description of the species. The fig. accompanying Summerhayes's description of *Bulbophyllum calvum* depicts the type specimen with a branched inflorescence. However, no trace of branches has been found in the type specimen itself. A branched inflorescence similar to that on Summerhayes's plate has been found in *Bulbophyllum cocoinum*, see under that species.

22 - **Bulbophyllum expallidum** J.J. Vermeulen - Fig. 37; Plate 5c.

Bulbophyllum expallidum J.J. Vermeulen, Bull. Soc. R. Bot. Belg. 54 (1984) 146.–Type: Leiden cult. Hort. 22269 (L).
Bulbophyllum mahonii auct. non Rolfe: Geerinck, Bull. Soc. R. Bot. Belg. 109 (1976) 176.
Bulbophyllum buntingii auct. non Rendle: Williamson, Orch. S. Centr. Afr. (1977) 138.
Bulbophyllum acutisepalum auct. non De Wild.: La Croix et al., Orch. Mal. 1 (1983) 41; Cribb, in F.T.E.A., Orch. 2 (1984) 309.

Roots 0.3–1 mm diam. *Rhizome* 2–3.5 mm diam., nodes not or hardly swollen. *Rhizome scales* tubular, tip acute or slightly acuminate; membranous, slightly woolly when old, with some nerves; not keeled abaxially. *Pseudobulbs* 1-leafed, ovoid to ellipsoid, occasionally narrowly ovoid, 0.5–1.3 cm apart, 1.5–4.5 by 1–1.5 cm, slightly flattened, obtusely to sharply 3–6-angled. *Petiole* 1–4 mm long. *Leaf blade* oblong to linear, 3–13 by 0.8–2.8 cm, tip obtuse, occasionally mucronate, either emarginate and/or slightly oblique or not; coriaceous, thick; midrib prominent abaxially. *Inflorescence* 12–35 cm long, 25–60-flowered. *Peduncle* usually erect, 3–10 cm by 1.2–1.9 mm, orbicular in section, glabrous, with 3–5 scales. *Peduncle scales* tubular, the longest 10–15 mm long; tip about acute; membranous, without nerves; slightly keeled and glabrous abaxially. *Rhachis* erect, terete, 8–26 cm long, glabrous. *Floral bracts* recurved, narrowly ovate-oblong to ovate-lanceolate, 4–6.5 by 1.5–1.7 mm, tip acute to acuminate; base rather broadly attached; membranous, without nerves; keeled and glabrous abaxially. *Flowers* scattered, in a dense to rather lax raceme, recurved to reflexed, secund or not, many open simultaneously, not fully open. *Pedicel and ovary* 2–6 mm long, glabrous. *Median sepal* recurved with a reflexed top part, narrowly ovate-lanceolate to broadly linear-lanceolate, 6–10.8 by 1.5–1.9 mm, tip about acute; margins entire, glabrous; base broadly attached; rather thin; glabrous; not keeled abaxially. *Lateral sepals* free, hardly falcate, oblique, 6.8–11 by 1.8–1.9 mm, otherwise as the median sepal. *Petals* not recurved, not falcate, broadly ovate-(oblong), or about elliptic, oblique, 1.5–2.5 by 1–1.5 mm, tip about obtuse; margins irregular and very finely papillose towards the tip; base narrowly attached; thin; very finely papillose towards the tip; not keeled abaxially. *Lip* slightly recurved near the base, distinctly recurved near the tip, narrowly ovate to broadly ovate-oblong in outline (not spread), 1.8–2.2 by 0.7–1.1 mm, tip obtuse; margins entire, glabrous to very finely papillose; thick; adaxially without ridges, with a slightly concave basal part, with a median slit starting at about the base and continuing over 3/4 of the lip, surface glabrous to very finely papillose; abaxially with a weak truncate ridge in the basal half, surface glabrous. *Column* from the base of the free part up to the tip of the stelidia 1–1.8 mm long, with distinct, gradually tapering, acute, slightly bent upwards stelidia of 0.4–0.7 mm long; column without teeth along its adaxial margins; stigma with a rather distinct, obtuse basal callus. *Anther* c. 0.5 by 0.7 mm; front not drawn out; front margin rounded, entire, glabrous; basally distinctly emarginate; abaxially with a weak, rounded, colliculate ridge near the top of the anther which overtops the front margin, surface glabrous, colliculate towards the top. *Pollinia* 2; ellipsoid to slightly ovoid, flattened on one side; a stipes has not been observed. *Fruit* not seen.

Colours. Pseudobulbs and leaves green. Sepals white or yellowish, occasionally tinged with pink. Petals white. Lip yellowish. Column white. Flowers rarely orange (RW. specimen). Distinctly fragrant.

Ecology. Hill forest to montane forest, Brachystegia woodland, at 800–2100 m alt. Epiphyte on the trunk and in the crown of trees. Often among lichens, associated with *Bulbophyllum josephii* var. *josephii* and *Bulbophyllum cochleatum* var. *gravidum* (La Croix, written comm.). Occasionally lithophytic. Flowering (VIII) XI–III.

Distribution. ZAI.: Haut Katanga Z11 (4 coll.). RW. (1). TANZ.: Southern Highlands T7 (2). ZAM.: West, Solwezi (2); Central, Mkushi (1); North, Kasama (2). MAL.: North (12). Material seen from BM, BR, K, L, WAG, Z.

Note. In general aspect *Bulbophyllum expallidum* is much like *Bulbophyllum josephii*. Living specimens, however, can be distinguished at first sight by the inflorescence which is always erect (La Croix, written comm.; see also note under *Bulbophyllum josephii*). Herbarium specimens of *Bulbophyllum expallidum* tend to have flowers which are more reflexed. The specimen Troupin 15681 from RW, however, which certainly belongs to *Bulbophyllum expallidum*, is in general aspect identical to specimens of *Bulbophyllum josephii* of the same area. Besides, the flowers are recorded to be orange.

23 - **Bulbophyllum ivorense** Cribb & Perez-Vera – Fig. 38.

Bulbophyllum ivorense Cribb & Perez-Vera, Adansonia, ser. 2, 15 (1975) 202, 206; Adam, Fl. Descr. Mt. Nimba 5 (1981) 1677.–Type: Perez-Vera 137 (Holo K; iso P).

Bulbophyllum flavidum Lindl. var. *elongatum* De Wild., Not. Pl. Ut. et Int. Fl. Congo 1 (1903) 120; id., Et. Fl. B. et Moy. Congo 1 (1904) 116; Durand & Durand, Syll. Fl. Congol. (1909) 516.–*Bulbophyllum elongatum* (De Wild.) De Wild., Pl. Beq. 1 (1921) 109; Vermeulen, Tijdschr. Ned. Orch. Ver. 44 (1982) 42.–Type: Laurent 124 (Holo BR; iso K).

Non *Bulbophyllum elongatum* (Bl.) Hassk., Cat. Bog. (1844) 39.

Bulbophyllum buntingii Rendle, Cat. Talb. S. Nig. Pl. (1913) 99, 146, pro sp. Talbot 935.

Bulbophyllum schimperanum auct. non Krzl.: Johansson, Acta Phyt. Suec. 59 (1974) 33.

Roots 0.3–1.4 mm diam. *Rhizome* 1.5–3.5 mm diam., nodes not or hardly swollen. *Rhizome scales* tubular, tip acute; membranous, slightly woolly when old, with few nerves; slightly keeled abaxially. *Pseudobulbs* 1-leafed, orbicular to ovoid, 0.6–2 cm apart, 1–2.3 by 0.6–1.8 cm, slightly flattened, obtusely 2–4-angled. *Petiole* 4–9 mm long. *Leaf blade* broadly lanceolate to narrowly linear-lanceolate, 3.5–14 by 0.7–1.8 cm, tip rounded, either slightly emarginate and/or oblique or not; coriaceous, thick; midrib not or hardly prominent abaxially. *Inflorescence* 7–28 cm long, 15–80-flowered. *Peduncle* usually about erect, 3.5–8 cm by 1.2–2.6 mm, orbicular in section, glabrous, with 5–7 scales. *Peduncle scales* tubular, the longest 9–17.5 mm long; tip acute; membranous, without nerves; slightly keeled towards the tip and glabrous abaxially. *Rhachis* about erect, terete, 7–21 cm long, glabrous. *Floral bracts* recurved, elliptic to ovate, 4–7 by 1.5–4 mm, tip acute; base broadly attached; membranous, without nerves; not or hardly keeled and glabrous abaxially. *Flowers* scattered, in a dense raceme, recurved to spreading, secund, many open simultaneously, not fully open. *Pedicel and ovary* 2.5–4 mm long, glabrous. *Median sepal* recurved towards the tip, narrowly ovate-oblong to broadly ovate-lanceolate, 4–6 by 1–1.6 mm, tip about obtuse to acute; margins entire, glabrous; base broadly attached; rather thin, slightly thickened towards the tip; finely papillose adaxially towards the tip; not keeled and glabrous abaxially. *Lateral sepals* free, hardly falcate, oblique, 3.7–5.8 by 1.4–1.8 mm, otherwise as the median sepal. *Petals* not recurved, not falcate, reduced to a pair of very small, about rectangular lobes at either side of the column, oblique, 0.25–0.3 by 0.3–0.4 mm, tip retuse to truncate; margins entire, finely papillose; base broadly attached; thin; finely papillose; not keeled abaxially. *Lip* slightly recurved near the base, about reflexed near the tip, lanceolate in outline (not spread), 1.8–2.8 by 0.5–0.7 mm, tip obtuse; margins entire, very finely papillose; thick; surface very finely papillose; adaxially without ridges, with a hardly concave basal part, with a median slit starting at about the base and continuing over about the entire length of the lip; abaxially with a weak, truncate ridge near the base. *Column* from the base of the free part up to the tip of the stelidia 1.2–2 mm long, abaxially papillose in the basal half, with distinct, narrowly triangular, either slightly falcate or not, acute stelidia of c. 0.8 mm long which may have a weak, obtuse tooth along their upper margin; column without teeth along its adaxial margins; stigma with a distinct, conical, obtuse, papillose basal tooth. *Anther* c. 0.4 by 0.6 mm; front not drawn out; front margin truncate, entire, papillose; basally distinctly emarginate; abaxially with a distinct, slightly conical, rounded, colliculate ridge near the top of the anther which hardly overtops the front margin, surface glabrous, colliculate near the top. *Pollinia* 4; the small ones more than half as long as the large ones, ellipsoid, distinctly laterally flattened on either side; the large ones semi-elliptic in dorsoventral view, elliptic in lateral view, distinctly dorsoventrally flattened on either side; a stipes has not been observed. *Fruit* not seen.

Colours. Pseudobulbs yellowish. Leaves pale green. Sepals white, cream coloured towards the tip, or yellowish green. Petals white. Lip cream coloured.

Ecology. Lowland and montane forest up to 700 m alt. Epiphytic. Flowering X–XI.

Distribution. LIB.: Nimba Mts., Bomi Hills (5 coll.). IV.C.: Mt. Tonkoui (1). NIG.: SE. part, Cross P. (1). CAM.: Coast (1). ZAI.: Forestier Central Z6 (2). Material seen from BM, BR, K, UPS, WAG.

24 - **Bulbophyllum mediocre** Summerh. - Fig. 39.

Bulbophyllum mediocre Summerh., Bull. Inst. Fr. d'Afr. N., ser. a, 21 (1959) 470; Vermeulen, Tijdschr. Ned. Orch. Ver. 44 (1982) 42.–Type: Monod 11842 (Holo K; iso BM).
Bulbophyllum recurvum auct. non Lindl.: Exell, Cat. Vasc. Pl. S. Tomé (1944) 327.
Bulbophyllum flavidum auct. non Lindl.: Exell, Cat. Vasc. Pl. S. Tomé (1944) 326.

Roots 0.6–1 mm diam. *Rhizome* 1–3 mm diam., nodes not or hardly swollen. *Rhizome scales* tubular, tip acute or slightly acuminate; membranous, slightly woolly when old, with many nerves; keeled abaxially towards the tip. *Pseudobulbs* 1-leafed, ovoid to narrowly ovoid, 2–3 cm apart, 1–3 by 0.5–1 cm, moderately flattened, (2 –) 4-angled. *Petiole* 8–15 mm long. *Leaf blade* narrowly lanceolate to broadly linear-lanceolate, 5.5–11.5 by 1.1–2.2 cm, tip obtuse, neither emarginate nor oblique; coriaceous, thick; midrib slightly prominent abaxially. *Inflorescence* 7–17 cm long, 5–30-flowered. *Peduncle* usually about erect, 4–11 cm by 0.7–1.2 mm, orbicular in section, glabrous, with 5–7 scales. *Peduncle scales* tubular, the longest 7.5–11 mm long;

tip about acute; membranous, with many nerves; keeled towards the tip and glabrous abaxially. *Rhachis* about erect, terete, either slightly zigzag bent or not, 2.2–9 cm long, glabrous. *Floral bracts* recurved to spreading, oblong to lanceolate, 3–7 by 1.3–2 mm, tip acute to acuminate; base rather broadly attached; membranous, with some nerves; keeled towards the tip and glabrous abaxially. *Flowers* scattered, in a rather lax raceme, recurved to spreading, either secund or not, many open simultaneously, not fully open. *Pedicel and ovary* 1.8–3 mm long, glabrous. *Median sepal* either only recurved towards the tip or not, ovate-lanceolate, 4.8–9.2 by 1.4–2.3 mm, tip acute to acuminate; margins entire; base broadly attached; thin; glabrous; not keeled abaxially. *Lateral sepals* free, hardly falcate, ovate-oblong, oblique, 5.6–9.3 by 2.2–3.4 mm, otherwise as the median sepal. *Petals* not recurved, hardly falcate, oblong to broadly lanceolate, oblique, 3–4.8 by 1–2.2 mm, tip obtuse to acute; margins entire, either very finely papillose towards the tip or not; base narrowly attached; very thin; glabrous; not keeled abaxially. *Lip* recurved, obovate in outline in natural position, 1.8–2.8 by 1.1–1.5 mm, tip rounded to subacute; margins entire, very finely papillose; rather thick, very thick towards the tip; surface entirely glabrous; adaxially without ridges, with a concave basal part, without a median slit; abaxially with a weak truncate ridge near the base. *Column* from the base of the free part up to the tip of the stelidia 1.7–2.2 mm long, with distinct, gradually tapering, falcate, acuminate stelidia of 1–1.4 mm long; column without teeth along its adaxial margins; stigma without a basal callus. *Anther* c. 0.7 by 0.5 mm; front drawn out, thin, concave; front margin rounded, entire, papillose; basally distinctly emarginate; abaxially without a ridge, surface glabrous, colliculate near the top. *Pollinia* 4, ellipsoid; the small ones about half as long as the large ones, distinctly flattened; the large ones flattened; a stipes has not been observed. *Fruit* not seen.

Colours. Flowers entirely orange.

Ecology. Montane vegetation at 700–1550 m alt. Epiphytic. Flowering VIII–IX.

Distribution. S.T. & PR.: Sao Tomé (3 coll.); Principe (2). Material seen from BM, K, P.

25 - Bulbophyllum schimperanum Krzl. – Fig. 40; Plate 5d.

Bulbophyllum schimperanum Krzl., Engl., Bot. Jahrb. 33 (1902) 71 ('*Bolbophyllum*'); De Wild., Pl. Beq. 1 (1921) 119; Summerh. in F.W.T.A. ed. 1, 2 (1936) 438; id., ed. 2, 3 (1968) 234; Adam, Fl. Descr. Mt. Nimba 5 (1981) 1684; Vermeulen, Tijdschr. Ned. Orch. Ver. 44 (1982) 45; Cribb, in F.T.E.A., Orch. 2 (1984) 311.–Type: Schimper 341 (B, lost); v.d. Laan 458 (Neo WAG, here chosen).

Non *Bulbophyllum schimperanum* auct.: Johansson, Acta Phyt. Suec. 59 (1974) 33 (= *Bulbophyllum ivorense*).

Bulbophyllum xanthoglossum Schltr, Engl., Bot. Jahrb. 38 (1906) 158; De Wild., Pl. Beq. 1 (1921) 121; Schltr, Fedde, Rep. Beih. 68 (1932) pl. 70.–Type: Schlechter 15755 (B, lost).

Bulbophyllum acutisepalum De Wild., Bull. J. Bot. Brux. 5 (1916) 179; id., Pl. Beq. 1 (1921) 101; Geerinck, Bull. Soc. R. Bot. Belg. 109 (1976) 175.–Syntypes: Laurent 1740, 1735, 2059 (all BR).

Non *Bulbophyllum acutisepalum* auct: La Croix et al., Orch. Mal. 1 (1983) 41; Cribb, in F.T.E.A., Orch. 2 (1984) 309 (= *Bulbophyllum expallidum*).

Roots 0.4–1.1 mm diam. *Rhizome* 1.2–4 mm diam., nodes not or hardly swollen. *Rhizome scales* tubular; membranous, with nerves; not or hardly keeled abaxially. *Pseudobulbs* 1-leafed, usually ovoid, occasionally orbicular to obreniform, 0.5–1.8 cm apart, 0.7–2.5 by 0.6–2.2 cm, slightly flattened, obtusely 2–4-angled. *Petiole* 2–10 mm long. *Leaf blade* narrowly oblong to linear-lanceolate, maximum width occasionally slightly above the middle, 3.5–16.5 by 0.9–2.5 cm, tip obtuse to acute, not or hardly emarginate, not oblique; coriaceous, thick; midrib hardly prominent abaxially. *Inflorescence* 6–36 cm long, 10–100-flowered. *Peduncle* usually about erect, 2.5–13 cm by 0.8–2.8 mm, orbicular in section, glabrous, with 3–6 scales. *Peduncle scales* tubular, the longest 5–16 mm long; tip acute to acuminate; membranous, without nerves; the lower keeled and all glabrous abaxially. *Rhachis* erect to arching, terete, 3.5–23 cm long, glabrous. *Floral bracts* recurved, narrowly elliptic to lanceolate, 2–7.5 by 0.9–2.5 mm, tip acute to acuminate; base rather narrowly attached; membranous, without nerves; not or hardly keeled and glabrous abaxially. *Flowers* scattered, in a rather dense raceme, recurved, either secund or not, many open simultaneously, not much open. *Pedicel and ovary* 1.5–2 mm long, glabrous. *Median sepal* not recurved, narrowly lanceolate to broadly linear-lanceolate, 6–8.8 by 1.2–2 mm, tip acute; margins entire, glabrous; base broadly attached; thin; glabrous; not keeled abaxially. *Lateral sepals* free, not falcate, (very) narrowly triangular, oblique, 5.5–10 by 2–3.5 mm, otherwise as the median sepal. *Petals* not recurved, hardly falcate, narrowly (ovate-)oblong, oblique, 2.5–3 by 0.8–1 mm, tip obtuse to acute; margins entire, very finely papillose; base (rather) broadly attached; thin; very finely papillose adaxially towards the margins; not keeled and glabrous abaxially. *Lip* recurved, elliptic in outline (not spread), 2.2–3.5 by 1.3–2 mm, tip rounded; margins entire, with fine, short, stiff hairs; rather thin; adaxially without ridges, with a hardly concave basal part, without a median slit, surface glabrous, either very finely papillose towards the margins or not; abaxially with a weak, obtuse ridge in the basal half, surface glabrous. *Column* from the base of the free part up to the tip of the stelidia c.1 mm long, with distinct, narrowly triangular, not falcate, acute stelidia of 0.5–0.6 mm long; column without teeth along its adaxial margins; stigma without a basal callus. *Anther* c. 0.4 by 0.3 mm; front slightly drawn out, thin, slightly concave; front margin rounded to truncate, entire, papillose; basally slightly emarginate; abaxially without a ridge, surface glabrous, colliculate towards the tip. *Pollinia* 2, ellipsoid, flattened on one side; a stipes has not been observed. *Fruit* ellipsoid, up to 0.7 by 0.4 cm, stalk 1 mm.

Colours. Pseudobulbs and leaves green. Sepals white, often with pink tips. Lip cream coloured to yellowish green.

Ecology. Lowland forest to montane forest, relict forest in savanna (Sanford, 1974), in UG. up to 2000 m alt. Epiphytic or lithophytic. Flowering observed in I–III, VI and VIII–XI.

Distribution. NIG.: Bendel P. (fide F.W.T.A. ed. 2). CAM.: S. part (7 coll.). C.A.R.: SW part, Lobaye P. (4). GAB. (1). ZAI.: Forestier Central Z6 (4); Lacs Edouard & Kivu Z9 (2). UG.: West U2, S. part (1). Material seen from BM, BR, K, L, P, WAG.

Note: A possible hybrid with *Bulbophyllum josephii* has been found, see under that species.

26 - Bulbophyllum oxychilum Schltr – Fig. 41; Plate 6a.

Bulbophyllum oxychilum Schltr, Westafr. Kautchuk Exp. (1901) 282, nom. nud.; Schltr, Engl., Bot. Jahrb. 38 (1905) 18; De Wild., Pl. Beq. 1 (1921) 115; Vermeulen, Tijdschr. Ned. Orch. Ver. 44 (1982) 167.– Type: Schlechter 13042 (B, lost); Bos 2336 (Neo WAG, here indicated).

Bulbophyllum buntingii Rendle, Cat. Talb. S. Nig. Pl. (1913) 99, 146, p.p., see below; De Wild., Pl. Beq. 1 (1921) 106; Summerh. in F.W.T.A. ed. 1, 2 (1936) 438; id., Kew Bull. 12 (1957) 112; id. in F.W.T.A. ed. 2, 3 (1968) 234; Johansson, Acta Phyt. Suec. 59 (1974) 29; Geerinck, Bull. Soc. R. Bot. Belg. 109 (1976) 175; Adam, Fl. Descr. Mt. Nimba 5 (1981) 1672, pl. 902, 948; Cribb, in F.T.E.A., Orch. 2 (1984) 309.–Syntypes: Bunting 29 (Lecto BM, here proposed; isolecto K), Talbot 935 (BM) (= *Bulbophyllum ivorense*, see below).

Non *Bulbophyllum buntingii* auct.: Williamson, Orch. S. Centr. Afr. (1977) 138 (= *Bulbophyllum expallidum*).

Bulbophyllum ellipticum De Wild., Rev. Zool. Afr. 9, suppl. bot. (1921) 29; id., Pl. Beq. 1 (1921) 108.– Type: Bequaert 2496 (BR).

Non *Bulbophyllum ellipticum* Schltr, Fedde, Rep., Beih. 1 (1912-13) 829.

Roots 0.5–1 mm diam. *Rhizome* 1–3 mm diam., nodes not or hardly swollen. *Rhizome scales* tubular, tip about acute; membranous, slightly woolly when old, with some nerves; slightly keeled abaxially. *Pseudobulbs* 1-leafed, ovoid to about orbicular, 0.8–2 cm apart, 0.6–2.3 by 0.5–1.6 cm, slightly flattened, obtusely 3-4-angled. *Petiole* 1–11 mm long. *Leaf blade* narrowly elliptic to broadly linear-lanceolate, maximum width either slightly above the middle or not, 1.6–12 by 0.7–2 cm, tip obtuse to acute, not or hardly emarginate, occasionally slightly oblique; coriaceous, thick; midrib not or hardly prominent abaxially. *Inflorescence* 6.5–31 cm long, 15–100-flowered. *Peduncle* erect to patent 1.8–7.5 cm by 1–2.5 mm, orbicular in section, glabrous, with 5–7 scales. *Peduncle scales* tubular, the longest 6–13 mm long; tip acute or acuminate, occasionally obtuse; membranous, without nerves; either slightly keeled towards the tip or not and glabrous abaxially. *Rhachis* about erect, terete, 4.5–24 cm long, glabrous. *Floral bracts* recurved, ovate-oblong to broadly ovate-lanceolate, 3.5–5 by 1.5–2 mm, tip acute; base broadly attached; membranous, without nerves; keeled and glabrous abaxially. *Flowers* scattered, in a dense to rather lax raceme, recurved or spreading, either secund or not, many open simultaneously, not fully open. *Pedicel and ovary* 1.5–5.5 mm long, glabrous. *Median sepal* not recurved, ovate-oblong to broadly ovate-lanceolate, 3.2–6.2 (– 7) by 1.2–2 mm, tip acute to slightly acuminate; margins entire, glabrous; base (rather) broadly attached; rather thick; glabrous; not keeled abaxially. *Lateral sepals* free, hardly falcate, oblique, 3.2–6.3 (– 7) by 1.5–2 mm, base broadly attached; otherwise as the median sepal. *Petals* not recurved, hardly falcate, about rhomboid, oblique, 1.5–3 by 0.6–1.2 mm, tip obtuse to acute; margins entire, finely papillose towards the tip; base narrowly attached; thin; finely papillose towards the tip; not keeled abaxially. *Lip* not or hardly recurved, about triangular in outline (not spread), cordate to hastate when spread (herbarium specimens), 1.5–2.3 by 0.5–1 mm, tip acute or subacute; margins finely papillose, in the basal half with about backwards directed, upwards folded, deltoid lateral lobes which have a rounded to subacute tip, entire and about glabrous margins, a broadly attached base and which are decurrent towards the tip of the lip and thick and glabrous; thick; adaxially without ridges, with a channeled median part, with a concave basal part, without a median slit, surface glabrous to very finely papillose; abaxially with a weak, rounded ridge towards the base, surface glabrous except for a semicircular, transverse papillose band of slightly farinose appearance about half-way the lip. *Column* from the base of the free part up to the tip of the stelidia 0.9–1.8 mm long, with distinct, gradually tapering, not or hardly falcate, acute stelidia of 0.5–1 mm long; column with weak, broadly rounded wings along its adaxial margins just below the base of the stelidia; stigma with two rather distinct obtuse basal teeth. *Anther* c. 0.4 by 0.4 mm; front not drawn out; front margin rounded, entire, papillose; basally distinctly emarginate; abaxially with a rather weak, conical, obtuse, colliculate ridge near the top of the anther which overtops the front margin, surface glabrous, papillose towards the top. *Pollinia* 4, ellipsoid; the small ones about half as long as the large ones, distinctly flattened; the large ones flattened; a stipes has not been observed. *Fruit* not seen.

Colours. Pseudobulbs yellowish brown, often suffused with red. Leaves dark green or glaucous green, often suffused with red. Sepals yellow, cream-coloured or white, in the latter case often with pink tips, or entirely pale pink. Petals white. Lip yellow, cream-coloured or white. White flowers may turn yellowish when a few days old. Whole inflorescence and, to a lesser extent, the leaves often slightly farinose.

Ecology. Lowland forest to hill forest, relict forest in savanna, at 0–500 m alt., at higher altitude in UG. Epiphyte in the crown of trees, on bark or in minor humus deposits (Johansson, 1974). Flowering VIII–XII.

Distribution. LIB. (10 coll.). IV.C. (6). GHANA (1). NIG.: SE. part, Cross P. (1). CAM. (2). C.A.R.: SW. part, Lobaye P. (1). GAB. (1). ZAI.: Forestier Central (2). UG.: West U2, S. part (2). Material seen from AMES, BM, BR, K, L, P, UPS, WAG.

Notes. 1 - According to the description, the holotype of *Bulbophyllum oxychilum* (Schlechter 13042, from CAM., not seen) has sepals of c. 7 mm long, which is slightly more than in the specimens seen by me. Otherwise Schlechter's description of *Bulbophyllum oxychilum* leaves no doubt about its identity.

2 - The specimen Bunting 29 from LIB. has been chosen as lectotype of *Bulbophyllum buntingii* because it is without doubt the specimen analysed by Rendle when preparing the description and the drawing of this species. Talbot 935 from NIG., the other syntype specimen, belonging to *Bulbophyllum ivorense*, has apparently never been analyzed by him.

27 - **Bulbophyllum nigritianum** Rendle – Fig. 42.

Bulbophyllum nigritianum Rendle, Cat. Talb. S. Nig. Pl. (1913) 99, 146; De Wild., Pl. Beq. 1 (1921) 114; Summerh. in F.W.T.A. ed. 1, 2 (1936) 438; id., Kew Bull. 12 (1957) 115; id. in F.W.T.A. ed. 2, 3 (1968) 234; Johansson, Acta Phyt. Suec. 59 (1974) 32; Geerinck, Bull. Soc. R. Bot. Belg. 109 (1976) 176; Adam, Fl. Descr. Mt. Nimba 5 (1981) 1681; Vermeulen, Tijdschr. Ned. Orch. Ver. 44 (1982) 167.–Type: Talbot 933 (Holo BM; iso K).

Bulbophyllum albidum De Wild., Rev. Zool. Afr. 9, suppl. bot. (1921) 27; De Wild., Pl. Beq. 1 (1921) 102.–*Bulbophyllum africanum* Hawkes, Lloydia 19 (1956) 92.–Type: Bequaert 6501 (BR).

Non *Bulbophyllum albidum* (Wight) Hook. f., Fl. Br. Ind. 5 (1890) 757.

Non *Cirrhopetalum africanum* Schltr, Engl., Bot. Jahrb. 53 (1915) 573 (= *Bulbophyllum longiflorum*).

Roots 0.5–1 mm diam. *Rhizome* 1.5–3 mm diam., nodes not or hardly swollen. *Rhizome scales* tubular, tip acute or shortly acuminate; membranous, slightly woolly when old, with some nerves; keeled abaxially towards the tip. *Pseudobulbs* 1-leafed, ovoid to narrowly ovoid, 0.3–2.5 cm apart, 1.4–4 by 0.6–1.5 cm, slightly flattened, obtusely 3–4-angled. *Petiole* 3–30 mm long. *Leaf blade* narrowly linear-lanceolate to linear, 7–21 by 0.5–1.5 cm, tip acute or subacute, neither emarginate nor oblique; coriaceous, thick; midrib prominent abaxially. *Inflorescence* 7–22 cm long, 35–150-

flowered. *Peduncle* erect to patent 1–5 cm by 1.5–2.5 mm, orbicular in section, glabrous, with 5–8 scales. *Peduncle scales* tubular, the longest 6–15 mm long; tip acute or acuminate; membranous, without nerves; keeled towards the tip and glabrous abaxially. *Rhachis* erect or slightly arching, terete, 4.5–18 cm long, glabrous. *Floral bracts* recurved, narrowly elliptic to lanceolate, 2.4–5.5 by 1.2–1.5 mm, tip acute to acuminate; base rather broadly attached; membranous, without nerves; keeled and glabrous abaxially. *Flowers* scattered, in a dense raceme, recurved or spreading, either secund or not, many open simultaneously, not much open. *Pedicel and ovary* 2.2–3.8 mm long, glabrous. *Median sepal* not recurved, very narrowly triangular, 3–5 by 0.8–1.4 mm, tip (sub)acute; margins entire, glabrous; base broadly attached; rather thin; glabrous; not keeled abaxially. *Lateral sepals* free, hardly falcate, oblique, 3–5 by 1–1.4 mm, otherwise as the median sepal. *Petals* not recurved, hardly falcate, oblong to broadly lanceolate, oblique, 0.9–1.5 by 0.3–0.5 mm, tip obtuse to acute; margins slightly irregular and finely papillose towards the tip; base rather broadly attached; thin; glabrous; not keeled abaxially. *Lip* recurved, oblong in outline (not spread), 0.6–1.2 by 0.4–0.7 mm, tip rounded; margins finely papillose to ciliate, near the base with forwars directed, obliquely semi-elliptic lateral lobes which have a rounded to obtuse tip, entire and ciliate margins, a broadly attached base and which are thin and finely papillose and separated from the median lobe by a more or less acute sinus; thick; surface finely papillose; adaxially without ridges, with a slightly concave basal part, without a median slit; abaxially with a weak, broadly rounded ridge near the base. Column from the base of the free part up to the tip of the stelidia 0.9–1.5 mm long, with distinct, gradually tapering, not falcate, subacute stelidia of 0.5–8 mm long; column with distinct, triangular, subacute teeth along its adaxial margins at the base of the stelidia; stigma with a rather weak to rather distinct, obtuse basal tooth. *Anther* c. 0.4 by 0.4 mm; front not drawn out; front margin rounded, entire, about glabrous; basally distinctly emarginate; abaxially with a rather weak, conical, obtuse, papillose ridge near the top of the anther which distinctly overtops the front margin, surface papillose, glabrous towards the top. *Pollinia* 4, ellipsoid; the small ones about half as long as the large ones, distinctly flattened; the large ones flattened; a stipes has not been observed. *Fruit* not seen.

Colours. Pseudobulbs dark green, often suffused with red. Leaves dark green. Flowers white or cream-coloured.

Ecology. Lowland to lower montane forest, at 0–1200 m alt., at higher altitude in UG. Epiphyte in the central part of the crown of trees, in humus deposits (Johansson, 1974). Flowering X–XII.

Distribution. S.L. (4 coll.). LIB. (4). IV.C. (4). GHANA (1). NIG.: SE. part, Cross P. (1). GAB. (2). ZAI.: Forestier Central Z6 (1). Material seen from AMES, BM, BR, E, K, L, UPS, Wag.

Note. The specimen Wageningen cult. Hort. 83-322 (= Leiden cult. Hort. 23617) from GAB. differs in having a distinctly ciliate lip.

28 - **Bulbophyllum humblotii** Rolfe – Fig. 43.

Bulbophyllum humblotii Rolfe, J. Linn. Soc. Bot. 29 (1891) 50; Grosvenor, Excelsa 6 (1976) 79; Ball, S. Afr. Ep. Orch. (1978) 80, pl.; La Croix et al., Orch. Mal. 1 (1983) 44; Cribb, in F.T.E.A., Orch. 2 (1984)

321.–S y n t y p e s: Scott Elliot s.n. (Lecto K, here chosen), Humblot 378, Meller s.n. (all K); Madagascar, resp. Fort Dauphin, Lake Nossive, Hirondio.

Bulbophyllum album Jum. & Perr., Ann. Fac. Sc. Marseille 21 (1912) 206.–T y p e: Perrier 1908 (P); Madagascar, Sambirano.

Bulbophyllum laggiarae Schltr, Fedde, Rep. 15 (1918) 330.–T y p e: Laggiara s.n. (?, n.v.); Madagascar, Ile de Sainte-Marie.

Bulbophyllum luteolabium Perr., Not. Syst. 14 (1951) 55.–S y n t y p e s: Decary 16996 (P), 17013 (P, n.v.); Madagascar, Tamatave.

Bulbophyllum linguiforme Cribb, Kew Bull. 32 (1977) 157.–T y p e: Ball 316 (Holo K; iso SRGH, n.v.).

Bulbophyllum sp., Schltr, Engl., Bot. Jahrb. 53 (1915) 572.–S p e c i m e n: Stolz 1475 (C, G, K, S, U, Z).

Bulbophyllum sp., Morris, Proc. Linn. Soc. London 179 (1968) 63; Morris, Ep. Orch. Mal. (1970) 45. S p e c i m e n: Morris 27 (K).

Roots 0.5–0.8 mm diam. *Rhizome* 1–1.8 mm diam., nodes not or hardly swollen. *Rhizome scales* tubular, tip shortly acuminate, the lower subacute or acute; membranous, the upper with some nerves; keeled abaxially towards the tip. *Pseudobulbs* 1-leafed, broadly ovoid or ellipsoid to orbicular, 1.2–3.5 cm apart, 0.6–1.1 by 0.6–1.2 cm, not or hardly flattened, obtusely 4-angled. *Petiole* c. 1 mm long. *Leaf blade* narrowly oblong to broadly linear-lanceolate, 1.8–5 by 0.6–1.1 cm, tip obtuse to acute, slightly emarginate, not oblique; coriaceous, thick to very thick; midrib not prominent abaxially. *Inflorescence* 1.5–3.5 cm long, 4–11-flowered. *Peduncle* erect to patent, often sturdy, slightly swollen below the nodes, 0.9–1.8 cm by 1–1.5 mm, orbicular in section, glabrous, with 4–5 scales. *Peduncle scales* tubular, the longest 3–3.5 mm long; tip acute; membranous, without nerves; not keeled and glabrous abaxially. *Rhachis* erect or arching, terete, 1–2 cm long, glabrous. *Floral bracts* recurved, ovate to triangular, 3–3.6 by 1.5–2.1 mm, tip acute to acuminate; base broadly attached; membranous, without nerves; slightly keeled and glabrous abaxially. *Flowers* scattered, in a rather lax raceme, appressed against the rhachis, many open simultaneously, not fully open. *Pedicel and ovary* c. 1.8 mm long, glabrous. *Median sepal* not recurved, elliptic, 4–5 by 2.8–3.3 mm, tip acute to slightly acuminate; margins entire, glabrous; base rather narrowly attached; rather thin; glabrous; not keeled abaxially. *Lateral sepals* free, slightly recurved, hardly falcate, triangular to narrowly triangular, oblique, 3.7–6 by 2–2.5 mm, tip acute; base broadly attached; slightly keeled towards the tip abaxially; otherwise as the median sepal. *Petals* not recurved, not falcate, ovate-oblong, oblique, 1.9–2.5 by 0.7–1 mm, tip acute; margins entire, glabrous; base broadly attached; rather thin; glabrous; not keeled abaxially. *Lip* recurved, oblong in outline (not spread), 2.5–3.1 by 1–1.2 mm, tip rounded; margins entire, glabrous to very finely papillose; thick; surface glabrous, locally very finely papillose; adaxially with two weak, rounded ridges in the basal half which diverge slightly towards the base and which have a short, weak, rounded, transverse ridge interconnecting them at about 1/3 of the lip, with a slightly concave basal part, without a median slit; abaxially with a weak, broadly rounded ridge near the base. *Column* from the base of the free part up to the tip of the stelidia c. 1.8 mm long, with distinct, gradually tapering, not or hardly falcate, acute stelidia of c. 0.9 mm long which have a weak, rounded wing along their upper margin; column with weak, broadly rounded wings along its adaxial margins just below the base of the stelidia; stigma without a basal callus. *Anther* c. 0.6 by 0.8 mm; front not drawn out; front margin rounded, entire, papillose; basally distinctly emarginate; abaxially with a rather weak, obtuse, colliculate ridge near the top of the anther which hardly overtops the front margin, surface glabrous but colliculate towards the top. *Pollinia* 4, ovoid in dorsoventral view, ellipsoid in lateral view; the small ones more than half as long as the large ones, distictly flattened; the large ones flattened; a stipes has not been observed. *Fruit* not seen.

Colours. Pseudobulbs yellowish or pale green. Leaves pale green, suffused with red. Rhachis very pale green. Sepals and petals white or cream coloured. Lip yellow or orange yellow.

Ecology. Brachystegia woodland, at 650–1200 m alt. Epiphyte in the crown of trees. Flowering IV.

Distribution. TANZ.: Southern Highlands T7 (1 coll.). MAL.: South (1). ZIM.: East (2). Material seen from CO, G, K, P, S, U, Z.

Distribution outside Africa: Seychelles (Archer 171 [K]); Madagascar (a.o. type specimens of *Bulbophyllum humblotii*, *B. album*, *B. laggiarae* and *B. luteolabium*).

Note. Specimens from Madagascar are identical but have slightly smaller flowers (median sepal 3.2–4 mm long).

29 - Bulbophyllum scariosum Summerh. – Fig. 44.

Bulbophyllum scariosum Summerh., Kew Bull. 8 (1953) 145; id. in F.W.T.A. ed. 2, 3 (1968) 239; Johansson, Acta Phyt. Suec. 59 (1974) 33; Cribb & Perez-Vera, Adansonia, ser. 2, 15 (1975) 206; Adam, Fl. Descr. Mt. Nimba 5 (1981) 1681, pl. 915.–Type: Jones 64 (K).
Bulbophyllum sp., Jaeger & Summerh., Kew Bull. 3 (1949) 482.–Specimen: Jaeger 568 (K).

Roots 0.5–1 mm diam. *Rhizome* 1.2–1.8 mm diam., nodes not or hardly swollen. *Rhizome scales* only tubular at their base, tip acuminate; fibrous, with many nerves close together; keeled abaxially. *Pseudobulbs* 2-leafed, narrowly ovoid, 2–4.6 cm apart, (1.2 –) 2.5–4 by 0.6–1.7 cm, slightly flattened, 4 (– 5)-angled. *Petiole* 0.5–3 mm long. *Leaf blade* narrowly lanceolate to linear-lanceolate, 3–9 by 0.6–1.3 cm, tip rounded, emarginate, oblique; coriaceous, rather thick; midrib prominent abaxially. *Inflorescence* 5.5–20 cm long, 2–8-flowered. *Peduncle* about erect, 4.3–15.5 cm by 1–1.8 mm, orbicular in section, glabrous, with 10–19 scales. *Peduncle scales* tubular, the longest 6.5–12 mm long; tip acuminate; fibrous, with many nerves nerves close together; keeled and glabrous abaxially. *Rhachis* arching or nodding, not or hardly swollen, 4-angled in section, with two slightly concave sides from which the flowers arise, zig-zag bent towards the tip, 1.2–5.5 cm long, edges about (?) sharp, entire, glabrous; surface glabrous. *Floral bracts* spreading, slightly concave, triangular to about deltoid, 5–8 by 3.5–7 mm, tip acuminate; base rather broadly attached; fibrous, with many nerves close together; keeled and glabrous abaxially. *Flowers* distichous, in a rather lax raceme, spreading, many open simultaneously, not fully open. *Pedicel and ovary* 1.5–3 mm long, glabrous. *Median sepal* not recurved, slightly concave, ovate, 4.5–7 by 3–4.4 mm, tip obtuse to slightly acuminate; margins entire, glabrous; base (rather) broadly attached; thin; glabrous; not keeled abaxially. *Lateral sepals* free, hardly recurved to about spreading, not concave, not falcate, triangular, oblique, 6–9 by 4–7 mm, base broadly attached; otherwise as the median sepal. *Petals* not recurved, not or hardly falcate, elliptic, oblique, 4–6 by 2.3–3.8 mm, tip rounded; margins slightly irregular towards the tip, glabrous; base (rather) broadly attached; thin; glabrous; not keeled abaxially. *Lip* recurved, elliptic in outline (not spread), 2–4.9 by

1.4–2.5 mm, tip broadly rounded; margins entire, glabrous; thick; adaxially with two weak, rounded ridges in the basal half which diverge towards the base of the lip, with a concave basal part, with a median slit which starts at the base and which continues over about 2/3 of the lip, surface glabrous to very finely papillose; abaxially with a rather distinct, broadly rounded ridge in the basal half, surface glabrous. *Column* from the base of the free part up to the tip of the stelidia 1.5–2.5 mm long, with rudimentary, deltoid, not falcate, rounded to acute stelidia; column usually with (rather) weak, deltoid, broadly rounded wings along its adaxial margins just below the base of the stelidia; stigma without a basal callus. *Anther* c. 1 by 1 mm; front not drawn out; front margin rounded, entire, papillose; basally distinctly emarginate; abaxially with a weak, conical, obtuse, colliculate ridge near the top of the anther which overtops the front margin, surface colliculate. *Pollinia* 4; the small ones much more than half as long as the large ones, about orbicular, distinctly flattened; the large ones ovoid, flattened; a stipes has not been observed. *Fruit* ellipsoid to obovoid, up to 0.8 by 0.4 cm, without a stalk.

Colours. Pseudobulbs pale yellowish green, often suffused with red. Leaves light green. Floral bracts straw yellow. Sepals and petals white. Lip white (?), yellow or orange.

Ecology. (Lower) montane forest at 1000–1800 m alt. Epiphyte in the outer part of the crown of trees, on bark or in humus deposits. Often very exposed (Johansson, 1974). Also lithophytic on boulders in montane grassland. Flowering IX–IV.

Distribution. GUIN.: Dalaba (3 coll.); Nimba Mts. (1). S.L.: Loma Mts, Tingi Hills (9). LIB.: Nimba Mts. (7). IV.C.: Nimba Mts, Mt. Tonkoui (2). EQ.GUIN.: Bioko (1). Material seen from AMES, BR, K, P, UPS, WAG.

Note. The specimen Sanford 4359 from Bioko has very inconspicuous wings along the adaxial margins of the column.

30 - **Bulbophyllum pandanetorum** Summerh. - Fig. 45.

Bulbophyllum pandanetorum Summerh., Kew Bull. 8 (1954) 580.–Type: Le Testu 5527 (Holo K; iso BM, BR, P).

Roots 0.4–1 mm diam. *Rhizome* 1.5–2.5 mm diam., nodes not or hardly swollen. *Rhizome scales* only tubular at their base, tip shortly acuminate; fibrous, with many nerves close together; keeled abaxially. *Pseudobulbs* 2-leafed, narrowly ovoid to ellipsoid, 2.5–4 cm apart, 2.2–4 by 0.8–1.3 cm, slightly flattened, 4-angled. *Petiole* 2–5 mm long. *Leaf blade* linear-lanceolate, 6–15.5 by 0.8–1 cm, tip obtuse, emarginate, oblique; coriaceous, rather thick; midrib prominent abaxially. *Inflorescence* 20–31 cm long, 8–16-flowered. *Peduncle* erect, slightly swollen towards the tip, 12–23.5 cm by 1.8–2 mm, elliptic in section, glabrous, with 19–22 scales. *Peduncle scales* tubular, the longest 16–18 mm long; tip acuminate; fibrous, with many nerves close together; keeled and glabrous abaxially. *Rhachis* erect, swollen, 4-angled in section, with two concave sides from which the flowers arise, only zig-zag bent towards the very tip, 6–8 cm by c. 4 mm, edges sharp, straight, irregularly dentate, glabrous; surface glabrous. *Floral bracts* recurved to spreading, distinctly concave, about deltoid, c. 14 by 12.5 mm, tip acuminate; base narrowly attached; fibrous, with many nerves close together; keeled and glabrous abaxially. *Flowers* distichous, in a rather lax raceme, recurved, (?) open simultaneously, not fully open. *Pedicel and ovary* c. 2 mm long, glabrous. *Median sepal* not recurved, slightly concave, ovate, 4.7–5.2 by 2.9–3.3 mm, tip about acuminate; margins entire, glabrous; base broadly attached; thin; glabrous; not keeled abaxially. *Lateral sepals* free, either recurved or not, not concave, not falcate, ovate-triangular, oblique, 5.7–6 by 3.8–4.4 mm, upper margin with a subacute tooth near the base, base rather broadly attached; otherwise as the median sepal. *Petals* not recurved, not falcate, obovate, oblique, 3.6–4 by 1.6–2 mm, tip rounded; margins slightly irregular towards the tip, glabrous; base rather broadly attached; very thin; glabrous; not keeled abaxially. *Lip* recurved, ovate in outline (not spread), c. 4 by 2–2.4 mm, tip obtuse; margins entire, glabrous; thick; surface glabrous; adaxially with two rather distinct, rounded ridges over almost the entire length which diverge towards the tip of the lip, with a concave basal part, without a median slit; abaxially with a distinct, truncate ridge in the basal half, towards the tip with a slightly concave median part. *Column* from the base of the free part up to the tip of the stelidia c. 2 mm long, with rudimentary, deltoid to triangular, not falcate, acute stelidia; column with weak, broadly rounded wings along its lower margins just below the base of the stelidia; stigma without a basal callus. *Anther* not seen (see note). *Pollinia* not seen. *Fruit* ellipsoid, up to 0.9 by 0.6 cm, without a stalk.

Colours. Flowers red.

Ecology. Lowland forest, both collections found growing as an epiphyte on Pandanus. Flowering IX (once recorded).

Distribution. GAB. (2 coll.). Material seen from BM, BR, K, P.

Note. Only one badly preserved anther could be studied which suggests a similarity with the anthers of species of group 4 rather than to that of *Bulbophyllum scariosum*.

GROUP 2 (sp. 31)

Pseudobulbs 2-leafed. Inflorescence a raceme, many-flowered. Rhachis terete, not thickened. Flowers scattered. Sepals slightly irregular abaxially; the lateral ones not connate along the upper margins, at most spreading. Lip thick, not soft and easy to compress, lateral lobes absent, margins glabrous. Stelidia long, rounded to truncate. Anther with a protrusion overtopping the front margin; protrusion not dorsoventrally flattened.

31 - Bulbophyllum denticulatum Rolfe – Fig. 46.

Bulbophyllum denticulatum Rolfe, Kew Bull. 1891 (1891) 197; Durand & Schinz, Consp. Fl. Afr. 5 (1895) 11; Rolfe in F.T.A. 7 (1897) 25; De Wild., Pl. Beq. 1 (1921) 108; Summerh. in F.W.T.A. ed. 1, 2 (1936) 438; id., ed. 2, 3 (1968) 236.–Type: O'Brien s.n. (K).

Roots 1–1.3 mm diam. *Rhizome* 3.4–7 mm diam., nodes not or hardly swollen. *Rhizome scales* tubular, tip (?) rounded; membranous, with many nerves; not keeled abaxially. *Pseudobulbs* 2-leafed, ovoid to narrowly ovoid, 1.5–1.8 cm apart, 1.2–3.5 by 1.2–2.3 cm, not or hardly flattened, obtusely 4-angled. *Petiole* 1–5 mm long. *Leaf blade* narrowly lanceolate to linear-lanceolate, 8.2–12 by 1–1.6 cm, tip obtuse, not emarginate, slightly oblique; coriaceous, thick; midrib slightly prominent abaxially. *Inflorescence* 22–34 cm long, 9–21-flowered. *Peduncle* erect to patent, 19–31 cm by c. 1.3 mm, orbicular in section, glabrous, with 7–9 scales. *Peduncle scales* tubular, the longest 5–9 mm long; tip obtuse to slightly retuse; membranous, with some nerves; not keeled and glabrous abaxially. *Rhachis* about nodding, terete, 3–4 cm long, glabrous. *Floral bracts* recurved, ovate, 2–3.5 by 1.7–2 mm, tip acute; base rather broadly attached; membranous, without nerves; not keeled and glabrous abaxially. *Flowers* scattered, in a rather dense raceme, recurved, many open simultaneously, not fully open. *Pedicel and ovary* 2.5–5 mm long, glabrous, jugae ending in an obtuse to acute tooth. *Median sepal* not recurved, ovate-oblong, 5.2–6 by 2.1–2.6 mm, tip acuminate; margins entire, finely papillose; base rather broadly attached; rather thin; glabrous adaxially; slightly keeled and with a slightly irregular surface and some papillae abaxially. *Lateral sepals* free, either slightly recurved or not, hardly falcate, ovate, oblique, 5–5.5 by 2.8–3 mm, base broadly attached, otherwise as the median sepal. *Petals* not recurved, not falcate, elliptic to oblong, oblique, 3.8–4.2 by 1.8–2.1 mm, tip acute; margins coarsely and irregularly denticulate, glabrous; base rather broadly attached; thin; glabrous; not keeled abaxially. *Lip* recurved near the base, elliptic in outline (not spread), 3–3.6 by 1.5–2.2 mm, tip rounded; edges entire, glabrous; thick; adaxially without ridges, with a slightly concave basal part, without a median slit but slightly channeled in the median part, surface slightly wrinkled and undulate in the median part, otherwise glabrous; abaxially without a ridge, surface glabrous, coarsely papillose near the base. *Column* from the base of the free part up to the tip of the stelidia c. 2.3 mm long, with distinct, oblong, not falcate, obliquely truncate to rounded stelidia of c. 1 mm long; column with distinct, triangular, obtuse teeth along its adaxial margins just below the base of the stelidia; stigma without a basal callus. *Anther* c. 0.8 by 0.7 mm; front not drawn out; front margin truncate, irregular, glabrous; basally distinctly emarginate; with a rather distinct, conical, rounded, colliculate ridge near the top of the anther which distinctly overtops the front margin; surface colliculate. *Pollinia* 4; the small ones about half as long as the large ones, almost orbicular, distinctly flattened on either side; the large ones ellipsoid, flattened on one side; no stipes has been observed. *Fruit* not seen.

Colours. Flowers brownish green, often striated purple. Lip orange.

Ecology. Lower montane forest at c. 900 m alt. (once recorded). Epiphytic. Flowering II (once recorded).

Distribution. S.L. (1 coll.). LIB.: Mt. Nimba (1). IV.C.: Mt. Tonkoui (1). Material seen from K, WAG.

GROUP 3 (sp. 32)

Pseudobulbs 1-leafed. Inflorescence a raceme, many-flowered. Rhachis distinctly swollen, spindle shaped. Flowers scattered. Sepals with long hairs abaxially, the lateral ones not connate along the upper margins, not or hardly recurved. Lip thick, not soft and easy to compress, either with very small lateral lobes or not, margins hirsute. Stelidia rather short, acute to retuse. Anther without a protrusion overtopping the front margin.

32 - Bulbophyllum comatum Lindl. - Fig. 47-48.

Literature: see under varieties.

Roots 0.5–1.2 mm diam. *Rhizome* 1.2–4 mm diam., nodes not or hardly swollen. *Rhizome scales* tubular, tip acuminate or apiculate; membranous, with many nerves; distinctly keeled abaxially. *Pseudobulbs* 1-leafed, about orbicular to ovoid or ellipsoid, 1–3.5 cm apart, (0.6 –) 1.5–4 by (0.5 –) 1–2.2 cm, slightly to moderately flattened, obtusely to sharply (2 –) 4-angled. *Petiole* 3–15 mm long. *Leaf blade* broadly oblong to linear, maximum width occasionally either slightly above or below the middle, (2.8–) 9–27 by 1.2–5 cm, tip acuminate, neither emarginate nor oblique; coriaceous, thick; midrib hardly prominent abaxially. *Inflorescence* 4.8–14 cm long, 25–65-flowered. *Peduncle* pendulous, swollen towards the tip, 2.3–9 cm by 1–2.5 mm, orbicular in section, glabrous, with 6–8 scales. *Peduncle scales* tubular, the longest 6–20 mm long; tip acuminate; membranous, either with some nerves or not; keeled and glabrous abaxially. *Rhachis* pendulous, distinctly swollen, spindle shaped, 2–6 cm by up to 18 mm, glabrous or woolly with fine ribbon-like hairs. *Floral bracts* recurved to spreading, ovate-oblong to broadly linear-lanceolate, 7–12 by 1.2–4 mm, tip acute to acuminate; base rather broadly atached, membranous, without nerves; keeled and about glabrous abaxially. *Flowers* arranged in a reticulate pattern, in a rather dense raceme, about appressed against the rhachis to recurved, many open simultaneously, not much open. *Pedicel and ovary* 1.5–2 mm long, with fine, ribbon-like hairs. *Median sepal* not recurved, narrowly (ovate-)oblong to (ovate-)lanceolate, 5.5–10 by 1.8–3 mm, tip acuminate to caudate; margins entire, glabrous or slightly papillose; base broadly attached; thick; glabrous adaxially; keeled and glabrous to papillose and with fine, long, ribbon-like hairs abaxially. *Lateral sepals* free, not falcate, narrowly ovate-oblong to ovate-lanceolate, oblique, 6.5–10 by 2–3.5 mm, otherwise as the median sepal. *Petals* not recurved, hardly falcate, narrowly obovate to broadly obovate-lanceolate, almost spathulate, oblique, 2.2–4 by 0.9–1.5 mm, tip obtuse; margins papillose towards the tip; base broadly attached; rather thin to rather thick; papillose adaxially towards the tip; with a low ridge and about glabrous abaxially. *Lip* recurved, narrowly elliptic to oblong in outline (not spread), 1.6–3.4 by 0.6–1.4 mm, tip rounded; margins entire, hirsute, just above the base either with very small, triangular, thin, glabrous lateral lobes with an obtuse tip, entire, glabrous margins and a broadly attached base or not; thick; adaxially with two weak, rounded ridges which are most prominent about half-way the lip, with a slightly concave basal part with or without a transverse, elliptic, glabrous callus just above the base (between the lateral lobes), without a median slit, surface densely hirsute; abaxially with a rather weak, rounded, truncate or retuse ridge over 2/3–3/4 of the lip, surface about glabrous to finely hirsute, densely hirsute towards the tip. *Column* from the base of the free part up to the tip of the stelidia 1.2–1.8 mm long, with distinct, triangular, not falcate, acute, truncate or retuse stelidia of 0.3–0.9 mm long; column with or without teeth along its adaxial margins (see under varieties); stigma without a basal callus. *Anther* 0.6–0.8 by 0.5–0.7 mm, front slightly drawn out, thin, slightly concave; front margin rounded, entire, glabrous; basally weakly emarginate; abaxially without a ridge, colliculate. *Pollinia* 4, ellipsoid; the small ones about half as long as the large ones, distinctly flattened on either side; the large ones flattened on one side; a stipes has not been observed. *Fruit* ellipsoid, up to 0.7 by 0.45 cm, without a stalk.

KEY TO THE VARIETIES

1 - Ovary thickly covered with long hairs; sepals abaxially entirely set with long, white, very fine, more or less ribbon-like hairs.............................var. **comatum** (a)
- Ovary sparsely covered with short hairs; sepals abaxially with long, white, ribbon-like hairs only on the keel of the median sepal, surface otherwise at most with much shorter hairs, lateral sepals often with a few longer hairs near the base along their upper margin..var. **inflatum** (b)

32a - var. comatum – Fig. 47.

Bulbophyllum comatum Lindl., J. Linn. Soc. Bot. 6 (1862) 128 ('*Bolbophyllum*'); Durand & Schinz, Consp. Fl. Afr. 5 (1895) 11; Rolfe in F.T.A. 7 (1897) 32; De Wild., Pl. Beq. 1 (1921) 107; Schltr in Mildbraed, 2e Deutsch. Zentr. Afr. Exp. 1910-11, 2 (1922) 180; Summerh. in F.W.T.A. ed. 1, 2 (1936) 437; id., ed. 2, 3 (1968) 236; Adam, Fl. Descr. Mt. Nimba 5 (1981) 1674.–*Phyllorchis comata* (Lindl.) Kuntze, Rev. Gen. Pl. 2 (1891) 677, incorr. name.–Type: Mann 642 (K).

Bulbophyllum hirsutissimum Krzl., Engl., Bot. Jahrb. 48 (1912) 392; De Wild., Pl. Beq. 1 (1921) 111.–Type: Ledermann 1214 (B, lost).

Bulbophyllum hirsutissimum Krzl., Engl., Bot. Jahrb. 51 (1914) 384.–*Bulbophyllum kraenzlinianum* De Wild., Pl. Beq. 1 (1921) 113.–Type: Ledermann 6328 (B, lost).

Peduncle 6–9 cm long. Rhachis a cylindrical body of up to 0.8 cm diam., surface glabrous or woolly with fine, more or less ribbon-like hairs. Floral bracts lanceolate to broadly linear-lanceolate, 9–12 by 1.2–3 mm. Flowers recurved. Median sepal 7.3–10 mm long, acuminate to caudate, abaxially entirely beset with fine, ribbon-like, long white hairs, surface otherwise glabrous. Abaxial surface of the lateral sepals as that of the median sepal. Petals 2.2–3 by 0.9–1.2 mm. Lip 1.6–2.5 by 0.6–1 mm, adaxially with a very inconspicuous basal callus. Column 1.2–1.5 mm long, stelidia truncate to retuse, 0.3–0.4 mm long; with distinct, forward directed, obliquely semi-elliptic to triangular, obtuse, thick teeth along the adaxial margins.

Colours. Flowers reddish.

Ecology. Hill forest to montane forest, 600–1300 m alt. Epiphytic. Flowering XII (once recorded).

Distribution. NIG.: SE. part, Cross P. (1 coll.). EQ.GUIN.: Bioko (2). Material seen from K.

32b - var. inflatum (Rolfe) J.J. Vermeulen – Fig. 48; Plate 66.

Bulbophyllum comatum Lindl. var. *inflatum* (Rolfe) J.J. Vermeulen, Bull. J. Bot. Nat. Belg. 56 (1986) 231.–*Bulbophyllum inflatum* Rolfe, Gard. Chron., ser. 3, 9, nr. 217 (1891) 234; Rolfe in F.T.A. 7 (1897) 33; De Wild., Pl. Beq. 1 (1921) 112; Summerh. in F.W.T.A. ed. 1, 2 (1936) 437; id., ed. 2, 3 (1968) 236; Johansson, Acta Phyt. Suec. 59 (1974) 31; Cribb & Perez-Vera, Adansonia, ser. 2, 15 (1975) 206; Adam, Fl. Descr. Mt. Nimba 5 (1981) 1677, pl. 910; Vermeulen, Tijdschr. Ned. Orch. Ver. 43 (1981) 88, 128.–Type: Kew cult. Hort. s.n. (K).

Peduncle 2.3–6 cm long. Rhachis an ellipsoid or obovoid, occasionally nearly cylindrical body of max 1.8 cm diam., surface more or less woolly with fine, more or less ribbon-like hairs. Floral bracts ovate-oblong to ovate-lanceolate, 7–12 by 2–4 mm. Flowers about appressed against the rhachis. Median sepal 5.5–8 mm long, acuminate, abaxially with ribbon-like, long white hairs on the crest of the keel, surface otherwise glabrous to papillose, only sparsely and locally beset with much shorter hairs of the same kind. Abaxial surface of the lateral sepals as that of th the median sepal. Petals 2.6–4 by 0.9–1.5 mm. Lip 2–3.4 by 0.9–1.4 mm, adaxially with a distinct basal callus. Column 1.5–1.8 mm long, stelidia acute, 0.6–0.9 mm long; usually with two rather inconspicuous edges along each lower margin, one of which or both may develop into an obtuse to retuse tooth, the front one forwards directed and occasionally about equaling the stelidia in size and shape, the back one either more or less forwards or downwards directed.

Colours. Plant light green or pale green. Inflorescence green, mottled with purple-red, or entirely purple-red. Sepals green, stained with dark purple-red in the centre, hairs white. Petals green, suffused with dark purple-red, or entirely dark purle-red. Lip dark purple-red. Column white.

Ecology. Lowland to lower montane forest, up to 1300 m alt., in RW. in montane forest at 2300 m alt. Epiphyte on shaded tree trunks and the basal part of large branches (Johansson, 1974). Flowering II–III, VIII and X–XII.

Distribution. S.L. (1 coll.). LIB. (8). IV.C.: Mt. Nimba, Mt. Momi (3). GAB. (1). RW. (1). Material seen from AMES, BR, E, K, L, P, UPS, WAG.

Notes. 1 - Most specimens differ from var. *comatum* in the shape and the position of the teeth along the adaxial margins of the column. However, a few specimens (Troupin 15766 from RW., Wageningen cult. Hort. 83-202 = Leiden cult. Hort. 23950 from GAB. and, to a lesser extent, Perez-Vera 259 and Geerling & Bokdam 1791, both from IV.C.) more or less approach var. *comatum* in this respect.

2 - The specimen Troupin 15766, the only one found in RW., is exceptionally small (the measurements in parenthesis in the species description). The flowers, however, are of normal size.

GROUP 4 (spp. 33 - 35)

Pseudobulbs 2-leafed. Inflorescence a raceme, many-flowered. Rhachis (slightly) flattened, flowers inserted along the margins. Flowers distichous. Sepals glabrous abaxially, the lateral ones not connate along the upper margins, reflexed and appressed against the rhachis. Lip thick, not very soft and easy to compress, lateral lobes absent, margins glabrous. Stelidia long and acuminate to short an bifid. Anther without a protrusion overtopping the front margin.

33 - **Bulbophyllum bifarium** Hook. f. - Fig. 49.

Bulbophyllum bifarium Hook. f., J. Linn. Soc. Bot. 7 (1864) 219 ('*Bolbophyllum*'); Durand & Schinz, Consp. Fl. Afr. 5 (1895) 10; Rolfe in F.T.A. 7 (1897) 27; De Wild., Pl. Beq. 1 (1921) 104; Summerh. in F.W.T.A. ed. 1, 2 (1936) 438; id., ed. 2, 3 (1968) 239, p.p.–*Phyllorchis bifaria* (Hook. f.) Kuntze, Rev. Gen. Pl. 2 (1891) 677, incorr. name.–Type: Mann 2121 (K).

Non *Bulbophyllum bifarium* auct.: Ake Assi, Et. Fl. C. Iv. (1964) 242; Johansson, Acta Phyt. Suec. 59 (1974) 29; Adam, Fl. Descr. Mt. Nimba 5 (1981) 1672, pl. 905 (= all *Bulbophyllum bidenticulatum* ssp. *bidenticulatum*); Cribb, in F.T.E.A., Orch. 2 (1984) 314 (= *Bulbophyllum bidenticulatum* ssp. *joyceae*).

Bulbophyllum pallescens Krzl., Engl., Bot. Jahrb. 51 (1914) 385; De Wild., Pl. Beq. 1 (1921) 115.–Type: Ledermann 6019 (B, lost).

Roots 0.4–1 mm diam. *Rhizome* 1.5–2 mm diam., nodes not or hardly swollen. *Rhizome scales* only tubular at their base, tip acute to acuminate; fibrous, with many nerves close together; more or less keeled abaxially. *Pseudobulbs* 2-leafed, (narrowly) ovoid to (narrowly) ellipsoid, 1.5–3.5 cm apart, 1.2–3.5 by 0.8–1.2 cm, slightly flattened, 4–5-angled. *Petiole* 0.5–2 mm long. *Leaf blade* lanceolate to linear-lanceolate, 3.2–10 by 0.7–1 cm, tip rounded to obtuse, emarginate, oblique; coriaceous, rather thick; midrib prominent abaxially. *Inflorescence* 16–19 cm long, 16–23-flowered. *Peduncle* about erect, slightly swollen towards the tip, 8.5–9.5 cm by 1.8–2.7 mm, elliptic in section, glabrous, with 9–14 scales. *Peduncle scales* tubular, the longest 8–11 mm long; tip acuminate; fibrous, with many nerves close together; keeled and glabrous abaxially. *Rhachis* erect or slightly nodding, swollen, 4-angled in section, with two concave sides from which the flowers arise, 6.5–9 cm by c. 3.5 mm, edges obtuse to about sharp, straight, entire, glabrous; surface glabrous. *Floral bracts* spreading, concave, ovate to ovate-oblong, 9–11 by 4.8–5 mm, tip acuminate; base broadly attached; fibrous, with many nerves close together; keeled and glabrous abaxially. *Flowers* distichous, in a rather lax raceme, recurved, many open simultaneously, fully open. *Pedicel and ovary* 1.5–2 mm long, glabrous. *Median sepal* not recurved, concave, elliptic, 5.6–6.4 by 3.7–5.4 mm, tip acuminate; margins entire, glabrous; base broadly attached; thin; glabrous; not keeled abaxially. *Lateral sepals* free, reflexed at the base, not concave, triangular to narrowly triangular, oblique, 6–6.8 by 2.3–3.9 mm, tip acute; otherwise as the median sepal. *Petals* not recurved, not falcate, obovate to about obrhomboid, oblique, 1.7–2.1 by 1.4–1.8 mm, tip broadly rounded; margins slightly irregular towards the tip, glabrous; base rather broadly attached; thin; glabrous; not keeled abaxially. *Lip* recurved, elliptic in outline (not spread), 1.8–2 by 1–1.4 mm, tip broadly rounded; margins entire, glabrous; thick; surface glabrous to very finely papillose; adaxially with two distinct, obtuse ridges continuing over 2/3–3/4 of the lip which diverge towars the base of the lip, with a concave basal part, without a median slit; abaxially with a weak, broadly rounded ridge near the base, slightly concave towards the top. *Column* from the base of the free part up to the tip of the stelidia 2–3 mm long, with distinct, triangular, not falcate, acute to acuminate stelidia of c. 1 mm long; column with distinct, deltoid, obtuse to subacute teeth along its adaxial margins just below the base of the stelidia; stigma without a basal callus. *Anther* c. 1 by 1 mm; front drawn out, thin, not concave; front margin truncate, with an apiculate tip, entire, glabrous; basally distinctly emarginate; abaxially with a distinct, short, obtuse, colliculate ridge near the base which does not overtop the front margin, surface colliculate but glabrous towards the top. *Pollinia* 4; all of about equal length; the inner pair ellipsoid, distinctly flattened on either side; the outer pair obovoid, flattened on one side; a stipes has not been observed. *Fruit* obovoid, up to 0.8 by 0.5 cm, without a stalk.

Colours. Floral bracts straw yellow. Sepals and petals whitish, pink or bluish pink. Lip yellow.

Ecology. Montane forest, 'semi montane grassland' (Sanford, 1974), at c. 1600 m alt. Epiphytic. Flowering XI.

Distribution. CAM.: West (3 coll.). Material seen from BR, K, P, WAG.

34 - Bulbophyllum bidenticulatum J.J. Vermeulen - Fig. 50–51.

Literature: see under the subspecies.

Roots 0.3–0.9 mm diam. *Rhizome* 0.9–3 mm diam., nodes not or hardly swollen. *Rhizome scales* only tubular at their base, tip acute or shortly acuminate; fibrous, with many nerves close together; more or less keeled abaxially. *Pseudobulbs* 2-leafed, (narrowly) ovoid to (narrowly) ellipsoid, occasionally about orbicular, 0.7–3 cm apart, 0.8–4 by 0.6–1.2 cm, slightly flattened, 4 (- 6)-angled. *Petiole* 0.5–2 mm long. *Leaf blade* narrowly lanceolate to linear, (1.6 -) 3–11 by 0.4–1 cm, tip rounded to obtuse, emarginate, oblique; coriaceous, rather thick; midrib prominent abaxially. *Inflorescence* 5–12 cm long, 6–22-flowered. *Peduncle* about erect, slightly swollen towards the tip, 2.8–6 cm by 1.2–4 mm, slightly elliptic in section, glabrous, with 9–13 scales. *Peduncle scales* tubular, the longest 7–14 mm long; tip acute to acuminate; fibrous, with many nerves close together; keeled and glabrous abaxially. *Rhachis* erect, swollen, 4-angled, with two concave sides from which the flowers arise, slightly zig-zag bent, 2–6.5 cm by 1.8–5 mm, edges sharp, straight, entire, glabrous; surface glabrous. *Floral bracts* spreading, either concave or not, ovate to about deltoid, 4.9–10 by 3.8–5 mm, tip acuminate; margins slightly rolled inwards; base rather broadly attached; fibrous, with many nerves close together; keeled and glabrous abaxially. *Flowers* distichous, in a rather lax raceme, recurved, many open simultaneously, fully open. *Pedicel and ovary* 1–2.2 mm long, glabrous. *Median sepal* (hardly) recurved, broadly elliptic, 3.9–6 by 2.5–4 mm, tip acute to acuminate; margins entire, glabrous; base broadly to rather narrowly attached; thin; glabrous; not keeled abaxially. *Lateral sepals* free, reflexed at the base, not falcate, (slightly ovate-)oblong, oblique, 4–7.5 by 2–3.2 mm, tip acuminate; base broadly attached; otherwise as the median sepal. *Petals* not recurved, not falcate, obrhomboid, oblique, 1.5–2.2 by 1.5–2.3 mm, top rounded to truncate, tip apiculate; margins entire, glabrous, the upper either with an obtuse to acute tooth or a distinct sinus near the base or not; base (rather) narrowly attached; rather thin; glabrous; not keeled abaxially. *Lip* either slightly recurved or not, broadly ovate to panduriform in outline (not spread), 1.5–2 by 1–1.8 mm, tip rounded to slightly retuse; margins entire, very finely papillose; thick; adaxially without ridges, with a slightly concave basal part, without a median slit, surface glabrous to very finely papillose; abaxially with a very weak, rounded ridge towards the base or not, surface about glabrous. *Column* from the base of the free part up to the tip of the stelidia 1.8–2.5 mm long, with distinct, triangular, not falcate, bifid (the upper tooth may be much smaller than the lower) stelidia; column with distinct, obliquely forwards directed, rounded wings along its adaxial margins just below the base of the stelidia which continue towards the column foot as narrow seams; stigma without a basal callus. *Anther* 0.7–1 by 0.7–1 mm; front drawn out, thin, not concave; front margin retuse to truncate, entire, glabrous; basally weakly emarginate (see note below var. joyceae); abaxially with or without a rather distinct, obtuse, colliculate ridge near the base of the anther which does not overtop the front margin, surface colliculate but glabrous towards the top. *Pollinia* 4 (see note below var. *joyceae*, otherwise not seen). *Fruit* obovoid, up to 0.5 by 0.4 cm, without a stalk.

KEY TO THE SUBSPECIES

1 - Median sepal 3.9–4.5 mm long.................................ssp. **bidenticulatum** (a)
- Median sepal c. 6 mm long (one specimen seen).......................ssp. **joyceae** (b)

34a - ssp. bidenticulatum – Fig. 50.

Bulbophyllum bidenticulatum J.J. Vermeulen, Bull. Soc. R. Bot. Belg. 54 (1984) 144.–Type: Morton & Gledhill 3157 (Holo WAG, iso K).

Bulbophyllum bifarium auct. non Hook. f.: Ake Assi, Et. Fl. C. Iv. (1964) 242; Summerh. in F.W.T.A. ed. 2, 3 (1968) 239, p.p.; Johansson, Acta Phyt. Suec. 59 (1974) 29; Adam, Fl. Descr. Mt. Nimba 5 (1981) 1672, pl. 905.

Bulbophyllum nigericum auct. non Summerh.: Summerh. in F.W.T.A. ed. 2, 3 (1968) 239, p.p.

Rhizome 0.9–1.2 mm diam. Pseudobulbs 0.7–2 cm distant, ovoid or ellipsoid to about orbicular, 0.8–1.8 cm long. Leaves (1.6 –) 3–7 by 0.4–0.7 cm. Peduncle scales up to 7–8 mm long. Floral bracts 4.9–9 by 3–5 mm. Median sepal 3.9–4.5 by 2.5–3.2 mm, basally broadly attached. Lateral sepals 4–6 by 2–2.3 mm. Petals 1.5–2 by 1.5–2 mm, top part truncate, tip apiculate. Lip 1.5–1.7 by 1–1.2 mm, rather thick, top part not recurved. Column 1.8–2 mm long, stelidia deeply bifid.

Colours. Pseudobulbs pale green, often spotted or suffused purple. Leaves light green, occasionally suffused purple. Floral bracts straw coloured to orange red. Sepals pale purplish to almost white, stained with slightly darker purple. Petals white, with dark purple spots. Lip yellow. Column white. Occasionally specimens with entirely white flowers occur.

Ecology. Lower montane forest at 900–1200 m alt. Epiphyte in the outer part of the crown of trees, in minor humus deposits (Johansson, 1974, sub *Bulbophyllum bifarium*). Also lithophytic on boulders in open vegetation. Often exposed. Flowering (III, V) X–XII.

Distribution. GUIN. (2 coll.). S.L.: Tingi Hills (1), Loma Mts. (2). LIB.: Nimba Mts. (2). IV.C.: Mt. Tonkoui, Mt. Momi (4). Material seen from AMES, K, P, UPS, WAG.

34b - ssp. joyceae J.J. Vermeulen ssp. nov. – Fig. 51.

Bulbophyllum bidenticulatum J.J. Vermeulen ssp. *joyceae* J.J. Vermeulen, ssp. nov. A ssp. bidenticulato in floribus maioribus differt.–Type: Stewart 1193 (K).

Bulbophyllum cf. *bifarium* Hook. f.: Stewart in Agnew, Upl. Ken. W. Fl. (1974) 763.
Bulbophyllum bifarium auct. non Hook. f.: Cribb, in F.T.E.A., Orch. 2 (1984) 314.

Rhizome 1–3 mm diam. Pseudobulbs 1–3 cm distant, ovoid to narrowly ovoid, 1.5–4 cm long. Leaves 3.7–11 by 0.5–1 cm. Peduncle scales up to 12–14 mm long. Floral bracts 9.5–10 by c. 5 mm. Median sepal c. 6 by 4 mm, basally rather narrowly attached. Lateral sepals c. 7.5 by 3.2 mm. Petals c. 2.2 by 2.3 mm, top part rounded, tip apiculate. Lip c. 2 by 1.8 mm, thick, top part slightly recurved. Column c. 2.5 mm long, stelidia bifid, but with a very inconspicuous upper tooth.

Colours. Pseudobulbs and leaves olive green. Floral bracts reddish, turning yellow. Sepals and petals yellowish, with purple spots. Lip red. Column yellow with purple spots.

Ecology. Montane forest at 2100 m alt. Flowering VII (once recorded).

Distribution. KEN.: Nyanza K5 (1 coll.).

Note. 1 - All flowers studied of the only available specimen have misshapen anthers and pollinia: the anther is immovably attached to the column by a thick strip of tissue; the thecae are cup-shaped towards the base, and the four pollinia are grown together at one end into two pairs so that two deeply cleft pollinia seem to be present.

It is not clear whether the specimen is a monstrosity or whether this is normal within the subspecies. The structure mentioned above does not at all impede pollination. Most flowers on the lower part of one inflorescence develop distinctly swollen ovaries with the perianth still fresh, only the lip shriveled.

2 - This subspecies is named after Mrs. Joyce Stewart (K) who discovered it.

35 - Bulbophyllum nigericum Summerh. - Fig. 52.

Bulbophyllum nigericum Summerh., Kew Bull. 16 (1962) 307; id. in F.W.T.A. ed. 2, 3 (1968) 239, p.p.– Type: King 124 (K).

Roots 0.5–1 mm diam. *Rhizome* 1.5–2.5 mm diam., nodes slightly swollen. *Rhizome scales* only tubular at their base, tip acute to acuminate; fibrous, with many nerves close together; keeled abaxially. *Pseudobulbs* 2-leafed, ovoid, 0.8–2.5 cm apart, 1.3–2.7 by 0.7–1.5 cm, slightly flattened, obtusely 4-angled. *Petiole* 1–2 mm long. *Leaf blade* lanceolate to linear-lanceolate, 3–7 by 0.5–1.2 cm, tip obtuse, slightly emarginate and oblique; coriaceous, rather thick; midrib prominent abaxially. *Inflorescence* 8–23 cm long, 7–30-flowered. *Peduncle* erect, swollen towards the tip, 4–9.5 cm by c. 3.5 mm, elliptic in section, glabrous, with 9–14 scales. *Peduncle scales* tubular, the longest 7–12 mm long; tip acuminate; fibrous, with many nerves close together, keeled and glabrous abaxially. *Rhachis* erect, swollen, 4-angled in section, with two concave sides from which the flowers arise, slightly zig-zag bent, 3.5–14 cm by c. 4 mm, edges sharp, straight, entire, glabrous; surface glabrous. *Floral bracts* spreading, concave, ovate, 7–12 by 4–6.5 mm, tip acute to slightly acuminate; base broadly attached; fibrous, with many nerves close together; keeled and glabrous abaxially. *Flowers* distichous, in a rather lax raceme, recurved, many open simultaneously, fully open. *Pedicel and ovary* 1.5–2 mm long, glabrous. *Median sepal* not recurved, concave, ovate, 4.5–7.2 by 3–4.5 mm, tip acuminate to cuspidate; margins entire, glabrous; base broadly attached; thin; glabrous; not keeled abaxially. *Lateral sepals* free, reflexed at the base, not concave, not falcate, ovate-oblong, oblique, 6–9 by 2.5–4.5 mm, otherwise as the median sepal. *Petals* not recurved, not falcate, rhomboid, oblique, 2–4 by 1.5–3 mm, top truncate, tip cuspidate to caudate (cauda 0.2–2 mm long); margins entire, glabrous, the upper with a distinct sinus near the base; base narrowly attached; thin; glabrous; not keeled abaxially. *Lip* recurved, either with a reflexed top part or not, slightly panduriform in outline (not spread), 2–4 by 1.3–3 mm, tip broadly rounded; margins entire, glabrous; thick; surface glabrous; adaxially without ridges, with a slightly concave basal part, without a median slit; abaxially with a weak, broadly rounded ridge near the base. *Column* from the base of the free part up to the tip of the stelidia 1.8–4 mm long, with distinct, gradually tapering, not falcate, acuminate stelidia of 0.9–2 mm long which may have a weak, obliquely forwards directed, obtuse tooth along their upper margins near the base; column with distinct, broadly rounded wings along its adaxial margins just below the base of the stelidia; stigma without a basal callus. *Anther* c. 1.2 by 1 mm, front hardly drawn out, thin, not concave, front margin truncate, entire, glabrous; basally hardly emarginate; abaxially with a very weak, rounded, colliculate ridge near the base, surface colliculate. *Pollinia* 4; all of about equal length; the inner pair narrowly ellipsoid, distinctly flattened; the outer pair narrowly obovoid, flattened. *Fruit* obovoid, up to 0.8 by 0.6 cm, without a stalk.

Colours. Sepals very pale green or yellow, occasionally stained purple. Petals purplish, spotted deep purple, or yellow. Lip bright yellow, suffused purplish brown near the base, or orange. Column purple, stelidia and wings along the adaxial margins white.

Ecology. Lithophytic at 1000–2050 m alt. Flowering X–XII.

Distribution. NIG.: Plateau P. (4 coll.). CAM., W. part (3). Material seen from K, P.

GROUP 5 (spp. 36 – 53; '*Megaclinium*')

Pseudobulbs 1–2 (– 3)-leafed. Inflorescence a raceme, 2–many-flowered. Rhachis either flattened or not, when flattened flowers inserted along the midnerve. Flowers distichous. Sepals either glabrous or with fine, dark hairs, the lateral ones not connate along the upper margins, either reflexed or not, not appressed against the rhachis. Lip usually thick, very soft and easy to compress (very thin when dried), lateral lobes absent, margins glabrous. Stelidia short, acute to bifid. Anther either with or without a not dorsoventrally flattened protrusion overtopping the front margin.

36 - **Bulbophyllum renkinianum** (Laurent) De Wild. – Fig. 53.

Bulbophyllum renkinianum (Laur.) De Wild., Pl. Beq. 1 (1921) 95; Summerh., Kew Bull. 12 (1957) 122.– *Megaclinium renkinianum* Laur., Bull. Agr. Congo Belge 5 (1914) 210.–Syntypes: Laurent 1772 (Lecto BR), 281 (BR, n.v.), 2054 (BR).

Bulbophyllum cercoglossum Summerh., Kew Bull. 8 (1954) 584.–Type: Le Testu 5772 (Holo K; iso BM, BR, P).

Roots 0.4–2 mm diam. *Rhizome* 2.8–5 mm diam., nodes not or hardly swollen. *Rhizome scales* tubular, tip obtuse to subacute; membranous, with many nerves; not or hardly keeled abaxially. *Pseudobulbs* 2-leafed, very narrowly ovoid to about cylindrical, 5.5–10.5 cm apart, 5.5–10 by 0.9–2 cm, slightly flattened, inconspicuously (?) 4-angled. *Petiole* 4–20 mm long. *Leaf blade* broadly lanceolate, maximum width occasionally slightly above the middle, 9.5–20 by 3–6 cm, tip acute to shortly acuminate, not emarginate, either slightly oblique or not; coriaceous, thick; midrib prominent abaxially. *Inflorescence* 20.5–40 cm long, 28–48-flowered. *Peduncle* about erect, slightly swollen and flattened towards the tip, 14–21.5 cm by 1.8–3 mm, orbicular in section, glabrous, with 8–11 scales. *Peduncle scales* tubular, the longest 6–9 mm long; tip about acute; membranous, with some nerves; hardly keeled and glabrous abaxially. *Rhachis* slightly nodding, widened and flattened, bladelike, 9–19 cm by 27–50 mm, edges sharp, about straight, irregularly dentate, locally slightly finely papillose; surface glabrous to very finely papillose, occasionally with fine dark hairs. *Floral bracts* reflexed, triangular, 1.8–3 by 1.4–2.2 mm, tip about acute; margins distinctly rolled up; base broadly attached; membranous, without nerves; not keeled and glabrous abaxially. *Flowers* distichous, placed along an excentric line on the rhachis, 4.5–7 mm apart, measured along one side of the rhachis, the lower up to 6–16 mm apart, recurved, secund, some open simultaneously, not fully open. *Pedicel and ovary* 6–8 mm long, about glabrous. *Median sepal* either with a recurved tip or not, slightly concave, ovate-lanceolate, 7–9.2 by 1.7–2.7 mm, tip acuminate; margins entire, glabrous; base rather broadly attached; rather thin; glabrous adaxially; not keeled and glabrous or with some fine dark hairs abaxially. *Lateral sepals* free, slightly recurved, not concave, falcate, obliquely ovate, 3.8–5 by 1.7–2.7 mm, tip caudate, (cauda 0.9–1.8 mm); base broadly attached; adaxially slightly and very finely papillose; otherwise as the median sepal. *Petals* not or hardly recurved, falcate, linear-lanceolate, oblique, 4–6 by 0.3–1 mm, tip acute; margins entire, glabrous; base broadly attached; rather thin; glabrous, not keeled abaxially. *Lip* slightly recurved near the base, top part reflexed, orbicular to broadly elliptic in outline (not spread), 3–4 by 2–3 mm, top part cuspidate to caudate (cauda 1–2 mm), tip acute; margins finely and irregularly denticulate, glabrous; rather thin, soft and easy to compress, with a rather thick base; surface glabrous; adaxially without ridges, with a slightly concave basal part, without a median slit; abaxially with a weak, rounded ridge near the base. *Column* from the base of the free part up to the tip of the stelidia 1.5–2 mm long, with distinct, narrowly triangular, upwards bent, subacute stelidia of c. 0.4 mm long; column with distinct, obliquely forwards directed, towards the top distinctly bent upwards (protruding far beyond the stelidia), about obliquely ax-shaped, truncate (with a slightly irregular front margin) wings along its adaxial margins, just below the base of the stelidia, which continue towards the column foot as narrow seams; stigma without a basal callus. *Anther* c. 1 by 1 mm; front distinctly drawn out, thin, not concave; front margin rounded with an apiculate tip, denticulate, glabrous; basally weakly emarginate; abaxially with a rather weak, obtuse, about glabrous ridge near the top of the anther which does not overtop the front margin, surface about glabrous. *Pollinia* 4; the small ones less than half as long as the large ones, about transversely placed, about semiorbicular, distinctly flattened on either side; the large ones ellipsoid, flattened; a stipes has not been observed. *Fruit* not seen.

Colours. Rhachis dark green with dark purple stains. Sepals yellow adaxially, finely spotted with purple; abaxially purple. Lip greenish with violet.

Ecology. Lowland forest, no altitude given. Epiphytic. Flowering observed in XI.

Distribution. GAB. (1 coll.). ZAI.: Forestier Central Z6 (3). Material seen from BM, BR, K.

37 - **Bulbophyllum dolabriforme** J.J. Vermeulen - Fig. 54.

Bulbophyllum dolabriforme J.J. Vermeulen, Kew. Bull. 42 (1987) 263–Type: Smith 570 (K).

Roots c. 0.5 mm diam. *Rhizome* c. 2 mm diam., nodes not or hardly swollen. *Rhizome scales* tubular, tip acute or shortly acuminate; membranous, with some nerves; keeled abaxially towards the tip. *Pseudobulbs* 1-leafed, ovoid, 0.3–0.6 cm apart, 1–1.8 by 0.6–0.9 cm, moderatedly flattened, obtusely 4-angled. *Petiole* 2–3 mm long. *Leaf blade* oblong to broadly lanceolate, 3.5–6.5 by 1.4–2.0 cm, tip obtuse, not or hardly emarginate, not oblique; coriaceous, thick; midrib hardly prominent abaxially. *Inflorescence* 4–7.5 cm long, 2–5-flowered. *Peduncle* about patent, wiry, swollen towards the tip, 2–4 cm by 0.5–0.7 mm, elliptic in section, glabrous, with c. 4 scales. *Peduncle scales* tubular, the longest c. 3 mm long; tip acute to acuminate; membranous, without nerves; keeled and glabrous abaxially. *Rhachis* patent, slightly swollen, spindle shaped, slightly flattened below the attachment of the flowers, 2–5 cm by 1.5–2 mm, glabrous. *Floral bracts* about spreading, narrowly ovate to broadly ovate-oblong, 3–4 by 1.5–2 mm, tip acuminate; base broadly attached; membranous, without nerves; keeled towards the tip and glabrous abaxially. *Flowers* distichous but the rows are not quite opposite (the whole rhachis is slightly twisted), in a lax raceme, 16–17 mm apart, measured along one side of the rhachis, the lower up to c. 17 mm apart, spreading, slightly secund, about all open simultaneously, not much open. *Pedicel and ovary* 3–4 mm long, glabrous. *Median sepal* only recurved at the very tip, slightly concave, oblong, c. 19.5 by 8.5 mm, tip cuspidate; margins entire, very finely papillose and with some ciliae; base rather broadly attached; thin; glabrous adaxially; not keeled and with scattered papillae and short ciliae abaxially. *Lateral sepals* free, either slightly recurved or not, not concave, falcate, rhomboid, oblique, c. 10 by 8 mm, tip acuminate; upper margin ciliate in the top half, lower margin glabrous; base broadly attached; with scattered ciliae abaxially in the top half; otherwise as the median sepal. *Petals* not recurved, slightly falcate, obovate-lanceolate, oblique, c. 8 by 2.2 mm, tip acuminate, with a seta; margins entire and glabrous towards the base, irregular and ciliate towards the tip; base rather broadly attached; thin; glabrous; not keeled abaxially. *Lip* not recurved, ovate in outline (not spread), slightly constricted near the tip, c. 5.5 by 3.5 mm, tip obtuse; margins entire, glabrous; thick but soft and easy to compress; surface glabrous; adaxially with two distinct, short, obtuse ridges near the tip which diverge towards the base of the lip, with a not or hardly concave basal part, without a median slit; abaxially with a rather distinct, very broad, truncate ridge over about the entire length of the lip which is marked off by two fine, lateral furrows. *Column* from the base of the free part up to the tip of the stelidia c. 2.5 mm long, with distinct, triangular, not falcate, subacute stelidia of c. 0.4 mm long; column with distinct, ax-shaped, truncate, finely papillose wings along its adaxial margins just below the base of the stelidia; stigma without a basal callus. *Anther* c. 0.5 by 0.4 mm; front slightly drawn out, thin, not concave; front margin rounded, irregularly denticulate, glabrous; basally distinctly emarginate; abaxially with a weak, slightly colliculate ridge near the top of the anther which does not overtop the front margin, surface glabrous but colliculate towards the top. *Pollinia* 4, about ellipsoid; the small ones about half as long as the large ones, distinctly flattened; the large ones flattened; a stipes has not been observed. *Fruit* not seen.

Colours. Pseudobulbs and leaves dark green, suffused with red. Inflorescence pale green, finely spotted pale red.

Ecology. Lowland forest, no altitude given. Epiphytic. Flowering VI in greenhouse K.

Distribution. NIG.: Bendel P. (1 coll.).

38 - Bulbophyllum curvimentatum J.J. Vermeulen - Fig. 55.

Bulbophyllum curvimentatum J.J. Vermeulen, Bull. Soc. R. Bot. Belg. 54 (1984) 137.–Type: Sanford 4035 (K).

Roots 0.4–0.8 mm diam. *Rhizome* 1–2 mm diam., nodes not or hardly swollen. *Rhizome scales* not seen. *Pseudobulbs* 1-leafed, ovoid, 0.3–0.6 cm apart, 0.6–1 by 0.5–0.6 cm, moderately flattened, 4-angled. *Petiole* c. 1 mm long. *Leaf blade* oblong to lanceolate, 2.5–3.5 by 0.6–1.3 cm, tip acute to subacute, neither emarginate nor oblique; coriaceous, thick; midrib slightly prominent abaxially. *Inflorescence* c. 12 cm long, 11–14-flowered. *Peduncle* erect to patent, wiry, 6–8 cm by 0.5–0.8 mm, orbicular in section, glabrous, with c. 6 scales. *Peduncle scales* tubular, the longest c. 4 mm long; tip acute; membranous, without nerves; keeled towards the tip and glabrous abaxially. *Rhachis* erect to slightly arching, terete, slightly zig-zag bent, 4.5–6 cm by 0.5–0.8 mm, glabrous. *Floral bracts* spreading, triangular, c. 1.8 by 1.4 mm, tip acute; base broadly attached; nerves; slightly keeled and glabrous abaxially. *Flowers* distichous, c. 10 mm apart, measured along one side of the rhachis, the lower up to c. 13 mm apart, spreading, (?) secund, probably cleistogamous (see note below). *Pedicel and ovary* not seen. *Median sepal* not recurved, concave, oblong, c. 5.3 by 2.5 mm, tip caudate (cauda incurved, c. 1.7 mm long); margins entire, glabrous; base broadly attached; thin, with slightly protruding nerves; glabrous adaxially; not keeled and with some fine dark hairs abaxially. *Lateral sepals* adhering (see note below), not recurved, not concave, falcate, deltoid, oblique, c. 3.8 by 3.5 mm, tip caudate (cauda c. 1.2 mm); otherwise as the median sepal. *Petals* not recurved, slightly falcate, broadly lanceolate, somehat constricted in the basal half, oblique, c. 3 by 0.8 mm, tip acuminate; margins denticulate towards the tip, glabrous; base broadly attached; thin; glabrous; keeled towards the base abaxially. *Lip* recurved, ovate in outline (not spread), c. 2 by 1.4 mm, tip obtuse; margins entire, glabrous; thick but soft and easy to compress; adaxially without ridges, with a slightly concave basal part, without a median slit, surface very finely papillose; abaxially with weak, rounded ridge near the base, surface bout glabrous. *Column* from the base of the free part up to the tip of the stelidia c. 1.5 mm long, with distinct, triangular, not falcate, acute stelidia of 0.2 mm long; column with distinct, deltoid, obtuse wings along its adaxial margins just below the base of the stelidia; stigma without a basal callus. *Anther* c. 0.8 by 0.6 mm; front slightly drawn out, thin, not concave; front margin truncate, entire, papillose; basally weakly emarginate; abaxially with a rather weak, obtuse, colliculate ridge near the base of the anther which does not overtop the front margin, surface colliculate. *Pollinia* not seen. *Fruit* obovoid, up to 0.8 by 0.4 cm, without a stalk.

Colours. Leaves green, suffused with dark red-purple. Petals with a few pale red dots (observed in dried material). Lip probably white. Fruiting XII.

Ecology. Abandoned coffee plantation at 1000 m alt. Epiphytic.

Distribution. EQ.GUIN.: Bioko (1 coll.).

Note. On the only specimen available, a fruiting one, the perianth is persistent on the mature fruits. All sepals are adherent along their margins, from base to tip. This, together with the fact that fruits are rather abundantly produced, indicates the species is cleistogamous.

39 - **Bulbophyllum colubrinum** (Rchb. f.) Rchb. f. - Fig. 56.

Bulbophyllum colubrinum (Rchb. f.) Rchb. f., Walp. Ann. 6 (1861) 257 ('*Bolbophyllum*'); De Wild., Pl. Beq. 1 (1921) 82; Summerh., Kew Bull. 6 (1952) 469; id., 16 (1962) 307; Summerh. in F.W.T.A. ed. 2, 3 (1968) 241; Geerinck, Bull. Soc. R. Bot. Belg. 109 (1976) 176.-*Megaclinium colubrinum* Rchb. f., Bonplandia 3 (1855) 223; Durand & Schinz, Consp. Fl. Afr. 5 (1895) 15; Finet, Bull. Soc. Bot. Fr. 57 (1910) 241; Rolfe in F.T.A. 7 (1897) 40.-*Phyllorchis colubrina* (Rchb. f.) Kuntze, Rev. Gen. Pl. 2 (1891) 677, incorr. name.-Type: Booth s.n. (W, n.v.; fragment in K).

Bulbophyllum inaequale Rchb. f., Flora 69 (1886) 553 (*Megaclinium inaequale* Rchb. f., nom. alt.); Durand & Schinz, Consp. Fl. Afr. 5 (1895) 12; De Wild., Pl. Beq. 1 (1921) 111.-*Phyllorchis inaequalis* (Rchb. f.) Kuntze, Rev. Gen. Pl. 2 (1891) 677, incorr. name.-Type: Paris cult. Hort. (?, n.v.).

Non *Bulbophyllum inaequale* (Bl.) Lindl., Gen. & Sp. Orch. Pl. (1830) 49.

Bulbophyllum imschootianum (Rolfe) De Wild., Pl. Beq. 1 (1921) 86; Summerh. in F.W.T.A. ed. 1, 2 (1936) 437.-*Megaclinium imschootianum* Rolfe, Kew Bull. 1895 (1895) 8; id. in F.T.A. 7 (1897) 39.-Type: Van Imschoot s.n. (K).

Bulbophyllum decipiens Schltr, Westafr. Kautchuk Exp. (1901) 281, nom. nud.; id., Engl., Bot. Jahrb. 38 (1905) 14; De Wild., Pl. Beq. 1 (1921) 83; Schltr in Mildbraed, 2e Deutsch. Zentr. Afr. Exp. 1910-11, 2 (1922) 30.-Type: Schlechter 12789 (B, lost).

Bulbophyllum gabunense Schltr, Westafr. Kautchuk Exp. (1901) 281, nom. nud.; id., Engl., Bot. Jahrb. 38 (1905) 15; De Wild., Pl. Beq. 1 (1921) 86.-Type: Schlechter 12819 (B, lost).

Bulbophyllum makakense Hansen, Am. Orch. Soc. Bull. 28 (1959) 278, pl., phot.-Type: Birket Smith s.n. (C, n.v.).

Roots 0.3–1.8 mm diam. *Rhizome* 2–5 mm diam., nodes not or hardly swollen. *Rhizome scales* tubular; membranous, with many nerves; not keeled abaxially. *Pseudobulbs* 1-leafed, about ellipsoid, 2.7–6 cm apart, 2.5–6.5 by 1–1.9 cm, hardly to moderately flattened, rather sharply 2–4-angled. *Petiole* 4–10 mm long. *Leaf blade* lanceolate to linear-lanceolate, 11–22 by 1.4–3.6 cm, tip obtuse, emarginate, oblique; coriaceous, thick; midrib slightly prominent abaxially. *Inflorescence* 14.5–51 cm long, 20–100-flowered. *Peduncle* erect, slightly swollen and flattened towards the tip, 8.5–22.5 cm by 1.8–3.2 mm, orbicular in section, glabrous, with 7–10 scales. *Peduncle scales* tubular, the longest 5.5–10 mm long; tip obtuse to acute; membranous, either with some nerves or not; hardly keeled and glabrous abaxially. *Rhachis* erect or slightly arching, widened and flattened, often bladelike, 6.5–38 cm by 4–24 mm, edges sharp, straight, entire to deeply and irregularly dentate, glabrous; surface glabrous. *Floral bracts* spreading to reflexed, oblong to broadly lanceolate, 3.5–6 by 1.5–3 mm, tip acuminate; margins distinctly folded inwards; base broadly attached; membranous, without nerves (apart from a fine venation); slightly keeled and glabrous abaxially. *Flowers* distichous, placed along a median to distinctly excentric line on the rhachis, 6.2–13 mm apart, measured along one side of the rhachis, the lower up to 12–32 mm apart, spreading, slightly secund, some open simultaneously, not fully open. *Pedicel and ovary* 2.5–3.8 mm long, glabrous. *Median sepal* with only a recurved tip, slightly concave, lanceolate to linear-lanceolate, 6–10.8 by 1.9–2.3 mm, tip cuspidate; margins entire, glabrous; base broadly attached; rather thin with slightly protruding nerves; glabrous; not keeled abaxially. *Lateral sepals* free, slightly recurved, not concave, falcate, ovate, oblique, 5–8.5 by 3.8–4.4 mm, tip cuspidate to caudate (cauda 1.5–3.5 mm); otherwise as the median sepal. *Petals* not recurved, falcate, (obovate-)oblong, oblique, 3–5 by 1–1.9 mm, tip acute to acuminate; margins entire to slightly irregular towards the tip, glabrous; base rather narrowly attached; thin; glabrous; not keeled abaxially. *Lip* recurved, elliptic to ovate in outline (not spread), 2.5–3 by 1.4–1.7 mm, tip obtuse; margins entire or irregularly denticulate towards the base, glabrous; thick but soft and easy to compress; surface glabrous; adaxially without ridges, with a slightly concave basal part, without a median slit but slightly channeled in the median part; abaxially with a weak, rounded ridge in the basal half. *Column* from the base of the free part up to the tip of the stelidia 2–3.8 mm long, with distinct, triangular, not falcate, slightly bifid to acute stelidia of c. 0.4 mm long; column with distinct, obliquely forwards directed, about semiorbicular, rounded wings along its adaxial margins just below the base of the stelidia which continue towards the column foot as narrow seams; stigma without a basal callus. *Anther* c. 1 by 1 mm; front drawn out, thin, not concave; front margin truncate, entire, glabrous; basally weakly emarginate; abaxially without a ridge, surface very finely colliculate. *Pollinia* 4; the small ones much more than half as long as the large ones, slightly transversely placed, obovoid, distinctly flattened on either side; the large ones ellipsoid, flattened on one side; a stipes has not been observed. *Fruit* ellipsoid, up to 1.8 by 0.7 cm, without a stalk.

Colours. Pseudobulbs (olive) green to yellowish brown. Leaves dark green. Rhachis pale to dark green, with purple spots, sometimes entirely purple. Sepals and petals yellowish, stained or spotted with purple, or entirely purple. Lip yellowish, stained purple.

Ecology. Lowland forest and lower montane forest up to c. 1000 m alt. Epiphytic. Flowering (I) VI–XI.

Distribution. S.L. (1 coll.). IV.C. (5). GHANA (1). NIG.: Ondo P. (1). CAM.: West, S. part; Coast (3). GAB. (4). CONGO (2). ZAI.: Forestier Central Z6 (5). ANG. (1). Material seen from BM, BR, E, K, P, W, WAG.

40 - Bulbophyllum maximum (Lindl.) Rchb. f. – Fig. 57–58; Plate 7a.

Bulbophyllum maximum (Lindl.) Rchb. f., Walp. Ann. 6 (1861) 259 ('*Bolbophyllum*'); Schltr, Die Orch. (1914) 328; De Wild., Pl. Beq. 1 (1921) 93; Summerh. in F.W.T.A. ed. 1, 2 (1936) 438; Exell, Cat. Vasc. Pl. S. Tomé (1944) 326; Summerh., Kew Bull. 12 (1957) 120; Ake Assi, Et. Fl. C. Iv. (1964) 245; Summerh. in F.W.T.A. ed. 2, 3 (1968) 241; Johansson, Acta Phyt. Suec. 59 (1974) 32; Adam, Fl. Descr. Mt. Nimba 5 (1981) 1680, pl. 912; c830 52; Segerbaeck, Orch. Nig. (1983) 67, col. phot.; Cribb, in F.T.E.A., Orch. 2 (1984) 317.–*Megaclinium maximum* Lindl., Gen. and Sp. Orch. Pl. (1830) 47; Durand & Schinz, Consp. Fl. Afr. 5 (1895) 16; Rolfe in F.T.A. 7 (1897) 38; De Wild., Et. Fl. B. et Moy. Congo 1 (1904) 116; Durand & Durand, Syll. Fl. Congol. (1909) 518.–*Phyllorchis maxima* (Lindl.) Kuntze, Rev. Gen. Pl. 2 (1891) 677, incorr. name.–Type: Smeathman s.n. (Holo BM; iso S, UPS).

Non *Megaclinium maximum* auct.: Lindl., Bot. Reg. 23 (1837) pl. 1959; Hook., Cent. Orch. Pl. (1849) 15, pl. 3 (*Megaclinium flaccidum* below plate) (= both *Bulbophyllum calyptratum*).

Non *Cirrhopetalum maximum* Ridl., J. Bot. 38 (1900) 70.

Bulbophyllum oxypterum (Lindl.) Rchb. f., Walp. Ann. 6 (1861) 258 ('*Bolbophyllum*'); Schltr, Westafr. Kautchuk Exp. (1901) 282; De Wild., Pl. Beq. 1 (1921) 94; Summerh., Kew Bull. 12 (1957) 121; id. in F.W.T.A. ed. 2, 3 (1968) 241; Piers, Orch. E. Afr. (1968) 105; Morris, Proc. Linn. Soc. London 179 (1968) 63; id., Ep. Orch. Mal. (1970) 41; Stewart & Campbell, Orch. Trop. Afr. (1970) 62, col. phot. 19; Stewart in Agnew, Upl. Ken. W. Fl. (1974) 763; Grosvenor, Excelsa 6 (1976) 79; Williamson, Orch. S. Centr. Afr. (1977) 132, excl. col. phot. (= *B. injoloense* ssp. *pseudoxypterum*); Ball, S. Afr. Ep. Orch. (1978) 82, col. phot.; Stewart & Hennesy, Orch. Afr., sel. rev. (1981) 88, pl. 21; Segerbaeck, Orch. Nig. (1983) 65.-*Megaclinium oxypterum* Lindl., Bot. Reg. 25 (1839) misc. 14; id., J. Linn. Soc. Bot. 6 (1862) 128; Durand & Schinz, Consp. Fl. Afr. 5 (1895) 16; Rolfe in F.T.A. 7 (1897) 39.-*Phyllorchis oxyptera* (Lindl.) Kuntze, Rev. Gen. Pl. 2 (1891) 677, incorr. name.-Type: Loddiges s.n. (K).

Non *Bulbophyllum oxypterum* auct: Cufodontis, En. Pl. Aeth. Sperm. (1972) 1618 (= *Bulbophyllum scaberulum* var. *scaberulum*).

Megaclinium purpuratum Lindl., J. Linn. Soc. Bot. 6 (1862) 128; Hook. f., Bot. Mag. 97 (1871) pl. 5936; Durand & Schinz, Consp. Fl. Afr. 5 (1895) 16.-Type: Barter 1854 (K).

Bulbophyllum platyrhachis (Rolfe) Schltr, Engl., Bot. Jahrb. 38 (1905) 15, sub *Bulbophyllum decipiens*; Schltr, Die Orch. (1914) 328; De Wild., Pl. Beq. 1 (1921) 94; Piers, Orch. E. Afr. (1968) 106; Morris, Proc. Linn. Soc. London 179 (1968) 63; id., Ep. Orch. Mal. (1970) 40; La Croix et al., Orch. Mal. 1 (1983) 55; Cribb, in F.T.E.A., Orch. 2 (1984) 319.-*Megaclinium platyrhachis* Rolfe in F.T.A. 7 (1897) 43; Hemsley in Bot. Mag. 130 (1904) pl. 7946.-Type: O'Brien s.n. (K).

Non *Bulbophyllum platirachis* De Wild., Miss. Laur. (1906) 223, pl. 55 (= *Bulbophyllum acutebracteatum* var. *acutebracteatum*).

Non *Bulbophyllum platyrrhachis* Ridl., J. Fed. Mal. St. Mus. 8 (1917) 96.

Bulbophyllum djumaense (De Wild.) De Wild., Pl. Beq. 1 (1921) 83.-*Megaclinium djumaense* De Wild., Not. Pl. Ut. et Int. Fl. Congo 1 (1903) 124; id., Et. Fl. B. et Moy. Congo 1 (1904) 116; Durand & Durand, Syll. Fl. Congol. (1909) 517.-Type: Gillet 2900 (BR).

Bulbophyllum ciliatum Schltr, Engl., Bot. Jahrb. 38 (1906) 156; De Wild., Pl. Beq. 1 (1921) 81; Schltr, Fedde, Rep. Beih. 68 (1932) pl. 65.-*Bulbophyllum cyrtopetalum* Schltr, Fedde, Rep. 9 (1911) 165.-*Bulbophyllum moireanum* Hawkes, Lloydia 19 (1956) 93, nom. superfl.-Type: Schlechter 15758 (B, lost).

Non *Bulbophyllum ciliatum* (Bl.) Lindl., Gen. and Sp. Orch. Pl. (1830) 48.

Megaclinium oxypterum Lindl. var. *mozambicense* Finet, Not. Syst. 1 (1910) 169.-*Bulbophyllum oxypterum* (Lindl.) Rchb. f. var. *mozambicense* (Fin.) De Wild., Pl. Beq. 1 (1921) 94.-Type: Le Testu 668 (Holo P; iso BM).

Bulbophyllum nyassanum Schltr, Engl., Bot. Jahrb. 53 (1915) 571; id., Fedde, Rep. Beih. 68 (1932) pl. 68.-Type: Stolz 1784 (Holo B, lost; iso BM, BR, K, P, S, Z).

Bulbophyllum sp. Schltr, Engl., Bot. Jahrb. 53 (1915) 572.-Specimens: Stolz 454 (G, K, L, S, U, WAG, Z), 667 (K, WAG, Z).

Bulbophyllum djumaense (De Wild.) De Wild. var. *grandifolium* De Wild., Pl. Beq. 1 (1921) 83.-Type: Seret 568 (BR).

Bulbophyllum subcoriaceum De Wild., Rev. Zool. Afr. 9, suppl. bot. (1921) 33 (*Megaclinium subcoriaceum* De Wild., nom. alt.); id., Pl. Beq. 1 (1921) 98.-Type: Bequaert 2310 (BR).

Roots 0.3–2.2 mm diam. *Rhizome* 2.2–11 mm diam., nodes not or hardly swollen. *Rhizome scales* tubular, tip obtuse, acute or shortly acuminate; membranous, with many nerves; not or hardly keeled abaxially. *Pseudobulbs* 2 (– 3)-leafed, (narrowly) ovoid, occasionally about ellipsoid, 2–10 cm apart, 3.5–10 by 1–3 cm, slightly flattened, obtusely to sharply 3–5-angled, edges occasionally with large knobs. *Petiole* 1–12 mm long. *Leaf blade* oblong to linear-lanceolate, maximum width usually slightly above the middle, 3.8–20 by 1.3–5.5 cm, tip rounded to obtuse, emarginate, slightly oblique; coriaceous, thick; midrib hardly prominent abaxially. *Inflorescence* 15–90 cm long, 16–120-flowered. *Peduncle* usually about erect, swollen and flattened towards the tip, 6–42 cm by 1.7–4.7 mm, orbicular in section, glabrous, with 5–14 scales. *Peduncle scales* tubular, the longest 7–15 mm long; tip obtuse to acute; membranous, with some nerves; not or hardly keeled and glabrous abaxially. *Rhachis* erect to patent, widened and flattened, often bladelike, 6–56 cm by 8–50 mm, edges sharp or nearly so, straight to undulate (the whole rhachis may be slightly twisted), about entire to regularly or very irregularly dentate to crenulate, glabrous; surface glabrous. *Floral bracts* spreading to reflexed, elliptic to ovate, 2.5–7 by 2–4 mm, tip acute to acuminate; margins distinctly rolled backwards; base broadly attached; membranous, without nerves; keeled and glabrous abaxially. *Flowers* distichous, placed along a median to distinctly excentric line on the rhachis, 2.5–9 (– 11) mm apart, measured along one side of the rhachis, the lower up to 5–22 mm apart, recurved to spreading, either secund or not, some open simultaneously, not fully open. *Pedicel and ovary* 2–4 mm long, either with some fine dark hairs or not. *Median sepal* with a recurved top part, slightly concave, ovate-lanceolate, 4.5–8 by 1.2–2 mm, tip acute to slightly acuminate; margins entire, glabrous; base broadly attached; thick, thickened towards the tip; glabrous to very finely papillose adaxially; not keeled and glabrous or with some fine dark hairs abaxially. *Lateral sepals* free, slightly recurved, tip spreading to reflexed, not concave, falcate, about rhomboid to broadly ovate, oblique, 3.7–6.2 by 1.9–3.7 mm, tip cuspidate to caudate (cauda 1.5–3 mm); base rather broadly attached; otherwise as the median sepal. *Petals* not recurved, falcate, lanceolate, oblique, 1.8–3.2 by 0.3–0.8 mm, tip acute to acuminate; margins entire, either slightly irregular towards the tip or not, glabrous; base (rather) broadly attached; (rather) thin; glabrous; not keeled abaxially. *Lip* recurved, ovate-oblong in outline (not spread), either slightly constricted about half-way or not, 1.5–3 by 0.7–1.4 mm, tip rounded; margins entire towards the tip, lacerate towards the base, glabrous; thick but rather soft and easy to compress; surface glabrous; adaxially without ridges, with a slightly concave basal part, without a median slit; abaxially with a rather distinct, truncate to retuse ridge up to 2/3 of the lip. *Column* from the base of the free part up to the tip of the stelidia 1.5–2.2 mm long, with rather distinct, about triangular, not falcate, emarginate to acute stelidia of c. 0.2 mm long; column with distinct, either obliquely forwards directed or not, deltoid to semiorbicular, rounded to obtuse wings along its adaxial margins just below the base of the stelidia; stigma without a basal callus. *Anther* c. 0.8 by 0.7 mm; front not or hardly drawn out; front margin rounded, entire, glabrous or slightly papillose; basally weakly emarginate; abaxially without a ridge, surface colliculate, glabrous towards the margins. *Pollinia* 4, ellipsoid, flattened on either side; the small ones much more than half as long as the large ones; a stipes has not been observed. *Fruit* ellipsoid, up to 2 by 0.9 cm, without a stalk.

Colours. Pseudobulbs pale green or yellowish green. Leaves green. Rhachis (pale) yellowish or greenish, often spotted with purple, or entirely purple or purplish brown with fine, pale yellow dots. Sepals yellowish or greenish adaxially, top part often bright yellow, with or without purple or brownish stains or spots, or entirely purple or brownish with pale yellow dots. Petals yellowish, usually with purple or brownish stains or dots. Lip yellowish or cream coloured, finely spotted purple. Column greenish, with or without purple stains.

Ecology. Lowland primary and secondary forest, montane forest, savanna (Brachystegia) woodland, at 0–1500 m alt. Epiphyte in the central part of the crown of trees, on bark or in minor humus deposits (Johansson, 1974); growing with lichens

(La Croix, pers. comm.). Also lithophytic on humus covered rocks. Flowering I–XII, but mainly VIII–II.

Distribution. GUIN. (1 coll.). S.L. (5). LIB. (8). IV.C. (11). GHANA (4). NIG.: S. & C. part (3). CAM.: North up to Adamawa Mts. (10). C.A.R.: SW. part (5). S.T. & PR.: Sao Tomé (1); Principe (1); Pagalu (1). GAB. (3). CONGO (1). ZAI.: Kasai Z4 (2); Forestier Central Z6 (3); Katanga Z11 (4). UG.: Buganda U4 (3). KEN.: Central K4, Thika (1); Coast K7, S. part (3). TANZ.: Tanga T3 (16); Zanzibar (1); East T6 (5); Southern Highlands T7 (11); South T8 (3). ANG. (3). ZAM.: West (3); Central (1). MAL.: South (8). ZIM.: Central (fide Grosvenor, 1976); East (11). MOZ.: Manica e Sofala (8); Sul do Save (1). Material seen from AMES, BM, BR, CO, E, G, HBG, K, L, P, S, U, UPS, WAG, Z.

Notes. 1 - Two entities may be recognized within *Bulbophyllum maximum*, one of which only occurs in part of the area of the species. Because of the existence of intermediate specimens these entities have not been granted taxonomical status.

a. - This entity includes the type specimens of *Bulbophyllum djumaense*, *Bulbophyllum djumaense* var. *grandifolium*, *Bulbophyllum maximum* and *Bulbophyllum purpuratum*.

Margins of the rhachis crenulate, regularly undulate with an amplitude of 1 cm or less. Column with slightly forwards directed, obliquely ovate wings along the adaxial margins, just below the base of the stelidia.

This entity is distributed only in W. Africa and C. Africa up to ZAI. It occurs together with entity b.; intermediates as well as specimens which show the characters mentioned above only indistinctly have been found.

The specimens Adamson 168 and Moreau 572 can also be considered intermediate but originate from E. Africa, MAL. and TANZ. resp., where entity a. does not occur. They have the column wings of entity a. but the rhachis of entity b.

b. - This entity includes all other synonyms listed under *Bulbophyllum maximum*.

Margins of the rhachis entire or finely or coarsely dentate, straight or more or less wavy with an amplitude of more than 1 cm. Column with downwards directed or only very slightly forwards or backwards directed, ovate or more or less deltoid wings along the lower margins just below the base of the stelidia.

2 - The name *platyrhachis* Rolfe applies to specimens with a very wide rhachis which are otherwise identical with *Bulbophyllum maximum*.

3 - The specimen Stolz 667 without flowers, from TANZ. (mentioned by Schlechter as *Bulbophyllum sp.*, Stolz 454 & 667) probably belongs here although in general habit it is reminiscent of *Bulbophyllum scaberulum* var. *scaberulum*. The BM specimen of Stolz 1784, also from TANZ. (type of *Bulbophyllum nyassanum*) consists of comparable though smaller specimens from the same locality which do have flowers and surely belong to *Bulbophyllum maximum*, together with typical specimens of *Bulbophyllum maximum*.

4 - The specimen Wageningen cult. Hort. 81-007 (= Leiden cult. Hort. 23646) from CAM. has a rhachis which is distinctly thickened along one margin. In section the rhachis is almost equilaterally triangular, with flowers on two sides.

5 - Some specimens, especially those from the Southern savannas, show pseudobulbs with knobby edges.

6 - The specimen Mavi 625 from Melsetter, ZIM., is intermediate between *Bulbo-*

phyllum scaberulum var. *scaberulum* and *Bulbophyllum maximum* in the shape of the floral parts. Both entities occur in the area mentioned, it may well be a natural hybrid between the two.

41 - Bulbophyllum injoloense De Wild. - Fig. 59-60.

Literature: see under the subspecies.

Roots 0.6–1.3 mm diam. *Rhizome* 4–6.5 mm diam., nodes not or hardly swollen. *Rhizome scales* tubular, tip obtuse to acute; membranous, with many nerves; the upper keeled abaxially towards the tip. *Pseudobulbs* 2-leafed, broadly to narrowly ovoid or ellipsoid, 2.5–12 cm apart, 3.5–9 by 2–4 cm, more or less flattened, 2–4-angled. *Petiole* 1–30 mm long. *Leaf blade* lanceolate to linear-lanceolate, 4–27 by 1.5–5.5 cm, tip obtuse, emarginate, either oblique or not; coriaceous, thick; midrib slightly prominent abaxially. *Inflorescence* 24–80 cm long, 25–65-flowered. *Peduncle* erect to about patent, usually rather sturdy, swollen and flattened towards the tip, 14–45 cm by 3–5 mm, orbicular in section, glabrous, with 10–18 scales. *Peduncle scales* tubular, the longest 14–25 mm long; tip obtuse to acute; membranous, with some nerves; slightly keeled towards the tip and glabrous abaxially. *Rhachis* erect to patent, widened and flattened, often bladelike, 10–45 cm by 4–15 mm, edges sharp or slightly obtuse, straight to slightly undulate, entire to slightly sinuous, glabrous; surface with fine dark hairs. *Floral bracts* reflexed, ovate, 3.5–8 by 3–7 mm, tip about acute; margins folded backwards; base broadly attached; membranous, without nerves; not keeled and with fine dark hairs abaxially. *Flowers* distichous, placed along a median to excentric line on the rhachis, 6–23 mm apart, measured along one side of the rhachis, the lower up to 18–35 mm apart, recurved, secund, some open simultaneously, not much open. *Pedicel and ovary* 3–5 mm long, with fine dark hairs. *Median sepal* either incurved or not, (ovate-)lanceolate, 5.8–9 by 2–2.7 mm, tip acute; margins entire, glabrous; base (rather) broadly attached; thick; glabrous to finely papillose adaxially; not keeled and glabrous to finely papillose and with some fine dark hairs abaxially. *Lateral sepals* free, not recurved, falcate, triangular, oblique, 5–6.5 by 3–4.8 mm, tip acute to slightly acuminate; base broadly attached; otherwise as the median sepal. *Petals* not recurved, falcate, narrowly (ovate-)lanceolate to (ovate) linear-lanceolate, oblique, 3.8–5.5 by 0.6–1 mm, tip acute; margins entire, glabrous to finely papillose; base rather broadly attached; thick; glabrous to rather coarsely papillose; slightly keeled abaxially. *Lip* recurved, ovate-oblong in outline (not spread), either slightly constricted about half-way or not, 2–3 by 1–2 mm, tip rounded; margins entire and either glabrous or papillose near the tip, very finely serrulate to coarsely lacerate in the basal 1/2 or 2/3 of the lip; thick but rather soft and easy to compress; adaxially without ridges, with a slightly concave basal part, without a median slit but slightly channeled in the median part, surface glabrous to very finely papillose or puberulous; abaxially with a rather distinct, truncate ridge in the basal 2/3 of the lip, surface glabrous, near the base either slightly papillose or not. *Column* from the base of the free part up to the tip of the stelidia 1.8–3 mm long, with distinct, triangular, not falcate, emarginate to acute stelidia of 0.5–1 mm long; column with distinct, either obliquely forwards directed or not, deltoid to semi-orbicular, broadly rounded to obtuse wings along its adaxial margins just below the base of the stelidia; stigma without a basal callus. *Anther* c. 1 by 0.8 mm; front usually (see note 3 under ssp. pseudoxypterum) distinctly drawn out, thin, concave; front margin truncate, entire, glabrous; basally weakly emarginate; abaxially with a rather distinct, obtuse, colliculate ridge near the top of the anther which hardly overtops the front margin, surface colliculate, glabrous towards the margins. *Pollinia* 4, all about detached, ovoid, flattened on either side; the small ones much more than half as long as the large ones; a stipes has not been observed. *Fruit* ellipsoid, up to 1.8 by 0.7 cm, without a stalk.

Notes. The variability within *Bulbophyllum injoloense* is comparable to that within *Bulbophyllum maximum*. In both species a division below species level can be made on account of a combination of two characters: one in the rhachis and one in the shape of the wings along the adaxial margins of the column.

Although the available material of *Bulbophyllum injoloense* was only less than one tenth of the material of *Bulbophyllum maximum* it seems safe to state that the differences between the entities within *Bulbophyllum injoloense*: 1 - are much better developed so that unambiguous description is possible; 2 - are geograpically as well as ecologically well separated. Therefore the entities have been given the rank of subspecies.

Intermediates between the entities have not been found.

KEY TO THE SUBSPECIES

1 - Leaves (13.5 –) 19.5–27 cm long; flowers c. half-way the fully grown rhachis 14–20 mm apart, measured along one side of the rhachis...............ssp. **injoloense** (a)
- Leaves 4–13.5 cm long; flowers c. half-way the fully grown rhachis 6–11 mm apart, measured along one side of the rhachis......................ssp. **pseudoxypterum** (b)

41a - ssp. injoloense – Fig. 59.

Bulbophyllum injoloense De Wild., Bull. J. Bot. Brux. 5 (1916) 175 (*Megaclinium injoloense* De Wild., nom. alt.); id., Pl. Beq. 1 (1921) 87.–Syntypes: Laurent 937 (Lecto BR, here proposed), 1784 (BR), 2056 (BR), Seret 809 (BR).

Pseudobulbs 4–12 cm distant, 3.5–9 by 2.5–4 cm, distinctly flattened, 2-3-angled. Petiole 15–30 mm. Leaf blade narrowly lanceolate to linear-lanceolate, (13.5 –) 19.5–27 by 2.5–5.5 cm. Rhachis 18–45 by 0.9–1.5 cm. Flowers (11 –) 14–23 mm apart, measured along one side of the rhachis, the lower 20–35 mm apart. Median sepal 5.8–9 by 2.2–2.7 mm. Lateral sepals 5–6.3 by 4–4.8 mm, top part rather abruptly bent downwards. Petals narrowly ovate-lanceolate to ovate-linear-lanceolate, 4–5.5 by 0.8–0.9 mm, either surface about glabrous. Lip c. 3 by 1.6–2 mm, margins finely serrulate up to over half-way the lip. Column c. 3 mm long, with forwards directed wings along the adaxial margins.

Colours. Rhachis greenish, stained dark purple. Sepals and petals yellowish green, striated dark purple. Lip yellowish, spotted purple.

Ecology. Lowland forest, no altitude given. Epiphytic. Flowering observed in III.

Distribution. ZAI.: Forestier Central Z6 (5 coll.). Material seen from BR.

Note. Insufficiently known. Most specimens seen are sterile, one contains a mature capsule with the badly preserved perianth still on top, another some rather mature buds. As a consequence, the measurements of the floral parts are not exact.

41b - ssp. pseudoxypterum (J.J. Vermeulen) J.J. Vermeulen.–Fig. 60.

Bulbophyllum injoloense De Wild. ssp. *pseudoxypterum* (J.J. Vermeulen) J.J. Vermeulen, Bull. J. Bot. Nat. Belg. 56 (1986) 232.–*Bulbophyllum injoloense* De Wild. var. *pseudoxypterum* J.J. Vermeulen, Bull. Soc. R. Bot. Belg. 54 (1984) 137.–Type: Fanshawe 12.116 (K).

Bulbophyllum oxypterum auct.: Williamson, Orch. S. Centr. Afr. (1977) col. phot. 114, not descr.

Pseudobulbs 2.5–9.5 cm distant, 5.5–8.5 by 2–2.7 cm, not much flattened, (3 -) 4-angled. Petiole 1–9 mm. Leaf blade lanceolate, 4–13.5 by 1.5–3.8 cm. Rhachis 10–35 by 0.4–0.8 (- 1.3) cm. Flowers 6–12 mm apart, measured along one side of the rhachis, the lower 18–32 mm apart. Median sepal 7–8 by 2–2.5 mm. Lateral sepals 5–6.5 by 3–3.2 mm, top part gradually bent downwards. Petals narrowly lanceolate to linear-lanceolate, 3.8–5 by 0.6–1 mm, either surface finely to coarsely papillose. Lip 2–2.5 by 1–1.4 mm, margins finely to coarsely serrulate or lacerate very near the base. Column 1.8–2 mm long, with downwards directed wings along the adaxial margins.

Colours. Pseudobulbs yellowish green to orange yellow. Leaves green. Rhachis pale yellowish to greenish, marked with brownish purple or purple, or entirely purple. Sepals yellow adaxially, marked with purple; abaxially purplish brown. Petals pale yellow, marked with purple.

Ecology. Brachystegia woodland, riverine forest. Epiphytic in the crown of trees, also lithophytic, altitude not given. Flowering VII–XI.

Distribution. ZAI.: Haut Katanga Z11 (1 coll.). ZAM.: North, Kasama, Mbala (7); West, Mwinilunga, Solwezi (6). Material seen from BM, BR, K.

Notes. 1 - Herbarium specimens without flowers can be distinguished in most cases from *Bulbophyllum maximum* because of the almost papery white peduncle scales.

2 - The specimen Greenway 6184 from ZAM. has an exceptionally wide rhachis of 1.3 cm wide.

3 - The specimens Greenway 6184 and Angus 459, both from ZAM., partly or entirely lack the concave top part of the anther.

42 - Bulbophyllum imbricatum Lindl. – Fig. 61; Plate 6c, d.

Bulbophyllum imbricatum Lindl., Bot. Reg. 27 (1841) misc. 37 ('*Bolbophyllum*'); Rchb. f., Walp. Ann. 6 (1861) 249 ('*Bolbophyllum*'); Durand & Schinz, Consp. Fl. Afr. 5 (1895) 12; Schltr, Engl., Bot. Jahrb. 38 (1906) 156; De Wild., Pl. Beq. 1 (1921) 87; Summerh. in F.W.T.A. ed. 1, 2 (1936) 438; id., ed. 2, 3 (1968) 242; Johansson, Acta Phyt. Suec. 59 (1974) 31; v.d. Burg, Tijdschr. Ned. Orch. Ver. 40 (1978) 152, col. phot.; Adam, Fl. Descr. Mt. Nimba 5 (1981) 1676.–*Phyllorchis imbricata* (Lindl.) Kuntze, Rev. Gen. Pl. 2 (1891) 677, incorr. name.–*Megaclinium imbricatum* (Lindl.) Rolfe in F.T.A. 7 (1897) 37.–Type: Loddiges s.n. (K).

Bulbophyllum strobiliferum Krzl., Mitth. Deutsch. Schutzgeb. 2 (1889) 155; Durand & Schinz, Consp. Fl. Afr. 5 (1895) 14; De Wild., Pl. Beq. 1 (1921) 98.–*Megaclinium strobiliferum* (Krzl.) Rolfe in F.T.A. 7 (1897) 38.–Type: Braun s.n. (B, lost).

Bulbophyllum leucorhachis (Rolfe) Schltr, Engl., Bot. Jahrb. 38 (1905) 17, sub *Bulbophyllum kamerunense*; Schltr, Die Orch. (1914) 326; De Wild., Pl. Beq. 1 (1921) 92; Summerh. in F.W.T.A. ed. 1, 2 (1936) 438;

id., ed. 2, 3 (1968) 242; Adam, Fl. Descr. Mt. Nimba 5 (1981) 1678, pl. 906.-*Megaclinium leucorhachis* Rolfe, Kew Bull. 1891 (1891) 198; Durand & Schinz, Consp. Fl. Afr. 5 (1895) 15; Rolfe in F.T.A. 7 (1897) 37; Hook. f., Bot. Mag. 127 (1901) pl. 7811.-Type: Trevor Lawrence s.n. (1890) (K).

Bulbophyllum triste (Rolfe) Schltr, Engl., Bot. Jahrb. 38 (1905) 17, sub *Bulbophyllum kamerunense*; De Wild., Pl. Beq. 1 (1921) 99.-*Megaclinium triste* Rolfe, Kew Bull. 1894 (1894) 362; id. in F.T.A. 7 (1897) 38; Anon., Gard. Chron., ser. 3, 53, no. 1366 (1913) 131, pl.-Type: Godefroy Lebeuf s.n. (K).

Non *Bulbophyllum triste* Rchb. f., Walp. Ann. 6 (1861) 253.

Bulbophyllum stenorhachis Krzl., Engl., Bot. Jahrb. 22 (1895) 25; De Wild., Pl. Beq. 1 (1921) 98.-Type: Preuss 1241 (B, lost, sketch of Summerhayes in K).

Bulbophyllum laurentianum Krzl. in Durand & De Wild., Mat. Fl. Congo 3 (1899) 50 ('*Bolbophyllum*'); De Wild., Not. Pl. Ut. et Int. Fl. Congo 1 (1903) 128; id., Et. Fl. B. et Moy. Congo 1 (1903) 22, pl. 22; id., (1904) 116; De Wild., Miss. Laur. (1905) 56; Durand & Durand, Syll. Fl. Congol. (1909) 518.-Type: Laurent s.n. (1895) (BR).

Bulbophyllum congolense (De Wild.) De Wild., Pl. Beq. 1 (1921) 82.-*Megaclinium congolense* De Wild., Et. Fl. B. et Moy. Congo 1 (1903) 21, pl. 8; id., Not. Pl. Ut. et Int. Fl. Congo 1 (1903) 127; id., Et. Fl. B. et Moy. Congo 1 (1904) 116; Durand & Durand, Syll. Fl. Congol. (1909) 517.-Type: Laurent s.n. (1892) (BR).

Bulbophyllum gilletii (De Wild.) De Wild., Pl. Beq. 1 (1921) 86.-*Megaclinium gilletii* De Wild., Et. Fl. B. et Moy. Congo 1 (1903) 22; Durand & Durand, Syll. Fl. Congol. (1909) 517.-Type: Gillet 2773b (BR).

Bulbophyllum kamerunense Schltr, Westafr. Kautchuk Exp. (1901) 281, nom. nud.; id., Engl., Bot. Jahrb. 38 (1905) 15; De Wild., Pl. Beq. 1 (1921) 89; Schltr, Fedde, Rep. Beih. 68 (1932) pl. 67; Summerh. in F.W.T.A. ed. 1, 2 (1936) 440; Ake Assi, Et. Fl. C. Iv. (1964) 244; Summerh. in F.W.T.A. ed. 2, 3 (1968) 242; Segerbaeck, Orch. Nig. (1983) 67.-Type: Schlechter 12430 (B, lost).

Bulbophyllum ledermannii (Krzl.) De Wild., Pl. Beq. 1 (1921) 91.-*Megaclinium ledermannii* Krzl., Engl., Bot. Jahrb. 48 (1912) 394.-Type: Ledermann 1215 (B, lost).

Megaclinium hebetatum Krzl., Vierteljahrreschr. Naturf. Ges. Zurich 68 (1923) 423.-Type: Spellenberg s.n. (B, lost).

Bulbophyllum linderi Summerh., Kew Bull. 1935 (1935) 202; id. in F.W.T.A. ed. 1, 2 (1936) 437; id., ed. 2, 3 (1968) 242; Johansson, Acta Phyt. Suec. 59 (1974) 31; Adam, Fl. Descr. Mt. Nimba 5 (1981) 1679, pl. 911.-Type: Linder 1325 (Holo K; iso AMES, P).

Roots 0.3–1.8 mm diam. *Rhizome* 3–10 mm diam., nodes not or hardly swollen. *Rhizome scales* tubular, tip rounded; membranous, usually with many nerves; not keeled abaxially. *Pseudobulbs* 1–2 (– 3)-leafed, ovoid to narrowly ovoid, (0.7 –) 1,5–6 cm apart, (0.7 –) 1.8–7 by (0.8 –) 1.2–4 cm, more or less flattened, obtusely to rather sharply 3–4 (– 5)-angled. *Petiole* 2–13 mm long. *Leaf blade* narrowly lanceolate to linear, (4.5 –) 6–25 by (0.6 –) 1–3.7 cm, tip obtuse, either emarginate and/or oblique or not; coriaceous, thick; midrib hardly prominent abaxially. *Inflorescence* (5 –) 17–60 (– 70) cm long, (8 –) 26–120-flowered. *Peduncle* erect to about patent, swollen and flattened towards the tip, (3 –) 9–51 cm by 2–5.5 mm, orbicular in section, glabrous, with (5 –) 7–15 (– 17) scales. *Peduncle scales* tubular, the longest 5–9 mm long; tip rounded; membranous, with many nerves; not keeled and glabrous abaxially. *Rhachis* about erect or slightly nodding, distinctly swollen, moderately widened and flattened, slightly dumb-bell shaped in section, (2 –) 4.5–26 cm by 5–13 mm by 1.5–5 mm, edges rounded, about straight, regularly crenulate, glabrous; surface glabrous. *Floral bracts* spreading to reflexed, concave, orbicular to elliptic, 4–6.5 by 3.5–8 mm, tip rounded; base broadly attached; membranous, with many nerves; not keeled and glabrous abaxially. *Flowers* distichous, placed along the median or a somewhat excentric line of the rhachis, 2.2–6 mm apart, measured along one side of the rhachis, the lower up to 3–8.5 mm apart, spreading, slightly secund, some open simultaneously, usually not fully open. *Pedicel and ovary* 1–3 mm long, glabrous. *Median sepal* either slightly recurved or not, ovate-lanceolate, 6–11 by 1.4–3 mm, tip acute; margins entire, about glabrous to papillose or puberulous; base broadly attached; thick; glabrous to papillose or finely puberulous; not keeled abaxially. *Lateral sepals* free, recurved, falcate, triangular to ovate, oblique, 3.2–7 by 1.8–5 mm, tip rounded to acute; otherwise as the median sepal. *Petals* not recurved, falcate, linear-lanceolate, oblique, 4–9 by 0.3–1.1 mm, tip acute; otherwise as the median sepal. *Lip* recurved, about triangular in outline (not spread), 1.2–3 by 0.5–2 mm, tip obtuse; margins entire towards the tip, deeply lacerate towards the base (sometimes with only some small teeth), glabrous; thick but rather soft and easy to compress; adaxially without ridges, with a concave basal part, without a median slit, surface glabrous; abaxially with a weak, rounded to retuse ridge towards the base, surface glabrous, slightly papillose near the base. *Column* from the base of the free part up to the tip of the stelidia 2–3.3 mm long, with distinct, triangular, not falcate, acute stelidia of 0.4–0.6 mm long; column usually with rather weak, obliquely forwards directed, triangular, acute teeth along its adaxial margins just below the base of the stelidia which continue towards the column foot as narrow seams; stigma without a basal callus. *Anther* c. 0.8 by 0.6 mm; front not or hardly drawn out, thin, slightly concave; front margin truncate to rounded, entire, about glabrous; basally not emarginate; abaxially with a rather distinct, obtuse, colliculate ridge near the top of the anther which hardly overtops the front margin, surface about glabrous. *Pollinia* 4, about ellipsoid, distinctly flattened; the small ones about half as long as the large ones; a stipes has not been observed. *Fruit* ellipsoid, up to 1.5 by 0.5 cm, without a stalk.

Colours. Pseudobulbs and leaves green. Rhachis yellowish white to greenish, often with purple stains, or entirely purple. Sepals and petals cream coloured, orange yellow or very pale greenish adaxially, often with large, dark purple blotches, or almost entirely dark purple; abaxially with or without dark purple stains. Lip white to pale greenish, often with pale purple spots.

Ecology. Lowland forest to lower montane forest, rarely above 800 m alt. Occasionally in littoral forest, plantations and secondary forest. Epiphytic in the crown of trees, on bark or in minor humus deposits (Johansson, 1974). Flowering mainly IX–II in W. Africa; I–XII in Central Africa.

Distribution. S.L. (5 coll.). LIB. (14). IV.C. (14). GHANA (2). NIG.: S. part (5). CAM.: S. & SW. part (11). C.A.R.: SW. part (3). EQ.GUIN.: Pagulu (1). S.T. & PR.: Sao Tomé (1). GAB. (7). CONGO (1). ZAI.: Bas Congo Z3 (1); Kasai Z4 (4); Forestier Central Z6 (23); Ubangi-Uele Z7 (3). Material seen from AMES, BM, BR, G, K, L, P, UPS, WAG.

Notes. 1 - The shape of the floral parts is very uniform in *Bulbophyllum imbricatum* but a considerable variability is present in the dimensions of all parts, the number of leaves per pseudobulb, the shape of the rhachis and the colour of the flowers.

The name *lindeni* Summerh. applies to rather slender plants with 1-leafed pseudobulbs and cream-coloured flowers. This entity is restricted to S.L., LIB. and IV.C. It occurs together with the 2-leafed entity with purple flowers.

The specimen Dekker 338 from IV.C. shows 1-leafed and 2-leafed pseudobulbs on the same specimen.

The specimen Gillet 2773b, type of *Bulbophyllum gilletii*, has 3-leafed pseudobulbs.

The name *leucorhachis* Rolfe applies to plants with 2-leafed pseudobulbs and cream coloured flowers.

2 - The type specimen of *Bulbophyllum kamerunense* from CAM. (Schlechter 12430) was lost at B. It is probably identical with *Bulbophyllum imbricatum* but, according to the description and Schlechter's sketch, it has slightly longer stelidia and a lip with entire margins. The latter character has never been observed in *Bulbophyllum imbricatum*, possibly Schlechter made a wrong observation.

43 - **Bulbophyllum magnibracteatum** Summerh. - Fig. 62; Plate 7b.

Bulbophyllum magnibracteatum Summerh., Kew Bull. 1935 (1935) 203; id. in F.W.T.A. ed. 1, 2 (1936) 437; id., ed. 2, 3 (1968) 242; Johansson, Acta Phyt. Suec. 59 (1974) 32; Cribb & Perez-Vera, Adansonia, ser. 2, 15 (1975) 206; Adam, Fl. Descr. Mt. Nimba 5 (1981) 1680.–Type: Miles 11 (K).

Roots 0.6–1.2 mm diam. *Rhizome* 1.5–3.5 mm diam., nodes not or hardly swollen. *Rhizome scales* tubular, tip acute to shortly acuminate, the lower obtuse; membranous, with nerves; the upper keeled abaxially. *Pseudobulbs* 1-leafed, (narrowly) ovoid or ellipsoid, 1.5–4.5 cm apart, 1–3 by 0.6–1.5 cm, hardly to moderately flattened, obtusely 3–4-angled. *Petiole* 3–9 mm long. *Leaf blade* lanceolate to broadly linear-lanceolate, 3.8–12 by 1–1.8 cm, tip obtuse, not or hardly emarginate or oblique, coriaceous, thick; midrib slightly prominent abaxially. *Inflorescence* 5.5–22 cm long, 10–24-flowered. *Peduncle* about erect, swollen and flattened towards the tip, 3–16 cm by 1.5–2 mm, orbicular in section, glabrous, with 6–12 scales. *Peduncle scales* tubular, the longest 4–9 mm long; tip acute to acuminate, the lower rounded; membranous, the upper with some nerves; keeled and glabrous abaxially. *Rhachis* erect to slightly arching, swollen, widened and flattened, slightly dumb-bell shaped in section, 1.8–7 cm by 7–15 mm by 3–4 mm (often much thinner when dry), edges obtuse (sharp when dry), straight, entire or slightly crenulate, glabrous; surface glabrous. *Floral bracts* recurved to reflexed, deltoid to broadly ovate, 5–11 by 6–11 mm, tip acuminate; base broadly attached; membranous, with some nerves; keeled and glabrous abaxially. *Flowers* distichous, placed along the median line of the rhachis, 1.8–4 mm apart, measured along one side of the rhachis, the lower up to 7 mm apart, about spreading, some open simultaneously, not fully open. *Pedicel and ovary* 3–4 mm long, glabrous or with some fine dark hairs. *Median sepal* either incurved or not, oblong to broadly lanceolate, 6.4–8.8 by 2.3–3 mm, tip acute; margins entire, glabrous to finely papillose; base rather broadly attached; rather thick; glabrous to finely papillose adaxially; not keeled, glabrous to finely papillose and either with some fine dark hairs or not abaxially. *Lateral sepals* free, either slightly recurved or not, falcate, broadly ovate, oblique, 3.5–5.5 by 3.2–4.5 mm, tip acute to finely acuminate; base broadly attached; otherwise as the median sepal. *Petals* not recurved, slightly falcate, ovate-oblong to ovate-lanceolate, oblique, 4–6 by 1.5–2 mm, tip acute; margins entire, either slightly irregular towards the tip or not, glabrous; base rather broadly to rather narrowly attached; rather thin to rather thick; glabrous; not keeled abaxially. *Lip* recurved, about triangular in outline, (not spread) slightly constricted below the middle, 1.8–2.5 by 1.4–2 mm, tip obtuse; margins entire, glabrous; thick but soft and easy to compress; surface glabrous; adaxially without ridges, with a slightly concave basal part, without a median slit but slightly channeled towards the tip; abaxially with a distinct, obtuse keel over almost the entire length of the lip. *Column* from the base of the free part up to the tip of the stelidia 2.2–3 mm long, with rather distinct, triangular, not falcate, subacute stelidia of c. 0.4 mm long; column without teeth along its adaxial margins; stigma without a basal callus. *Anther* c. 1 by 0.7 mm; front not drawn out; front margin about rounded, entire, papillose; basally distinctly emarginate; abaxially with a weak, obtuse, colliculate ridge near the top of the anther which overtops the front margin, surface about glabrous. *Pollinia* 4; the small ones less than half as long as the large ones, ellipsoid, distinctly flattened; the large ones narrowly ellipsoid, flattened; a stipes has not been observed. *Fruit* not seen.

Colours. Peduncle dark green, stained with dark purple towards the tip. Rhachis yellowish, stained with purple or blackish purple. Sepals cream coloured, stained with dark purple towards the tip or almost entirely dark purple, slightly paler or cream coloured with purple dots towards the base. Petals cream coloured, slightly stained with purple towards the pase. Lip yellowish, stained with purple, or purple.

Ecology. Lowland forest and hill forest, up to 800 m alt. Epiphytic in the central part of the crown of trees, on bark (Johansson, 1974). Flowering I–XII.

Distribution. LIB. (3 coll.). IV.C. (4). GHANA (1). NIG. (1). CAM. (2). C.A.R. (1). EQ.GUIN.: Mbini (1). GAB. (1). ZAI.: Forestier Central Z6 (2). Material seen from BM, BR, HBG, K, P, UPS.

44 - Bulbophyllum sandersonii (Hook. f.) Rchb. f. – Fig. 63–64; Plate 7c, d.

Literature: see under the subspecies.

Roots 0.2–1.2 mm diam. *Rhizome* 1.5–6 mm diam., nodes not or hardly swollen. *Rhizome scales* tubular, tip acute; membranous, with some nerves; not keeled abaxially. *Pseudobulbs* 2-leafed, (narrowly) ovoid, occasionally ellipsoid, 1.2–6.5 cm apart, 1.6–7 by 0.7–2.5 cm, hardly to moderately flattened, obtusely to rather sharply 3–6-angled, edges occasionally with large knobs. *Petiole* 1–7 mm long. *Leaf blade* lanceolate to linear, 3.5–26 by 0.5–2.5 cm, tip rounded to obtuse, emarginate, oblique; coriaceous, thick; midrib slightly prominent abaxially. *Inflorescence* 5.5–30 cm long, 4–60-flowered. *Peduncle* erect to patent, swollen and flattened towards the tip, 3.5–17 cm by 1–3.5 mm, orbicular in section, glabrous, with 6–10 (– 12) scales. *Peduncle scales* tubular, the longest 5–13 mm long; tip rounded to acute; membranous, with nerves; not or hardly keeled and glabrous abaxially. *Rhachis* erect or slightly nodding, either slightly swollen or not, widened and flattened, often about bladelike, 1.5–14.5 cm by 2–11 mm by 1–3 mm, edges sharp to obtuse (sharp when dry), straight to slightly undulate, entire to irregularly dentate, glabrous to papillose; surface glabrous to papillose or puberulous. *Floral bracts* recurved to reflexed, ovate, 3.5–7 by 2–5 mm, tip acute; base broadly attached; membranous, with nerves; not keeled and glabrous abaxially. *Flowers* distichous, placed along a median to excentic line on the rhachis, 3–11 mm apart, measured along one side of the rhachis, the lower up to 5–15 mm apart, spreading, either slightly secund or not, some open simultaneously, fully or wide open. *Pedicel and ovary* 2.5–4.5 mm long, glabrous. *Median sepal* hardly recurved to spreading, ovate-lanceolate to about linear, 6.4–15 by 0.8–2.8 mm, tip obtuse to acute;

margins entire, glabrous to finely papillose, occasionally with some ciliae; base broadly attached; rather thin to rather thick; glabrous to finely papillose or puberulous adaxially; not keeled and glabrous to finely papillose abaxially. *Lateral sepals* free, spreading, falcate, broadly ovate to triangular, oblique, 3.5–9 by 2.5–4.5 mm, otherwise as the median sepal. *Petals* not recurved, falcate, either zig-zag bent towards the tip or not, (thinly) ensiform, subulate towards the tip, towards the base broadened and flattened, base oblique, 5–14 by 0.4–1 mm, top part slightly thickened and club-shaped, (easily breaking off in herbarium specimens), tip obtuse; margins entire, glabrous; base broadly attached; thick; glabrous; slightly keeled towards the base abaxially. *Lip* slightly recurved in the basal half, broadly ovate to ovate-oblong in outline (not spread), triangular when flattened (the basal edges may protrude distinctly), 1.5–3 by 1–2.5 mm, tip rounded to obtuse; margins entire, glabrous; thick but soft and easy to compress; glabrous; adaxially with two weak to very distinct, rounded to sharp ridges in the top half or over about the entire length of the lip which diverge towards the base of the lip, with a concave basal part, without a median slit; abaxially with a distinct, obtuse to retuse keel in the basal half or over about the entire length of the lip. *Column* from the base of the free part up to the tip of the stelidia 2–3.6 mm long, with rather distinct, triangular, not falcate, obtuse to acute stelidia of 0.3–0.4 mm long; column usually with rather weak, obliquely forwards directed, deltoid, subacute teeth along its lower margins just below the base of the stelidia which continue towards the column foot as narrow seams; stigma without a basal callus. *Anther* 0.7–1 by 0.5–0.7 mm; front slightly drawn out, thin, slightly concave; front margin obtuse, entire, glabrous to papillose; basally weakly emarginate; abaxially with a weak, rounded, colliculate ridge near the top of the anther which hardly overtops the front margin, surface glabrous but colliculate towards the top. *Pollinia* 4, flattened; the small ones less than half as long as the large ones, narrowly ellipsoid; the large ones slightly obovoid; a thin stipes has been observed in some specimens. *Fruit* ellipsoid, up to 1.5 by 0.5 cm, without a stalk.

KEY TO THE SUBSPECIES

1 - Floral bracts distinctly narrower than the fully developed part of the rhachis; flowers usually placed along an excentric line on the rhachis; lip adaxially with at most weak, obtuse keels in the top half (flowers usually with large, purple blotches or entirely purple; occurring throughout Tropical Africa up to CAM., not further westwards)..ssp. **sandersonii** (a)

- Floral bracts as wide as the fully developed part of the rhachis or wider; flowers usually placed along the median line of the rhachis; lip adaxially with two high and sharp keels over the entire length of the lip, very close to the margins (flowers with white, green or orange sepals, not or hardly marked with purple; occurring in ZAI. and westwards)..ssp. **stenopetalum** (b)

44a - ssp. sandersonii – Fig. 63; Plate 7c.

Bulbophyllum sandersonii (Hook. f.) Rchb. f., Flora 61 (1878) 78; Bolus, J. Linn. Soc. 25 (1889) 181; Bolus, Ic. Orch. Austro-Afr. Extra Tr. 1, 1 (1893) pl. 3; De Wild., Pl. Beq. 1 (1921) 96; Harrison, Ep. Orch.

S. Afr. (1972) 33, pl. 13; Grosvenor, Excelsa 6 (1976) 79; Williamson, Orch. S. Centr. Afr. (1977) 134, col. phot.; Ball, S. Afr. Ep. Orch. (1978) 84; Stewart & all., Wild Orch. S. Afr. (1982) 219, col. phot.; La Croix et all., Orch. Mal. 1 (1983) 55; Cribb, in F.T.E.A., Orch. 2 (1984) 320.-*Megaclinium sandersonii* Hook. f., Bot. Mag. 97 (1871) sub pl. 5936, nom. nud.; Durand & Schinz, Consp. Fl. Afr. 5 (1895) 16; Rolfe, Fl. Capensis 5, 3 (1913) 10.-Type: Sanderson 898 (K).

Bulbophyllum mellerii (Hook. f.) Rchb. f., Flora 61 (1878) 78; De Wild., Pl. Beq. 1 (1921) 93.-*Megaclinium mellerii* Hook. f., Bot. Mag. 97 (1871) sub pl. 5936, nom. nud.; Durand & Schinz, Consp. Fl. Afr. 5 (1895) 16; Engler, Pflanzenw. O-Afr. 2, c (1895) 157; Rolfe in F.T.A. 7 (1897) 42.-Type: Meller s.n. (K).

Bulbophyllum tentaculigerum Rchb. f., Flora 61 (1878) 77; De Wild., Pl. Beq. 1 (1921) 99; Summerh. in F.W.T.A. ed. 1, 2 (1936) 438; id., ed. 2, 3 (1968) 239; Stewart in Agnew, Upl. Ken. W. Fl. (1974) 763.-*Megaclinium tentaculigerum* (Rchb. f.) Durand & Schinz, Consp. Fl. Afr. 5 (1895) 16; Rolfe in F.T.A. 7 (1897) 42.-Type: Kalbreyer s.n. (W).

Bulbophyllum pusillum (Rolfe) De Wild., Pl. Beq. 1 (1921) 95.-*Megaclinium pusillum* Rolfe, Kew Bull. 1894 (1894) 362; Engler, Pflanzenw. O-Afr. 2, c (1895) 157; Rolfe in F.T.A. 7 (1897) 42.-*Bulbophyllum mooreanum* Robyns & Tournay in Robyns, Fl. Sp. P. Alb. 3 (1955) 496.-Type: Glasnevin cult s.n. (K).

Bulbophyllum bibundiense Schltr, Engl., Bot. Jahrb. 38 (1906) 155; De Wild., Pl. Beq. 1 (1921) 104; Schltr, Fedde, Rep. Beih. 68 (1932) pl. 64; Summerh. in F.W.T.A. ed. 1, 2 (1936) 440; id., ed. 2, 3 (1968) 241.-Type: Schlechter 12784 (B, lost, sketch of Summerhayes in K).

Bulbophyllum sp. Schltr, Engl., Bot. Jahrb. 53 (1915) 572.-Specimen: Stolz 1023 (G, L, S, U, WAG, Z).

Rhizome 1.5-6 mm diam. Pseudobulbs (narrowly) ovoid. Leaves lanceolate to linear. Rhachis 1.5-9 (- 14) by 0.3-1.1 cm, margins narrowly obtuse to sharp (sharp when dried), straight to slightly undulate, irregular or slightly dentate, more or less papillose, surface occasionally glabrous, usually densely papillose or puberulous. Flowers placed along an eccentric line, 3-10 mm apart, measured along one side of the rhachis, the lower 5-12 mm apart. Floral bracts narrower than the fully developed part of the rhachis. Median sepal ovate lanceolate, 6.4-11.4 by 1.2-2.8 mm, margins glabrous to finely and densely papillose, adaxially usually finely and densely papillose to puberulous (rarely glabrous), abaxially glabrous to finely papillose. Lateral sepals 3.5-6 by 2.5-4 mm, top part gradually to abruptly bent downwards, margins and surface as the median sepal. Petals ensiform, subulate towards the tip, 5-10 by 0.4-1 mm. Lip broadly ovate, adaxially with two weak, rounded keels in the top part.

Colours. Pseudobulbs green, yellowish or brownish. Leaves green. Rhachis green or yellowish, usually suffused with purple, or entirely purple. Sepals and petals yellowish or greenish, heavily suffused or spotted with purple or entirely purple. Lip purple or cream-coloured, spotted purple. Plants with entirely cream coloured flowers occasionally occur.

Ecology. Lowland to montane forest, Brachystegia woodland, rarely in lowland forest, 200-2100 m alt. Epiphytic, occasionally lithophytic. Usually growing together with ferns (La Croix, written comm.). Flowering I-XII.

Distribution. CAM.: S. and SW. part (7 coll.). GAB. (2). ZAI.: Lacs Edouard & Kivu Z9 (2). UG.: West U2, S. part (3). RW. (3). KEN.: Central K4, Aberdare Mts (2). TANZ.: Tanga T3 (8); East T6 (2); Southern Highlands T7 (11). ZAM.: Barotseland (1); Central, NE part (3); North (5). MAL.: North (3); Central (2); South (9). ZIM.: Central (fide Grosvenor 1976); East (4). MOZ.: Zambezia (1). R.S.A.: Transvaal (5); Natal, S. to Alexandria (5). Material seen from BM, BR, C, E, G, K, L, P, S, U, W, WAG, Z.

44b - ssp. stenopetalum (Krzl.) J.J. Vermeulen – Fig. 64; Plate 7d.

Bulbophyllum sandersonii (Hook. f.) Rchb. f. ssp. *stenopetalum* (Krzl.) J.J. Vermeulen, Bull. J. Bot. Nat. Belg. 56 (1986) 234.–*Bulbophyllum stenopetalum* Krzl., Engl., Bot. Jahrb. 22 (1895) 25; Schltr, Westafr. Kautchuk Exp. (1901) 282; De Wild., Pl. Beq. 1 (1921) 98.–Type: Preuss 1217 (B, lost; sketch Summerhayes in K).

Megaclinium minor De Wild., Not. Pl. Ut. et Int. Fl. Congo 1 (1903) 125; id., Et. Fl. B. et Moy. Congo 1 (1904) 116; Durand & Durand, Syll. Fl. Congol. (1909) 518.–*Bulbophyllum minor* De Wild., Pl. Beq. 1 (1921) 94 ('*minus*').–Type: Gentil s.n. (BR).

Bulbophyllum rhodopetalum Krzl., Engl., Bot. Jahrb. 36 (1905) 114; De Wild., Pl. Beq. 1 (1921) 117.–Type: Zenker 2895 (Holo B, lost; iso G).

Rhizome 1.8–3.5 mm diam. Pseudobulbs narrowly ellipsoid to narrowly ovoid. Leaves narrowly lanceolate to linear-lanceolate. Rhachis 6–14.5 by 0.2–0.5 cm, margins narrowly rounded, straight, entire, glabrous, surface glabrous. Flowers placed along the median line, 5.5–11 mm apart, measured along one side of the rhachis, the lower 10–15 mm apart. Floral bracts as broad as the fully developed part of the rhachis or broader. Median sepal linear-lanceolate, nearly linear, 7–15 by 0.8–1.5 mm, margins glabrous, occasionally very finely papillose or slightly ciliate, either surface glabrous or at most very finely papillose. Lateral sepals 4.2–9 by 2.6–4.5 mm, top part gradually bent downwards, margins and surface as the median sepal. Petals very thinly ensiform, nearly filiform, very thinly subulate towards the tip, 7–14 by 0.4–0.7 mm. Lip narrowly ovate to ovate-oblong, adaxially with two distinct, sharp keels over the entire length which are confluent with the margins towards the edges near the base.

Colours. Rhachis (pale) green. Sepals yellowish green to cream coloured, turning orange yellow after a few days. Median sepal occasionally with a few very small purple dots. Petals yellowish green, suffused with purple, top entirely purple. Lip yellowish with purple dots.

Ecology. Lowland forest to lower montane forest, occasionally in secondary forest, up to 1000 m alt. Epiphytic. Flowering III–V (IX).

Distribution. LIB.: E. part (1 coll.). IV.C. (3). GHANA (fide F.W.T.A.). NIG.: SE part (1). CAM.: S. part (21). GAB. (1). ZAI.: Forestier Central Z6 (5). Material seen from AMES, BM, BR, C, E, G, K, L, P, S, U, W, WAG.

Notes. 1 - The specimen Cole 38 from CAM. has a rhachis which is extraordinarily thickened at one of its margins so that it is about triangular in section, with flowers along two sides.

2 - In addition to the characters used in the key above, E. African specimens of ssp. *sandersonii* differ distinctly from ssp. *stenopetalum* in having a more compact habit (wider leaves etc.), more abruptly downwards curved lateral sepals, wider (not really filiform) petals and a shorter and wider column (compare the illustrated specimens). CAM. and GAB. specimens of ssp. *sandersonii*, however, approach ssp. *stenopetalum* in these characters, e.g. the type of *Bulbophyllum bibundiense* (according to a sketch by Summerhayes in K) and the type of *Bulbophyllum tentaculigerum*. They only differ from ssp. *stenopetalum* in the characters mentioned in the key, especially those of the rhachis. Although, in the area mentioned, the ranges of the taxa show a slight overlap, and the taxa approach eachother morphologically, I prefer to maintain them at subspecies level.

3 - Due to the softness of the lip the shape of the keels on the lip can only be observed in fresh or spirit material. Thus many specimens were not checked so far as this character is concerned. It should be used with some reserve in the key to the subspecies.

45 - **Bulbophyllum lizae** J.J. Vermeulen - Fig. 65; Plate 8a.

Bulbophyllum lizae J.J. Vermeulen, Bull. Soc. R. Bot. Belg. 54 (1984) 141.-Type: De Wilde, Arends & Groenendijk 129 (WAG).

Roots 1-1.5 mm diam. *Rhizome* 5-7 mm diam., nodes not or hardly swollen. *Rhizome scales* tubular, tip acute, the lower obtuse; membranous, with some nerves; slightly keeled abaxially. *Pseudobulbs* 2-leafed, narrowly ovoid, 5-7 cm apart, 4-7 by 1.4-2 cm, hardly to moderately flattened, about sharply 5-6-angled. *Petiole* 5-15 mm long. *Leaf blade* narrowly lanceolate to linear-lanceolate, 10-23 by 1.3-2.5 cm, tip obtuse, emarginate, slightly oblique; coriaceous, thick; midrib hardly prominent abaxially. *Inflorescence* 17-25 cm long, 10-12-flowered. *Peduncle* erect, swollen and flattened towards the tip, 12.5-18 cm by 3-4 mm, orbicular in section, glabrous, with 7-10 scales. *Peduncle scales* tubular, the longest 11-14 mm long; tip obtuse; membranous, with some nerves; not keeled and glabrous abaxially. *Rhachis* about erect, slightly swollen, widened and flattened, slightly dumb-bell shaped in section, 6-8 cm by c. 7 mm, edges rounded, straight, about entire, glabrous; surface glabrous. *Floral bracts* spreading, about elliptic, 6-9 by 4-6 mm, tip acute; base broadly attached; membranous, without nerves; not keeled and glabrous abaxially. *Flowers* distichous, placed along the median line of the rhachis, 12-17 mm apart, measured along one side of the rhachis, the lower up to c. 30 mm apart, spreading, some open simultaneously, wide open. *Pedicel and ovary* 7-10 mm long, glabrous. *Median sepal* rolled up backwards, narrowly ovate-lanceolate, c. 28 by 6 mm, tip acute; margins entire, glabrous; base broadly attached; thick; glabrous; not keeled abaxially. *Lateral sepals* free, spreading with a reflexed top part, falcate, broadly ovate, oblique, c. 17 by 13 mm, otherwise as the median sepal. *Petals* spreading, falcate, linear, towards the tip slightly thickened, ensiform-subulate in section, slightly broadened towards the base, oblique, c. 23 by 2 mm, top part slightly club-shaped, tip obtuse; margins entire, glabrous; base rather narrowly attached; rather thin; glabrous; not keeled abaxially. *Lip* recurved, broadly ovate to about deltoid in outline (not spread), c. 7 by 8 mm, tip obtuse to subacute; margins entire, glabrous; thick but soft and easy to compress; surface glabrous; adaxially without ridges, with a hardly concave basal part, without a median slit; abaxially with a weak, rounded ridge near the base. *Column* from the base of the free part up to the tip of the stelidia c. 9 mm long, with rounded stelidia; column with weak, obliquely forwards directed, broadly rounded teeth along its adaxial margins, just below the base of the stelidia, which continue towards the column foot as narrow seams; stigma without a basal callus. *Anther* c. 1.3 by 1.1 mm; front not or hardly drawn out; rather thin, slightly concave; front margin rounded, entire, papillose; basally not emarginate; abaxially without a ridge, surface glabrous. *Pollinia* 4; the small ones less than half as long as the large ones, narrowly ellipsoid, distinctly flattened; the large ones obovoid, flattened; a stipes has not been observed. *Fruit* not seen.

Colours. Plant light green. Rhachis light green. Flowers entirely cream-coloured or slightly greenish white.

Ecology. Montane forest at c. 1400 m alt. Epiphytic. Flowering I (once observed).

Distribution. S.T. & PR.: Sao Tomé (1 coll.). Material seen from L, WAG.

46 - Bulbophyllum purpureorhachis (De Wild.) Schltr – Fig. 66; Plate 8b, c.

Bulbophyllum purpureorhachis (De Wild.) Schltr, Die Orch. (1914) 328; De Wild., Pl. Beq. 1 (1921) 241; Summerh. in F.W.T.A. ed. 2, 3 (1968) 241.–*Megaclinium purpureorhachis* De Wild., Not. Pl. Ut. et Int. Fl. Congo 1 (1903) 126; id., Et. Fl. B. et Moy. Congo 1 (1904) 116; id. 2 (1907) 123, pl. 61; id. 2 (1908) 237; Durand & Durand, Syll. Fl. Congol. (1909) 518; Fowler, J. R. Hort. Soc. 35 (1909) proc. LXXIII, pl.; Rolfe, Bot. Mag. 135 (1909) pl. 8273; Anon., Gard. Chron., ser. 3, 45, no. 1167 (1909) 293, pl.– Type: Gentil s.n. (1903) (BR).

Roots 0.5–2 mm diam. *Rhizome* 5–13 mm diam., nodes not or hardly swollen. *Rhizome scales* tubular, tip acute to acuminate; membranous, with many nerves; not keeled abaxially. *Pseudobulbs* 2-leafed, ovoid to about rectangular, 4–8 cm apart, 4.5–11.5 by 2–5.5 cm, distinctly flattened, rather sharply 2–3-angled. *Petiole* 6–15 mm long. *Leaf blade* (obovate)-lanceolate, 12–30 by 2.8–8 cm, tip rounded to obtuse, emarginate, usually oblique; coriaceous, thick; midrib prominent abaxially. *Inflorescence* 28–95 cm long, 20–80-flowered. *Peduncle* erect, sturdy, swollen and flattened towards the tip, 13–45 cm by 5–8 mm, orbicular in section, locally with dark hairs, with 8–10 scales. *Peduncle scales* tubular, the longest 8–15 mm long; tip acute to acuminate; membranous, with many nerves; not keeled and with dark hairs abaxially. *Rhachis* erect, widened and flattened, bladelike, 11–55 cm by 11–45 mm, edges about sharp, straight or slightly undulate, about entire and glabrous; surface with dark hairs. *Floral bracts* reflexed, rhomboid to slightly ovate, 6–8 by 6–9 mm, tip acute; margins slightly rolled inwards; base broadly attached; membranous, with many nerves; not keeled and with some dark hairs abaxially. *Flowers* distichous, placed along the median line of the rhachis, 9–21 mm apart, measured along one side of the rhachis, the lower up to 12–30 mm apart, spreading but with an incurved top part, some or many open simultaneously, not much open. *Pedicel and ovary* c. 5 mm long, locally with dark hairs. *Median sepal* incurved, narrowly triangular, 7–14 by 2.2–4 mm, tip obtuse; margins folded inwards, entire, glabrous; base broadly attached; thick; glabrous adaxially; not keeled and thickly beset with dark hairs abaxially. *Lateral sepals* free, either recurved or not, falcate, triangular, oblique, 6–13 by 4–7 mm, only the upper margin folded inwards; finely papillose towards the base adaxially; otherwise as the median sepal. *Petals* not recurved, falcate, lanceolate, oblique, 5–9 by 0.7–2.2 mm, tip caudate; margins entire, glabrous; base rather broadly attached; rather thin; glabrous; not keeled abaxially. *Lip* slightly recurved, ovate in outline (not spread), slightly constricted in the top half, 3.3–6 by 2.8–4.3 mm, tip rounded; margins entire to finely denticulate, glabrous; rather thin and soft and easy to compress; surface glabrous; adaxially without ridges, with a concave basal part, without a median slit; abaxially with a distinct, obtuse keel over almost the entire length of the lip. *Column* from the base of the free part up to the tip of the stelidia 2–2.5 mm long, with distinct, triangular, not falcate, acute stelidia of c. 0.5 mm long; column with distinct, obliquely forwards directed, triangular, bifid to acute teeth along its adaxial margins just below the base of the stelidia; stigma without a basal callus. *Anther* c. 0.8 by 0.6 mm; front slightly drawn out, thin, not or hardly concave; front margin rounded, slightly irregular, glabrous; basally not emarginate; abaxially with a rather weak, obtuse, finely colliculate ridge near the top of the anther which hardly overtops the front margin, surface finely colliculate. *Pollinia* 4, flattened; the small ones about half as long as the large ones, transversely placed, about crescent shaped; the large ones ovoid; a stipes has not been observed. *Fruit* ellipsoid, up to 2.2 by 0.7 cm, without a stalk.

Colours. Pseudobulbs olive green or yellowish brown. Leaves dark green. Peduncle dark green, heavily spotted dark purple. Rhachis very pale greenish in the centre, mouldy white towards the margins, heavily spotted with purple. Sepals very pale greenish to white, striated and spotted with purple or purplish brown adaxially, suffused with purple abaxially. Petals white with a purple midnerve. Lip white with purple dots. Column purple.

Ecology. Lowland forest, once recorded from secondary forest. Epiphytic. Flowering I–XII.

Distribution. IV.C. (6 coll.). CAM.: S. part (2). GAB. (6). CONGO (3). ZAI.: Forestier Central Z6 (4); Ubangi-Uele Z7 (2). Material seen from BM, BR, K, L, P, WAG.

47 - **Bulbophyllum falcipetalum** Lindl. - Fig. 67.

Bulbophyllum falcipetalum Lindl., J. Linn. Soc. Bot. 6 (1862) 128 ('*Bolbophyllum*'); Durand & Schinz, Consp. Fl. Afr. 5 (1895) 12; Rolfe in F.T.A. 7 (1897) 26; De Wild., Pl. Beq. 1 (1921) 110; Summerh. in F.W.T.A. ed. 1, 2 (1936) 440; id., ed. 2, 3 (1968) 239.–*Phyllorchis falcipetala* (Lindl.) Kuntze, Rev. Gen. Pl. 2 (1891) 677, incorr. name.–Type: Mann 526 (K).

(?) *Bulbophyllum braunii* Krzl., Mitth. Deutsch. Schutzgeb. 2 (1889) 154; Durand & Schinz, Consp. Fl. Afr. 5 (1895) 10; Rolfe in F.T.A. 7 (1897) 25; De Wild., Pl. Beq. 1 (1921) 105.–Type: Braun 81 (?B, lost).

Bulbophyllum lutescens (Rolfe) De Wild., Pl. Beq. 1 (1921) 93.–*Megaclinium lutescens* Rolfe, Kew Bull. 1910 (1910) 158.–Type: Band s.n. (K).

Roots 0.3–0.8 mm diam. *Rhizome* 0.9–1.8 mm diam., nodes not or hardly swollen. *Rhizome scales* tubular, tip about acute; membranous, with some nerves; hardly keeled abaxially. *Pseudobulbs* 2-leafed, (narrowly) ovoid to (narrowly) ellipsoid, 0.3–1 cm apart, 1–2.2 by 0.4–0.8 cm, moderately to distinctly flattened, 3–5-angled. *Petiole* 1–2 mm long. *Leaf blade* linear lanceolate, 2.1–4.5 by 0.3–0.8 cm, tip obtuse, emarginate, oblique; coriaceous, rather thick; midrib prominent abaxially. *Inflorescence* 3.4–8 cm long, 7–15-flowered. *Peduncle* about erect, swollen and flattened towards the tip, 1.3–3.5 cm by 0.7–1.2 mm, orbicular in section, with some fine dark hairs, with 6–8 scales. *Peduncle scales* tubular, the longest 4–7 mm long; tip acute to acuminate; membranous, with many nerves; keeled and glabrous abaxially. *Rhachis* erect or slightly nodding, usually slightly widened and flattened, rarely terete, usually slightly bladelike, zig-zag bent, 1.8–3.8 cm by up to 1.5 mm, edges (when flattened) sharp, about straight, entire and glabrous; surface with fine dark hairs. *Floral bracts* spreading to reflexed, elliptic, 2–5 by 1.2–3 mm, tip acute; base rather broadly attached; membranous, with many fine nerves; slightly keeled and glabrous abaxially. *Flowers* distichous, placed along the median line of the rhachis, 3.5–6.5 mm apart, measured along one side of the rhachis, the lower up to 6–11 mm apart, spreading, many open simultaneously, fully open. *Pedicel and ovary* 1.5–2 mm long, with some fine dark hairs. *Median sepal* hardly recurved to spreading, lanceolate, 4.3–5 by 1.1–1.5 mm, tip acute; margins entire, glabrous; base broadly attached; rather thin; glabrous adaxially; not keeled and with some fine dark hairs abaxially. *Lateral sepals* free, spreading to reflexed at the base, falcate, triangular, oblique, 3.4–4 by 1.8–2.1 mm, otherwise as the median sepal. *Petals* not recurved, falcate, ovate-lanceolate, oblique, 3–4 by 0.6 mm, tip acute; margins entire, glabrous; base broadly attached; thin; glabrous; not keeled abaxially. *Lip* recurved, narrowly ovate in outline (not spread), constricted in the lower half, near the middle, and gradually tapering towards the tip, 1.6–1.8 by 0.9–1.1 mm, tip obtuse to subacute; margins entire, glabrous; rather thick but soft and easy to compress; surface glabrous; adaxially without ridges, with a concave basal part, without a median slit; abaxially with a weak, rounded ridge in the basal half. *Column* from the base of the free part up to the tip of the stelidia c. 0.8 mm long, with rudimentary, subacute to acute stelidia; column with folded inwards, distinct, obliquely forwards directed, about semi-elliptic, obtuse wings along its adaxial margins just below the base of the stelidia which continue towards the column foot as narrow seams; stigma without a basal callus. *Anther* c. 0.5 by 0.4 mm; front moderately drawn out, thin, slightly concave; front margin obtuse, entire, papillose; basally not emarginate; abaxially with a distinct, obtuse, colliculate ridge near the top of the anther which does not overtop the front margin, surface glabrous but colliculate towards the top. *Pollinia* 4; the small ones about half as long as the large ones, about orbicular, distinctly flattened; the large ones ellipsoid, flattened; a thick stipes has been observed. *Fruit* ellipsoid, up to 0.5 by 0.4 cm, without a stalk.

Colours. Flowers yellow or yellowish green.

Ecology. Lowland forest, also in mangrove forest. Epiphytic. Flowering observed in II and X–XI.

Distribution. IV.C. (4 coll.). GHANA (1). NIG.: Bendel P., Rivers P. (2). CAM.: Coast (1). GAB. (3). Material seen from K, P, WAG.

48 - Bulbophyllum scaberulum (Rolfe) Bolus – Fig. 68–70.

Literature: see under the varieties.

Roots 0.3–2 mm diam. *Rhizome* 2–6 mm diam., nodes not or hardly swollen. *Rhizome scales* tubular, tip about acute; membranous, with nerves; not keeled abaxially. *Pseudobulbs* 2-leafed, about orbicular to narrowly ovoid or narrowly ellipsoid, 2.3–13 cm apart, 1.8–9 by 0.8–3.6 cm, more or less flattened, obtusely to sharply 3–5-angled, edges occasionally with large knobs. *Petiole* 1–60 mm long. *Leaf blade* elliptic to linear, 3–28 by 0.7–6 cm, tip obtuse, emarginate, oblique; coriaceous, thick; midrib usually slightly prominent abaxially. *Inflorescence* 6–55 cm long, 12–100-flowered. *Peduncle* erect to patent, swollen and flattened towards the tip, 3.5–28 cm by 1.2–5 mm, orbicular in section, glabrous, with 7–14 scales. *Peduncle scales* tubular, the longest 5–17 mm long; tip obtuse to acute; membranous, with some nerves; hardly keeled and glabrous abaxially. *Rhachis* erect to patent, widened and flattened, often about bladelike, 3–30 cm by 2–14 mm, edges sharp or obtuse, about straight, entire to irregularly dentate, glabrous; surface glabrous or finely papillose, either with some fine dark hairs or not. *Floral bracts* spreading to reflexed, deltoid to ovate, 2–9 by 2–13 mm, tip acute; margins either rolled inwards or not; base broadly attached; membranous, without nerves; not keeled and glabrous abaxially. *Flowers* distichous, placed along a median or excentric line on the rhachis, 2.5–18 mm apart, measured along one side of the rhachis, the lower up to 4–18 mm apart, recurved to spreading, either secund or not, some or many open simultaneously, either fully open or not. *Pedicel and ovary* 2–5 mm long, either with some fine dark hairs or not. *Median sepal* either slightly incurved and/or with a recurved top part or not, (ovate-)oblong to broadly linear-lanceolate, 4–9 by 1.2–3.8 mm, tip subacute to acute; margins entire, glabrous to finely papillose; base broadly to rather narrowly attached; thick; glabrous adaxially; not keeled, glabrous, either with some fine dark hairs or not abaxially. *Lateral sepals* free, not recurved or with a recurved top part, either falcate or not, triangular, oblique, 3.5–7 by 2.3–5 mm, tip acute to cuspidate; base (rather) broadly attached; otherwise as the median sepal. *Petals* not recurved, falcate, oblong to broadly lanceolate or broadly ovate-lanceolate to linear (in the latter case the top part may be slightly swollen), oblique, 2.3–7 by 0.4–1.5 mm, tip subacute to acute; without fine dark hairs abaxially; otherwise as the median sepal. *Lip* recurved, ovate in outline (not spread), slightly constricted about half-way, 1.2–3 by 1–2.2 mm, tip obtuse; margins entire or slightly irregular towards the base, glabrous; rather thick but soft and easy to compress; surface glabrous; adaxially without ridges, with a slightly concave basal part, without a median slit; abaxially with a distinct, rounded to truncate ridge over almost the entire length of the lip. *Column* from the base of the free part up to the tip of the stelidia 1.4–3 mm long, with distinct, triangular, not falcate, acute stelidia of 0.2–0.5 mm long; column with rather distinct, about downwards or slightly obliquely forwards directed, deltoid, obtuse to subacute teeth along its adaxial margins, just below the base of the stelidia, which continue towards the column foot as narrow seams; stigma without a basal callus. *Anther* c. 0.7 by 0.5 mm; front slightly drawn out, thin, slightly concave; front margin obtuse to acute, entire, slightly papillose; basally weakly emarginate; abaxially with a weak, rounded, colliculate ridge near the top of the anther which overtops the front margin, surface colliculate. *Pollinia* 4; all of about equal length, obovoid, flattened; a stipes has not been observed. *Fruit* ellipsoid, up to 1.5 by 0.7 cm, without a stalk.

Note. *Bulbophyllum scaberulum* and *Bulbophyllum resupinatum* are very similar. Some of the varieties of each species are rather constant in their characters whereas others are very variable, especially in the vegetative parts. The flowers of all varieties differ only in small details.

The decision to maintain two species is based on the size of the vegetative parts and the shape of the rhachis. In *Bulbophyllum scaberulum* three varieties have been put together usually consisting of large plants with a widened and distinctly flattened rhachis; in *Bulbophyllum resupinatum* two varieties consisting of small plants with a rhachis which is neither widened nor flattened. Some additional differences between the two species are present in the flowers: in *Bulbophyllum resupinatum* the stelidia are smaller but the wings below the adaxial margins of the column are distinctly larger and tend to be directed more forwards. In addition, the inner pair of pollinia is distinctly smaller.

Some overlap in size of pseudobulbs and leaves is present: small specimens of *Bulbophyllum scaberulum* may approach *Bulbophyllum resupinatum* in this respect. However, such specimens differ in having a very sturdy appearance probably caused by less favourable environmental circumstances. They have hardly flattened, almost orbicular pseudobulbs and thick leaves whereas *Bulbophyllum resupinatum* of comparable size has narrowly ovoid, distinctly flattened pseudobulbs and much thinner leaves.

KEY TO THE VARIETIES

1 - Floral bracts about half-way the rhachis 7–13 mm wide; petals 6–7 mm long.... ....var. **crotalicaudatum** (c)
- Floral bracts about half-way the rhachis 2–4 mm wide; petals 2.3–5.5 mm long. 2

2 - Median sepal narrowly oblong to broadly lanceolate; petals oblong to broadly lanceolate..var. **fuerstenbergianum** (b)
- Median sepal ovate-lanceolate to broadly linear-lanceolate; petals broadly ovate-lanceolate to linear-lanceolate..........................var. **scaberulum** (a)

48a - var. scaberulum – Fig. 68.

Bulbophyllum scaberulum (Rolfe) Bolus, J. Linn. Soc. 25 (1889) 181; Schelpe, An Intr. to the S. Afr. Orch. (1966) 22, pl. 2, col. phot.; Harrison, Ep. Orch. S. Afr. (1972) 35, pl. 14; Grosvenor, Excelsa 6 (1976) 79; Ball, S. Afr. Ep. Orch. (1978) 86, pl.; Hennesy, Am. Orch. Soc. Bull. 50 (1981) 402, pl.; Stewart & all., Wild Orch. S. Afr. (1982) 220, col. phot.; La Croix et all., Orch. Mal. 1 (1983) 56; Cribb, in F.T.E.A., Orch. 2 (1984) 320.–*Megaclinium scaberulum* Rolfe, Gard. Chron., ser. 3, 4, nr. 80 (1888) 6; Durand & Schinz, Consp. Fl. Afr. 5 (1895) 16; Rolfe, Fl. Capensis 5, 3 (1913) 10.–Type: Tillett s.n. (K).

Bulbophyllum clarkei (Rolfe) Schltr, Engl., Bot. Jahrb. 38 (1905) 13, sub *Bulbophyllum bakossorum*; De Wild., Pl. Beq. 1 (1921) 81.–*Megaclinium clarkei* Rolfe, Kew Bull. 1891 (1891) 198; Durand & Schinz, Consp. Fl. Afr. 5 (1895) 15; Rolfe in F.T.A. 7 (1897) 41.–*Bulbophyllum summerhayesii* Hawkes, Lloydia 19 (1956) 93.–Type: Clarke s.n. (K).

Non *Bulbophyllum clarkei* Rchb. f., Flora 71 (1888) 155.

Bulbophyllum congolanum Schltr, Westafr. Kautchuk Exp. (1901) 281, nom. nud.; id., Engl., Bot. Jahrb. 38 (1905) 14; Durand & Durand, Syll. Fl. Congol. (1909) 517; De Wild., Pl. Beq. 1 (1921) 82; Schltr in Mildbraed, 2e Deutsch. Zentr. Afr. Exp. 1910-11, 2 (1922) 20; Summerh., Kew Bull. 12 (1957) 117; id. in F.W.T.A. ed. 2, 3 (1968) 239; Piers, Orch. E. Afr. (1968) 99; Williamson, Orch. S. Centr. Afr. (1977) 133, col. phot.; Adam, Fl. Descr. Mt. Nimba 5 (1981) 1674, pl. 906; Segerbaeck, Orch. Nig. (1983) 63, phot., col. phot.–Syntypes: Schlechter 12693 (AMES, BM, BR, G, L, P, Z), 12781 (BR).

Bulbophyllum eburneum (Pfitz. ex Krzl.) De Wild., Pl. Beq. 1 (1921) 85.–*Megaclinium eburneum* Pfitz. ex Krzl., Orchis 2 (1908) 134.–Type:? (?, lost according to Summerhayes 1957).

Bulbophyllum pobeguinii (Finet) De Wild., Pl. Beq. 1 (1921) 95; Summerh. in F.W.T.A. ed. 1, 2 (1936) 440.–*Megaclinium pobeguinii* Finet, Not. Syst. 1 (1910) 167.–Syntypes: Pobeguin 636 (P), 925 (P).

Bulbophyllum bambiliense De Wild., Bull. J. Bot. Brux. 5 (1916) 174 (*Megaclinium bambiliense* De Wild., nom. alt.); id., Pl. Beq. 1 (1921) 78.–Type: Seret 208 (BR).

Bulbophyllum ealaense De Wild., Bull. J. Bot. Brux. 5 (1916) 175 (*Megaclinium ealaense* De Wild., nom. alt.); id., Pl. Beq. 1 (1921) 85.–Type: Laurent 685 (BR).

Bulbophyllum jespersenii De Wild., Bull. J. Bot. Brux. 5 (1916) 176 (*Megaclinium jespersenii* De Wild., nom alt.); id., Pl. Beq. 1 (1921) 89.–Type: Jespersen s.n. (1910) (BR).

Bulbophyllum zobiaense De Wild., Bull. J. Bot. Brux. 5 (1916) 178 (*Megaclinium zobiaense* De Wild., nom. alt.); id., Pl. Beq. 1 (1921) 100.–Type: Seret 874 (BR).

Bulbophyllum chevalierii De Wild., Pl. Beq. 1 (1921) 80 (*Megaclinium chevalierii* De Wild., nom. alt.- Type: Chevalier 12838 (Holo BR; iso BM, L, P, Z).
Bulbophyllum sp. Piers, Orch. E. Afr. (1968) 108.-Specimen: Ombler s.n. (1963) (?).
Bulbophyllum sp. Williamson, Orch. S. Centr. Afr. (1977) 135.-Specimen: Williamson 15 (K).
Bulbophyllum oxypterum auct. non (Lindl.) Rchb.: Cufodontis, En. Pl. Aeth. Sperm. (1922) 1618).

Rhachis not or only slightly thickened. Floral bracts 2–6.5 by 2–4 mm. Median sepal ovate-lanceolate to broadly linear-lanceolate, 4–9 by 1.2–2 mm, margins usually hardly folded inwards. Lateral sepals 3.5–7 by 2.3–5 mm, top part distinctly but gradually bent downwards, tip acuminate to about cuspidate. Petals broadly ovate to linear, 2.3–5.5 by 0.4–1.5 mm, rarely with a slightly thickened top part.

Colours. Plant green. Sepals and petals cream-coloured, yellow or greenish, often striated, spotted or suffused with purple adaxially and suffused with purple abaxially. Lip cream-coloured or yellow, often spotted with purple, or almost entirely pale purple.

Ecology. Lowland rainforest and semideciduous forest, montane forest, Brachystegia woodland, occasionally in secondary forest, up to 2300 m alt. Epiphyte in the crown of trees, on bark or in minor humus deposits, often very exposed (Johansson, 1974). Also on tree trunks, a.o. among lichens (La Croix, written comm.). Occasionally lithophytic. Flowering mainly VII–III.

Distribution. GUIN. (13 coll.). S.L. (5). LIB. (6). IV.C. (11). GHANA (7). NIG.: S. part (6); Plateau P. (2). CAM.: N to Adamawa Mts. (6). C.A.R. (7). SUDAN: Equatoria (1). AETH.: E part, Wollega P., Kaffa P. (3). GAB. (1). CONGO (2). ZAI.: Bas Congo Z3 (1); Kasai Z4 (1); Forestier Central Z6 (16); Ubangi-Uele Z7 (3); Lacs Edouard & Kivu Z9 (1); Haut Katanga Z11 (11). KEN.: Coast K7, S. part (5). TANZ.: Lake T1 (1); Tanga T3 (Fide F.T.E.A); East T6 (4); Southern Highlands T7 (1). ANG. (3). ZAM.: West, Mwinilunga (1); Central, N. part (3); North (2). MAL.: South (1). ZIM.: East (5); South (3). MOZ.: Niassa (2); Manica e Sofala (2). R.S.A.: Natal (7); Cape, S. to East London (2). Material seen from AMES, BM, BR, E, G, HBG, K, L, NY, P, S, UPS, WAG, Z.

Notes. 1 - This variety is rather constant in the shape of the floral parts (except for a variability in the l/w-ratio, see below) but shows an extreme diversity in shape and size of all other parts. Three entities are distinguished here which are linked by numerous intermediate specimens.

a. - Relatively large plants, usually with closely placed pseudobulbs, wide leaves and a long and wide rhachis. Specimens of this entity tend to have very short and wide floral parts (e.g. Wubben sn., Bates 1130, Louis 6720 and 13729). They occur throughout the area of the variety except for the its Southern (drier) part.

b. - Plants with widely spaced, narrow pseudobulbs, c. linear leaves and a short, narrow rhachis with closely placed flowers. This entity seems to be restricted to ZAI. (around Eala and Yangambi). Comparable specimens which show these characters less pronounced occur in W. Africa.

c. - Relatively small and sturdy plants with closely placed, ovoid or almost orbicular, only hardly flattened pseudobulbs and fleshy, elliptic or oblong leaves. The rhachis, however, is often relatively long. This entity occurs throughout the range of the variety, but in its most typical form in the drier regions: AETH., ANG, ZAM., ZIM., MAL., MOZ. and further south. Some specimens show pseudobulbs with crenulate edges.

2 - The specimen Moreau 385b from TANZ. has petals with a slightly swollen tip. A possible hybrid with *B. maximum* has been found, see under that species.

48b - var. fuerstenbergianum (De Wild.) J.J. Vermeulen – Fig. 69.

Bulbophyllum scaberulum (Rolfe) Bolus var. *fuerstenbergianum* (De Wild.) J.J. Vermeulen, Bull. J. Bot. Nat. Belg. 56 (1986) 234.–*Bulbophyllum fuerstenbergianum* (De Wild.) De Wild., Pl. Beq. 1 (1921) 86; Summerh. in F.W.T.A. ed. 2, 3 (1968) 239.–*Megaclinium fuerstenbergianum* De Wild. in Cogn. & Goossens, Dict. Ic. Orch. 2, Megacl. (1905) pl. 1.–Type: Fuerstenberg s.n. (Holo BR; iso K).

Rhachis not or only slightly thickened. Floral bracts 4–6.5 by 2–3 mm. Median sepal narrowly oblong to broadly lanceolate, 5.5–6.5 by 1.8–2.1 mm, margins usually distinctly folded inwards. Lateral sepals 5–5.5 by 3.2–4 mm, top part straight or only slightly bent downwards, tip acute, at most very slightly acuminate. Petals oblong to broadly lanceolate, 3.2–3.8 by 0.9–1.4 mm, top part not thickened.

Colours. Flowers yellowish or greenish, suffused with purple.

Ecology. Lowland forest and montane forest. Epiphytic. Flowering observed in V and XII.

Distribution. NIG.: SE part, Ogoja P. (1 coll.). CAM.: W. part (1). EQ.GUIN.: Bioko (1). ZAI.: Haut Katanga Z11 (1). Material seen from BR, K.

48c - var. crotalicaudatum J.J. Vermeulen, var. nov. – Fig. 70.

Bulbophyllum scaberulum (Rolfe) Bolus var. *crotalicaudatum* J.J. Vermeulen, var. nov. A var. scaberulo fuerstenbergianoque in bracteis floriferibus maioribus latioribusque differt.–Type: Richards 15656b (K).

Rhachis distinctly thickened (in one herbarium specimen only slightly thickened, but this may be due to drying). Floral bracts 6–9 by 6–13 mm. Median sepal ovate-oblong to ovate-lanceolate, c. 9 by 2–3.8 mm, margins usually hardly folded inwards. Lateral sepals 3.8–4.8 by 3.5–4.5 mm, top part distinctly and very abruptly bent downwards, tip acute. Petals broadly linear-lanceolate to linear, 6–7 by 0.7–1.2 mm, top part not thickened.

Colours. Flowers white or deep purple.

Ecology. Montane forest.

Distribution. TANZ.: Tanga T3 (2); Southern Highlands T7 (1). Material seen from K.

49 - Bulbophyllum resupinatum Ridl. – Fig. 71–73; Plate 8d.

Literature: see under the varieties.

Roots 0.2–0.9 mm diam. *Rhizome* 1–2.2 mm diam., nodes not or hardly swollen. *Rhizome scales* tubular, tip acute; membranous, with nerves; keeled abaxially. *Pseudobulbs* 2-leafed, (narrowly) ovoid to (narrowly) ellipsoid, 0.5–3.5 cm apart, 0.8–4 (–5) by 0.5–1.4 cm, flattened, obtusely to rather sharply 2–4-angled. *Petiole* 1–3 mm long. *Leaf blade* oblong to linear-lanceolate, maximum width either slightly above the middle or not; 1.3–5 (– 8.5) by 0.5–1.6 cm, tip rounded to obtuse, emarginate, oblique; coriaceous, thin to rather thick; midrib hardly to moderately prominent abaxially. *Inflorescence* 1.8–40 cm long, 8–60-flowered. *Peduncle* erect to patent, 0.9–23 cm by 0.8–1.5 mm, orbicular in section, glabrous or with fine dark hairs, with 5–10 scales. *Peduncle scales* tubular, the longest 4.5–9 mm long; tip acute to acuminate; membranous, either with some nerves or not; keeled and glabrous abaxially. *Rhachis* erect to about pendulous, either slightly flattened below the attachment of the flowers and/or slightly zig-zag bent or not, 0.9–28 cm long, with fine dark hairs. *Floral bracts* recurved to reflexed, elliptic to (ovate-)oblong, 2–5.5 by 1.5–3 mm, tip acute to acuminate; base rather broadly attached; membranous, without nerves; keeled and glabrous abaxially. *Flowers* distichous, 1.8–11 mm apart, measured along one side of the rhachis, the lower up to 2–15 mm apart, spreading, either secund or not, many open simultaneously, not much to about fully open. *Pedicel and ovary* 1–4.1 mm long, with fine dark hairs. *Median sepal* slightly incurved to recurved, ovate-oblong to narrowly ovate-lanceolate, 2.2–5 by 0.7–1.2 mm, tip acute; margins entire, glabrous; base broadly attached; thick; glabrous adaxially; not keeled, glabrous to papillose and either with fine dark hairs or not abaxially. *Lateral sepals* free, either recurved or not, slightly falcate, triangular, oblique, 2–3.9 by 1.2–2.8 mm, otherwise as the median sepal. *Petals* either recurved and/or slightly falcate or not, narrowly lanceolate to linear-lanceolate, oblique, 1.4–3 by 0.25–0.6 mm, tip acute; margins entire, glabrous; base broadly attached; rather thin; glabrous adaxially; not keeled and either glabrous or with some fine dark hairs near the base abaxially. *Lip* recurved, broadly ovate in outline (not spread), 0.9–1.6 by 0.7–1.1 mm, tip rounded; margins entire, glabrous; thick but soft and easy to compress; surface glabrous; adaxially without ridges, with a concave basal part, without a median slit; abaxially with a rather distinct, obtuse keel near the base. *Column* from the base of the free part up to the tip of the stelidia 0.8–1.6 mm long, with about rudimentary, rounded to obtuse stelidia; column with folded inwards, distinct, obliquely forwards directed, deltoid, subacute to acute teeth along its adaxial margins, just below the base of the stelidia, which continue towards the column foot as narrow seams; stigma without a basal callus. *Anther* c. 0.5 by 0.4 mm; front moderately drawn out, thin, slightly concave; front margin obtuse, entire, papillose; basally weakly emarginate; abaxially with a rather weak, rounded, colliculate ridge near the top of the anther which does not overtop the front margin, surface glabrous but colliculate towards the top and either slightly papillose towards the base or not. *Pollinia* 4, more or less obovoid; the small ones less than half as long as the large ones, distinctly flattened; the large ones flattened; a thin stipes has been observed. *Fruit* ellipsoid, up to 0.7 by 0.3 cm, without a stalk.

Note. *Bulbophyllum resupinatum* is very similar to *Bulbophyllum scaberulum*, see note under the latter species.

KEY TO THE VARIETIES

1 - Flowers abaxially glabrous or papillose, without dark hairs (ovary may have dark hairs)..var. **resupinatum** (a)
- Flowers abaxially with fine, dark hairs..............................var. **filiforme** (b)

49a - var. resupinatum – Fig. 71; Plate 8d.

Bulbophyllum resupinatum Ridley, Bol. Soc. Brot. 5 (1888) 198; Durand & Schinz, Consp. Fl. Afr. 5 (1895) 14; Rolfe in F.T.A. 7 (1897) 26; De Wild., Pl. Beq. 1 (1921) 117; Exell, Cat. Vasc. Pl. S. Tomé (1944) 328.–Syntypes: Quintas 10 (Lecto BM, here proposed) = Quintas s.n. (COI, n.v.), Moller s.n. (COI, n.v.).

Bulbophyllum rhizophorae auct. non Lindl.: Summerh. in F.W.T.A. ed. 1, 2 (1936) 440, p.p.; id., ed. 2, 3 (1968) 239, p.p.; Vermeulen, Tijdschr. Ned. Orch. Ver. 43 (1981) 90.

Inflorescence 1.8–8.6 cm long. Peduncle 0.9–4 cm long. Rhachis 0.9–6.5 cm long, not flattened. Flowers 1.8–3 mm apart, measured along one side of the rhachis, the lower 2–4.2 mm apart. Floral bracts 2–4 by 1.5–2.5 mm. Pedicel and ovary 1.5–2.5 mm long. Flowers not much opening. Median sepal incurved, 2.5–3 by 1–1.1 mm, narrowly ovate-oblong; abaxially glabrous to very finely papillose, without fine, dark hairs. Lateral sepals 2–2.7 by 1.3–2 mm, abaxially as the median sepal. Petals not recurved, c. 1.4 by 0.3–0.4 mm.

Colours. Pseudobulbs green, often with some purple spots. Leaves green. Flowers entirely dark red, sepals occasionally slightly greenish towards the base.

Ecology. Lowland forest, montane forest. Altitude not given. Epiphytic. Flowering VI and XII–I.

Distribution. IV.C. (3 coll.). GHANA (1). NIG.: SE part, Cross P. (1). CAM. (1). S.T. & PR.: Sao Tomé (1). ZAI.: Forestier Central Z6 (1). Material seen from BM, BR, K, L, P, WAG.

49b - var. filiforme (Krzl.) J.J. Vermeulen – Fig. 72–73.

Bulbophyllum resupinatum var. *filiforme* (Krzl.) J.J. Vermeulen, Bull. J. Bot. Nat. Belg. 56 (1986) 235.– *Bulbophyllum filiforme* Krzl., Engl., Bot. Jahrb. 22 (1895) 25 (‘*Bolbophyllum*’); Rolfe in F.T.A. 7 (1897) 27; Schltr, Westafr. Kautchuk Exp. (1901) 281; De Wild., Pl. Beq. 1 (1921) 110; Summerh. in F.W.T.A. ed. 1, 2 (1936) 440; id., ed. 2, 3 (1968) 239.–Type: Preuss 1242 (B, lost); Leiden cult. Hort. 25160 (neo L).

Bulbophyllum longispicatum Krzl. & Schltr, Orchis 2 (1908) 98; De Wild., Pl. Beq. 1 (1921) 113.–Type: Schlechter 15759 (Holo B, lost; iso BR).

Non *Bulbophyllum longispicatum* Cogn., J. des Orch. 4 (1893) 265.

Bulbophyllum rubroviolaceum De Wild., Rev. Zool. Afr. 9, suppl. bot. (1921) 32; De Wild., Pl. Beq. 1 (1921) 118.–Type: Bequaert 6705 (BR).

Bulbophyllum daloaense Cribb & Perez-Vera, Adansonia, ser. 2, 15 (1975) 202, 206.–Type: Perez-Vera 270 (Holo K; iso P).

Bulbophyllum victoris Cribb & Perez-Vera, Adansonia, ser. 2, 15 (1975) 202, 206.–Type: Perez-Vera 553 (Holo K; iso P, UCJ, n.v.).

Bulbophyllum rhizophorae auct. non Lindl.: Summerh. in F.W.T.A. ed. 2, 3 (1968) 239, p.p.

Inflorescence 3.5–40 cm long. Peduncle 1.5–12 cm long. Rhachis 2–28 cm long, occasionally slightly flattened below the attachment of the flowers. Flowers 2.8–11 mm apart, measured along one side of the rhachis, the lower 4–15 mm apart. Floral bracts 3.2–5.5 by 2–3 mm. Pedicel and ovary 1–4.1 mm long. Flowers (not) fully opening. Median sepal recurved to incurved, 4–5.5 by 1–1.2 mm, ovate-oblong to narrowly lanceolate; abaxially glabrous, with fine, dark hairs. Lateral sepals 3–3.9 by 1.8–2.8 mm, abaxially as the median sepal. Petals recurved, 2.2–3 by 0.3–0.6 mm.

Colours. Pseudobulbs yellowish green to olive green, often with a few purple spots. Leaves green. Sepals yellow, spotted with (brownish-)purple, or almost entirely (brownish-)purple except for the basal part. Petals (brownish-)purple. Lip purple with a whitish base.

Ecology. Mangrove forest, lowland forest, lower montane forest. Epiphytic. Flowering III, IX - I.

Distribution. S.L. (1 coll.). LIB. (1). IV.C. (4). CAM. (1). GAB. (4). ZAI.: Forestier Central Z6 (1). Material seen from BR, K, L, P, WAG.

Note. 1 - The plates of *Bulbophyllum daloaense* and *B. victoris*, accompanying the original descriptions are highly inaccurate in the details of the flowers.

2 - The length of the inflorescence is very variable. Using this character, in combination with a few others, two entities can be distinguished:

a. - Inflorescence 4.5–40 cm long. Flowers 4–11 mm apart, measured along 1 side of the rhachis, the lower 4–15 mm apart. Median sepal straight to incurved. This entity includes the type specimens of *B. rubroviolaceum* and *B. victoris*.

b. - Inflorescence 3.5–11 cm long. Flowers 2.8–5 mm apart, measured along 1 side of the rhachis, the lower 5–7 mm apart. Median sepal straight to recurved. This entity includes the type specimens of *B. filiforme*, *B. longispicatum* and *B. daloaense*.

An intermediate specimen is e.g. Leiden cult. Hort. 25160 from GAB.

50 - **Bulbophyllum parvum** Summerh. - Fig. 74.

Bulbophyllum parvum Summerh., Kew Bull. 12 (1957) 122; id. in F.W.T.A. ed. 2, 3 (1968) 239.–Type: Jones 235 (K).

Roots c. 0.2 mm diam. *Rhizome* c. 1 mm diam., nodes not or hardly swollen. *Rhizome scales* tubular, tip acute; membranous, with some nerves; slightly keeled abaxially. *Pseudobulbs* 2-leafed, about orbicular, 0.2–0.4 cm apart, 0.6–0.8 by 0.5–0.8 cm, slightly flattened, very obtusely 4-angled. *Petiole* c. 0,5 mm long. *Leaf blade* elliptic to broadly oblong, 0.6–0.9 by 0.3–0.5 cm, tip rounded, emarginate, not oblique; coriaceous, very thick; midrib not prominent abaxially. *Inflorescence* 1.2–1.6 cm long, 6–8-flowered. *Peduncle* about erect, sturdy, swollen towards the tip and below the nodes, 0.8–1.1 cm by 0.5–0.8 mm, orbicular in section, with some fine dark hairs, with c. 5 scales. *Peduncle scales* tubular, the longest c. 3.5 mm long; tip acute to acuminate; membranous, with some nerves; slightly keeled and glabrous abaxially. *Rhachis* erect, swollen, slightly flattened below the attachment of the flowers, about spindle-shaped, 0.4–0.6 cm by c. 1 mm, with some fine dark hairs. *Floral bracts* recurved to spreading, elliptic to ovate, c. 3 by 1.9 mm, tip acute; base broadly attached; membranous, without nerves; slightly keeled and with some fine dark hairs abaxially. *Flowers* distichous, c. 1.2 mm apart, measured along one side of the rhachis, the lower up to c. 1.9 mm apart, spreading, secund, many open simultaneously, not much open. *Pedicel and ovary* c. 1.5 mm long, with some fine dark hairs. *Median sepal* not recurved, lanceolate, c. 3.4 by 1 mm, tip acute; margins entire, glabrous; base rather broadly attached; thick; glabrous adaxially; not keeled and with fine dark hairs abaxially. *Lateral sepals* free, slightly falcate, triangular, oblique, c. 3.3 by 1.8 mm, otherwise as the median sepal. *Petals* not recurved, slightly falcate, ovate linear-lanceolate, oblique, c. 2.3 by 0.25 mm, tip acute to slightly acuminate; margins entire, glabrous; base broadly attached; rather thin; glabrous; not keeled abaxially. *Lip* recurved, elliptic in outline (not spread), c. 1.2 by 0.8 mm, tip rounded; margins entire, glabrous; thick but soft and easy to compress; surface glabrous; adaxially without ridges, with a slightly concave basal part, without a median slit; abaxially with an inconspicuous, obtuse ridge in the basal half. *Column* from the base of the free part up to the tip of the stelidia c. 1.5 mm long, with rather distinct, triangular, not falcate, acute stelidia of c. 0.1 mm long; column with slightly folded inwards, distinct, obliquely forwards directed, triangular, subacute teeth along its lower margins, just below the base of the stelidia, which continue towards the column foot as narrow seams; stigma without a basal callus. *Anther* c. 0.7 by 0.4 mm, front drawn out into a beak with a c. truncate tip; front margin slightly irregular, c. glabrous; basally hardly emarginate; abaxially hardly keeled, surface finely papillose. *Pollinia* 2; about ellipsoid with a slightly truncate base, flattened on either side. *Fruit* not seen.

Colours. Flowers purple.

Ecology. On a dead tree among lichens, 700 m alt. Flowering II (once recorded).

Distribution. S.L.: Picket Hill (1 coll.).

51 - Bulbophyllum falcatum (Lindl.) Rchb. f. - Fig. 75–77; Plate 9a-d.

Literature: see under the varieties.

Roots 0.25–1.5 mm diam. *Rhizome* 1–4 mm diam., nodes not or hardly swollen. *Rhizome scales* tubular, tip obtuse to acute, occasionally mucronate; membranous, with nerves; hardly to moderately keeled abaxially. *Pseudobulbs* 2-leafed, broadly to very narrowly ovoid, occasionally about ellipsoid, 0.4–5 cm apart, 0.9–7 by 0.6–2 cm, more or less flattened, obtusely to about sharply 2-4-angled. *Petiole* 1–25 mm long. *Leaf blade* broadly lanceolate to linear, maximum width either slightly above the middle or not, 2–21 by 0.5–3 cm, tip obtuse, usually emarginate, either oblique or not; coriaceous, thin to rather thick; midrib prominent abaxially. *Inflorescence* 3–40 cm long, 8–60-flowered. *Peduncle* erect to pendulous, either swollen and flattened towards the tip or not, 1–16 cm by 0.5–3.2 mm, orbicular in section, glabrous, with 4–8 scales. *Peduncle scales* tubular, the longest 4–14 mm long; tip acute to acuminate, membranous, either with some nerves or not; slightly keeled and glabrous abaxially. *Rhachis* erect to pendulous, terete to widened and flattened, either bladelike and/or zig-zag bent or not, 2–17 cm by up to 18 mm, edges (when flattend) sharp to obtuse, straight or slightly undulate, entire to irregularly dentate, glabrous; surface glabrous or with fine dark hairs. *Floral bracts* spreading to reflexed, ovate to broadly ovate-oblong, 1.3–6.2 by 1–3.5 mm, tip about acute; margins either folded inwards or not, base broadly attached; membranous, without nerves; not keeled and glabrous abaxially. *Flowers* distichous, placed along a median to excentric line on the rhachis, 2–20 (- 30) mm apart, measured along one side of the rhachis, the lower up to 4–30 mm apart, spreading, either secund or not, many open simultaneously, fully open. *Pedicel and ovary* 1.5–5 mm long, with fine dark hairs. *Median sepal* not recurved or with a recurved top part, narrowly obovate to broadly obovate-lanceolate, or more or less spatulate, 2.3–9 by 0.8–4 mm, tip obtuse to acuminate; margins entire, glabrous; base broadly attached; thick, top half with a fleshy callus along both margins; glabrous adaxially, occasionally with some fine dark hairs near the base; not keeled and with fine dark hairs abaxially. *Lateral sepals* free, basal half recurved and concave, top half spreading to reflexed; slightly falcate, broadly ovate to ovate oblong, or triangular, oblique, 1.8–9 by 0.8–7 mm, tip subacute to acute; rather thick, without calli; finely papillose to coarsely puberulous adaxially in the basal half; otherwise as the median sepal. *Petals* either recurved or not, either falcate or not, ovate to broadly linear-lanceolate, oblique, 1.3–3.2 by 0.25–1 mm, tip rounded to acuminate; margins entire, glabrous; base (rather) broadly attached; thin, either with a thickened top part or not; glabrous; not keeled abaxially. *Lip* recurved in the lower half, deltoid to broadly ovate in outline (not spread), 0.7–4 by 0.5–3 mm, tip obtuse; margins entire, glabrous; thick but soft and easy to compress; surface glabrous; adaxially without ridges, with a slightly concave basal part, without a median slit; abaxially with a distinct obtuse keel in the basal half. *Column* from the base of the free part up to the tip of the stelidia 0.5–1.5 mm long, with rather distinct to rudimentary, deltoid to triangular, not falcate, obtuse to acute stelidia of at most 0.4 mm long; column with folded inwards, distinct, obliquely forwards directed, deltoid, obtuse wings along its adaxial margins, just below the base of the stelidia, which continue towards the column foot as narrow seams; stigma without a basal callus. *Anther* 0.4–0.8 by 0.3–0.6 mm; front slightly drawn out, thin, slightly concave; front margin truncate to subacute, entire, papillose; basally weakly emarginate; abaxially with a rather weak, rounded, colliculate ridge near the top of the anther which does not overtop the front margin, surface glabrous but colliculate towards the top. *Pollinia* 4; the small ones much more than half as long as the large ones, ellipsoid, distinctly flattened; the large ones deltoid to ovoid, flattened; a thick stipes has been observed. *Fruit* ellipsoid, up to 1.5 by 0.5 cm, stalk c. 1 mm.

Note. In the present concept *Bulbophyllum falcatum* shows extensive variability in many parts. Several species accepted by previous authors have been included here. Typical specimens of these 'species' are easily recognizable indeed and locally even well separated morphologically (see e.g the note on *Bulbophyllum arnoldianum* below var. *velutinum*). However, when studying all material available many intermediate specimens appear to disrupt this distinctness.

Three varieties can be recognized, based on the shape of the petals. Generally spoken, these varieties are also slightly different in their vegetative parts.

KEY TO THE VARIETIES

1 - Top part of petals more or less thickened, rounded to subacute var. **falcatum** (a)
- Top part of petals not thickened, acute to acuminate, or subacute................2

2 - Ratio length median sepal/petal 2.4–5; Median sepal (6 –) 6.5–9 mm long, usually rather distinctly spatulate; petals ovate to ovate-oblong, occasionally ovate-lanceolate, tip acute or subacute..var. **bufo** (b)
- Ratio length median sepal/petal 1.3–2.3; Median sepal 2.3–6 (– 7) mm long, usually obovate to broadly obovate-lanceolate; petals very narrowly triangular, gradually narrowing towards the sharply acute to acuminate tip..........var. **velutinum** (c)

51a - var. falcatum – Fig. 75; Plate 9a.

Bulbophyllum falcatum (Lindl.) Rchb. f., Walp. Ann. 6 (1861) 258 ('*Bolbophyllum*'); Schltr, Die Orch. (1914) 326; De Wild., Pl. Beq. 1 (1921) 86; Schltr in Mildbraed, 2e Deutsch. Zentr. Afr. Exp. 1910-11, 2 (1922) 180; Summerh. in F.W.T.A. ed. 1, 2 (1936) 440; Exell, Cat. Vasc. Pl. S. Tomé (1944) 326; Summerh., Kew Bull. 13 (1958) 75; id., Bull. Inst. Fr. d'Afr. N., ser. a, 21 (1959) 473; Ake Assi, Et. Fl. C. Iv. (1964) 243; Summerh. in F.W.T.A. ed. 2, 3 (1968) 241; Piers, Orch. E. Afr. (1968) 101; Stewart & Campbell, Orch. Trop. Afr. (1970) 60, col. phot.; Johansson, Acta Phyt. Suec. 59 (1974) 31; Stewart in Agnew, Upl. Ken. W. Fl. (1974) 763; Folkersma, Tijdschr. Ned. Orch. Ver. 38 (1976) 238, col. phot.; Adam, Fl. Descr. Mt. Nimba 5 (1981) 1675; Vermeulen, Tijdschr. Ned. Orch. Ver. 43 (1981) 124, pl.; Segerbaeck, Orch. Nig. (1983) 65, col. phot.; Cribb, in F.T.E.A., Orch. 2 (1984) 317.–*Megaclinium falcatum* Lindl., Bot. Reg. 12 (1826) pl. 989; id., Gen. and Sp. Orch. Pl. (1830) 47; Durand & Schinz, Consp. Fl. Afr. 5 (1895) 15; Rolfe in F.T.A. 7 (1897) 41.–*Phyllorchis falcata* (Lindl.) Kuntze, Rev. Gen. Pl. 2 (1891) 677, incorr. name.–Type: Afzelius s.n. (K).

Bulbophyllum oxyodon Rchb. f., Gard. Chron., ser. 3, 4, nr. 80 (1888) 91, nom. alt.; Schltr, Beih. Bot. Zentralbl. 33 (1915) 419.–*Megaclinium oxyodon* Rchb. f., Gard. Chron., ser. 3, 4, nr. 80 (1888) 91; Durand & Schinz, Consp. Fl. Afr. 5 (1895) 16.–Type: Lendy s.n. (W).

Bulbophyllum leptorhachis Schltr, Westafr. Kautchuk Exp. (1901) 281, nom. nud.; Schltr, Engl., Bot. Jahrb. 38 (1905) 17; De Wild., Pl. Beq. 1 (1921) 92; Schltr, Fedde, Rep. Beih. 68 (1932) pl. 67.–Type: Schlechter 12992 (B, lost).

Megaclinium endotrachys Krzl., Engl., Bot. Jahrb. 36 (1905) 115.–Type: Dinklage 1852 (Holo B, lost; iso AMES, K, WAG, Z).

Bulbophyllum hemirhachis (Pfitz.) De Wild., Pl. Beq. 1 (1921) 86.–*Megaclinium hemirhachis* Pfitz., Orchis 2 (1908) 135.–Type: Schneider s.n. (?).

Bulbophyllum ugandae (Rolfe) De Wild., Pl. Beq. 1 (1921) 100.–*Megaclinium ugandae* Rolfe, Kew Bull. 1913 (1913) 338.–Type: Glasnevin cult. (K, n.v.).

Bulbophyllum dahlemense Schltr, Notizbl. Berl.-Dahl. 7 (1919) 325.–Type: Berlin cult. (Deistel s.n.) (B, lost).

Bulbophyllum bufo auct. non Lindl.: Vermeulen, Tijdschr. Ned. Orch. Ver. 43 (1981) 124.

Rhizome 1.5–3 mm diam. Pseudobulbs 1.8–6 cm long. Leaves 2–14 by 0.5–3 cm. Inflorescence 7–40 cm long. Peduncle 4–17 cm long. Rhachis 3–23 by 0.4–1.3 cm, margins sharp. Flowers 5–18 mm apart, measured along one side of the rhachis, the lower up to 32 mm apart. Median sepal narrowly obovate to broadly obovate-lanceolate, 3.5–8 by 1.5–2.5 mm, tip rounded to obtuse; abaxially not or sparsely beset with dark hairs. Lateral sepals 3–5.5 by 2.2–3.5 mm, adaxially finely papillose towards the base, abaxially as the median sepal. Petals ovate-lanceolate to broadly linear-lanceolate, 1.8–3 by 0.4–0.8 mm, top thickened, rounded to subacute. Lip 1.5–2 by 1–1.8 mm.

Colours. Pseudobulbs (olive) green. Leaves green. Rhachis greenish, often spotted with dark red-purple, occasionally entirely dark red-purple. Median sepal with bright yellow or red calli. Sepals otherwise yellowish adaxially, the lateral ones often spotted with dark red; abaxially whitish, yellowish or greenish, stained with dark red. Petals yellow, with a bright yellow tip. Lip cream-coloured, spotted with dark red, or almost entirely dark red with a yellow base.

Ecology. Lowland forest, lower montane forest, relict forest in savanna, up to 1400 m alt. Epiphyte in the central part of the crown of trees, in minor humus deposits (Johansson, 1974). Flowering VIII–V.

Distribution. GUIN. (2). S.L. (14 coll.). LIB. (10). IV.C. (21). GHANA (4). TOGO (fide F.W.T.A. ed. 2). NIG.: S. part (16). CAM.: mainly W. part (8).

EQ.GUIN.: Bioko (2). ZAI.: Kasai Z4 (1); Forestier Central Z6 (2). UG.: West U2 (1); Buganda U4 (3). Material seen from AMES, BM, BR, E, HBG, K, L, P, UPS, WAG.

Notes. 1 - The rhachis is always more or less flattened.

2 - Dinklage, the collector of the type specimen of *Megaclinium endotrachys*, is mentioned as 'the botanist D.' in Graham Greene's 'Journey Without Maps' (Penguin, Harmondsworth, ed. 1980, p. 66).

51b - var. bufo (Lindl.) J.J. Vermeulen - Fig. 76; Plate 9b.

Bulbophyllum falcatum (Lindl.) Rchb. f. var. *bufo* (Lindl.) J.J. Vermeulen, Bull. J. Bot. Nat. Belg. 56 (1986) 235.-*Bulbophyllum bufo* (Lindl.) Rchb. f., Walp. Ann. 6 (1861) 258 ('*Bolbophyllum*'); De Wild., Pl. Beq. 1 (1921) 80; Summerh. in F.W.T.A. ed. 1, 2 (1936) 440; id., ed. 2, 3 (1968) 241; Johansson, Acta Phyt. Suec. 59 (1974) 29; Adam, Fl. Descr. Mt. Nimba 5 (1981) 1672.-*Megaclinium bufo* Lindl., Bot. Reg. 27 (1841) misc. 42; id., Gard. Chron. 22 (1841) 348; Durand & Schinz, Consp. Fl. Afr. 5 (1895) 15; Rolfe in F.T.A. 7 (1897) 43.-*Phyllorchis bufo* (Lindl.) Kuntze, Rev. Gen. Pl. 2 (1891) 677, incorr. name.-Type: Loddiges s.n. (K).

Non *Bulbophyllum bufo* auct.: Vermeulen, Tijdschr. Ned. Orch. Ver. 43 (1981) 124 (= var. *falcatum*).

Bulbophyllum deistelianum (Krzl.) Schltr, Engl., Bot. Jahrb. 38 (1905) 17; De Wild., Pl. Beq. 1 (1921) 83.-*Megaclinium deistelianum* Krzl., Engl., Bot. Jahrb. 33 (1902) 72.-Type: Berlin cult. (Deistel s.n.) (B, lost, sketch Summerhayes in K).

Megaclinium gentilii De Wild., Bel. Col. 8 (1902) 425, n.v.; id., Et. Fl. B. et Moy. Congo 1 (1903) 23; id., 3 (1909) 59, pl. 26; Durand & Durand, Syll. Fl. Congol. (1909) 517.-*Bulbophyllum lubiense* De Wild., Pl. Beq. 1 (1921) 92.-Type: Gentil 1-159 (1902) (BR).

Non *Bulbophyllum gentilii* Rolfe, Gard. Chron., ser. 3, 35, no. 929 (1904) 266 (= *Bulbophyllum schinzianum* var. *schinzianum* and var. *phaeopogon*).

Bulbophyllum bakossōrum Schltr, Westafr. Kautchuk Exp. (1901) 280, nom. nud.; id., Engl., Bot. Jahrb. 38 (1905) 13; De Wild., Pl. Beq. 1 (1921) 78.-Type: Schlechter 12398 (B, lost, sketch Summerhayes in K).

Bulbophyllum longibulbum Schltr, Westafr. Kautchuk Exp. (1901) 281, nom. nud.; id., Engl., Bot. Jahrb. 38 (1905) 17; De Wild., Pl. Beq. 1 (1921) 92; Summerh. in F.W.T.A. ed. 1, 2 (1936) 438; id., ed. 2, 3 (1968) 239.-Type: Schlechter 12893 (B, lost; sketch Summerh. in K).

Bulbophyllum sereti De Wild., Bull. J. Bot. Brux. 5 (1916) 177 (*Megaclinium sereti* De Wild., nom. alt.); id., Pl. Beq. 1 (1921) 96.-Syntypes: Seret 654 (BR), Boone 116 (BR).

Rhizome 2–4 mm diam. Pseudobulbs 3.5–8 cm long. Leaves 8.5–21 by 1.3–4.5 cm. Inflorescence 11–40 cm long. Peduncle 3–12 cm long. Rhachis 7–28 by 0.2–1.8 cm, margins obtuse to sharp. Flowers 8–30 mm apart, measured along one side of the rhachis, the lower up to 30 mm apart. Median sepal spatulate, rarely obovate, (5 -) 6–8 by 2.5–4 mm, tip rounded to subacute to shortly acuminate; abaxially usually sparsely beset with dark hairs. Lateral sepals 5–9 by 3–7 mm, adaxially finely papillose, abaxially as the median sepal. Petals ovate to ovate-oblong, occasionally ovate-lanceolate, 1.5–2.5 by 0.5–1 mm, top not thickened, subacute to acute. Lip 1.5–4 by 1.2–3 mm.

Colours. Pseudobulbs olive green. Leaves dark green. Rhachis green, spotted red-purple. Sepals abaxially whitish with dark red-purple dots; the median one adaxially yellow, the lateral ones whitish with few and small red-purple spots. Petals whitish with very small red-purple spots. Lip whitish, partly or almost entirely dark red-purple.

Ecology. Lowland forest and hill forest, up to 800 m alt. Epiphyte in the central part of the crown of trees, in humus deposits (Johansson, 1974). flowering IV and VIII–XI.

Distribution. GUIN. (2 coll.). S.L. (1). LIB. (10). IV.C. (5). GHANA (1). NIG.: S. part (3). CAM.: Coast, S. part (3). ZAI.: Bas Congo Z3 (1); Kasai Z4 (1); Forestier Central Z6 (3). Material seen from AMES, BM, BR, K, P, S, UPS, W, WAG.

Notes. 1 - Specimens with a flattened rhachis occur together with specimens with a terete rhachis. Otherwise these specimens are entirely identical and intermediates are frequently found.

2 - The specimen W. de Wilde & de Wilde-Duyfjes 3732 from CAM. has slightly larger vegetative parts and an acuminate median sepal.

The specimen Leeuwenberg 5438 from CAM. has slightly larger flowers (median sepal 10 mm long). The wings along the adaxial margins of the column are not folded inwards and have a slightly serrulate tip.

51c - var. velutinum (Lindl.) J.J. Vermeulen - Fig. 77; Plate 9c, d.

Bulbophyllum falcatum (Lindl.) Rchb. f. var. *velutinum* (Lindl.) J.J. Vermeulen, Bull. J. Bot. Nat. Belg. 56 (1986) 235.-*Bulbophyllum velutinum* (Lindl.) Rchb. f., Walp. Ann. 6 (1861) 258 ('*Bolbophyllum*'); De Wild., Pl. Beq. 1 (1921) 100; Summerh. in F.W.T.A. ed. 1, 2 (1936) 440; Robyns, Fl. Sp. P. Alb. 3 (1955) 496; Summerh., Bull. Inst. Fr. d'Afr. N., ser. a, 21 (1959) 472; id. in F.W.T.A. ed. 2, 3 (1968) 239, pl.; Adam, Fl. Descr. Mt. Nimba 5 (1981) 1685, pl. 914; Vermeulen, Tijdschr. Ned. Orch. Ver. 43 (1981) 126.-*Megaclinium velutinum* Lindl., Bot. Reg. 33 (1847) sub pl. 32; Durand & Schinz, Consp. Fl. Afr. 5 (1895) 16; Rolfe in F.T.A. 7 (1897) 39.-*Phyllorchis velutina* (Lindl.) Kuntze, Rev. Gen. Pl. 2 (1891) 678, incorr. name.-Type: Loddiges s.n. (K).

Bulbophyllum rhizophorae Lindl., J. Linn. Soc. Bot. 6 (1862) 125 ('*Bolbophyllum*'); Hook., Bot. Mag. 88 (1862) pl. 5309 ('*Bolbophyllum*'); Durand & Schinz, Consp. Fl. Afr. 5 (1895) 14; Rolfe in F.T.A. 7 (1897) 26; Schltr, Westafr. Kautchuk Exp. (1901) 282; Malmquist, Orchis 8 (1914) 69; Schltr, Die Orch. (1914) 329; De Wild., Pl. Beq. 1 (1921) 117; Summerh. in F.W.T.A. ed. 1, 2 (1936) 440, p.p.; id., Kew Bull. 12 (1957) 123; id. in F.W.T.A. ed. 2, 3 (1968) 239, p.p.; Johansson, Acta Phyt. Suec. 59 (1974) 32; Adam, Fl. Descr. Mt. Nimba 5 (1981) 1683.-*Phyllorchis rhizophorae* (Lindl.) Kuntze, Rev. Gen. Pl. 2 (1891) 677, incorr. name.-Type: Barter 2048 ('20118') (K).

Non *Bulbophyllum rhizophorae* auct.: Vermeulen, Tijdschr. Ned. Orch. Ver. 43 (1981) 90 (= *Bulbophyllum resupinatum* var. *resupinatum*).

Bulbophyllum melanorrhachis Rchb. f., Gard. Chron., ser. 2, 4, no. 84 (1875) 162, nom. alt.; De Wild., Pl. Beq. 1 (1921) 93; Summerh. in F.W.T.A. ed. 1, 2 (1936) 440; id., Kew Bull. 6 (1952) 470; id. in F.W.T.A. ed. 2, 3 (1968) 241; Cribb & Perez-Vera, Adansonia, ser. 2, 15 (1975) 206; Adam, Fl. Descr. Mt. Nimba 5 (1981) 1680; Vermeulen, Tijdschr. Ned. Orch. Ver. 43 (1981) 128.-*Megaclinium melanorrhachis* Rchb. f., Gard. Chron., ser. 2, 4, no. 84 (1875) 162; Durand & Schinz, Consp. Fl. Afr. 5 (1895) 16; Rolfe in F.T.A. 7 (1897) 41.-Type: Hb. Reichenbach f. (W).

Bulbophyllum minutum (Rolfe) Engler, Pflanzenw. Afr. 2 (1908) 417.-*Megaclinium minutum* Rolfe, Kew Bull. 1893 (1893) 5; Hook. f., Bot. Mag. 119 (1893) pl. 7314; Rolfe in F.T.A. 7 (1897) 40.-*Bulbophyllum kewense* Schltr, Die Orch. (1914) 328; De Wild., Pl. Beq. 1 (1921) 89.-Type: Scott Elliot s.n. (K).

Non *Bulbophyllum minutum* Thou., Orch. Isles Austr. Afr. (1822) pl. 109.

Bulbophyllum millenii (Rolfe) Schltr, Engl., Bot. Jahrb. 38 (1905) 13, sub *Bulbophyllum bakossorum*; De Wild., Pl. Beq. 1 (1921) 93.-*Megaclinium millenii* Rolfe in F.T.A. 7 (1897) 40; Rendle, Cat. Talb. S. Nig. Pl. (1913) 146.-Type: Millen s.n. (K).

Bulbophyllum arnoldianum (De Wild.) De Wild., Pl. Beq. 1 (1921) 77.-*Megaclinium arnoldianum* De Wild., Miss. Laur. (1905) 55; Durand & Durand, Syll. Fl. Congol. (1909) 517.-Type: Laurent s.n. (1903) (BR).

Bulbophyllum minutum (Rolfe) Engler var. *purpureum* (De Wild.) De Wild., Pl. Beq. 1 (1921) 89.-*Megaclinium minutum* Rolfe var. *purpureum* De Wild., Bull. J. Bot. Brux. 3 (1911) 255; id., Et. Fl. B. et Moy. Congo 3 (1912) 376.-Type: Laurent s.n. (1905) (BR).

Megaclinium lasianthum Krzl., Engl., Bot. Jahrb. 48 (1912) 393.-*Bulbophyllum simonii* Summerh., Kew Bull. 1935 (1935) 204; id. in F.W.T.A. ed. 1, 2 (1936) 438; id., ed. 2, 3 (1968) 239; Lodewijk, Tijdschr.

Ned. Orch. Ver. 38 (1976) 234, col. phot.; Vermeulen, Tijdschr. Ned. Orch. Ver. 43 (1981) 91, pl.– Type: Simon 11 (B, lost, sketch of Summerhayes in K).

Non *Bulbophyllum lasianthum* Lindl., Gard. Chron. (1855) 53.

Bulbophyllum fractiflexum Krzl., Engl., Bot. Jahrb. 48 (1912) 392; De Wild., Pl. Beq. 1 (1921) 110.–Type: Zenker 3910 (B, lost).

Non *Bulbophyllum fractiflexum* J.J. Smith, Bull. Dep. Agr. Ind. Neerl. 19 (1908) 6; nec *Bulbophyllum fractiflexum* Pabst, Arq. do J. Bot. do R. de Jan. 14 (1956) 23.

Bulbophyllum solheidii De Wild., Bull. J. Bot. Brux. 5 (1916) 178 (*Megaclinium solheidii* De Wild., nom. alt.); id., Pl. Beq. 1 (1921) 97.–Type: Solheid 78 (BR).

Bulbophyllum brixhei De Wild., Bull. J. Bot. Brux. 5 (1916) 174 (*Megaclinium brixhei* De Wild., nom. alt.); id., Pl. Beq. 1 (1921) 79.–Type: Pynaert 1456 (BR).

Bulbophyllum lanuriense De Wild., Rev. Zool. Afr. 9, suppl. bot. (1921) 30 (*Megaclinium lanuriense* De Wild., nom. alt.); id., Pl. Beq. 1 (1921) 90.–Syntypes: Bequaert 4489 (BR), 4493 (BR).

Megaclinium angustum Rolfe, Kew Bull. 1922 (1922) 26.–Type: Holland s.n. (K).

Rhizome 1–3.5 mm diam. Pseudobulbs 0.9–4.2 cm long. Leaves 2–15.5 by 0.5–2 cm. Inflorescence 3–28 cm long. Peduncle 1–7 (– 10) cm long. Rhachis 2–20 by 0.1–1 cm, margins obtuse to sharp. Flowers 2–15 (– 30) mm apart, measured along one side of the rhachis, the lower up to 30 mm apart. Median sepal obovate to broadly obovate-lanceolate, 2.3–6 (– 7) by 0.8–2.2 mm, tip obtuse to acute; abaxially sparsely to densely set with dark hairs. Lateral sepals 1.8–5.2 by 0.8–3.5 mm, adaxially finely papillose to coarsely puberulous, abaxially as the median sepal. Petals very narrowly triangular, 1.3–3.2 by 0.25–0.8 mm, top not thickened, gradually tapering into a sharply acute to acuminate tip. Lip 0.7–2.5 by 0.5–1.8 mm.

Colours. Pseudobulbs (pale) green or olive green. Leaves green. Rhachis green, often spotted or stained with dark red or entirely dark red. Median sepal with yellow, orange or red calli. Sepals otherwise yellowish or greenish, the lateral ones often partly dark red.Petals yellow or dark red. Lip yellow or dark red with a yellow base. Occasionally flowers entirely dark red.

Ecology. Lowland forest, relict forest in savanna, montane forest, littoral forest, mangrove forest, also recorded from coffee trees; up to 1800 m alt. Epiphytic, rarely lithophytic. Flowering I–XII.

Distribution. S.L. (6 coll.). LIB. (10). IV.C. (26). GHANA (7). NIG.: S. part (16). CAM.: S. part (19). EQ.GUIN.: Mbini (3); Bioko (2); Pagalu (1). S.T. & PR.: Sao Tomé (7); Principe (1). GAB. (16). ZAI.: Forestier Central Z6 (18); Lacs Edouard & Kivu Z9 (6). Material seen from AMES, BM, BR, E, G, K, L, P, UPS, W, WAG, Z.

Notes. Within var. *velutinum* two entities may be recognized on account of the shape of the rhachis. The flowers of these entities are identical.

a. - Rhachis erect to patent, distinctly widened and flattened.

This entity occurs throughout the area of var. *velutinum*. The name *velutinum* applies to specimens with large flowers (median sepal 4–6 mm long); the name *melanorhachis* to specimens with small flowers (median sepal 2.3–4 mm long). A single specimen from GHANA (Hall GC 39020) and most specimens of Sao Tomé (e.g. Wageningen cult. Hort. 80-98, = v.d. Laan 492) resemble var. *falcatum* in general aspect, however, the shape of the petals is of var. *velutinum*. They occur together with typical specimens of var. *velutinum* entity a.

The specimen Groot 785 from ZAI. has exceptionally wide sepals.

b. - Rhachis arching, patent to pendulous, not widened nor flattened but terete in section when flowering, occasionally slightly thickened and flattened with rounded margins when developing fruits.

This entity only occurs in NIG., CAM. (especially shrub vegetations near the coast), EQ.GUIN., Principe, GAB. and ZAI. In most of its area it is morphologically well separated from entity a., only in ZAI. intermediates between a. and b. occur (e.g. the type specimen of *Bulbophyllum arnoldianum*).

The name *simonii* applies for specimens with large flowers. Specimens with small flowers are very rare (e.g. the types of *Bulbophyllum brixhei* and *Bulbophyllum rhizophorae*).

The specimen Sanford 5862 from EQ.GUIN. has a much elongated inflorescence with widely spaced flowers (30 mm apart, measured along one side of the rhachis). This variability is comparable to that within *B. resupinatum* var. *filiforme*.

52 - Bulbophyllum calyptratum Krzl. - Fig. 78–80; Plate 10a, b.

Literature: see under the varieties.

Roots 0.8–1.5 mm diam. *Rhizome* 2–5 mm diam., nodes not or hardly swollen. *Rhizome scales* tubular, tip obtuse to acute; membranous, with nerves; slightly keeled abaxially towards the tip. *Pseudobulbs* 2-leafed, (narrowly) ovoid to (narrowly) ellipsoid, 1.3–5 cm apart, 1.8–6 by 0.7–2 cm, more or less flattened, (2 –) 3–4-angled. *Petiole* 5–25 mm long. *Leaf blade* linear-lanceolate to narrowly linear, 7.2–26 by 0.2–1.6 cm, tip obtuse, emarginate, oblique; coriaceous, rather thick to thick; midrib not or hardly prominent abaxially. *Inflorescence* 8–60 cm long, 8–50-flowered. *Peduncle* erect to about pendulous, swollen and flattened towards the tip, 3.5–30 cm by 1–2.8 mm, orbicular in section, glabrous, with 6–9 scales. *Peduncle scales* tubular, the longest 8–13 mm long; tip obtuse to acuminate; membranous, either with some nerves or not; slightly keeled towards the tip and glabrous or with some fine black hairs abaxially. *Rhachis* erect, patent or reflexed, about terete to widened and flattened, either bladelike or not, either slightly zig-zag bent or not, 4–31 cm by 2–17 mm, edges obtuse to sharp, straight to undulate, entire to irregularly dentate, glabrous; surface glabrous or with fine dark hairs. *Floral bracts* spreading to reflexed, ovate to ovate-oblong, 2.5–7 by 1.2–3 mm, tip acute to acuminate; margins either folded backwards or not; base broadly attached; membranous, without nerves; not keeled and glabrous or with fine dark hairs abaxially. *Flowers* distichous, placed along a median to excentric line on the rhachis, 5–20 mm apart, measured along one side of the rhachis, the lower up to 8–28 mm apart, incurved to spreading, either secund or not, some to many open simultaneously, fully open. *Pedicel and ovary* 2–7 mm long, glabrous or with fine dark hairs. *Median sepal* incurved to recurved, spathulate, 5–8.3 by 2.2–4.3 mm, tip emarginate to acute; margins entire, glabrous; base rather broadly attached; thick, top half with a fleshy callus along both margins; glabrous adaxially, occasionally with some fine dark hairs near the base; not keeled and with fine dark hairs abaxially. *Lateral sepals* free, basal half recurved and concave, top half spreading to reflexed; slightly falcate, triangular, oblique, 3.8–8 by 2.5–5 mm, tip acute; base broadly attached; rather thick, without calli; finely papillose adaxially in the basal half; not keeled and with fine dark hairs abaxially; otherwise as the median sepal. *Petals* not recurved, hardly falcate, (ovate-)oblong to broadly ovate-lanceolate, oblique, 1.1–2.5 by 0.3–0.8 mm, tip obtuse to acute; margins entire, glabrous; base (rather) broadly attached; thin; glabrous; not keeled abaxially. *Lip* recurved, ovate to broadly ovate-oblong in outline (not spread), 1–2.5 by 0.8–2 mm, tip rounded; margins entire, glabrous; thick but rather soft and easy to compress; surface glabrous; adaxially without ridges, with a not or hardly concave basal part, without a median slit; abaxially with a distinct, truncate to retuse keel. *Column* from the base of the free part up to the tip of the stelidia 0.8–2.1 mm long, with distinct, triangular, not falcate, (sub)acute stelidia of 0.2–0.4 mm long; column with (rather) weak, deltoid, obtuse to subacute wings along its adaxial margins, just below the base of the stelidia, which continue towards the column foot as narrow seams; stigma without a basal callus. *Anther* c. 1 by 0.5 mm; front distinctly drawn out, thin, slightly concave; front margin truncate to rounded, entire, glabrous; basally weakly emarginate; abaxially with a rather weak, rounded, colliculate ridge near the top of the anther which does not overtop the front margin, surface glabrous but colliculate towards the top. *Pollinia* 4; obovoid, flattened, the small ones about half as long as the large ones; a stipes has not been observed. *Fruit* ellipsoid, up to 1.4 by 0.4 cm, stalk 2 mm.

KEY TO THE VARIETIES

1 - Flowers with the median sepal curved towards the rhachis only just after opening, flowers spreading afterwards and becoming even reflexed (nb: Pollinated flowers and fruits resume a recurved position again); pedicel and ovary (unpollinated flowers of dried specimens!) 4–7 mm long .... var. **calyptratum** (a)
- Flowers with the median sepal curved towards the rhachis all the time they are open; pedicel and ovary (unpollinated flowers of dried specimens!) 2–4 mm long 2

2 - Inflorescence erect; rhachis terete or nearly so; petals 1.7–2.5 mm long .... var. **graminifolium** (b)
- Inflorescence curved downwards; rhachis distinctly widened and flattened, more or less reflexed below the substratum on which the plant grows; petals 1.1–1.5 mm long .... var. **lucifugum** (c)

52a - var. calyptratum – Fig. 78; Plate 10a.

Bulbophyllum calyptratum Krzl., Engl., Bot. Jahrb. 22 (1895) 24 ('*Bolbophyllum*'); Rolfe in F.T.A. 7 (1897) 26; Schltr, Westafr. Kautchuk Exp. (1901) 281; De Wild., Pl. Beq. 1 (1921) 107; Summerh. in F.W.T.A. ed. 1, 2 (1936) 440; id., Kew Bull. 12 (1957) 116; id., Bull. Inst. Fr. d'Afr. N., ser. a, 21 (1959) 473; id. in F.W.T.A. ed. 2, 3 (1968) 241; Segerbaeck, Orch. Nig. (1983) 65, col. phot.–Type: Preuss 1215 (Holo B, lost; iso BM, P).

Bulbophyllum lindleyi (Rolfe) Schltr, Die Orch. (1914) 327; De Wild., Pl. Beq. 1 (1921) 92; Summerh. in F.W.T.A. ed. 1, 2 (1936) 440; Ake Assi, Et. Fl. C. Iv. (1964) 244.–*Megaclinium lindleyi* Rolfe in F.T.A. 7 (1897) 43.–*Megaclinium maximum* auct. non Lindl. (1830): Lindl., Bot. Reg. 23 (1837) pl. 1959; Hook., Cent. Orch. Pl. (1849) 15, pl. 3 (*Megaclinium flaccidum* on plate).–Type: Loddiges s.n. (K).

Bulbophyllum buchenavianum (Krzl.) De Wild., Pl. Beq. 1 (1921) 106.–*Megaclinium buchenavianum* Krzl., Engl., Bot. Jahrb. 36 (1905) 115.–Type: Zenker 2894 (Holo B, lost; iso G).

Megaclinium lepturum Krzl., Ann. Nat. Hofmus. Wien 30 (1916) 61.–Type: Ludwigs s.n. (B, lost).

Leaves linear-lanceolate to linear, 7.2–20 by 0.5–1.6 cm. Inflorescence 22–47 cm long, with 8–50 flowers. Peduncle erect to about patent, 11–21 cm long. Rhachis erect or arching, about terete to widened and flattened, either bladelike or not, edges obtuse to sharp. Flowers with the median sepal curved towards the rhachis only just after opening, when fully open spreading or reflexed, recurved again after pollination. Pedicel and ovary 4–7 mm long. Petals 1.2–1.8 by 0.3–0.7 mm. Lip 1–2 by 0.4–0.8 mm.

Colours. Pseudobulbs and leaves usually rather dark green. Rhachis greenish to almost white, with or without purplish dots. Median sepal with yellow or brownish calli. Sepals otherwise whitish or slightly greenish, with dark red dots and stains. Petals white. Lip white with or without dark red dots or stains, or entirely dark red.

Ecology. Lowland rainforest and semideciduous forest, lower montane forest up to 1100 m. Also in mangrove forest and quite frequently found in secondary forest and plantations. Epiphytic. Flowering mainly II–VI and X–XI.

Distribution. GUIN. (1 coll.). S.L. (4). LIB. (2). IV.C. (3). GHANA (2). NIG.: S. part (3). CAM.: S. part (14). EQ.GUIN.: Mbini (3). GAB. (9). CONGO (1). ZAI.: Forestier Central Z6 (3); Lacs Edouard et Kivu Z9 (1). Material seen from BM, BR, G, K, L, P, WAG, Z.

Notes. 1 - Included in var. *calyptratum* are specimens with an almost terete rhachis next to specimens with a bladelike rhachis. This difference is not supported by any other character and intermediates between the two extremes occur frequently.

2 - Remarkable is that in the extreme W. and E. part of the area of var. *calyptratum* (GUIN., S.L., LIB., IV.C. and ZAI.) specimens with a bladelike rhachis are present almost exclusively, whereas in the central part (GHANA, NIG., CAM., EQ.GUIN. and GAB.) specimens with a terete rhachis and specimens with a bladelike rhachis occur together, the latter often in distinctly fewer numbers.

52b - var. graminifolium (Summerh.) J.J. Vermeulen - Fig. 79; Plate 10b.

Bulbophyllum calyptratum Krzl. var. *graminifolium* (Summerh.) J.J. Vermeulen, Bull. J. Bot. Nat. Belg. 56 (1986) 235.–*Bulbophyllum graminifolium* Summerh., Kew Bull. 12 (1957) 117; id. in F.W.T.A. ed. 2, 3 (1968) 241; Adam, Fl. Descr. Mt. Nimba 5 (1981) 1676.–Type: Deighton 5744 (K).

Bulbophyllum intermedium De Wild., Bull. J. Bot. Brux. 5 (1916) 176 (*Megaclinium intermedium* De Wild., nom. alt.); id., Pl. Beq. 1 (1921) 88.–Type: Seret 335 (BR).

Non *Bulbophyllum intermedium* F.M. Bail., Bot. Bull. Dep. Agr. Qd. 14 (1896) 12.

Leaves (narrowly) linear, 11–26 by 0.3–1.2 cm. Inflorescence 8–28 cm long, with 11–40 flowers. Peduncle erect, 3.5–11 cm long. Rhachis erect, about terete or slightly flattened, not bladelike, edges obtuse. Flowers with the median sepal curved towards the rhachis all the time they are open. Pedicel and ovary 2–4 mm long. Petals 1.7–2.5 by 0.5–0.8 mm. Lip 1.8–2.5 by 0.9–2 mm.

Colours. Pseudobulbs and leaves usuallly medium or light green. Rhachis light green. Sepals cream coloured, occasionally with a few small pale purple dots, the median with yellowish calli. Petals white. Lip white, suffused with pale purple or pinkish. Column white.

Ecology. Lowland forest. Also in orchards and plantations. Altitude not given. Epiphytic. Flowering (II, VI) X–XII.

Distribution. GUIN.: Central part (2 coll.). S.L. (3). LIB. (3). IV.C. (1). GHANA (5). ZAI.: Forestier Central Z6 (4). Material seen from BM, BR, K, L, P, WAG.

Notes. 1 - Var. *graminifolium* has roughly the same distribution as the specimens of var. *calyptratum* with a bladelike rhachis.

2 - Some specimens of var. *graminifolium* approach var. *calyptratum* in the shape of the flower parts.

52c - var. lucifugum (Summerh.) J.J. Vermeulen, stat. nov.-Fig. 80.

Bulbophyllum lucifugum Summerh., Kew Bull. 12 (1957) 119; id. in F.W.T.A. ed. 2, 3 (1968) 241; Johansson, Acta Phyt. Suec. 59 (1974) 32; Cribb & Perez-Vera, Adansonia, ser. 2, 15 (1975) 206; Adam, Fl. Descr. Mt. Nimba 5 (1981) 1679, pl. 954.-Type: Deighton 4377 (K).

Leaves narrowly linear, 19–24 by 0.2–0.8 cm. Inflorescence 10–18 cm long, with 12–14 flowers. Peduncle curved downwards, 4.5–10 cm long. Rhachis pendulous or reflexed below the substratum on which the plant grows, widened and flattened, bladelike, edges sharp. Flowers with the median sepal curved towards the rhachis all the time they are open. Pedicel and ovary c. 4 mm long. Petals 1.1–1.5 by 0.5 mm. Lip 1.8–2.2 by 1.2 mm.

Colours. Rhachis with small purple dots. Sepals cream coloured, with small purple dots, the median one with yellowish calli. Petals white. Lip suffused with purple.

Ecology. Lowland forest and hill forest up to 700 m alt. Epiphyte in the central part of the crown of trees, on bark or in minor humus deposits (Johansson, 1974).

Distribution. S.L. (2 coll.). LIB. (3). IV.C. (3). Material seen from AMES, K, UPS.

Note. Var. *lucifugum* and var. *graminifolium* are kept as separate varieties because they differ in two independent characters: the length of the petals and the shape and peculiar position of the rhachis (tucked away below the branch on which the plant grows). The latter character suggests a system of pollination which is quite different from that of var. *graminifolium*.

53 - Bulbophyllum carnosilabium Summerh. - Fig. 81.

Bulbophyllum carnosilabium Summerh., Kew Bull. 8 (1954) 582.-Type: Le Testu 6473 (Holo K; iso BM, P).

Roots 0.4–1.2 mm diam. *Rhizome* 1.7–4 mm diam., nodes not or hardly swollen. *Rhizome scales* tubular, tip subacute; membranous, with nerves; not keeled abaxially. *Pseudobulbs* 2-leafed, narrowly ovoid to about cylindrical, 2–3.5 cm apart, 3.7–7 by 0.8–1.8 cm, flattened, (2 -) 4-angled. *Petiole* 0.8–2 cm long. *Leaf blade* linear lanceolate to linear, 11–22.5 by 0.8–2.4 cm, tip obtuse, emarginate, oblique; coriaceous, rather thin to rather thick; midrib slightly prominent abaxially. *Inflorescence* 10–40 cm long, 8–34-flowered. *Peduncle* erect to patent, swollen and flattened towards the tip, 5–9.5 cm by 1–2 mm, orbicular in section, glabrous, with 5–8 scales. *Peduncle scales* tubular, the longest 5–10 mm long; tip obtuse to subacute; membranous, with some nerves; slightly keeled towards the tip and glabrous abaxially. *Rhachis* erect, widened and flattened, about bladelike, 6–30 cm by 2.5–8 mm, edges sharp, about straight, entire, glabrous; surface glabrous. *Floral bracts* reflexed, oblong, 3–6.5 by 1.5–2.5 mm, tip acute to acuminate; margins folded inwards; base broadly attached; membranous, without nerves; hardly keeled and glabrous abaxially. *Flowers* distichous, placed along the median line of the rhachis, 10–18 mm apart, measured along one side of the rhachis, the lower up to 17–22 mm apart, spreading, either secund or not, some open simultaneously, not fully open. *Pedicel and ovary* 2.5–4 mm long, glabrous or with some fine dark hairs. *Median sepal* either slightly recurved or not, elliptic, 4–6.8 by 2.2–3.3 mm, tip acuminate; margins entire, finely papillose; base rather broadly attached; thick, top half adaxially with a fleshy callus along both margins; adaxially finely papillose, basal part with some fine dark hairs; abaxially not keeled, slightly papillose and with some fine, dark hairs. *Lateral sepals* free, basal half recurved, concave, top half entirely reflexed; not or hardly falcate, triangular, oblique, 4.2–6.8 by 2.2–4 mm, tip acute; margins glabrous; base broadly attached; rather thick, without calli; finely papillose adaxially only in the basal half; otherwise as the median sepal. *Petals* not recurved, hardly falcate, ovate(-oblong), oblique, 1.2–2 by 0.4–1 mm, tip acute; margins entire, glabrous; base rather broadly attached; thin; glabrous; not keeled abaxially. *Lip* slightly recurved, ovate in outline (not spread), c. 3 by 1.7 mm, tip rounded; margins entire, glabrous; thick and slightly soft and easy to compress; surface glabrous; adaxially with two weak, rounded ridges over almost the entire length of the lip which diverge towards the base of the lip, with a concave basal part, without a median slit, surface slightly corrugate (due to drying?); abaxially with a weak, rounded ridge over almost the entire length of the lip. *Column* from the base of the free part up to the tip of the stelidia 1.5–2 mm long, with distinct, about triangular, not falcate, acute stelidia of 0.4–0.6 mm long which may have a weak, obliquely forwards directed tooth along their upper margin; column with rather distinct, deltoid, acute wings along its lower margins just below the base of the stelidia; stigma without a basal callus. *Anther* c. 1 by 0.8 mm; front moderately drawn out, thin, hardly concave; front margin retuse, entire, glabrous; basally weakly emarginate; abaxially with a rather distinct, obtuse, colliculate ridge near the top of the anther which does not overtop the front margin, surface glabrous but colliculate towards the top. *Pollinia* 4, obovoid; the small ones about half as long as the large ones, distinctly flattened, the large ones flattened; a stipes has not been observed. *Fruit* not seen.

Colours. Flowers reddish.

Ecology. Lowland and hill forest up to 850 m alt. Epiphytic. Flowering IX–XII.

Distribution. GAB. (4 coll.). ZAI.: Lacs Edouard & Kivu Z9 (1). Material seen from BM, BR, K, P.

GROUP 6 (sp. 54)

Pseudobulbs 1-leafed. Inflorescence a raceme, 3–many-flowered. Rhachis flattened, flowers inserted along the margins. Flowers distichous. Sepals glabrous abaxially, the lateral ones not connate along the upper margins, not recurved. Lip thick, not very soft and easy to compress, lateral lobes absent, margins c. glabrous. Stelidia long, acute to sharply acuminate. Anther with a slightly dorsoventrally flattened protrusion overtopping the front margin.

54 - Bulbophyllum carnosisepalum J.J. Vermeulen - Fig. 82.

Bulbophyllum carnosisepalum J.J. Vermeulen, Bull. J. Bot. Nat. Belg. 56 (1986) 236.-Type: Bruneel 465 (BR).

Roots 0.5–0.8 mm diam. *Rhizome* 1.5–4 mm diam., nodes not or hardly swollen. *Rhizome scales* tubular, tip acute to acuminate; membranous, with nerves; keeled abaxially towards the tip. *Pseudobulbs* 1-leafed, orbicular to ellipsoid, occasionally narrowly obovoid, 0.4–0.8 cm apart, 0.9–1.5 by 0.4–1 cm, flattened, obtusely 2–4-angled. *Petiole* 1–9 mm long. *Leaf blade* (linear-)lanceolate, 4–8.5 by 0.8–1.6 cm, tip obtuse, neither emarginate nor oblique; coriaceous, thick; midrib slightly prominent abaxially. *Inflorescence* 4.8–7 cm long, 3–12-flowered. *Peduncle* erect, slightly swollen towards the tip, 1.5–4 cm by 1–1.5 mm, orbicular in section, glabrous, with c. 6 scales. *Peduncle scales* tubular, the longest 6–8 mm long; tip acuminate; membranous, without nerves; keeled and glabrous abaxially. *Rhachis* erect, slightly swollen and flattened, 4-angled in section, with two concave sides from which the flowers arise, slightly zig-zag bent, 1.5–4 cm by up to 4 mm, edges slightly obtuse to sharp, straight, entire, glabrous; surface glabrous. *Floral bracts* recurved to spreading, concave, triangular, 6.1–8 by 4–6 mm, tip acute to slightly acuminate; base broadly attached; rather thin, without nerves; distinctly keeled and glabrous abaxially. *Flowers* distichous, 6–9 mm apart, measured along one side of the rhachis, slightly recurved, some open simultaneously, not much open. *Pedicel and ovary* 2–3 mm long, either with some fine dark papillae or not. *Median sepal* not recurved, ovate(-oblong), 5–5.8 by 2,5–3.7 mm, tip acuminate to cuspidate (cusp up to 1.4 mm long); margins entire, glabrous; base broadly attached; thick, thickened towards the tip; glabrous adaxially; keeled and with some fine dark papillae abaxially. *Lateral sepals* free, not falcate, narrowly triangular, oblique, 5–5.8 by 2–2.5 mm, distinctly keeled abaxially; otherwise as the median sepal. *Petals* straight, either with a recurved tip or not, not or hardly falcate, linear-lanceolate, oblique, 2–2.2 by 0.3–0.4 mm, tip obtuse to subacute; margins entire, glabrous; base broadly attached; rather thin; glabrous; not keeled abaxially. *Lip* hardly recurved, broadly oblong in outline (not spread), slightly constricted about half-way, 1.6–2 by 0.7–1 mm, tip rounded; margins entire towards the base, irregularly crenulate towards the tip, glabrous; thick; adaxially without ridges, with a slightly concave basal part, without a median slit but slightly channeled in the median part about half-way the lip, surface glabrous towards the base, densely hirsute-aculeate towards the tip; abaxially with a distinctly convex surface, glabrous, either slightly irregular towards the tip or not. *Column* from the base of the free part up to the tip of the stelidia 1.8–2.2 mm long, with distinct, either gradually tapering or in the middle slightly widened, slightly falcate, narrowly acute to acuminate stelidia of 1–1.2 mm long; column with rather distinct, deltoid, acute teeth along its adaxial margins just below the base of the stelidia; stigma without a basal callus. *Anther* c. 0.8 by 0.6 mm; front not drawn out; front margin rounded, irregular, glabrous; basally distinctly emarginate; abaxially without a ridge but with a thick, not concave but slightly flattened, conical, rounded beak near the top of the anther which distinctly overtops the front margin, surface glabrous but colliculate towards the top. *Pollinia* 4; ellipsoid, distinctly flattened on one side, the small ones less than half as long as the large ones; a stipes has not been observed. *Fruit* ellipsoid, up to 0.6 by 0.4 cm, without a stalk.

Colours. Pseudobulbs and leaves medium green, either suffused with some red or not. Rhachis greenish. Floral bracts yellowish, often suffused with red. Flowers greenish, often stained with red-purple, or entirely red-purple.

Ecology. Lowland forest, montane forest. No altitude given. Epiphytic. Flowering X–XII.

Distribution. IV.C. (1 coll.). CAM.: Coast (2). GAB. (2). ZAI.: Forestier Central Z6 (1). UG. (1). Material seen from BR, K, P, WAG.

GROUP 7 (spp. 55 - 56)

Pseudobulbs 1-2-leafed. Inflorescence a raceme, many-flowered. Rhachis slightly swollen, spindle shaped. Flowers scattered. Sepals slightly irregular abaxially, the lateral ones not connate along the upper margins, not recurved. Lip thick, not very soft and easy to compress, with seamlike lateral lobes which have a long-haired margin. Stelidia long, narrowly acute. Anther with a dorsoventrally flattened, rounded to bifid protrusion overtopping the front margin.

55 - **Bulbophyllum ballii** Cribb - Fig. 83.

Bulbophyllum ballii Cribb, Kew Bull. 32 (1977) 159; Ball, S. Afr. Ep. Orch. (1978) 78, pl.–*Bulbophyllum* sp. Grosvenor, Excelsa 6 (1976) 79.–Type: Ball 2 (Holo K; iso SRGH, n.v.).

(?) *Bulbophyllum* sp. Williamson, Orch. S. Centr. Afr. (1977) 138.–Specimen: Williamson 141 (?, n.v., identity uncertain).

Roots 0.3–1 mm diam. *Rhizome* 1–2 mm diam., nodes not or hardly swollen. *Rhizome scales* tubular, tip acute to acuminate; membranous, without nerves; keeled abaxially. *Pseudobulbs* 1-2-leafed, ovoid to ellipsoid, 0.3–0.6 cm apart, 0.8–1.4 by 0.6–0.9 cm, hardly to moderately flattened, 4-angled. *Petiole* 1.5–6 mm long. *Leaf blade* lanceolate to linear-lanceolate, 2.5–7.5 by 0.8–1.2 cm, tip obtuse, emarginate, usually oblique; coriaceous, thick; midrib slightly prominent abaxially. *Inflorescence* 7–16 cm long, 15–27-flowered. *Peduncle* erect to patent, slightly swollen towards the tip, 3.5–7.5 cm by c. 1.5 mm, orbicular in section, glabrous, with 3–4 scales. *Peduncle scales* tubular, the longest 4.5–6 mm long; tip acute, the lower obtuse; membranous, without nerves; slightly keeled and glabrous abaxially. *Rhachis* erect to patent, slightly swollen, spindle-shaped, 3.5–7.5 cm by 1.5–5 (?) mm, glabrous. *Floral bracts* recurved, triangular to ovate, 2–3.5 by 1.5–2 mm, tip acute; base broadly attached; membranous, without nerves; either keeled or not and with scattered irregular papillae abaxially. *Flowers* placed in a regularly reticulate pattern, in a lax raceme, appressed against the rhachis to slightly recurved, some open simultaneously, not much open. *Pedicel and ovary* c. 1.2 mm long, with scattered, irregular papillae and some fine dark hairs, jugae ending in an acute tooth. *Median sepal* not recurved, oblong to broadly lanceolate, 3–3.2 by 1–1.5 mm, tip acuminate; margins entire, glabrous; base broadly attached; rather thick; glabrous adaxially; not keeled, slightly irregularly papillose and with some fine dark hairs abaxially. *Lateral sepals* free, slightly falcate, triangular to narrowly triangular, oblique, 2.8–3.7 by 1.2–1.8 mm, tip acute; otherwise as the median sepal. *Petals* not recurved, hardly falcate, ovate-oblong to ovate-lanceolate, oblique, 1.1–1.8 by 0.3–0.8 mm, tip acute; margins entire, glabrous; base broadly attached; rather thick; glabrous; not keeled abaxially. *Lip* recurved, elliptic to oblong in outline (not spread), 1.3–1.5 by 0.5–0.8 mm, tip rounded; margins with a narrow, thin seam with some scattered, hairlike lacerae along the edges; thick; surface glabrous; adaxially without ridges, except for two very small and short ridges along the margins, just above the base, with a slightly concave basal part, without a median slit but slightly channeled in the median part; abaxially with a weak retuse ridge in the basal 2/3 of the lip. *Column* from the base of the free part up to the tip of the stelidia 1.4–1.8 mm long, with distinct, distinctly falcate, gradually tapering, acute stelidia of 0.8–1 mm long; column with weak, broadly rounded wings along its adaxial margins, just below the base of the stelidia; stigma without a basal callus. *Anther* c. 0.5 by 0.5 mm; front hardly drawn out, thin, concave; front margin rounded, about entire, glabrous; basally distinctly emarginate; abaxially with a distinct, deeply bifid, glabrous ridge near the top of the anther which distinctly overtops the front margin, surface finely papillose. *Pollinia* 4; see note below. *Fruit* ellipsoid, up to 0.7 by 0.3 cm, without a stalk.

Colours. Sepals yellowish green, marked with purple-red, or entirely purple-red. Petals white with purple. Lip dark purple-red.

Ecology. Lower montane forest at c. 1000 m alt. Epiphytic.

Distribution. ZIM.: East (4 coll.). Material seen from K, material also present in SRGH.

Note. The species is probably autogamous. In all flowers studied the pollinia could not be removed from the top of the column. The anther, however, easily falls off. In spite of this fructification is abundant; the few specimens studied all bear old inflorescences with many well developed fruits.

56 - **Bulbophyllum elliotii** Rolfe – Fig. 84.

Bulbophyllum elliotii Rolfe, J. Linn. Soc. Bot. 29 (1891) 51; etc.–Type: Scott Elliot (1890) s.n. (K), Madagascar, Fort Dauphin.

Bulbophyllum malawiense Morris, Proc. Linn. Soc. London 179 (1968) 63; Morris, Ep. Orch. Mal. (1970) 45; Grosvenor, Excelsa 6 (1976) 79; Williamson, Orch. S. Centr. Afr. (1977) 136, col. phot.; Ball, S. Afr. Ep. Orch. (1978) 90, col. phot.; Vermeulen, Tijdschr. Ned. Orch. Ver. 43 (1981) 88, pl.; Stewart & all., Wild Orch. S. Afr. (1982) 219, col. phot.; La Croix et al., Orch. Mal. 1 (1983) 52; Cribb, in F.T.E.A., Orch. 2 (1984) 312.–Type: Morris 172 (K).

Bulbophyllum oreonastes auct. non Rchb. f.: Harrison, Ep. Orch. S. Afr. (1972) 37, pl. 15.

Roots 0.4–1.2 mm diam. *Rhizome* 1.5–3 mm diam., nodes not or hardly swollen. *Rhizome scales* tubular, tip obtuse, mucronate; membranous, the upper with some nerves; not keeled abaxially. *Pseudobulbs* 2-leafed, orbicular to ovoid or ellipsoid, 0.3–2 cm apart, 0.5–2.2 by 0.4–2.2 cm, not or hardly flattened, very obtusely 4-angled. *Petiole* 0.05–2 mm long. *Leaf blade* about orbicular to narrowly oblong, rarely broadly lanceolate, 1–4.4 by 0.6–1.7 cm, tip rounded to obtuse, emarginate, either oblique or not; coriaceous, thick to very thick; midrib hardly prominent abaxially. *Inflorescence* 4–15.5 cm long, 8–47-flowered. *Peduncle* patent to pendulous, sturdy, swollen towards the tip and below the nodes, 2–7.5 cm by 1.2–3 mm, orbicular in section, usually slightly granulate towards the tip, with 5–9 scales. *Peduncle scales* tubular, the longest 5–9 mm long; tip acute to acuminate; membranous, either with some nerves or not; keeled and glabrous abaxially. *Rhachis* patent to about pendulous, swollen, spindle-shaped, 2.2–7.5 cm by up to 4 mm, finely papillose, either with fine dark hairs or not. *Floral bracts* recurved, broadly ovate, 4–6 by 3–4 mm, tip acute; base broadly attached; membranous, without nerves; slightly keeled and either about glabrous or with some fine dark hairs abaxially. *Flowers* placed in a reticulate pattern or in 3–5 rows, in a rather dense raceme, about appressed against the rhachis or slightly recurved, many open simultaneously, not much open. *Pedicel and ovary* 1.3–2.5 mm long, glabrous or with some fine, dark hairs; jugae ending in a short acute tooth. *Median sepal* not recurved, narrowly elliptic to oblong, 3.5–5 by 1.5–2.2 mm, tip acute to acuminate; margins entire, glabrous; base (rather) broadly attached; thick; glabrous adaxially; not keeled, slightly irregularly papillose and either with or without some fine dark hairs abaxially. *Lateral sepals* free, slightly recurved, slightly falcate, ovate to broadly ovate-oblong, oblique, 3.5–4.5 by 1.7–2.8 mm, base broadly attached; otherwise as the median sepal. *Petals* not recurved, slightly falcate, ovate lanceolate to narrowly triangular or linear, oblique, 1.9–2.8 by 0.2–0.9 mm, tip acute to acuminate; margins either entire or irregularly denticulate or lacerate near the base, glabrous; base broadly attached; thin; glabrous; not keeled abaxially. *Lip* recurved, elliptic in outline (not spread), 1.6–2 by 0.8 - 1 mm, tip rounded to about truncate; margins with a narrow, thin seam with edges which are densely beset with very long hairs; thick; surface glabrous; adaxially without ridges, with a slightly concave basal part, without a median slit; abaxially with a weak, rounded ridge in the basal half, top part convex. *Column* from the base of the free part up to the tip of the stelidia 1.5–2.2 mm long, with distinct, gradually tapering, more or less falcate, acute stelidia of 0.6–1 mm long; column with at most very inconspicuous, broadly rounded wings along its adaxial margins, just below the base of the stelidia; stigma without a basal callus. *Anther* c. 0.5 by 0.6 mm; front not drawn out; front margin rounded, slightly irregular, about glabrous; basally distinctly emarginate; abaxially with a distinct, bifid or truncate to rounded, colliculate ridge near the top of the anther which distinctly overtops the front margin, surface colliculate. *Pollinia* 4, ellipsoid to ovoid; the small ones much more than half as long as the large ones, distinctly flattened; the large ones flattened; a stipes has not been observed. *Fruit* ellipsoid, up to 0.8 by 0.45 cm, without a stalk.

Colours. Pseudobulbs light green. Leaves light or medium green. Vegetative parts often heavily suffused with red. Rhachis dull green, suffused with brownish purple, or entirely dark reddish purple or brownish purple. Sepals greenish or yellow, adaxially spotted or stained dark red or almost entirely dark red except for the central part; abaxially stained or suffused with dark red. Petals yellowish white, with dark red margins or midnerve, or entirely dark red. Lip very dark red, often with a white base; hairs along margins white. Column pale yellow.

Ecology. Dry Brachystegia woodland, less common in montane forest, at 300–1800 m alt. Epiphyte in the central part of the crown of trees. Usually growing among lichens (La Croix, written comm.). Occasionally lithophytic. Flowering II–V and VII–X.

Distribution. ZAI.: Haut Katanga Z11 (1 coll.). BU. (1). TANZ.: East T6 (2); Southern Highlands T7 (2); South T8 (2). ZAM.: West, Mwinilunga (1); Central (2); North (2). MAL.: North (4); South (3). ZIM.: Central (fide Grosvenor, 1976); East (2); South (1). R.S.A.: E. Transvaal (Fide Harrison 1972, as *B. oreonastes*). Material seen from BM, BR, K, L, S, WAG.

Distribution: Madagascar.

Notes. 1 - No important differences exist between the type of *Bulbophyllum malawiense* from MAL. and the type of *Bulbophyllum elliotii* of Madagascar.

2 - *Bulbophyllum elliotii* is variable in the shape of the petals. Some specimens have narrowly triangular to linear petals, gradually narrowing from a wide and slightly thickened base, with glabrous margins. Others have ovate-lanceolate petals with slightly and irregularly denticulate or lacerate margins near the base. Drawings of a series of intermediate petals have been added to fig. 56; this variability could not be used to distinguish varieties.

The species is also variable in the shape of the abaxial ridge of the anther, see the description above. Here, too, intermediates exist. No relation has been found between the variability of the petals and that of the anther.

3 - The specimen Leedal 7252 from TANZ. probably is a natural hybrid between *Bulbophyllum elliotii* and *Bulbophyllum encephalodes*. The specimen has the distichous inflorescence and the warty top part of the lip of *Bulbophyllum encephalodes* but the ciliate rim along the margins of the lip of *Bulbophyllum elliotii*.

GROUP 8 (spp. 57 - 59)

Pseudobulbs 1-2-leafed. Inflorescence a raceme, many-flowered. Rhachis flattened, flowers inserted along the margins. Flowers distichous. Sepals glabrous or with fine, dark hairs abaxially, the lateral ones not connate along the upper margins, at most spreading. Lip thick, not very soft and easy to compress, with seamlike lateral lobes with denticulate margins. Stelidia short, acute. Anther with a dorsoventrally flattened protrusion overtopping the front margin.

57 - **Bulbophyllum porphyrostachys** Summerh. - Fig. 85; Plate 10 c-d.

Bulbophyllum porphyrostachys Summerh., Bot. Mus. Leafl. Harv. Un. 14 (1951) 230; id., Hook. Ic. Pl. 36 (1956) pl. 3547; id. in F.W.T.A. ed. 2, 3 (1968) 239.-Type: Keay FHI 22711 (K).

Non *Bulbophyllum porphyrostachys* auct.: Williamson, Orch. S. Centr. Afr. (1977) 136 (= *Bulbophyllum lupulinum*) (Williamson, written comm.).

Roots 0.5–1.5 mm diam. *Rhizome* 2.5–3.5 mm diam., nodes not or hardly swollen. *Rhizome scales* tubular, tip obtuse to subacute; membranous, with nerves; keeled abaxially towards the tip *Pseudobulbs* 2-leafed, ellipsoid to narrowly ellipsoid, occasionally slightly ovoid, 3.5–7 cm apart, 3–6 by 1.2–2 cm, slightly flattened, sharply 4–5-angled. *Petiole* 3–25 mm long. *Leaf blade* lanceolate to linear-lanceolate, 9–19.5 by 1.5–3 cm, tip obtuse, emarginate, oblique; coriaceous, thick; midrib slightly prominent abaxially. *Inflorescence* 7.5–22 cm long, 18–50-flowered. *Peduncle* erect to patent, rather sturdy, slightly swollen towards the tip and below the nodes, 4–13 cm by 2.5–5 mm, orbicular in section, glabrous, with 6–9 scales. *Peduncle scales* tubular, the longest 6–10 mm long; tip acute, the lower obtuse to acute; membranous, without nerves; keeled towards the tip and glabrous abaxially. *Rhachis* about erect, slightly swollen and flattened, 4-angled in section, with 2 slightly concave sides from which the flowers arise, slightly zig-zag bent, 3.5–11 cm by up to 6 mm, edges about sharp, straight, entire, glabrous; surface glabrous. *Floral bracts* recurved, distinctly concave, broadly ovate, 7–12 by 7–12 mm, tip acuminate; base rather broadly attached; membranous, without nerves; keeled towards the tip and glabrous abaxially. *Flowers* distichous, 3.5–5 mm apart, measured along one side of the rhachis, recurved, some open simultaneously, not much open. *Pedicel and ovary* 1.5–2 mm long, glabrous. *Median sepal* not recurved, ovate to broadly ovate oblong, 3–4 by 1.7–2.5 mm, top subacute, tip apiculate; margins entire, glabrous; base broadly attached; rather thick; glabrous; not keeled abaxially. *Lateral sepals* free, hardly falcate, oblique, 3–4 by 1.8–3 mm, otherwise as the median sepal. *Petals* not recurved, slightly falcate, narrowly linear-lanceolate, oblique, 1.8–2.5 by 0.2–0.3 mm, tip obtuse to subacute; margins entire, glabrous; base broadly attached; rather thin, slightly thickened towards the tip; glabrous; keeled abaxially. *Lip* with a reflexed top part, ovate in outline (not spread), 1.5–2 by 0.8–1.7 mm, top part a reniform thick body with a rounded to retuse tip and some large warts on its adaxial surface; margins with a narrow, thin seam with irregularly denticulate edges over the entire length of the lip except for the top part; thick; surface glabrous; adaxially with two weak, rounded ridges over about the entire length of the lip which diverge towards the base, with a slightly concave basal part, without a median slit; abaxially without a ridge, surface distinctly convex. *Column* from the base of the free part up to the tip of the stelidia 1.8–2 mm long, with distinct, triangular, not falcate, acute stelidia of c. 0.8 mm long; column with weak, broadly rounded wings along its lower margins just below the base of the stelidia which continue towards the column foot as narrow seams; stigma with two distinct obtuse basal teeth. *Anther* c. 1 by 0.8 mm; front not drawn out; front margin rounded, irregular, glabrous; basally distinctly emarginate; abaxially without ridge, but with a distinct, thick, concave, rounded, colliculate protrusion which distinctly overtops the front margin, surface glabrous but colliculate towards the top. *Pollinia* 4, ovoid, flattened on one side; the small ones about half as long as the large ones; a stipes has not been observed. *Fruit* ellipsoid, up to 1 by 0.4 cm, without a stalk.

Colours. Plant dull green. Rhachis green, often suffused with some red. Floral bracts yellowish or dark red. Sepals and petals yellow with dark red spots. Lip dark red or greenish.

Ecology. Rainforest and semideciduous forest, relict forest in savanna (Sanford, 1974). No altitude given. Epiphytic. Flowering (IV) X–XII.

Distribution. NIG.: S. part (5 coll.). CAM.: W. part (2). Material seen from E, K, L.

58 - Bulbophyllum lupulinum Lindl. – Fig. 86; Plate 11a.

Bulbophyllum lupulinum Lindl., J. Linn. Soc. Bot. 6 (1862) 126 ('Bolbophyllum); Durand & Schinz, Consp. Fl. Afr. 5 (1895) 12; Rolfe in F.T.A. 7 (1897) 28; Schltr, Westafr. Kautchuk Exp. (1901) 281; De Wild., Pl. Beq. 1 (1921) 113; Summerh. in F.W.T.A. ed. 1, 2 (1936) 438; id., Kew Bull. 6 (1952) 470; id. in F.W.T.A. ed. 2, 3 (1968) 239, p.p.; Cribb & Perez-Vera, Adansonia, ser. 2, 15 (1975) 206; Adam, Fl.

Descr. Mt. Nimba 5 (1981) 1679; Segerbaeck, Orch. Nig. (1983) 63, phot.-*Phyllorchis lupulina* (Lindl.) Kuntze, Rev. Gen. Pl. 2 (1891) 677, incorr. name.-Type: Mann 783 (K).

Non *Bulbophyllum lupulinum* auct.: Johansson, Acta Phyt. Suec. 59 (1974) 32 (= *Bulbophyllum tetragonum*).

Bulbophyllum urbanianum Krzl., Engl., Bot. Jahrb. 28 (1900) 163 ('*Bolbophyllum*'); De Wild., Pl. Beq. 1 (1921) 121.-Type: Deistel 79 (B, lost).

Bulbophyllum ituriense De Wild., Rev. Zool. Afr. 9, suppl. bot. (1921) 29; De Wild., Pl. Beq. 1 (1921) 112.-Type: Bequaert 2497 (BR).

Bulbophyllum porphyrostachys auct. non Summerh.: Williamson, Orch. S. Centr. Afr. (1977) 136 (Williamson, written comm.).

Roots 0.5–1.7 mm diam. *Rhizome* 2–5 mm diam., nodes not or hardly swollen. *Rhizome scales* tubular, tip obtuse to acute; membranous, the upper with nerves; not keeled abaxially. *Pseudobulbs* 1–2-leafed, ovoid to narrowly ovoid or ellipsoid, 3–13 cm apart, 2.7–7.5 by 1.2–2.5 cm, slightly flattened, sharply 4-angled. *Petiole* 2–15 mm long. *Leaf blade* lanceolate to linear-lanceolate, maximum width usually slightly above the middle, 8–23 by 1.2–5 cm, tip obtuse, emarginate, oblique; coriaceous, thick; midrib prominent abaxially. *Inflorescence* 15–38 cm long, 28–68-flowered. *Peduncle* about erect, rather sturdy, 6.5–19 cm by 2.5–5 mm, orbicular in section, with fine dark hairs towards the tip, with 8–14 scales. *Peduncle scales* tubular, the longest 10–14 mm long; tip rounded to about acute; membranous, without nerves; not keeled and with fine dark hairs abaxially, the lower glabrous. *Rhachis* erect, hardly swollen, slightly flattened, 4-angled in section, with two slightly concave sides from which the flowers arise, 5–23 cm long, edges about sharp, straight, entire, glabrous; surface with dots of fine dark hairs. *Floral bracts* recurved, distinctly concave, about triangular, 9.5–16 by 7–13 mm, tip obtuse to acute; base broadly attached; membranous, without nerves; not keeled and with fine dark hairs abaxially. *Flowers* distichous, 5–7 mm apart, measured along one side of the rhachis, recurved, some open simultaneously, not much open. *Pedicel and ovary* 1.3–2.2 mm long, with fine dark hairs. *Median sepal* not recurved, broadly to narrowly triangular, 4–5.2 by 1.5–3 mm, tip acute; margins entire, glabrous; base broadly attached; thick; glabrous adaxially; not keeled and with some fine dark hairs abaxially. *Lateral sepals* free, either slightly recurved or not, slightly falcate, oblique, 3.8–5.2 by 1.5–2.9 mm, otherwise as the median sepal. *Petals* not recurved, slightly falcate, obovate-lanceolate to linear, oblique, 2.5–3.2 by 0.25–0.5 mm, tip obtuse; margins entire, glabrous; base broadly attached; rather thick; glabrous; not keeled abaxially. *Lip* with a reflexed top part, about rectangular in outline (not spread), 1.5–2.4 by 0.8–1.7 mm, tip rounded; margins with a narrow, thin seam with denticulate edges over the entire length of the lip except for the reflexed top part, which has entire to slightly irregular, glabrous margins; thick; surface glabrous; adaxially without ridges, with a hardly concave basal part, without a median slit; abaxially with a distinctly convex surface. *Column* from the base of the free part up to the tip of the stelidia 1.8–2.6 mm long, with rather distinct, deltoid, not falcate, acute stelidia of 0.3–0.5 mm long; column with weak, broadly rounded wings along its adaxial margins just below the base of the stelidia; stigma with two weak, rounded basal teeth. *Anther* c. 1 by 0.8 mm; front not drawn out; front margin rounded, about entire, glabrous; basally weakly emarginate; abaxially with a distinct, rounded, colliculate ridge near the top of the anther which ends in a thick, concave, truncate, colliculate protrusion which distinctly overtops the front margin, surface glabrous. *Pollinia* 4, ellipsoid, flattened on one side; the small ones less than half as long as the large ones, slightly transversely placed; a stipes has not been observed. *Fruit* ellipsoid, up to 0.9 by 0.5 cm, without a stalk.

Colours. Pseudobulbs green or olive green. Leaves green. Rhachis very pale green or greyish white with brown appendages. Floral bracts whitish, often stained with pale purple or pink, abaxially with brown hairs. Sepals yellow, adaxially often spotted or striated with dark red; abaxially stained red and with brown hairs. Petals pale yellow, often striated with dark red. Lip yellow, often stained or spotted with dark red.

Ecology. Lowland rainforest to montane forest, semideciduous forest, up to 1700 m alt. Epiphyte, also frequently found lithophytic, even on exposed places such as rocky mountain slopes. Flowering (V, VIII) XI–III.

Distribution. GUIN.: W. part (2 coll.). S.L. (5). IV.C. (4). NIG.: S. part (4). CAM.: N. up to Adamawa Mts. (16). ZAI.: Forestier Central Z6 (2); Haut Katanga Z11 (1). AETH.: East, Welega P., Ilubador P. (2). ZAM.: West, Mwinilunga (1). Material seen from AMES, BM, BR, C, E, G, K, P, S, UPS, WAG, Z.

59 - Bulbophyllum tetragonum Lindl. – Fig. 87.

Bulbophyllum tetragonum Lindl., Gen. and Sp. Orch. Pl. (1830) 53 ('*Bolbophyllum*'); Durand & Schinz, Consp. Fl. Afr. 5 (1895) 14; Rolfe in F.T.A. 7 (1897) 32; De Wild., Pl. Beq. 1 (1921) 120.–Type: Drawing in Hb. Lindley (K).

Bulbophyllum wrightii Summerh., Kew Bull. 16 (1962) 309; id. in F.W.T.A. ed. 2, 3 (1968) 239; v.d.Burg, Tijdschr. Ned. Orch. Ver. 40 (1978) 152, col. phot.–Type: Wright 58/18 (K).

Bulbophyllum lupulinum auct. non Lindl.: Summerh. in F.W.T.A. ed. 2, 3 (1968) 239, p.p.; Johansson, Acta Phyt. Suec. 59 (1974) 32.

Roots 0.8–1.4 mm diam. *Rhizome* 2.8–5 mm diam., nodes not or hardly swollen. *Rhizome scales* tubular, tip about acute, the lower emarginate; membranous, the upper with nerves; not or hardly keeled abaxially. *Pseudobulbs* 1–2-leafed, narrowly ovoid to narrowly ellipsoid, 1.5–9 cm apart, 4.5–7.5 by 1.2–2.5 cm, slightly flattened, about sharply 4–6-angled. *Petiole* 4.5–10 mm long. *Leaf blade* linear-lanceolate to broadly linear, 10–25 by 1.2–2.6 cm, tip obtuse, emarginate, oblique; coriaceous, thick; midrib slightly prominent abaxially. *Inflorescence* 15.5–28 cm long, 24–50-flowered. *Peduncle* erect, rather sturdy, slightly swollen towards the tip and below the nodes, 8.5–16 cm by 2.5–3.5 mm, orbicular in section, with fine dark hairs towards the tip, with 7–9 scales. *Peduncle scales* tubular, the longest 8–10 mm long; tip obtuse to about acute; membranous, without nerves; hardly keeled and with fine dark hairs abaxially towards the tip. *Rhachis* erect, slightly swollen, 4-angled in section, with two slightly concave sides from which the flowers arise, 7–15 cm by 2.5–4 mm, edges obtuse, straight, entire, glabrous; surface locally densely covered with fine dark hairs. *Floral bracts* spreading, hardly concave, elliptic, 9–13 by 5–7 mm, tip acute; margins slightly rolled inwards; base rather broadly attached; membranous, without nerves; slightly keeled and with fine dark hairs abaxially. *Flowers* distichous, 5–7 mm apart, measured along one side of the rhachis, recurved, some open simultaneously, fully open. *Pedicel and ovary* 1.5–2 mm long, with fine dark hairs. *Median sepal* either slightly recurved or not, ovate, 3.5–4.5 by 2.2–2.5 mm, tip acuminate; margins entire, glabrous, with a distinct edge just above the base; base rather broadly attached; thick; glabrous adaxially; not keeled and fine dark hairs abaxially. *Lateral sepals* free, spreading, hardly falcate, oblique, 3.2–4 by 2–2.5 mm, tip about acute, only the upper margin with an inconspicuous edge just above the base, base broadly attached; otherwise as the median sepal. *Petals* not recurved, slightly falcate, linear-lanceolate to linear, oblique, 2–3 by 0.2–0.4 mm, tip acute; margins entire, glabrous; base broadly attached; rather thick, thickened towards the base; glabrous; not keeled abaxially. *Lip* with a reflexed top part, elliptic in outline (not spread), 1.3–2 by 1–1.5 mm, top part convex, tip rounded; margins with a narrow, towards the tip slightly broadened, thin seam with denticulate margins over the entire length of the lip except for the top part which has entire, glabrous margins; thick; surface glabrous; adaxially without ridges, with a hardly concave basal part, without a median slit; abaxially with a distinctly convex surface. *Column* from the base of the free part up to the tip of the stelidia 1.6–1.8 mm long, with rather distinct, deltoid, not falcate, subacute to acute stelidia of c. 0.4 mm long; column with weak, broadly rounded wings along its adaxial margins just below the base of the stelidia; stigma with a distinct obtuse, transverse basal callus. *Anther* c. 1 by 0.6 mm; front not drawn out; front margin truncate, irregular, glabrous; basally distinctly emarginate; abaxially with a weak, rounded, colliculate ridge near the top of the anther which ends into a thick, concave, rounded, colliculate protrusion which distinctly overtops the front margin, surface glabrous but colliculate towards the top. *Pollinia* 4, ellipsoid, flattened on one side; the small ones less than half as long as the large ones, slightly transversely placed; a stipes has not been observed. *Fruit* ellipsoid, up to 0.8 by 0.4 cm, without a stalk.

Colours. Pseudobulbs yellowish green. Leaves green. Rhachis very pale greenish with dark brown hairs. Floral bracts whitish, suffused with brown-purple towards the tip, abaxially with dark brown hairs. Sepals yellowish white, adaxially with large dark red spots; abaxially stained with red and with some brown hairs. Petals yellow. Lip yellow, slightly stained with dark red. Column yellow.

Ecology. Lowland and hill forest up to 900 m alt. Epiphyte in the central part of the crown of trees, on bark or in minor humus deposits (Johansson, 1974, under *Bulbophyllum lupulinum*). Flowering XI–II.

Distribution. S.L. (3 coll.). LIB. (2). IV.C. (1). GHANA (2). CAM.: W. part (1). ZAI.: Forestier Central Z6 (1). Material seen from AMES, BR, K, UPS, WAG.

GROUP 9 (spp. 60 - 66)

Pseudobulbs 1–2-leafed. Inflorescence a raceme, many-flowered. Rhachis flattened, flowers inserted along the margins. Flowers distichous. Sepals glabrous to papillose abaxially, the lateral ones not connate along the upper margins, usually at most spreading (rarely reflexed but not appressed against the rhachis). Lip thick, not very soft and easy to compress, semi-orbicular to obovate lateral lobes present or not, margins glabrous to papillose. Stelidia short, rounded to acute. Anther with a dorsoventrally flattened protrusion overtopping the front margin.

60 - Bulbophyllum horridulum J.J. Vermeulen – Fig. 88.

Bulbophyllum horridulum J.J. Vermeulen, Bull. J. Bot. Nat. Belg. 56 (1986) 238.–Type: Christiaensen 2458 (BR).

Roots 0.7–1.2 mm diam. *Rhizome* 0.9–1.8 mm diam., nodes slightly swollen. *Rhizome scales* tubular, tip acute; membranous, with some nerves; keeled abaxially. *Pseudobulbs* 2-leafed, very narrowly ellipsoid or obovoid, 13.5–16 cm apart, 3–4.5 by 0.7–1 cm, slightly flattened, about sharply 4-angled. *Petiole* 2–5 mm long. *Leaf blade* oblong to broadly lanceolate, 4.5–6.5 by 1.5–2.3 cm, tip obtuse, emarginate, slightly oblique; coriaceous, thick; midrib slightly prominent abaxially. *Inflorescence* 8.5–14.5 cm long, 12–30-flowered. *Peduncle* erect to patent, slightly swollen towards the tip, 3.5–5 cm by 0.8–1 mm, orbicular in section, glabrous, with 5–7 scales. *Peduncle scales* tubular, the longest 5–6 mm long; tip acute; membranous, without nerves; keeled and glabrous abaxially. *Rhachis* about erect, slightly swollen and flattened, 4-angled in section, with two concave sides from which the flowers arise, slightly zig-zag bent towards the tip, 4–10 cm by 1.5–2 mm, edges sharp, straight, entire, glabrous; surface glabrous. *Floral bracts* spreading, concave, triangular, c. 5 by 3 mm, tip (?); base broadly attached; membranous, without nerves; keeled and glabrous abaxially. *Flowers* distichous, 6–8 mm apart, measured along one side of the rhachis, recurved, hardly open. *Pedicel and ovary* c. 1.8 mm long, glabrous. *Median sepal* not recurved, very narrowly triangular, c. 6 by 1.9 mm, tip acute; margins entire, glabrous; base broadly attached; rather thin; glabrous; abaxially with a keel which is irregularly dentate towards the tip. *Lateral sepals* free, slightly falcate, ovate-oblong, oblique, c. 6 by 2.2 mm, abaxially with a winglike keel which is deeply and irregularly lacerate; otherwise as the median sepal. *Petals* not recurved, falcate, linear-lanceolate, oblique, c. 2.8 by 0.3 mm, tip subacute; margins entire, glabrous; base broadly attached; thin; glabrous; abaxially with a winglike keel. *Lip* slightly recurved in the basal half, elliptic in outline (not spread), c. 2.1 by 1.5 mm, tip broadly rounded; edges rounded, entire, glabrous, near the base with large, semi-orbicular lateral lobes which have a rounded tip, entire, slightly papillose margins, a broadly attached base and which are thin and finely papillose; thick; adaxially with two distinct, obtuse ridges in the basal half near the edges which are most prominent about half-way the lip and which diverge slightly towards the base and a very weak, obtuse median ridge which starts above half-way the lip and protrudes slightly over its top, with a concave basal part, without a median slit, surface finely papillose; abaxially with a weak, rounded ridge in the basal half, surface glabrous, finely papillose towards the edges. *Column* from the base of the free part up to the tip of the stelidia c. 1.3 mm long, with about rudimentary, deltoid, not falcate, subacute stelidia; column with distinct, obliquely forwards directed, triangular, obtuse teeth along its adaxial margins just below the base of the stelidia; stigma with a weak, rounded basal callus. *Anther* c. 0.9 by 0.7 mm; front not drawn out; front margin rounded, entire, glabrous; basally distinctly emarginate; abaxially with a concave, broadly rounded, thick protrusion near the top which distinctly overtops the front margin and which bears on its abaxial side an obliquely backwards directed, conical, rounded, colliculate protrusion; surface glabrous but colliculate towards the top. *Pollinia* 4, about ellipsoid; the small ones about half as long as the large ones, distinctly flattened; the large ones flattened on one side; a stipes has not been observed. *Fruit* not seen.

Colours. Flowers red.

Ecology. Lower montane forest at c. 1400 m alt. Epiphytic. Flowering VIII (once observed).

Distribution. ZAI.: Lacs Edouard & Kivu Z9 (1 coll.).

Note. Only the type specimen was available, consisting of a sterile specimen and two detached flowers. It is most probable but not entirely sure that these parts belong together.

61 - Bulbophyllum acutebracteatum De Wild. - Fig. 89–90; Plate 11b.

Literature: see under the varieties.

Roots 0.3–0.8 mm diam. *Rhizome* 0.7–1.8 mm diam., nodes not or hardly swollen. *Rhizome scales* tubular, tip acute to shortly acuminate; membranous, with nerves; slightly keeled and often very finely papillose abaxially. *Pseudobulbs* 2-leafed, (narrowly) ovoid to (narrowly) ellipsoid, 1–8 cm apart, 0.8–4 by 0.4–1.2 cm, hardly to moderately flattened, rather obtusely 4–6-angled. *Petiole* 0.5–4 mm long. *Leaf blade* elliptic to lanceolate, maximum width occasionally slightly above the middle, 0.8–7 by 0.5–2 cm, tip rounded to obtuse, emarginate, oblique; coriaceous, thick; midrib slightly prominent abaxially. *Inflorescence* 5–16 cm long, 6–46-flowered. *Peduncle* erect to patent, slightly swollen towards the tip, 2.5–8 cm by 0.7–1.8 mm, orbicular in section, glabrous, with 5–10 scales. *Peduncle scales* tubular, the longest 5–9 mm long; tip acute to acuminate; membranous, without nerves; keeled towards the tip and glabrous to very finely papillose abaxially. *Rhachis* erect or nodding, slightly swollen and flattened, 4-angled in section, with two concave sides from which the flowers arise, slightly zig-zag bent towards the tip, 1.8–11 cm by 1.2–3.3 mm, edges obtuse, sharp or winged, straight, entire, glabrous to coarsely papillose; surface glabrous to coarsely papillose. *Floral bracts* recurved to spreading, concave, broadly triangular to ovate, 4–5.5 by 2–4 mm, tip acute to acuminate; base broadly attached; membranous, without nerves; keeled towards the tip and glabrous to slightly papillose abaxially. *Flowers* distichous, 4–8 mm apart, measured along one side of the rhachis, recurved, some to many open simultaneously, not much open. *Pedicel and ovary* 1.7–2 mm long, glabrous to papillose. *Median sepal* not recurved, ovate-oblong to narrowly triangular, 3.5–5.2 by 1.3–2 mm, tip acute; margins entire, glabrous, either papillose towards the tip or not; base broadly attached; rather thick; glabrous adaxially; not keeled and glabrous to coarsely papillose abaxially. *Lateral sepals* free, slightly recurved, hardly falcate, oblique, 3.5–6.2 by 2–2.8 mm, otherwise as the median sepal. *Petals* not recurved, hardly falcate, lanceolate to linear, oblique, 1.7–3.2 by 0.2–0.3 mm, tip obtuse; margins entire, glabrous; base broadly attached; thin; glabrous; not keeled abaxially. *Lip* not recurved, elliptic in outline (not spread), 1.2–2.5 by 0.7–1.2 mm, tip obtuse; edges entire, glabrous to finely papillose; thick; adaxially with two weak, obtuse ridges in the lower part near the edges which diverge towards the base of the lip and with a weak to rather distinct, rounded to obtuse median ridge starting just above half-way the lip and protruding over its top, with a concave basal part, with a convex top part, with a median slit near the base, surface glabrous, either finely papillose towards the tip or not; abaxially with a rather distinct, broadly rounded ridge near the base ending in a more or less developed lump half way, distinctly concave towards the tip, surface glabrous. *Column* from the base of the free part up to the tip of the stelidia 0.7–1.8 mm long, with rudimentary, at most about deltoid, not falcate, emarginate or rounded to obtuse stelidia; column with rather weak to very distinct, obliquely forwards directed, deltoid to narrowly triangular, obtuse teeth along its lower margins just below the base of the stelidia; stigma with or without an inconspicuous basal callus. *Anther* c. 0.8 by 0.6 mm; front not drawn out; front margin rounded, about entire, glabrous; basally distinctly emarginate; abaxially with a weak, rounded, colliculate ridge about half-way the anther which ends into a thick, not or hardly concave, distinctly flattened, broadly rounded, colliculate protrusion which distinctly overtops the front margin; surface colliculate. *Pollinia* 4; the small ones about half as long as the large ones, orbicular to ellipsoid, distinctly flattened; the large ones ellipsoid to ovoid, flattened on one side; a stipes has not been observed. *Fruit* ellipsoid, up to 0.7 by 0.2 cm, without a stalk.

Note. In *Bulbophyllum acutebracteatum* as well as in *B. unifoliatum* infraspecific taxa have been based on the absence/presence, and on the distribution of papillae on the rhachis and the flower parts. Contrary to the infraspecific taxa within *B. acutebracteatum*, those within *B. unifoliatum* are different in more than one character and they are geographically separated. I think that use of the variety concept within *B. acutebracteatum* next to the use of the subspecies concept within *B. unifoliatum* shows due appreciation of this difference between the two species.

KEY TO THE VARIETIES

1 - Rhizome scales, inflorescence and inflorescence scales about glabrous to very finely papillose; rhachis, abaxial surface of the floral bracts and sepals partly or entirely coarsely papillose............................var. **rubrobrunneopapillosum** (b)
- All these parts glabrous..var. **acutebracteatum** (a)

61a - var. acutebracteatum – Fig. 89; Plate 11b.

Bulbophyllum acutebracteatum De Wild., Pl. Beq. 1 (1921) 101.–*Bulbophyllum platirachis* De Wild., Miss. Laur. (1906) 223, pl. 55; id., Et. Fl. B. et Moy. Congo 2 (1908) 237.–Type: Laurent s.n. (1903) (BR).
Non *Megaclinium platyrhachis* Rolfe in F.T.A. 7 (1897) 43 (= B. maximum).
Non *Bulbophyllum platyrrhachis* Ridl., J. Fed. Mal. St. Mus. 8 (1917) 96.
Bulbophyllum fuscum auct. non Lindl., Rolfe in F.T.A. 7 (1897) 24, p.p.
Bulbophyllum oreonastes auct. non Rchb. f., Summerh. in F.W.T.A. ed. 1, 2 (1936) 438, p.p.; id., ed. 2, 3 (1968) 236, p.p.

Rhizome scales, peduncle, peduncle scales, rhachis and the abaxial surface of the floral bracts and the sepals glabrous or nearly so. Column with inconspicuous, deltoid, obtuse, thin teeth along the adaxial margins, just below the base of the stelidia.

Colours. Rhizome and rhizome scales dark brownish red. Pseudobulbs pale green to olive green. Leaves green. Rhachis olive green, suffused with red or entirely dark red. Floral bracts yellow or dark red. Sepals yellow to orange red. Petals yellow with an orange tip. Lip bright orange or orange yellow, yellow at the base. Column pale yellow.

Ecology. Lowland forest. No altitude given. Epiphytic. Flowering in VII–XI.

Distribution. S.L. (2 coll.). LIB. (3). GAB. (2). ZAI.: Forestier Central Z6 (4). Material seen from AMES, BR, K, L, P.

Note. In the vegetative parts and the inflorescence var. *acutebracteatum* is rather like *Bulbophyllum fuscum*. The main difference is found in the shape of the lip. Herbarium specimens are occasionally difficult to identify due to shriveling of that part: during the drying process specimens of *Bulbophyllum fuscum* may develop a ridge towards the tip of the lip which can be mistaken for that of *Bulbophyllum acutebracteatum* var. *acutebracteatum*. In such cases the latter can still be recognized by the rather hunch-backed appearance of the lip when observed from aside which is caused by the more or less developed lump about half-way the abaxial side of the lip.

61b - var. rubrobrunneopapillosum (De Wild.) J.J. Vermeulen–Fig. 90.

Bulbophyllum acutebracteatum De Wild. var. *rubrobrunneopapillosum* (De Wild.) J.J. Vermeulen, Bull. J. Bot. Nat. Belg. 56 (1986) 238.–*Bulbophyllum rubrobrunneopapillosum* De Wild., Rev. Zool. Afr. 9, suppl. bot. (1921) 31; id., Pl. Beq. 1 (1921) 117.–Type: Bequaert 6706 (BR).

Bulbophyllum fuscoides Petersen, Bot. Tidsskr. 49 (1952) 163; Summerh. in F.W.T.A. ed. 2, 3 (1968) 239.–Type: Birket Smith s.n. (C, n.v., photograph in K).

Rhizome scales, peduncle and peduncle scales about glabrous to finely papillose, rhachis and the abaxial surface of the floral bracts and the sepals partly or entirely coarsely papillose. Column usually (not always) with distinct, obliquely forwards directed, ovate-oblong, obtuse, thin teeth of 0.3–0.6 mm long along the adaxial margins, just below the base of the stelidia.

Colours. Rhizome and rhizome scales dark reddish brown. Pseudobulbs and leaves green, often suffused with red. Rhachis and flower bracts dark red. Sepals greenish, suffused with dark red or entirely dark red. Petals yellow or dark red. Lip yellow or red.

Ecology. Lowland forest and hill forest up to 1000 m alt. Epiphyte in moss cushions (once recorded). Flowering in XI–I.

Distribution. CAM.: S. part (4 coll.). EQ.GUIN.: Mbini (1). ZAI.: Forestier Central Z6 (1). R.S.A. Natal (1, probably wrongly labeled). Material seen from BM, BR, K, P.

62 - Bulbophyllum fuscum Lindl. – Fig. 91–92; Plate 11c.

Literature: see under the varieties.

Roots 0.2–1.1 mm diam. *Rhizome* 0.9–2.2 mm diam., nodes not or hardly swollen. *Rhizome scales* tubular, tip about acute; membranous, the upper with nerves; slightly keeled abaxially. *Pseudobulbs* 2-leafed, (narrowly) ovoid or ellipsoid, (0.8 –) 2.5–13 cm apart, 1–5 by 0.4–1.4 cm, slightly flattened, obtusely 4-angled. *Petiole* 1–9 mm long. *Leaf blade* broadly elliptic to linear-lanceolate, maximum width occasionally slightly above the middle, tip obtuse, emarginate, oblique; coriaceous, thick; midrib slightly prominent abaxially. *Inflorescence* 4–23 cm long, 6–34 (–64) flowered. *Peduncle* erect to patent, slightly swollen towards the tip, 2.5–10 cm by 0.7–2 mm, orbicular in section, glabrous, with 4–8 scales. *Peduncle scales* tubular, the longest 3–8 mm long; tip acute to acuminate; membranous, without nerves; slightly keeled towards the tip and glabrous abaxially. *Rhachis* erect or nodding, slightly swollen and flattened, 4-angled in section, with two concave sides from which the flowers arise, either zig-zag bent or not, 1.6–18 cm by up to 4 mm, edges sharp, straight, about entire, glabrous; surface glabrous. *Floral bracts* recurved to spreading, concave, broadly triangular to ovate, 2–6 by 1.6–4 mm, tip acute to acuminate; base broadly attached; membranous, without nerves; keeled and glabrous abaxially. *Flowers* distichous, 3–9 mm apart, measured along one side of the rhachis, recurved, many open simultaneously, not much open. *Pedicel and ovary* 1–2.5 mm long, glabrous. *Median sepal* not recurved, ovate to ovate-oblong, 2–5.2 by 0.9–2 mm, tip acute to acuminate; margins entire, glabrous; base broadly attached; rather thin to rather thick; glabrous; slightly keeled abaxially. *Lateral sepals* free, either recurved or not, hardly falcate, ovate to (narrowly) triangular, oblique, 2.2–6 by 1.3–2.7 mm, tip subacute to acuminate; otherwise as the median sepal. *Petals* not recurved, either slightly falcate or not, lanceolate to linear-lanceolate, oblique, 1.3–4 by 0.2–0.5 mm, tip rounded to subacute; margins entire, glabrous; base broadly attached; thin; glabrous; not keeled abaxially. *Lip* hardly recurved, elliptic in outline (not spread), 0.9–3 by 0.5–1.4 mm, top part rounded with a bent downwards, about apiculate tip; edges entire, glabrous, near the base either with or without small to very large, semicircular to oblique broadly obovate lateral lobes which have a rounded truncate tip, towards the tip irregularly denticulate (rarely entire) margins, a narrowly to broadly attached base and which are thin and glabrous; thick; adaxially with two weak to rather distinct rounded to obtuse ridges in the basal half near the edges which diverge towards the base of the lip, with a concave basal part, with a convex top part, either with a median slit near the base or not, surface glabrous; abaxially with a rather distinct, broadly rounded ridge near the base which rather abruptly ends in a convex top part, surface glabrous. *Column* from the base of the free part up to the tip of the stelidia 0.6–1.4 mm long, with about rudimentary, at most deltoid, not falcate, obtuse to acute stelidia; column with weak, either obliquely forwards directed or not, deltoid, rounded to subacute teeth along its adaxial margins just below the base of the stelidia; stigma without a basal callus. *Anther* c. 0.3 by 0.3 mm; front not drawn out; front margin rounded, entire, glabrous; basally distinctly emarginate; abaxially without a ridge but with a thick, slightly concave, distinctly flattened, rounded, colliculate protrusion near the top of the anther which distinctly overtops the front margin; surface colliculate. *Pollinia* 4, ellipsoid; the small ones about half as long as the large ones, distinctly flattened; the large ones slightly flattened on one side; a stipes has not been observed. *Fruit* ellipsoid, up to 0.7 by 0.2 cm; without a stalk.

Colours. Rhizome and rhizome scales often dark reddish brown. Pseudobulbs and leaves green. Inflorescence green, occasionally suffused with some red. Floral bracts yellowish, often suffused with red. Sepals yellow, greenish or ochrish, often suffused with dark red or reddish brown or entirely dark red or reddish brown. Petals yellow or dark red. Lip yellow, orange or dark red.

Note. In the vegetative parts and the inflorescence this species is rather like *Bulbophyllum acutebracteatum* var. *acutebracteatum*. The main difference is found in the shape of the lip. However, herbarium specimens are occasionally difficult to identify due to shriveling of that part: see note under *Bulbophyllum acutebracteatum* var. *acutebracteatum*.

KEY TO THE VARIETIES

1 -Lip with distinct, usually denticulate lateral lobes near its basevar. **fuscum** (a)
-Lip without lateral lobes..var. **melinostachyum** (b)

N.b.: Due to shriveling, some dried specimens of var. *melinostachyum* seem to have thin, rather large, obliquely ovate, rounded lateral lobes near the base of the lip. In such specimens confusion with var. *fuscum* may occur. The difference is, however, that the anterior margins of these 'pseudowings' of var. *melinostachyum* constitute the ridges on the adaxial side of the lip which diverge towards the base and become winglike. They gradually emerge from the median lobe of the lip. The wings of var. *fuscum*, which are superimposed on these 'pseudowings', arise from them with a distinct sinus.

62a - var. fuscum - Fig. 91; Plate 11c.

Bulbophyllum fuscum Lindl., Bot. Reg. 25 (1839) misc. 3 ('*Bolbophyllum*'); Rchb. f., Walp. Ann. 6 (1861) 256 ('*Bolbophyllum*'); Durand & Schinz, Consp. Fl. Afr. 5 (1895) 12; Rolfe in F.T.A. 7 (1897) 24, p.p.; Schltr, Westafr. Kautchuk Exp. (1901) 281; De Wild., Pl. Beq. 1 (1921) 110; Summerh. in F.W.T.A. ed. 1, 2 (1936) 438; id., ed. 2, 3 (1968) 236; Adam, Fl. Descr. Mt. Nimba 5 (1981) 1676.–*Phyllorchis fusca* (Lindl.) Kuntze, Rev. Gen. Pl. 2 (1891) 677, nom incorr.–Type: Loddiges s.n. (K).

Non *Bulbophyllum fuscum* auct.: Rendle, Cat. Talb. S. Nig. Pl. (1913) 146 (= *Bulbophyllum oreonastes*); Piers, Orch. E. Afr. (1968) 104 (= var. *melinostachyum*).

Bulbophyllum ogoouense Guillaumin, Bull. Mus. Hist. Nat. Paris 30 (1924) 522.–Type: Pobeguin s.n. (P).

Leaves 1.2–5.5 (– 9) cm long. Inflorescence 4–12 cm long. Peduncle 2.5–6.5 cm long. Rhachis 1.6–7 cm long, with 6–20 flowers. Floral bracts 2–4 mm long. Median sepal 2–3.2 mm long. Lateral sepals 2.3–3.2 mm long. Petals 1.3–2 mm long. Lip 0.9–2 mm long, with lateral lobes which usually are denticulate.

Colours. See under the species description.

Ecology. Lowland forest and hill forest up to 1000 m alt. Epiphytic. Flowering IX–III.

Distribution. GUIN.: Central part (2 coll.). S.L. (3). LIB. (2). IV.C. (2). NIG.: S. part (1). CAM.: Coast (2). C.A.R.: SW part (2). GAB. (2). ZAI.: Forestier Central Z6 (4); Lacs Edouard & Kivu Z9 (1). ANG. (2). Material seen from BM, BR, CO, K, P, WAG.

Note. The specimen Exell & Mendonça 1086 from ANG. has entire lateral lobes on the lip.

62b - var. melinostachyum (Schltr) J.J. Vermeulen - Fig. 92.

Bulbophyllum fuscum Lindl. var. *melinostachyum* (Schltr) J.J. Vermeulen, Bull. J. Bot. Nat. Belg. 56 (1986) 240.-*Bulbophyllum melinostachyum* Schltr, Engl., Bot. Jahrb. 26 (1899) 342; De Wild., Pl. Beq. 1 (1921) 114.-Type: Schlechter 12250 (Holo B, lost; iso BM, G, K, Z).

Bulbophyllum obanense Rendle, Cat. Talb. S. Nig. Pl. (1913) 101, 146; De Wild., Pl. Beq. 1 (1921) 115.-Type: Talbot 932 (BM).

Bulbophyllum caespitosum auct. non Thou.: Durand & Schinz, Consp. Fl. Afr. 5 (1895) 10; Rolfe in F.T.A. 7 (1897) 25, excl. descr. (see note).

Bulbophyllum oreonastes auct. non Rchb. f.: Summerh. in F.W.T.A. ed. 1, 2 (1936) 438, p.p.; id., ed. 2, 3 (1968) 236, p.p.; Morris, Proc. Linn. Soc. London 179 (1968) 63; Morris, Ep. Orch. Mal. (1970) 43; Grosvenor, Excelsa 6 (1976) 79; Williamson, Orch. S. Centr. Afr. (1977) 136; Ball, S. Afr. Ep. Orch. (1978) 88, pl.; Adam, Fl. Descr. Mt. Nimba 5 (1981) 1681, pl. 914; Segerbaeck, Orch. Nig. (1983) 61, phot.; La Croix et al., Orch. Mal. 1 (1983) 53; Cribb, in F.T.E.A., Orch. 2 (1984) 312.

Bulbophyllum fuscum auct. non Lindl.: Piers, Orch. E. Afr. (1968) 104.

Leaves 1.8–11 cm long. Inflorescence 4.5–23 cm long. Peduncle 2.5–10 cm long. Rhachis 2–18 cm long, with 6–34 (- 64) flowers. Floral bracts 2.5–6 mm long. Median sepal 2.5–5.2 mm long. Lateral sepals 2.2–6 mm long. Petals 1.5–4 mm long. Lip 1.2–4 mm long, without lateral lobes.

Colours. See under the species description.

Ecology. Lowland forest to montane forest, Brachystegia woodland, up to 2100 m alt. Epiphytic on tree trunks; lithophytic on lichen-covered rocks (La Croix, written comm.). Flowering I–XII; in W. Africa up to GAB. mainly IX–I, in E Afr. mainly X–XI and II–VI.

Distribution. S.L. (2). LIB. (2 coll.). IV.C. (2). NIG.: SE part, N. to Adamawa Mts. (4). CAM.: S. and SW. part (7). EQ.GUIN.: Bioko (1). GAB. (11). ZAI.: Forestier Central Z6 (1); Lacs Edouard & Kivu Z9 (1); Haut Katanga Z11 (5). UG.: West U2, S. part (1). TANZ.: Tanga T3 (2); East T6 (1); Southern Highlands T7 (2). ZAM.: Northwest, Mwinilunga, Solwezi, Copperbelt (5); North (12). MAL.: Central (2), South (2). ZIM.: Central (fide Grosvenor, 1976, sub *Bulbophyllum oreonastes*); East (3); South (2). MOZ.: Niassa (2); Manica e Sofala (2). Material seen from BM, BR, C, E, G, K, L, NY, P, S, UPS, WAG, Z.

Notes. 1 - The specimens studied from S.L. and LIB. are conspicuously larger than the other specimens, having long inflorescences with up to 64 flowers (usually no more than 30). Specimens from ZIM. tend to be slightly shorter and wider in the vegetative parts.

2 - The specimens Wageningen cult. Hort. 83-339 (= v.d. Laan 616) and Wageningen cult. Hort. 83-351 (= v.d. Laan 737 = v.d. Laan 839) both from GAB. have sepals which are reflexed half-way.

The specimen Christiaensen 593 from ZAI. has slightly more acuminate sepals. The other floral parts are too decayed for detailed study.

3 - Several authors have used the name *Bulbophyllum oreonastes* for *Bulbophyllum fuscum* var. *melinostachyum*. This was caused by wrong interpretation of the very poor type specimen of *Bulbophyllum oreonastes*.

4 - Specimens of var. *melinostachyum* have been wrongly identified by Kraenzlin as *Bulbophyllum caespitosum* Thou. This view was followed by Durand & Schinz and Rolfe, hence the records of *Bulbophyllum caespitosum* from S.L.

63 - **Bulbophyllum oreonastes** Rchb. f. - Fig. 93.

Bulbophyllum oreonastes Rchb. f., Otia Hamb. 2 (1881) 118; Durand & Schinz, Consp. Fl. Afr. 5 (1895) 13; Rolfe in F.T.A. 7 (1897) 24; Schltr, Westafr. Kautchuk Exp. (1901) 282; De Wild., Pl. Beq. 1 (1921) 115; Summerh. in F.W.T.A. ed. 1, 2 (1936) 438, p.p.; Summerh., Kew Bull. 6 (1952) 471; Summerh. in F.W.T.A. ed. 2, 3 (1968) 236, p.p.; Johansson, Acta Phyt. Suec. 59 (1974) 32; Lodewijk, Tijdschr. Ned. Orch. Ver. 38 (1976) 235, col. phot.-*Phyllorchis oreonastes* (Rchb. f.) Kuntze, Rev. Gen. Pl. 2 (1891) 677, incorr. name.-*Buibophyllum* sp. Hook. f., J. Linn. Soc. Bot. 7 (1864) 220.-Type: Mann 2122 (K).

Non *Bulbophyllum oreonastes* auct.: Morris, Proc. Linn. Soc. London 179 (1968) 63; Morris, Ep. Orch. Mal. (1970) 43 (= both *Bulbophyllum fuscum* var. *melinostachyum*); Harrison, Ep. Orch. S. Afr. (1972) 37, pl. 15 (= *Bulbophyllum elliotii*); Grosvenor, Excelsa 6 (1976) 79; Williamson, Orch. S. Centr. Afr. (1977) 136; Ball, S. Afr. Ep. Orch. (1978) 88, pl.; Adam, Fl. Descr. Mt. Nimba 5 (1981) 1681, pl. 914; La Croix et al., Orch. Mal. 1 (1983) 53; Cribb, in F.T.E.A., Orch. 2 (1984) 312; Segerbaeck, Orch. Nig. (1983) 61, phot. (= all *Bulbophyllum fuscum* var. *melinostachyum*).

Bulbophyllum hookerianum Krzl., Engl., Bot. Jahrb. 17 (1893) 49; Krzl., Gard. Chron., ser. 3, 15, no. 388 (1894) 685; De Wild., Pl. Beq. 1 (1921) 111.-Type: Preuss 943 (Holo B, lost; iso K).

Bulbophyllum zenkerianum Krzl., Engl., Bot. Jahrb. 48 (1912) 391; De Wild., Pl. Beq. 1 (1921) 121; Summerh., Kew Bull. 12 (1957) 116; id. in F.W.T.A. ed. 2, 3 (1968) 239 ('*zenkeranum*'); Williamson, Orch. S. Centr. Afr. (1977) 136, col. phot. ('*zenkeranum*'); Adam, Fl. Descr. Mt. Nimba 5 (1981) 1685; Segerbaeck, Orch. Nig. (1983) 61, phot.; Cribb, in F.T.E.A., Orch. 2 (1984) 313 ('*zenkeranum*').-Type: Zenker 3902 (Holo B, lost; iso BM, E, G, K, L, S).

Bulbophyllum rhopalochilum Krzl., Engl., Bot. Jahrb. 51 (1914) 384; De Wild., Pl. Beq. 1 (1921) 117.-Type: Kassner 2380 (Holo B, lost; iso BM, BR, E, P, Z).

Bulbophyllum infundibuliflorum Petersen, Bot. Tidsskr. 49 (1952) 164.-Type: Birket Smith s.n. (C).

Bulbophyllum planiaxe Petersen, Bot. Tidsskr. 49 (1952) 166.-Type:? (C, n.v., sketch Summerh. in K)

Roots 0.3–1 mm diam. *Rhizome* 0.8–3 mm diam., nodes not or hardly swollen. *Rhizome scales* tubular, tip about acuminate, the lower obtuse to acute, mucronate; membranous, the upper with some nerves; keeled abaxially. *Pseudobulbs* 2-leafed, about orbicular to narrowly ovoid or ellipsoid, 0.7–4 cm apart, 0.4–3.5 by 0.4–1.2 cm, hardly to moderately flattened, obtusely 4-angled. *Petiole* 0.5–8 mm long. *Leaf blade* broadly elliptic to linear-lanceolate, maximum width occasionally slightly above the middle, 0.6–8.2 by 0.4–2 cm, tip rounded to obtuse, emarginate, either slightly oblique or not; coriaceous, thick; midrib prominent abaxially. *Inflorescence* 1.5–17.5 cm long, 5–36-flowered. *Peduncle* erect to patent, slightly swollen towards the tip, 0.7–6.5 cm by 0.9–2 mm, orbicular in section, glabrous, with 2–5 scales. *Peduncle scales* tubular, the longest 4–9 mm long; tip acuminate; membranous, without nerves; keeled and glabrous abaxially. *Rhachis* usually slightly nodding, swollen and flattened, 4-angled in section, with two concave sides from which the flowers arise, either zig-zag bent or not, 0.7–16 cm by 1.2–6 mm, edges sharp or winged, straight, entire to irregularly dentate, glabrous; surface glabrous. *Floral bracts* about spreading, concave, narrowly triangular, 3–7 by 1.7–3.6 mm, tip acute; base broadly attached; membranous, without nerves; keeled (keel either irregularly dentate or not) and glabrous abaxially. *Flowers* distichous, 4–12 mm apart, measured along one side of the rhachis, spreading to about reflexed, many open simultaneously, not much open. *Pedicel and ovary* 1.8–3 mm long, glabrous. *Median sepal* not recurved, ovate(-oblong), 3–6.2 by 1–1.8 mm, tip acute; margins entire, glabrous; base broadly attached; rather thin; glabrous; with an entire to irregularly dentate keel abaxially. *Lateral sepals* free, either slightly recurved or not, hardly falcate, (narrowly) triangular, oblique, 3–7 by 1.5–3.2 mm, not keeled abaxially; otherwise as the median sepal. *Petals* not recurved, hardly falcate, (linear-)lanceolate, oblique, 1–3 by 0.2–0.6 mm, tip obtuse to subacute; margins entire, glabrous; base broadly attached; thin; glabrous; not keeled abaxially. *Lip* recurved about half-way, narrowly elliptic to broadly oblong in outline (not spread), 1.4–2.5 by 0.6–1.2 mm, tip rounded; edges entire, glabrous; thick; adaxially with two weak, obtuse ridges in the basal half near the edges which may diverge slightly towards the base of the lip, with a concave basal part, with a convex top part, without a median slit near the base, surface glabrous; abaxially with a rather distinct, rounded ridge near the base, ending abruptly about half-way the lip, top part of the lip distinctly swollen and convex, separated from the basal part by a transverse furrow or a slit, surface glabrous. *Column* from the base of the free part up to the tip of the stelidia 0.8–1.2 mm long, with slightly rudimentary, deltoid to triangular, not falcate, obtuse to acute stelidia; column usually with weak, at most deltoid, broadly rounded to subacute teeth or wings along its lower margins just below the base of the stelidia; stigma usually with a rather distinct obtuse basal tooth. *Anther* c. 0.8 by 0.6 mm; front not drawn out; front margin truncate to rounded, slightly irregular, glabrous; basally distinctly emarginate; abaxially with a rather distinct, rounded, colliculate ridge near the top of the anther which ends into a thick, hardly concave, distinctly flattened, truncate to slightly retuse, colliculate protrusion which distinctly overtops the front margin; surface glabrous but colliculate towards the top. *Pollinia* 4; the small ones about half as long as the large ones, about orbicular, distinctly flattened; the large ones about ovoid, flattened on one side; a stipes has not been observed. *Fruit* ellipsoid to slightly obovoid, up to 0.9 by 0.5 cm, without a stalk.

Colours. Rhizome and rhizome scales dark brownish red. Pseudobulbs olive green. Leaves green. Rhachis green. Floral bracts yellowish, often suffused with red. Sepals and petals pale yellowish, bright yellow or bright orange, usually striated dark red-purple, occasionally entirely dark red-purple. Lip yellow, brownish red or purple.

Ecology. Lowland forest to montane forest and other montane vegetation, up to 2300 m alt. Occasionally in Brachystegia forest. Epiphyte in the central part of the crown of trees, on bark or in minor humus deposits (Johansson, 1974). Also in bushes and lithophytic on rocks. Flowering mainly III–VI and VIII–XII.

Distribution. GUIN. (2). S.L. (1 coll.). LIB. (7). IV.C. (3). GHANA (2). NIG. S. part, mainly SE (8). CAM.: S. part (16). C.A.R. (1). EQ.GUIN.: Bioko (1). GAB. (11). ZAI.: Forestier Central Z6 (3); Lacs Edouard & Kivu Z9 (1); Haut Katanga Z11 (3). UG.: West U2, S. part (3). RW.: (3). ZAM.: Northwest, Mwinilunga (5), Copperbelt (3). MOZ. (1). Material seen from AMES, BM, BR, C, E, G, HBG, K, L, P, UPS, WAG.

Notes. 1 - The plant: Wageningen cult. Hort. 79-82 (= Leiden cult. Hort. 23640 = v. Setten 388 & 494) from GAB. has only produced perfectly cleistogamous flowers during the years it was cultivated in Wageningen and Leiden. The flowers hardly open and all developed a seed pod afterwards.

2 - Several authors have used the name *Bulbophyllum oreonastes* for *Bulbophyllum fuscum* var. *melinostachyum*. This was caused by wrong interpretation of the very poor type specimen of *Bulbophyllum oreonastes*. *Bulbophyllum oreonastes*, as understood here, has frequently been referred to as *Bulbophyllum zenkerianum*, a junior synonym of *Bulbophyllum oreonastes*.

64 - **Bulbophyllum encephalodes** Summerh. - Fig. 94.

Bulbophyllum encephalodes Summerh., Bot. Mus. Leafl. Harv. Un. 14 (1951) 228; Summerh., Hook. Ic. Pl. 36 (1956) pl. 3546; Morris, Proc. Linn. Soc. London 179 (1968) 63; Piers, Orch. E. Afr. (1968) 101; Morris, Ep. Orch. Mal. (1970) 44; Stewart in Agnew, Upl. Ken. W. Fl. (1974) 760; Geerinck, Bull. Soc. R. Bot. Belg. 109 (1976) 176; Grosvenor, Excelsa 6 (1976) 79; Williamson, Orch. S. Centr. Afr. (1977) 139; Ball, S. Afr. Ep. Orch. (1978) 70, pl.; La Croix et al., Orch. Mal. 1 (1983) 44; Cribb, in F.T.E.A., Orch. 2 (1984) 307.-Type: Moreau 6 (K).

Roots 0.6–1.1 mm diam. *Rhizome* 2–3.5 mm diam., nodes not or hardly swollen. *Rhizome scales* tubular, tip about acute; membranous, the upper with nerves; keeled abaxially. *Pseudobulbs* 1-leafed, ovoid to ellipsoid, (1.5 -) 3–8 cm apart, 1.2–3.5 by 1–2.2 cm, slightly flattened, about sharply 4-angled. *Petiole* 4–6 mm long. *Leaf blade* lanceolate, 3–14.5 by 1.2–3.2 cm, tip rounded, emarginate, not or hardly oblique; coriaceous, thick; midrib hardly prominent abaxially. *Inflorescence* 13–43 cm long, 14–36-flowered. *Peduncle* about erect, slightly swollen towards the tip, 11–32 cm by 1.5–2 mm, orbicular in section, glabrous, with 4–11 scales. *Peduncle scales* tubular, the longest 10–15 mm long; tip obtuse to acute; membranous, with some nerves; keeled towards the tip and glabrous abaxially. *Rhachis* nodding, slightly swollen and flattened, 4-angled in section, with two concave sides from which the flowers arise, either zig-zag bent or not, 2–11 cm by 2–3.5 mm, edges sharp, straight, entire to irregularly dentate, glabrous; surface glabrous. *Floral bracts* about spreading, concave, ovate to triangular, 3–5 by 2.6–4 mm, tip acuminate; base broadly attached; membranous, without nerves; keeled and glabrous abaxially. *Flowers* distichous, 4–6 mm apart, measured along one side of the rhachis, recurved, many open simultaneously, not fully open. *Pedicel and ovary* 1.7–3 mm long, glabrous. *Median sepal* not recurved, ovate(-oblong), 4–6 by 2–2.9 mm, tip acuminate; margins entire, glabrous; base (rather) broadly attached; rather thick; glabrous; not keeled abaxially. *Lateral sepals* free, either slightly recurved or not, hardly falcate, ovate to ovate-oblong or narrowly triangular, oblique, 4–6 by 2.2–3 mm, base broadly attached; otherwise as the median sepal. *Petals* not recurved, not or hardly falcate, oblong to linear-lanceolate, oblique, 2.2–3 by 0.6–1 mm, tip acute; margins entire, glabrous; base broadly attached; rather thin; glabrous; not keeled abaxially. *Lip* not recurved, ovate to broadly ovate-oblong in outline (not spread), (1.5 -) 2.2–3 by (0.6 -) 1.4–1.8 mm, tip irregularly warty; edges entire, glabrous; thick; adaxially with two rather weak, rounded, slightly irregular ridges which are most prominent about half-way the lip, which diverge slightly towards the base and which develop into a number of rather large, irregularly shaped warts towards the tip of the lip, with a slightly concave basal part, without a median ridge, surface glabrous; abaxially with a distinct, lumplike, broadly rounded ridge near the base abruptly ending in a large, convex lump with a brainlike surface half-way, slightly concave and slightly warty towards the tip, surface glabrous. *Column* from the base of the free part up to the tip of the stelidia 1.5–2 mm long, with rather distinct, triangular, not falcate, subacute stelidia of 0.2–0.3 mm long; column usually without teeth along the adaxial margins (see note below), stigma without a basal callus. *Anther* c. 0.8 by 0.7 mm; front not drawn out; front margin about truncate, entire, glabrous; basally distinctly emarginate; abaxially with a thick, not concave, retuse, colliculate protrusion towards the top of the anther which distinctly overtops the front margin, and a rather weak, rounded ridge about half-way the anther; surface almost entirely colliculate. *Pollinia* 4, ellipsoid; the small ones about half as long as the large ones, distinctly flattened; the large ones flattened; a stipes has not been observed. *Fruit* ellipsoid, up to 0.9 by 0.5 cm, without a stalk.

Colours. Rhizome and rhizome scales reddish brown. Leaves dark green. Rhachis reddish brown. Sepals greenish, suffused, spotted or striated with purple or entirely purple. Petals pale greenish or purple. Lip very dark purple or white with purple markings.

Ecology. Montane forest, occasionally in Brachystegia woodland up to 1500 m alt. Also in plantations. Epiphytic, often on the bare bark of tree trunks (La Croix, written comm.). Flowering I–XII.

Distribution. CAM.: SW part (1 coll.). ZAI.: Haut Katanga Z11 (9). UG.: West U2, S. part (1). BUR. (2). KEN.: Nyanza K5 (2). TANZ.: Tanga T3 (5); Southern Highlands T7 (1). ZAM.: West, Mwinilunga (3), Copperbelt (1). MAL.: South (5). ZIM.: East (2). Material seen from AMES, BM, BR, E, K, L, NY, S, UPS, WAG.

Notes. 1 - The specimen Sanford 5655, the only one from NIG., has slightly longer and narrower sepals and petals and a very small lip.

The specimen Brass 17871 from MAL. has very small, deltoid, forwards directed wings along the adaxial margins of the column, just below the stelidia.

2 - A possible hybrid between *Bulbophyllum encephalodes* and *Bulbophyllum elliotii* has been found, see note 3 under the latter species.

65 - **Bulbophyllum concatenatum** Cribb & Taylor – Fig. 95.

Bulbophyllum concatenatum Cribb & Taylor, Kew Bull. 35 (1980) 433; Cribb, in F.T.E.A., Orch. 2 (1984) 305.–Type: Schlieben 3155 (holo K; iso BR).

Roots 0.3–0.9 mm diam. *Rhizome* 0.9–2.3 mm diam., nodes not or hardly swollen. *Rhizome scales* tubular, tip acute, slightly mucronate; membranous, the upper with some nerves; keeled abaxially. *Pseudobulbs* 1-leafed, (narrowly) ovoid to ellipsoid, 1.4–8.8 cm apart, 1–3.4 by 0.6–1.3 cm, slightly flattened, 4-angled. *Petiole* 1–3 mm long. *Leaf blade* broadly lanceolate to linear-lanceolate, maximum width slightly above the middle; 2.8–9.5 by 0.5–1.5 cm, tip rounded, emarginate, either oblique or not; coriaceous, thick; midrib slightly prominent abaxially. *Inflorescence* 7–16 cm long, 5–16-flowered. *Peduncle* erect to patent, slightly swollen towards the tip, 5.4–12.2 cm by 0.7–1.2 mm, orbicular in section, glabrous, with 6–9 scales. *Peduncle scales* tubular, the longest 5–8.5 mm long; tip acute to acuminate; membranous, without nerves; keeled and glabrous abaxially. *Rhachis* nodding, slightly swollen and flattened, 4-angled in section, with two concave sides from which the flowers arise, either slightly zig-zag bent or not, 1.6–4 cm by 1.3–2.5 mm, edges winged, straight, entire to irregularly dentate, glabrous; surface glabrous. *Floral bracts* recurved to spreading, concave, ovate, 3.5–5 by 2–3 mm, tip acute; base rather broadly attached; membranous, without nerves; keeled and glabrous abaxially. *Flowers* distichous, 4.5–7 mm apart, measured along one side of the rhachis, recurved, some open simultaneously, not fully open. *Pedicel and ovary* c. 1.6 mm long, glabrous. *Median sepal* not recurved, ovate, c. 5 by 2.8 mm, tip acute; margins entire, glabrous; base broadly attached; rather thick; glabrous; not keeled abaxially. *Lateral sepals* free, about spreading, not falcate, narrowly ovate, oblique, c. 5.2 by 2.7 mm, otherwise as the median sepal. *Petals* not recurved, hardly falcate, linear-lanceolate, oblique, c. 3.2 by 0.4 mm, tip obtuse; margins entire, glabrous; base rather broadly attached; thin; glabrous; not keeled abaxially. *Lip* recurved, narrowly ovate in outline (not spread), c. 2.5 by 1.3 mm, tip rounded; edges entire, glabrous or nearly so, in the basal half with obliquely deltoid lateral lobes which have a rounded tip, slightly irregularly denticulate, about glabrous margins, a broadly attached base and which are decurrent towards the tip of the lip, thin and about glabrous; thick; adaxially with to rather distinct, obtuse ridges in the basal half, with a concave basal part, without a median slit, surface finely puberulous; abaxially with a distinct, broadly rounded ridge, surface puberulous. *Column* from the base of the free part up to the tip of the stelidia c. 1.4 mm long, with rather distinct, triangular, not falcate, obtuse to subacute stelidia of c. 0.4 mm long; column without teeth along its adaxial margins; stigma without a basal callus. *Anther* c. 0.7 by 0.6 mm; front not drawn out; front margin truncate, irregular, about glabrous; basally distinctly emarginate; abaxially without a ridge but with a thick, hardly concave, flattened, retuse, glabrous protrusion near the top of the anther which distinctly overtops the front margin; surface glabrous. *Pollinia* 4, ellipsoid; the small ones about half as long as the large ones, distinctly flattened; the large ones flattened; a stipes has not been observed. *Fruit* not seen.

Colours. Sepals and petals yellowish or greenish, stained with purple. Lip yellow, spotted purple.

Ecology. Montane forest, 1400–2000 m alt. Epiphyte on trees and liana (1 rec.). Flowering observed in I, III, XII.

Distribution. TANZ.: East T6 (6 coll.). Southern Highlands T7 (1). Material seen from BR, K.

66 - Bulbophyllum unifoliatum De Wild. - Fig. 96–98.

Literature: see under the subspecies.

Roots 0.5–1 mm diam. *Rhizome* 1–2.2 mm diam., nodes not or hardly swollen. *Rhizome scales* tubular, tip about acute, the lower rounded; membranous, the upper with some nerves; slightly keeled abaxially. *Pseudobulbs* 1-leafed, (narrowly) ellipsoid, occasionally slightly ovoid, 0.6–9 cm apart, 1.2–4 by 0.5–1.4 cm, slightly flattened, 4-angled. *Petiole* 1–7 mm long. *Leaf blade* oblong to linear, 1.8–17 by 0.6–1.5 cm, tip rounded to acute, either emarginate and/or oblique or not; coriaceous, thick; midrib slightly prominent abaxially. *Inflorescence* 5.5–20 cm long, 5–38-flowered. *Peduncle* erect or patent, slightly swollen and flattened towards the tip, 3–14 cm by 0.8–1.8 mm, orbicular in section, glabrous, with 3–8 scales. *Peduncle scales* tubular, the longest 3–12 mm long; tip obtuse to acute, occasionally mucronate; membranous, without nerves; keeled and glabrous abaxially. *Rhachis* erect to arching or nodding, slightly swollen and flattened, 4-angled in section, with two concave sides from which the flowers arise, either zig-zag bent or not, 1.5–8.5 cm by up to 3.2 mm, edges sharp, straight, entire to irregularly dentate, glabrous; surface glabrous to slightly papillose. *Floral bracts* recurved to spreading, concave, about triangular to ovate, 3.5–7.2 by

1.8–5.2 mm, tip acute; margins straight to undulate; base broadly attached; membranous, without nerves; slightly keeled and glabrous abaxially. *Flowers* distichous, 4–8 mm apart, measured along one side of the rhachis, recurved to about spreading, many open simultaneously, not much open. *Pedicel and ovary* 1–2 mm long, about glabrous. *Median sepal* not recurved, ovate to ovate-oblong, 2.9–5.8 by 1.3–2.5 mm, tip acute to acuminate; margins entire, glabrous; base broadly attached; rather thin to rather thick; glabrous adaxially; not keeled and glabrous to coarsely papillose abaxially. *Lateral sepals* free, recurved, hardly falcate, oblique, 2.8–6 by 1.5–2.8 mm, otherwise as the median sepal. *Petals* not recurved, not or hardly falcate, oblong to linear-lanceolate, oblique, 1.5–3 by 0.25–0.9 mm, tip obtuse to acute; margins entire, glabrous; base broadly to rather narrowly attached; thin; glabrous; not keeled abaxially. *Lip* either recurved or not, elliptic in outline (not spread), 1.4–2.6 by 0.8–1.7 mm, tip rounded to obtuse; margins entire, glabrous to coarsely papillose; thick; adaxially with two very weak to rather distinct, rounded to obtuse ridges starting at the base and continuing over half-way the lip which diverge towards the base and occasionally towards the tip, with a more or less concave basal part, without a median slit, surface either glabrous or finely to coarsely papillose, especially towards the margins, except for a well defined, about glabrous strip along the median line, occasionally partly puberulous in stead of papillose; abaxially with a weak to distinct, broadly rounded ridge in the basal half, surface glabrous to finely papillose, either coarsely papillose towards the margins or not. *Column* from the base of the free part up to the tip of the stelidia 0.8–1.5 mm long, with about rudimentary, at most deltoid, not falcate, obtuse to acute stelidia; column with (rather) distinct, deltoid to triangular, rounded to obtuse teeth along its lower margins just below the base of the stelidia; stigma without a basal callus. *Anther* c. 0.6–1 by 0.6–0.8 mm; front not drawn out; front margin rounded, entire, glabrous; basally weakly to distinctly emarginate; abaxially without a ridge but with a thick, hardly concave, flattened, rounded to truncate, colliculate protrusion which distinctly overtops the front margin; surface glabrous but colliculate towards the top. *Pollinia* 4; the small ones about half as long as the large ones, orbicular to ellipsoid, distinctly flattened on either side; the large ones ellipsoid, flattened on one side; a stipes has not been observed. *Fruit* ellipsoid to ovoid, up to 0.6 by 0.25 cm, without a stalk.

Note: See note under the species description of *B. acutebracteatum*.

KEY TO THE SUBSPECIES

1 - Lip glabrous or at most very finely papillose................ssp. **infracarinatum** (b)
  - Lip coarsely papillose, especially the ridges on its adaxial surface................2
2 - Sepals abaxially glabrous or at most very finely papillose...ssp. **unifoliatum** (a)
  - Sepals abaxially coarsely papillose, especially towards the tip....ssp. **flectens** (c)

66a - ssp. unifoliatum – Fig. 96.

Bulbophyllum unifoliatum De Wild., Rev. Zool. Afr. 9, suppl. bot. (1921) 34; De Wild., Pl. Beq. 1 (1921) 121; Geerinck, Bull. Soc. R. Bot. Belg. 109 (1976) 176; Williamson, Orch. S. Centr. Afr. (1977) 138.– Type: Bequaert 6616 (BR).

Non *Bulbophyllum unifoliatum* auct.: Morris, Ep. Orch. Mal. (1970) 47 (= ssp. *infracarinatum*).

Pseudobulbs 5.5–9 cm apart, 1.4–4 cm long, narrowly ellipsoid. Leaves linear-lanceolate to linear, 6–17 by 0.6–1.4 cm, tip obtuse to acute. Rhachis glabrous. Floral bracts 3.5–4.2 by 1.8–3.5 mm, margins straight. Median sepal ovate, 3–5.2 by 1.5–2.4 mm, glabrous. Lateral sepals ovate, 3.3–6 by 1.9–2.8 mm, glabrous. Petals narrowly oblong to linear-lanceolate, 1.8–3 by 0.25–0.9 mm, base rather narrowly attached. Lip 1.8–2.5 by 1–1.6 mm, adaxially about entirely coarsely papillose except for a well-defined, glabrous strip along the median line. Anther with a rather thin, rounded to truncate, finely papillose protrusion on its abaxial side near the tip which is narrower than the anther cap itself.

Colours. Rhizome and rhizome scales reddish brown. Rhachis greenish. Sepals and petals yellowish or brownish orange. Lip orange red, purple near the base.

Ecology. Lowland forest, montane forest, 400–1800 m alt. Epiphyte, growing with ferns and mosses (La Croix, written comm.). Flowering X and I–III.

Distribution. ZAI.: Forestier Central Z6 (4 coll.). RW. (2). TANZ.: Eastern P. T6 (1). ANG.: Huambo P. (1). ZAM.: Northwest, Copperbelt (1). Widespread but only locally. Material seen from BR, K.

Note. In general habit the specimen Exell & Mendonça 1882, the only from ANG., resembles ssp. *infracarinatum*, but the lip is more like that of ssp. *unifoliatum*. The specimen Leonard 1420 from ZAI. may be similar but this cannot be verified because all the lips have fallen off.

66b - ssp. infracarinatum (Williamson) J.J. Vermeulen, stat. nov. – Fig. 97.

Bulbophyllum infracarinatum Williamson, J. S. Afr. Bot. 47 (1981) 133; La Croix et al., Orch. Mal. 1 (1983) 41.–*Bulbophyllum carinatum* Williamson, J. S. Afr. Bot. 46 (1980) 333.–Type: Drewe 33 (SRGH, n.v.). Non *Cirrhopetalum carinatum* Teysm. & Binnend., Ned. Kruidk. Arch. 3 (1855) 397; nec *Bulbophyllum carinatum* Cogn., Journ. des Orch. 6 (1895) 216; nec *Bulbophyllum carinatum* Ames, Phil. J. Sc. 7 (1912) 140.

Bulbophyllum sp. Grosvenor, Excelsa 6 (1976) 79; Ball, S. Afr. Ep. Orch. (1978) 74, pl.–Specimen: Ball 986 (BR, K).

Bulbophyllum unifoliatum auct. non De Wild.: Morris, Ep. Orch. Mal. (1970) 47.

Pseudobulbs 0.6–3 cm apart, 1.2–2.6 cm long, ellipsoid to narrowly ellipsoid. Leaves linear-lanceolate, 3–10 by 0.6–1.5 cm, tip obtuse. Rhachis glabrous. Floral bracts 3.5–4.8 by 3–4 mm, margins straight. Median sepal ovate to ovate oblong, c. 3 by 1.3–2 mm, abaxially glabrous or nearly so. Lateral sepals ovate to ovate-oblong, 2.8–4.2 by 1.5–2.3 mm, abaxially as the median sepal. Petals linear-lanceolate, 1.5–2 by 0.25–0.35 mm, base broadly attached. Lip 1.4–1.8 by 0.8–1.3 mm, adaxially about glabrous, very finely papillose towards the margins. Anther with a rather fleshy, rounded, coarsely papillose protrusion on its abaxial side near the tip which is narrower than the anther cap itself.

Colours. Rhizome and rhizome scales reddish brown. Pseudobulbs yellowish green. Rhachis green. Floral bracts brown. Sepals yellowish, greenish or pale purplish, blotched with purple. Petals cream coloured, with some purple dots. Lip purple.

Ecology. Lower montane forest at c. 1100 m alt. Epiphytic. Flowering XII–II.

Distribution. MAL.: South, Mlanje Mts. (1 coll.). ZIM.: East (2). MOZ.: Manica e Sofala (1). Material seen from BM, K. Also material present in SRGH.

66c - ssp. flectens (Cribb & Taylor) J.J. Vermeulen, stat. nov. – Fig. 98.

Bulbophyllum flectens Cribb & Taylor, Kew Bull. 35 (1980) 436; La Croix et al., Orch. Mal. 1 (1983) 44; Cribb, in F.T.E.A., Orch. 2 (1984) 307.–Type: Cribb, Grey-Wilson & Mwasumbi 11446 (K).

Pseudobulbs 1.8–5 cm apart, 1.5–3.3 cm long, ellipsoid to narrowly ellipsoid to slightly ovoid. Leaves oblong to linear-lanceolate, 1.8–9 by 0.8–1.3 cm, tip rounded to obtuse. Rhachis glabrous, occasionally slightly papillose. Floral bracts 5–7.2 by 4–5.2 mm, margins more or less undulate. Median sepal ovate-oblong, 4–5.8 by 2–2.5 mm, abaxially coarsely papillose, especially towards the tip. Lateral sepals ovate-oblong, 4.2–5.5 by 1.9–2.5 mm, abaxially as the median sepal. Petals lanceolate to linear-lanceolate, 2.2–3 by 0.3–0.7 mm, base broadly attached. Lip 2–2.6 by 1.2–1.7 mm, adaxially coarsely papillose along the margins and on the ridges, puberulous elsewhere, except for a well-defined, glabrous strip along the median line. Anther with a fleshy, rounded, coarsely papillose protrusion on its abaxial side near the tip which is about as wide as the anther cap itself.

Colours. Pseudobulbs brownish. Rhachis dark brown-purple. Flowers entirely dark purple.

Ecology. Montane forest at 1700–2100 m alt. Epiphytic. Flowering I–III.

Distribution. TANZ.: Southern Highlands T7 (8 coll.); South T8, Songea (1). MAL.: Central (1). Material seen from AMES, K.

GROUP 10 (sp. 67)

Pseudobulbs 2-leafed. (Adaxial surface of the leaves finely papillose). Inflorescence a raceme, 1–3-flowered. Rhachis slighlty swollen, spindle-shaped. Flowers scattered. Sepals with a few fine, dark hairs abaxially, the lateral ones not connate along the upper margin, not recurved. Lip rather thick, not very soft and easy to compress, lateral lobes absent, margins ciliate. (Column distinctly elongated.) Stelidia short, acuminate. Anther with a dorsoventrally flattened protrusion overtopping the front margins.

Note. Characters not mentioned in the descriptions of the other groups have been placed between brackets.

67 - Bulbophyllum gilgianum Krzl. – Fig. 99.

Bulbophyllum gilgianum Krzl., Engl., Bot. Jahrb. 28 (1900) 162 ('*Bolbophyllum*'); De Wild., Pl. Beq. 1 (1921) 111; Cribb, in F.T.E.A., Orch. 2 (1984) 311.–Type: Goetze 614 (holo B, lost); Moreau 464 (neo K).

Roots 0.8–1.2 mm diam. *Rhizome* 1–2.2 mm diam., nodes not or hardly swollen. *Rhizome scales* tubular, tip obtuse to acute; membranous, without nerves; keeled abaxially towards the tip. *Pseudobulbs* 2-leafed, ovoid to about orbicular, 1.5–5 cm apart, 0.5–1.3 by 0.5–1.1 cm, slightly flattened, 4-angled. *Petiole* 1–2 mm long. *Leaf blade* elliptic, 0.8–2.2 by 0.6–1.3 cm, tip rounded, emarginate, slightly oblique; coriaceous, very thick; surface finely and regularly papillose, midrib not or hardly prominent abaxially. *Inflorescence* 1.5–2.1 cm long, 2–3-flowered. *Peduncle* about patent, rather sturdy, slightly swollen towards the tip and below the nodes, 1-1.4 cm by c. 1.5 mm, orbicular in section, locally with fine dark hairs, with 4-6 scales. *Peduncle scales* tubular, the longest c. 4 mm long; tip acute to acuminate; membranous, without nerves; slightly keeled towards the tip and with some fine dark hairs abaxially. *Rhachis* erect to patent, slightly swollen and flattened below te attachment of the flowers, 0.5-1 cm by c. 1 mm, with some fine dark hairs. *Floral bracts* recurved, slightly concave, broadly elliptic, 5-5.9 by 3-4 mm, tip acuminate; base broadly attached; membranous, without nerves; slightly keeled and with some fine dark hairs abaxially. *Flowers* (?) scattered, in a lax raceme, about appressed against the rhachis, about all open simultaneously, not much open. *Pedicel and ovary* 2-3.5 mm long, with some fine dark hairs, jugae ending in an acute tooth. *Median sepal* not recurved, broadly ovate-oblong, 8.2-10 by c. 4 mm, tip acute; margins entire, glabrous; base rather narrowly attached; thick; with very fine dark hairs and two rather inconspicuous swellings near the base adaxially; not or hardly keeled and with rather coarse dark hairs abaxially. *Lateral sepals* free, hardly falcate, broadly lanceolate, oblique, c. 10 by 3 mm, tip acuminate; with an obtuse ridge adaxially near the base; slightly and obtusely keeled abaxially; otherwise as the median sepal. *Petals* not recurved, hardly falcate, broadly (obovate)-lanceolate, oblique, 6-7 by c. 2 mm, tip acute; margins entire, finely papillose; base rather narrowly attached; rather thin but with a thick base; finely papillose towards the tip and with a distinct, obtuse, fleshy ridge in the basal half adaxially; keeled and finely papillose abaxially. *Lip* slightly recurved, narrowly oblong in outline, maximum width slightly above the middle, 5.3-5.5 by 1.6-1.8 mm, tip obtuse; margins entire, ciliate; rather thick; adaxially without ridges, with a slightly concave basal part, with a distinct, sharply cut median furrow in the basal half which widens towards the tip, surface finely papillose, slightly pubescent near the base; abaxially with a weak, rounded ridge near the base, surface finely papillose. *Column* from the base of the free part up to the tip of the stelidia 3.7-4 mm long, with distinct, gradually tapering, slightly falcate, acuminate stelidia of c. 0.4 mm long which have a rather distinct, obliquely forwards directed, deltoid, obtuse tooth along their upper margin; column without teeth along its lower margins; stigma without a basal callus. *Anther* c. 1 by 0.7 mm; front not drawn out; front margin truncate, irregular, glabrous; basally distinctly emarginate; abaxially without a ridge but with a very thick, flattened, rounded, colliculate top part which distinctly overtops the front margin; surface colliculate. *Pollinia* 4, flattened on one side; the small ones much more than half as long as the large ones, orbicular; the large ones ellipsoid; a stipes has not been observed. *Fruit* not seen.

Colours. Leaves suffused with purple. Flowers yellowish green, mottled purple. Lip yellow or purple. Column green.

Ecology. Montane forest, at c. 1900 m alt. Epiphytic. Flowering observed in II.

Distribution. TANZ.: East T6 (2 coll.), Southern Highlands T7 (fide Kraenzlin). Material seen from K.

GROUP 11 (sp. 68; '*Cirrhopetalum*')

Pseudobulbs 1-leafed. Inflorescence subumbellate, 3-many-flowered. Rhachis not swollen. Flowers scattered. Sepals glabrous abaxially, the lateral ones connate along their upper margins, not recurved. Lip thick, not very soft and easy to compress, lateral lobes absent, margins glabrous. Stelidia long, acuminate. Anther without a protrusion overtopping the front margin.

68 - Bulbophyllum longiflorum Thouars - Fig. 100; Plate 11d.

Bulbophyllum longiflorum Thouars, Orch. Il. Austr. Afr. (1822) pl. 97; Seidenf., Dansk Bot. Ark. 29 (1973) 126; La Croix et al., Orch. Mal. 1 (1983) 46; Cribb, in F.T.E.A., Orch. 2 (1984) 304.-*Phyllorchis longiflora* (Thouars) Kuntze, Rev. Gen. Pl. 2 (1891) 675, incorr. name.-*Cirrhopetalum longiflorum* (Thouars) Schltr, Beih. Bot. Zentralbl. 33 (1915) 420.-*Cirrhopetalum thouarsii* Lindl., Gen. and Sp. Orch. Pl. (1830) 58, incorr. name; Lindl., Bot. Reg. 24 (1838) pl. 11; Krzl., Engl., Bot. Jahrb. 34 (1904) 60.-*Phyllorchis thouarsii* (Lindl.) Kuntze, Rev. Gen. Pl. 2 (1891) 677, incorr. name.-Type: Du Petit Thouars s.n. (Holo P; iso L), Mauritius.

Non *Bulbophyllum longiflorum* Rchb. f., Naves, Nov. App. (1880) 235; nec *Bulbophyllum longiflorum* Ridl., J. Linn. Soc. Bot. 32 (1896) 268.

Epidendrum umbellatum Forst. f.; Fl. Ins. Austr. Prod. (1786) 60.-*Cirrhopetalum umbellatum* (Forst. f.) Hook & Arn., Bot. Beech. Voy. (1832) 71; Hunt & Summerh., Kew Bull. 20 (1966) 59; Morris, Proc. Linn. Soc. London 179 (1968) 63; Piers, Orch. E. Afr. (1968) 110; Morris, Ep. Orch. Mal. (1970) 53; Stewart & Campbell, Orch. Trop. Afr. (1970) 6, col. phot.; Grosvenor, Excelsa 6 (1976) 79; Ball, S. Afr. Ep. Orch. (1978) 96, pl.; Stewart & Hennesy, Orch. Afr., sel. rev. (1981) 90, pl.; Stewart & all., Wild Orch. S. Afr. (1982) 220, pl.-*Phyllorchis umbellata* (Forst. f.) Kuntze, Rev. Gen. Pl. 2 (1891) 675, incorr. name.-Type: Forster 147 (Holo?; iso M, W, n.v., fide Seidenfaden 1973), Tahiti.

Non *Epidendrum umbellatum* Sw., Kgl. Sv. Vet. Ac. Nya Handl. 21 (1800) 241.

Non *Bulbophyllum umbellatum* Lindl., Gen. and Sp. Orch. Pl. (1830) 56.

Cirrhopetalum africanum Schltr, Engl., Bot. Jahrb. 53 (1915) 573; Schltr, Fedde, Rep. Beih. 68 (1932) pl. 70.-Syntypes: Stolz 453 (G, L, S, U, WAG); 1411 (G, L, S, U, WAG); 2046 (BM, BR,P).

Non *Bulbophyllum africanum* Hawkes, Lloydia 19 (1956) 92 (= *Bulbophyllum nigritianum*).

Roots rising along the entire rhizome, 1-2 mm diam. *Rhizome* 2-5 mm diam., nodes either slightly swollen or not. *Rhizome scales* tubular, tip acuminate; membranous, with many nerves which remain as coarse fibres after decay of the scales; keeled abaxially towards the tip. *Pseudobulbs* 1-leafed, (narrowly) ovoid, 2.6-9 cm apart, 1.5-4.5 by 0.6-1.8 cm, slightly flattened, obtusely 4-6-angled. *Petiole* 4-23 mm long. *Leaf blade* broadly (linear-)lanceolate, maximum width occasionally slightly above the middle, 6.8-21 by 1.5-4 cm, tip rounded to obtuse, emarginate, usually oblique; coriaceous, thick; midrib hardly prominent abaxially. *Inflorescence* 8.5-21 cm long, 3-6-flowered. *Peduncle* erect to patent, rather wiry, 8.3-20.5 cm by 1-1.8 mm, orbicular in section, glabrous, with 4-5 scales. *Peduncle scales* tubular, the longest 8-10 mm long; tip acuminate; membranous, with some nerves; keeled and glabrous abaxially. *Rhachis* erect, terete, 0.2-0.7 cm long, glabrous. *Floral bracts* spreading, ovate-oblong, 4-8 by 1.8-3 mm, tip acuminate; base rather broadly attached; membranous, with some nerves; not keeled and glabrous abaxially. *Flowers* scattered, in a dense, subumbellate raceme, spreading, either slightly secund or not, all open simultaneously, not fully open. *Pedicel and ovary* 12-18 mm long, glabrous. *Median sepal* not recurved, concave, orbicular to elliptic, 9-10 by 7-9 mm, top obtuse, tip with a seta of 4.5-8 mm long which is slightly thickened towards the tip, another much shorter seta may be present at each side of the tip; margins entire near the base, increasingly irregular and dentate towards the tip, glabrous; base narrowly attached; rather thin; glabrous; not keeled abaxially. *Lateral sepals* connate along their upper margins, twisted at their base, not recurved, together convex, not falcate, each one linear-lanceolate, oblique, 28–38 by 3.8–4.2 mm, tip about acute; margins entire, glabrous; base broadly attached; thin; glabrous, not keeled abaxially. *Petals* not recurved, slightly falcate, oblong to ovate-oblong, oblique, 5.5–8 by 2.2–2.5 mm, tip acute to acuminate, with a seta of 3–5 mm long; margins coarsely and irregularly denticulate towards the tip, each tooth with a short seta; base rather broadly attached; rather thin; glabrous; not keeled abaxially. *Lip* recurved, narrowly triangular in outline (not spread), c. 5 by 2.2–2.5 mm, tip rounded to obtuse; margins entire, glabrous; thick; adaxially without ridges, with a slightly concave basal part, without a median slit, surface glabrous; abaxially with a weak, retuse ridge near the base, surface glabrous. *Column* from the base of the free part up to the tip of the stelidia 5.5–7 mm long, with distinct, gradually tapering, more or less falcate, acuminate stelidia of 3–4.5 mm long which have a distinct, obliquely forwards and inwards directed, triangular, acute tooth along their upper margin about half-way and which usually have a weak, at most deltoid, obtuse tooth near the base; column with weak, obtuse wings along its lower margins; stigma without a basal callus. *Anther* c. 1.3 by 1.3 mm; front slightly drawn out, thin, slightly concave; front margin truncate, denticulate, glabrous; basally distinctly emarginate; abaxially with a rather weak, rounded, glabrous ridge near the top of the anther which does not overtop the front margin; surface glabrous. *Pollinia* 4; all of about equal length, obovoid, the outer flattened on one side, the inner pair flattened on either side; a stipes has not been observed. *Fruit* ellipsoid, up to 6 by 0.8 cm, stalk 2 mm.

Colours. Pseudobulbs olive green. Leaves green. Peduncle pale green, finely stained with purple. Sepals cream coloured or yellowish; the median one heavily spotted dark red, especially towards the tip; the lateral ones only slightly stained with red. Petals as the median sepal. Lip yellow, suffused with dark red or entirely dark red. Column pale yellow, spotted dark red. Top of median sepal and petals with scattered, enlarged cells which are slightly swollen, hyaline and almost black. Flowers either scentless or with an unpleasant smell.

Ecology. Hill and montane forest, 500–1400 m alt. Epiphytic, usually on tree trunks (La Croix, written comm.). Flowering (IV, VII) X–II.

Distribution. ZAI.: Forestier Central Z6 (1). UG. (1 coll.). TANZ.: Tanga T3 (11); East T6 (2); Southern Highlands T7 (5). MAL.: North (3); South (2). ZIM.: East (3). Material seen from BR, C, E, G, K, L, S, U, UPS, WAG.

Distribution outside Africa: Madagascar, Seychelles, Réunion, Mauritius, Indonesia, Philippines, Papua New Guinea, Australia, New Caledonia, Fiji, Cook Isl., Tubuai Isl., Society Isl. (See Seidenfaden, 1973, p. 126).

Notes. 1 - For further synonymy see: Seidenfaden l.c.

2 - Applying the key for identification of some herbarium specimens may lead to mistakes due to the fact that the lateral sepals may occasionally get loose during the drying process. Besides, the seta of the median sepal easily breaks off in dried specimens, so that this character cannot be confirmed.

69 - Bulbophyllum nummularia (Wendl. & Krzl.) Rolfe.

Bulbophyllum nummularia (Wendl. & Krzl.) Rolfe in F.T.A. 7 (1897) 35; De Wild., Pl. Beq. 1 (1921) 115.– *Megaclinium nummularia* Wendl. & Krzl., Gard. Chron., ser. 3, 15, no. 388 (1844) 685.–Type: Braun s.n. (?, n.v.).

Distribution. CAM.

Note. No material has been seen by me. The combination of characters ascribed to the species is unique among continental African *Bulbophyllum*; it is impossible to include it in any of the groups.

70 - Bulbophyllum sangae Schltr

Bulbophyllum sangae Schltr, Westafr. Kautchuk Exp. (1901) 282, n.n.; id., Engl., Bot. Jahrb. 38 (1905) 19; id., Fedde, Rep. Beih. 68 (1932) pl. 69; De Wild., Pl. Beq. 1 (1921) 119.–Type: Schlechter 12687 (B, lost).

Distribution. CONGO.

Note. All type material of this species is lost and it has not been collected again. The shape of the lip is similar to that of *Bulbophyllum renkinianum* (Group 5) but *Bulbophyllum sangae* is a very small species with 1-leafed pseudobulbs and not flattened, only 1–2-flowered inflorescences.

71 - Bulbophyllum sessiliflorum Wendl. & Krzl.

Bulbophyllum sessiliflorum Wendl. & Krzl., Xen. Orch. 3 (1896) 156, pl. 290 ('*Bolbophyllum*'); De Wild., Pl. Beq. 1 (1921) 120.–Type: Braun s.n. '*Bolbophyllum* n. 20' (?, n.v.).

Distribution. CAM.

Note. No material of this species has been seen by the author. The plate in Xenia Orch. shows a plant which is similar to *Bulbophyllum scariosum* and *Bulbophyllum pandanetorum* (both group 1), although the flowers are quite differently coloured. Neither of the latter mentioned species is known to occur in CAM. where the type specimen from *Bulbophyllum sessiliflorum* is said to have come from.

72 - Bulbophyllum teretifolium Schltr

Bulbophyllum teretifolium Schltr, Westafr. Kautchuk Exp. (1901) 282, nom. nud.; Schltr, Engl., Bot. Jahrb. 38 (1905) 18; De Wild., Pl. Beq. 1 (1921) 120; Summerh. in F.W.T.A. ed. 1, 2 (1936) 437; id., ed. 2, 3 (1968) 234.–Type: Schlechter 12362 (B, lost).

Distribution. CAM.: Bibundi, Man Of War Bay. Common (fide Schlechter).

Notes. Unknown to the present author. The character 'terete leaves' is unique among continental African *Bulbophyllum*. Otherwise the species is difficult to place. It is amazing that a species which has so frequently been observed by Schlechter at a much visited place was never collected again.

NOMINA NUDA

N1 - Bulbophyllum apiferum Batem. ex Steudel, nom. nud.

Bulbophyllum apiferum Batem. ex Steudel, Nom., ed. 2, 1 (1840) 214, nom. nud.
Non *Bulbophyllum apiferum* Carr, Gard. Bull. s.s. (1930) 133.

Distribution. 'Senegambia'.
Note. As far as could be ascertained a description has never been published.

EXCLUDED SPECIES

E1 - Bulbophyllum apetalum Lindl.

Bulbophyllum apetalum Lindl., J. Linn. Soc. Bot. 6 (1862) 127; Krzl., Mitth. Deutsch. Schutzgeb. 2 (1889) 156.–Type: Barter 72 (K).

Note. 1 - Is *Genyorchis apetala* (Lindl.) J.J. Vermeulen, nov. comb.

In 1901 (Westafr. Kautchuk Exp.: 280) Schlechter established the generally accepted genus *Genyorchis*, including among others *Genyorchis pumila* (Sw.) Schltr., basionym: *Dendrobium pumilum* Sw. (= *Bulbophyllum pumilum* (Sw.) Lindl., see species 16 of this monograph).

However, Schlechter wrongly assumed that the specimens he studied of what he called *Genyorchis pumila* (Schlechter 12702 and 12737) belonged to the same species as the type specimen of *Dendrobium pumilum* Sw. (Afzelius in herb. Swartz s.n.), which in fact is a *Bulbophyllum* species. See species 17 of this revision.

As a consequence the name *Genyorchis pumila* (Sw.) Schltr is a homotypic synonym of *Bulbophyllum pumilum* (Sw.) Lindl. Schlechter's specimens of '*Genyorchis pumila*' should be renamed. The oldest name for this species is *Bulbophyllum apetalum* Lindl., the new combination in *Genyorchis* has been made above.

A proposal for the lectotypification of *Genyorchis* is due to appear in Taxon in 1987.

E2 - Bulbophyllum cupuligerum Krzl.

Bulbophyllum cupuligerum Krzl., Engl., Bot. Jahrb. 43 (1909) 342; id. in Mildbraed, Deutsch. Zentr. Afr. Exp. 1907-08, 2 (1914) 85; De Wild., Pl. Beq. 1 (1921) 107.–Type: Mildbraed 951 (B, lost).

Note. Is *Stolzia cupuligera* (Krzl.) Summerh., see Cribb, Kew Bull. 33 (1978) 83.

E3 - Bulbophyllum densum Thouars

Bulbophyllum densum Thouars, Orch. Isles Austr. Afr. (1822) pl. 108; Cribb, in F.T.E.A., Orch. 2 (1984) 322.

Note. The specimen Boivin s.n. (1854) from K belongs to this species and is labeled 'Zanzibar'. The species has never been found there again and is probably wrongly labeled. It belongs to a group which occurs on Madagascar and which has never been recorded from continental Africa.

E4 - **Bulbophyllum elaidum** Lindl.

Bulbophyllum elaidum Lindl., J. Linn. Soc. Bot. 6 (1862) 127; Rolfe in F.T.A. 7 (1897) 29; De Wild., Pl. Beq. 1 (1921) 108; Summerh. in F.W.T.A. ed. 1, 2 (1936) 438.-Syntypes: Barter 73 (Lecto K) & 1841 (K).

Note. Is *Stolzia elaidum* (Lindl.) Summerh., see Cribb, Kew Bull. 33 (1978) 84.

E5 - **Bulbophyllum galeatum** (Sw.) Lindl.

Bulbophyllum galeatum (Sw.) Lindl., Gen. and Sp. Orch. Pl. (1830) 57.-Dendrobium galeatum Sw., Schrad. Neu. J. Bot. 1, part 1 (1805) 97.-Type: Afzelius s.n.

Note. Is *Polystachya galeata* (Sw.) Rchb., see Podzorsky & Cribb, Kew Bull. 34 (1979) 165.

E6 - **Bulbophyllum inopinatum** W.W. Smith.

Bulbophyllum inopinatum W.W. Smith, Not. R. Bot. Gard. Edinb. 8 (1915) 346; Summerh., Kew Bull. 10 (1956) 587.-Type: Edinburgh cult. s.n. (E).

Note. This species is conspecific with the Asiatic *Bulbophyllum penicillium* Par. & Rchb. f. It is not of African origin, as was supposed by W.W. Smith.

E7 - **Bulbophyllum micropetalum** Lindl.

Bulbophyllum micropetalum Lindl., J. Linn. Soc. Bot. 6 (1862) 27.-Type: Mann 644 (K).

Note. Is *Genyorchis micropetala* (Lindl.) Schltr, see note under Excluded Species no. 1.

E8 - **Bulbophyllum nigripetalum** Rolfe.

Bulbophyllum nigripetalum Rolfe, Kew Bull. 1891 (1891) 197; id. in F.T.A. 7 (1897) 31; De Wild., Pl. Beq. 1 (1921) 114; Summerh. in F.W.T.A. ed. 1, 2 (1936) 437; id., Kew Bull. 6 (1952) 471; id. 10 (1956) 587.-Type: Sander s.n. (K).

Note. Wrongly supposed to be of African origin, is conspecific with *Bulbophyllum secundum* Hook. f. See Seidenfaden, Dansk Bot. Ark. 33, 3 (1979) 103.

E9 - **Bulbophyllum nutans** Thouars

Bulbophyllum nutans Thouars, Orch. Isles Austr. Afr. (1822) pl. 106; Cribb, in F.T.E.A., Orch. 2 (1984) 322.

Note. The specimen Boivin s.n. (1854) from K belongs to this species and is labeled 'Zanzibar'. The species has never been found there again and is probably wrongly labeled. It occurs a.o. on Madagascar and has never been recorded from continental Africa.

E10 - Bulbophyllum peperomioides Krzl.

Bulbophyllum peperomioides Krzl., Engl., Bot. Jahrb. 43 (1909) 341; id. in Mildbraed, Deutsch. Zentr. Afr. Exp. 1907-08, 2 (1914) 84; De Wild., Pl. Beq. 1 (1921) 116.–Type: Mildbraed 2814 (B, lost).

Note. Is *Stolzia peperomioides* (Krzl.) Summerh., see Cribb, Kew Bull. 33 (1978) 85.

NOT VALIDLY PUBLISHED SPECIES OF UNKNOWN AFFINITY

V1 - Bulbophyllum eurhachis Schltr

Bulbophyllum eurhachis Schltr in Mildbraed, 2e Deutsch. Zentr. Afr. Exp. 1910-11, 2 (1922) 74.–Type: Mildbraed 5020 (B, lost).

Distribution. CAM.?
Note. The name was published without description.

V2 - Bulbophyllum fernandopoanum Schltr

Bulbophyllum fernandopoänum Schltr in Mildbraed, 2e Deutsch. Zentr. Afr. Exp. 1910-11, 2 (1922) 180.–Type: Mildbraed 6892 (B, lost).

Distribution. EQ. GUIN.: Bioko.
Note. The name was published without description. Included in sect. *Megaclinium* (group 5) by Schlechter.

V3 - Bulbophyllum insulanum Schltr

Bulbophyllum insulanum Schltr in Mildbraed, 2e Deutsch. Zentr. Afr. Exp. 1910-11, 2 (1922) 180.–Type: Mildbraed 6955 (B, lost).

Distribution. EQ. GUIN.: Bioko.
Note. The name was published without description.

V4 - Bulbophyllum schultzeanum Schltr

Bulbophyllum schultzeanum Schltr in Mildbraed, 2e Deutsch. Zentr. Afr. Exp. 1910-11, 2 (1922) 180.–Type: Mildbraed 6951 (B, lost).

Distribution. EQ. GUIN.: Bioko.
Note. The name was published without description.

2 - **CHASEELLA** SUMMERH.

Chaseella Summerh., Kirkia 1960-61 (1961) 88.-Type species: *Chaseella pseudohydra* Summerh.

Description of the genus: Pseudobulbs 6–12-leafed. Pollinia 2.

Note. Monotypic genus.

1 - **Chaseella pseudohydra** Summerh. - Fig. 101.

Chaseella pseudohydra Summerh., Kirkia 1960-61 (1961) 88; Grosvenor, Excelsa 6 (1976) 79; Ball, S. Afr. Ep. Orch. (1978) 94, pl.; Cribb, in F.T.E.A., Orch. 2 (1984) 323.-Type: Chase SRGH 22738 (Holo K, iso SRGH, n.v.).

Chaseella sp. Stewart in Agnew, Upl. Ken. W. Fl. (1974) 763.-Specimen: Stewart 1194 (K).

Plant creeping, epiphytic. *Roots* 0.4–0.6 mm diam. *Rhizome* 0.7–1.1 mm diam., nodes slightly swollen. *Rhizome scales* tubular, tip acuminate, the lower obtuse to acute; membranous, with many nerves; slightly keeled abaxially towards the tip. *Pseudobulbs* 6–12-leafed, orbicular to ovoid, 0.3–2.2 cm apart, 0.4–0.8 by 0.3–0.7 cm, not or hardly flattened, obtusely 4–8-angled. *Petiole* at most 0.4 mm long. *Leaf blade* lanceolate to linear-lanceolate, 0.2–0.9 by 0.05–0.1 cm, tip obtuse to acute, neither emarginate nor oblique; coriaceous; very thick (orbicular to semi-orbicular in section); midrib not prominent abaxially. *Inflorescence* c. 2.2 cm long, 1 (– ?2)-flowered. *Peduncle* about patent, wiry, c. 1.9 cm by c. 0.6 mm, orbicular in section, slightly flattened and with 2 opposite, slightly denticulate ridges towards the tip, with some fine dark hairs, with c. 4 scales. *Peduncle scales* tubular, the longest c. 3 mm long; tip acute to acuminate; membranous, without nerves; keeled towards the tip and with some fine dark hairs abaxially. *Rhachis* erect or slightly nodding, terete, 0.3–0.4 cm long, with some fine dark hairs. *Floral bracts* recurved, elliptic, c. 3 by 2.5 mm, tip acute to acuminate; base rather narrowly attached; membranous; without nerves; keeled towards the tip and with some fine dark hairs abaxially. *Flowers* recurved, not fully open. *Pedicel and ovary* c. 3 mm long, with some fine dark hairs. *Median sepal* not recurved, ovate, c. 4 by 3 mm, tip subacute; margins entire, glabrous; base rather broadly attached; rather thin; glabrous adaxially; slightly keeled and with some fine dark hairs abaxially. *Lateral sepals* free, recurved, not falcate, about deltoid, oblique, c. 4 by 3.7 mm, keeled abaxially; otherwise as the median sepal. *Petals* not recurved, not falcate, elliptic, oblique, c. 1.8 by 1.2 mm, tip rounded; margins entire, glabrous; base rather broadly attached; thin; glabrous; not keeled abaxially. *Lip* recurved half-way, broadly ovate in outline (not spread), c. 1.8 by 1.2 mm, tip obtuse; margins entire, glabrous; thick; adaxially without ridges, with a slightly concave basal part, without a median slit, surface glabrous; abaxially with a weak, rounded ridge near the base, surface glabrous. *Column* from the base of the free part up to the tip of the stelidia c. 1.8 mm long, with distinct, triangular, not falcate, truncate to slightly bifid stelidia of c. 1 mm long; column with rather weak, rounded, finely serrulate teeth along its adaxial margins just below the base of the stelidia; stigma without a basal callus. *Anther* c. 0.8 by 0.7 mm; front not drawn out; front margin rounded, entire, papillose; basally distinctly emarginate; abaxially with a rather weak, rounded, colliculate ridge near the top of the anther which overtops the front margin, surface glabrous but colliculate towards the top. *Pollinia* 2, triangular, distinctly flattened on one side, a stipes has not been observed. *Fruit* not seen.

Colours. Flowers greenish.

Ecology. Montane forest, altitude not given.

Distribution. KEN.: Nyanza K5 (1 coll.). ZIM.: East (1).

Notes. The KEN. specimen is sterile and therefore of somewhat uncertain identity.

IDENTIFICATION LIST

Note. This identification list includes all collections examined for this revision. The identifications are given as numbers between brackets, which correspond with the species numbers in this revision. To facilitate the use of this list it is preceded by a list of the taxa with their numbers.

1a -*B. schinzianum* Krzl. var. *schinzianum*.
1b -var. *irigaleae* (Cribb & Perez-Vera) J.J. Vermeulen.
1c -var. *phaeopogon* (Schltr) J.J. Vermeulen.
2 -*B. subligaculiferum* J.J. Vermeulen.
3a -*B. saltatorium* Lindl. var. *saltatorium*.
3b -var. *albociliatum* (Finet) J.J. Vermeulen.
3c -var. *calamarium* (Lindl.) J.J. Vermeulen.
4 -*B. barbigerum* Lindl.
5 -*B. vulcanicum* Krzl.
6a -*B. cochleatum* Lindl. var. *cochleatum*.
6b -var. *gravidum* (Lindl.) J.J. Vermeulen.
6c -var. *bequaertii* (De Wild.) J.J. Vermeulen.
6d -var. *tenuicaule* (Lindl.) J.J. Vermeulen.
6e -var. *brachyanthum* (Lindl.) J.J. Vermeulen.
7 -*B. burttii* Summerh.
8 -*B. kivuense* J.J. Vermeulen.
9 -*B. inornatum* J.J. Vermeulen.
10 -*B. bavonis* J.J. Vermeulen.
11 -*B. stolzii* Schltr.
12 -*B. prorepens* Summerh.
13 -*B. intertextum* Lindl.
14 -*B. vanum* J.J. Vermeulen.
15 -*B. capituliflorum* Rolfe.
16 -*B. pipio* Rchb. f.
17 -*B. pumilum* (Sw.) Lindl.; entity a.
18 -*B. coriscense* Rchb. f.
19 -*B. cocoinum* Lindl.
20a -*B. josephii* (Kuntze) Summerh. var. *josephii*.
20b -var. *mahonii* (Rolfe) J.J. Vermeulen.
21 -*B. calvum* Summerh.
22 -*B. expallidum* J.J. Vermeulen.
23 -*B. ivorense* Cribb & Perez Vera.
24 -*B. mediocre* Summerh.
25 -*B. schimperanum* Krzl.
26 -*B. oxychilum* Schltr.
27 -*B. nigritianum* Rendle.
28 -*B. humblotii* Rolfe.
29 -*B. scariosum* Summerh.
30 -*B. pandanetorum* Summerh.
31 -*B. denticulatum* Rolfe.
32a -*B. comatum* Lindl. var. *comatum*.
32b -var. *inflatum* (Rolfe) J.J. Vermeulen.
33 -*B. bifarium* Hook. f.
34a -*B. bidenticulatum* J.J. Vermeulen ssp. *bidenticulatum*
34b -ssp. *joyceae* J.J. Vermeulen.
35 -*B. nigericum* Summerh.
36 -*B. renkinianum* (Laurent) De Wild.
37 -*B. dolabriforme* J.J. Vermeulen.
38 -*B. curvimentatum* J.J. Vermeulen.
39 -*B. colubrinum* (Rchb. f.) Rchb. f.
40 -*B. maximum* (Lindl.) Rchb. f.
41a -*B. injoloense* De Wild. ssp. *injoloense*.
41b -ssp. *pseudoxypterum* (J.J. Vermeulen) J.J. Vermeulen.
42 -*B. imbricatum* Lindl.
43 -*B. magnibracteatum* Summerh.
44a -*B. sandersonii* (Hook. f.) Rchb. f. ssp. *sandersonii*.
44b -ssp. *stenopetalum* (Krzl.) J.J. Vermeulen.
45 -*B. lizae* J.J. Vermeulen.
46 -*B. purpureorhachis* (De Wild.) Schltr.
47 -*B. falcipetalum* Lindl.
48a -*B. scaberulum* (Rolfe) Bolus var. *scaberulum*.
48b -var. *fuerstenbergianum* (De Wild.) J.J. Vermeulen.
48c -var. *crotalicaudatum* J.J. Vermeulen.
49a -*B. resupinatum* Ridl. var. *resupinatum*.
49b -var. *filiforme* (Krzl.) J.J. Vermeulen.
50 -*B. parvum* Summerh.
51a -*B. falcatum* (Lindl.) Rchb. f. var. *falcatum*.
51b -var. *bufo* (Lindl.) J.J. Vermeulen.
51c -var. *velutinum* (Lindl.) J.J. Vermeulen.
52a -*B. calyptratum* Krzl. *calyptratum*.
52b -var. *graminifolium* (Summerh.) J.J. Vermeulen.
52c -var. *lucifugum* (Summerh.) J.J. Vermeulen.
53 -*B. carnosilabium* Summerh.
54 -*B. carnosisepalum* J.J. Vermeulen.
55 -*B. ballii* Cribb.
56 -*B. elliotii* Rolfe.
57 -*B. porphyrostachys* Summerh.
58 -*B. lupulinum* Lindl.
59 -*B. tetragonum* Lindl.
60 -*B. horridulum* J.J. Vermeulen.
61a -*B. acutebracteatum* De Wild. var. *acutebracteatum*.
61b -var. *rubrobrunneopapillosum* (De Wild.) J.J. Vermeulen.
62a -*B. fuscum* Lindl. var. *fuscum*.
62b -var. *melinostachyum* (Schltr) J.J. Vermeulen.
63 -*B. oreonastes* Rchb. f.
64 -*B. encephalodes* Summerh.
65 -*B. concatenatum* Cribb & Taylor.
66a -*B. unifoliatum* De Wild. ssp. *unifoliatum*.
66b -ssp. *infracarinatum* (Williamson) J.J. Vermeulen.
66c -ssp. *flectens* (Cribb & Taylor) J.J. Vermeulen.
67 -*B. gilgianum* Krzl.
68 -*B. longiflorum* Thouars
C1 -*Chaseella pseudohydra* Summerh.

Abbayes 1326 (29); Achten 664 (48a); 743 (3b); 774 (13); 774b (13); Adam (1954-55-56) s.n. (29); (1959) s.n. (63); 16033 (51a); 20685 (42); 20686 (1a); 20820 (40); 20869 (6a); 20870 (32b); 20887 (29); 21164 (32b); 21164b (32b); 22219 (29); 22346 (29); 22347 (6a); 22591b (34a); 23281 (62b); 23624 (51c); 25356 (39); 25524 (63); 26199 (20b); 26392 (23); Adames 46 (17); 84 (52a); 98 (52a); 99 (16); 100 (16); 105 (51a); 106 (16); 109 (40); 110 (49b); 128 (39); 194 (17); 198 (62a); 200 (51a); 209 (51c); 550 (1a); 557 (1b); 588 (23); 640 (20b); 657 (51b); 657a (51b); 664 (26); 680 (13); 685 (42); 686 (3b); 801 (29); 828 (62a); 832 (6a); 851 (40); 852 (51c); Adamson 168 (40); 169 (68); Afzelius in herb. Swartz s.n. (17); Ake Assi 5202 (40); 5602 (40); 5606 (34a); 5639 (46); 5667 (20b); 5672 (27); 6035 (51a); 7149 (51c); 8826 (52a); 9036 (4); 9122 (51c); 9123 (51a); 9124 (27); 9126 (51c); 9128 (27); 9129 (42); 9137 (62b); 9138 (51b); 9139 (51b); 9140 (51a); 9192 (51c); 9193 (26); 9195 (51b); 9196 (51a); 9218 (34a); 9223 (4); 9224 (3b); 9333 (40); 9336 (48a); 9338 (46); 9343 (26); 9410 (6a); 9767 (51c); 9776 (39); 9786 (51c); 9797 (16); 9868 (32b); 9870 (51c); 6943 (6a); 9335 (51a); Alleizette, d' s.n. (48a); Andrews A 2033 (6a); Angus 459 (41b); 3844 (22); Annet 345 (25); Archbold 1355 (48c); 2586 (44a); Arends c.s. 329 (54); Armstrong & Brown cult. s.n. (48a); Arnstein Lye 4175 (6c); Ash 2248 (13); Auquier 3478 (6e); 4145 (56).

Bain (1957) s.n. (64); Baker 39 (47); Bakshi 31 (4); 192 (63); Baldwin 9881 (17); 9882 (17); 10036 (52b); 10079 (4); 10100 (1a); 10173 (52b); 10174 (51a); 10239 (51a); 10298 (61a); 10301 (13); 10307 (27); 10313 (61a); 10339 (51c); 10343 (61a); 10389 (62b); 10454 (62a); 11380 (13); 11382 (6a); 11478 (6a); Ball 150 (66b); 258 (55); 316 (28); 454 (68); 651 (64); 696 (40); 712 (48a); 713 (40); 843 (48a); 881 (55); 882 (55); 986 (66b); 1055 (13); 1308 (13); 1323 (55); 1366 (44a); Bally 11501 (48a); Bamps 1717 (40); 2049 (1a); 2051 (16); 2859 (20a); 3225 (6d); 3286 (63); Bamps, Symoens & vanden Berghen 823 (44a); Band (1909) s.n. (47); Barbosa 1404 (40); Barter s.n. (17); 1482 (1a); 1854 (40); 2026 (40); 2048 (51c); Bates 1130 (48a); 1374 (1a); 1503 (51b); Beckton (1940) s.n. (3b); Beentje 1457 (52a); 1633b (15); Benoit 549 (17); Bequaert (1919) s.n. (63); 1020 (3b); 1461 (52b); 2210 (49a); 2310 (40); 2496 (26); 2497 (58); 2650 (17); 4489 (51c); 4493 (51c); 6297 (13); 6413 (6c); 6498 (3b); 6500 (51c); 6501 (27); 6616 (66a); 6705 (49b); 6706 (61b); Biegel 2775 (40); Bigger 2119 (11); Birket Smith s.n. (63); Birket Smith s.n. (6a); Blayney s.n. (27); Bodenghien 295 (48b); Boivin (1854) s.n. (excl. sp. 3); (1854) s.n. (excl. sp. 9); Bonardi (1965) s.n. (39); (1965) s.n. (51b); (1970) s.n. (48a); 61 (51c); 63 (16); 64 (47); 230 (26); 234 (51c); Boom 32565 (4); Boone (1911) s.n. (40); 104 (46); 105 (19); 116 (51b); Booth s.n. (39); Bos 1070 (44a); 2327 (51c); 2336 (26); 2390 (29); 3613 (40); 3948 (1c); 3949 (42); 4219 (51c); 4473 (51c); 4474 (42); 5688 (47); 6423 (3b); 6529 (63); 6530 (63); 6630 (51c); 6680 (1c); 6708 (51c); Botte 513 (63); Bouxin 166 (7); 172 (7); 175 (5); 194 (63); 220 (7); 264 (63); 1320 (5); Bouxin & Radoux 687 (6d); Bowden (1923) s.n. (51a); 135/22 (17); 135/23 (27); 112 (17); Bowling GC 36570 (1c); GC 36613 (51c); GC 36615 (40); GC 36619 (52b); GC 38107 (39); Brass 17871 (64); Braun s.n. (16); Brenan 9350 (13); 9568 (63); 9569 (44a); FHI 25420 (17); Brenan, Jones, Onochie & Keay 8643 (58); Breteler 2015 (42); 2016 (1a); 2809 (44b); Breteler & de Wilde 83 (20 x 25); 207 (63); 209 (63); 215 (25); 321 (13); 545 (62b); 642 (14); 643 (51c); Breteler, de Wilde & Leeuwenberg 2252 (58); Breteler, de Wilde & de Bruyn 454 (40); Bridson 386 (6c); 391 (66a); 443 (20a); 450 (20a); 497a (6d); 497b (5); 499 (7); Brummitt 9408 (44a); 9480 (62b); Brummitt & Pawek 11097 (56); Brummitt, Banda, Seyani & Patel 14367 (6b); Brummitt, Chisumpa & Polhill 13837 (63); Brummitt, Seyani & Patel 14939 (62b); Bruneel 465 (54); Brunt 930 (58); 1202 (6b); 1238 (63); 1245 (6b); 1252 (6c); 1252 (6d); Brussel cult. Hort. (1911) s.n. (23); (1911) s.n. (51c); (1913) s.n. (23); (1913) s.n. (51b); (1915) s.n. (3b); 78.849 (6c); Bullock 1098 (41b); 2732 (62b); Bunting 10 (17); 10 (19); 11 (3c); 25 (51a); 29 (26); 31 (58); 74 (17); 102 (51b); Burg, v.d. 111 (51a); 137 (51c); 188 (51c); 208 (51a); 370 (51a); 370 (51c); 373 (51a); 376 (58); 379 (26); 380 (51a); 388 (51a); 803 (51c); 837 (51a); 844 (51c); 892 (51a); 1075 (6a); 1156a (39); 1240 (42); 1241 (1a); 1246 (42); 1253 (52b); 1256 (17); 1258 (26); 1259 (59); 1265 (51c); 1266 (51c); Burgess 4 (62b); Burtt 2995 (48a); 3136 (5); 3152 (12); 3164 (7); 4829 (40); 4872 (20a); 6325 (62b); 6596 (48a).

Caille 2 (48a); Cambridge exp. 193 (6d); Cameron 14863 (44a); Carmichael 351 (66c); Carroll in Moreau 663 (6a); 663a (6a); 670 (20a); 675 (13); Carroll in Tweedie 623 (13); Chandler 1885 (40); 2452 (52a); 2770 (25); 2791 (17); 2797 (19); Chapman 211 (22); 307 (11); 4397 (6a); Chase 992 (40); 994 (62b); 4079 (56); 4081 (40); 7180 (40); 22738 (C1); Chevalier 412 (48a); 4006 (48a); 10563 (48a); 12838 (48a); 13657b (6d); 13750 (24); 16527 (39); 16839 (48a); 18817 (6a); 19696 (63); 20280 (6a); 20470 (48a); 20569 (58); 20569 (58); 20669 (40); 20670 (52b); 20670b (52a); 26928 (47); 26967 (44a); 26982b (47); 27135 (3c); 27138 (3b); 27979 (1a); 27980 (3c); 27981 (42); Chew 12 (1c); Chillon 1266 (48a); Christiaensen 172 (52a); 593 (62b); 996 (52a); 1333 (6d); 1363 (63); 1383 (5); 1498 (6e); 1714 (6c); 1965 (52a); 1966 (3b); 1969 (44a); 2042

(25); 2082 (62a); 2458 (60); 6473 (53); Claessens 393 (41a); 483 (44b); Cole 38 (44b); Cole in King 153 (52a); Copley 54 (6d); 74 (6a); 111 (48a); Corbisier 500 (1a); Corbusier-Baland 1526 (42); Cox 25 (19); 129 (16); Cribb & Grey Wilson 10014 (40); 10032 (13); 10087 (40); 10128 (44a); 10171 (6e); 10172 (13); 10244 (44a); 10255 (64); 10256 (13); 10273 (6c); 10274 (40); 10285 (68); 10368 (65); 10420 (67); 10626 (20a); 10666 (68); 10818 (11); 11494 (44a); Cribb, Grey Wilson & Mwasumbi 10646 (48a); 11197 (40); 11283 (40); 11300 (66c); 11315 (20a); 11446 (66c); Crockewit 604 (6d); Croix, la (1979) s.n. (11); 93 (40); 98 (22); 138a (20b); 139 (20a); 194 (40); 201 (44a); 230 (66c); 249 (22); 286 (40); 337 (20b); Crowley cult. (1890) s.n. (13); Cunningham v. Someren 12 (40); 16 (6e); Cuthbert 4 (44a); 58 (6b).

Davies 911 (22); Dawkins 661 (51a); Decary 16996 (28); 17013 (28); Deighton 1148 (17); 2569 (51a); 3823 (27); 4273 (52b); 4274 (48a); 4327 (17); 4329 (13); 4371 (3c); 4375 (51c); 4376 (51a); 4377 (52c); 4392 (42); 4638 (3a); 4640 (17); 4649 (27); 4652 (17); 4657 (4); 4658 (51a); 4685 (51a); 4687 (51a); 4688 (51a); 4836 (17); 4921 (42); 4954 (42); 5026 (52a); 5029 (48a); 5244 (62a); 5648 (61a); 5653 (51a); 5714 (51c); 5718 (59); 5719 (6a); 5744 (52b); 5745 (52b); 5910 (51a); 5913 (17); 6011 (51c); 6012 (51a); 6022 (13); Deistel 54 (20a); Dekker 79 (17); 97 (3c); 333 (42); 334 (51a); 338 (42); 339 (42); 340 (51c); 431 (48a); 434 (3b); 445 (51c); Delarge 74 (64); 93 (6e); Demeuse s.n. (1a); Denton (1890) s.n. (51c); Depsem 18 (64); Deuse 1153 (6e); 1155 (5); 1165 (7); Deville 300 (3b); Devred 1471 (48a); 3562 (51a); Dewulf 85 (1a); 216 (42); 310 (52b); 347 (62a); 452 (52b); 915 (42); Dinklage 1852 (51a); 2940 (51a); Don Atwell 6 (40); 8 (17); Dowsett-Lemaire 369 (22); 381 (44a); 484 (6a); 524 (11); 660 (11); 674 (10); 776 (11); Drake 1243 (17); Dralle s.n. (17); Drummond & Hemsley 1690 (65); 3463 (20a); Drummond & Rutherford-Smith 7118 (62b); Dubois (1931) s.n. (44b); Dummer 2664 (68); 2997 (51a); 4323 (51a); Dybowski 11 (17); Dyer 4504 (48a).

Edinburgh cult. Hort. s.n. (52a); (1898) s.n. (17); (1899) s.n. (32b); (1915) s.n. (17); Edwards 680 (62b); Eggeling 2133 (48a); 3191 (6c); 3196 (64); 3242 (63); 3244 (6d); 3270 (20a); 3359 (48a); 3574 (6c); 3669 (3b); 3816 (6c); 4046 (26); 4141 (20a); 4148 (44a); 4192 (6c); 5206 (40); Ekema 26 (51a); 232 (58); Evrard 1433 (39); 1532 (19); 1685 (1a); 2744 (1a); 2829 (51c); 3781 (48a); 4050 (51c); 4888 (48a); 5188 (51c); 5574 (62a); Exell 390 (6d); Exell & Mendonça 1085 (13); 1086 (62a); 1882 (66a); Exell, Mendonça & Wild 1390 (62b).

Faden, Evans & Rathburn 71/682 (40); Fanshawe 4211 (62b); 8227 (63); 8812 (41b); 12116 (41b); Farron 4209 (46); 4422 (46); 4423b (52a); 4598 (46); 4711 (39); Faulkner 4515 (68); Feith s.n. (40); Felix 663 (62a); Fielding s.n. (3c); Fleckwood 11486 (6a); Flock 482 (64); 841 (56); Fredericq in De Witte 8030 (51c); Fries, Norlindh & Weimarck 2199 (48a); Frils, Aweke, Rasmussen & Vollesen 2360 (58); Fuerstenberg s.n. (48b).

Galt 15 (62b); Gardner 2548 (6e); Gassner 16 (44a); Gassner & Cribb 178 (11); 199 (11); Geerling & Bokdam 1714 (29); 1742 (32b); 1743 (6a); 1791 (32b); 2270 (46); 2272 (44b); 2273 (52a); Geneve cult. Hort. (1896) s.n. (17); Gentil (1903) s.n. (19); (1903) s.n. (42); (1903) s.n. (44b); (1903) s.n. (46); 1/159 (51b); Gerard 1524 (48a); 5330 (48a); Germain 108 (39); 1370 (5); 2996 (7); 3511 (5); 3535 (5); 3895 (5); 4232 (20a); Ghesquière 4893 (6e); Gillet (1936) s.n. (1c); 7 (52b); 18 (42); 2476 (40); 2773 (42); 2900 (40); Gilliland 1541 (40); 1544 (40); 1647 (66b); Glasnevin cult. Hort. (1894) s.n. (44a); (1900) s.n. (15); (1905) s.n. (3b); (1918) s.n. (54); Godefroy-Lebeuf s.n. (42); Goetze 1480 (40); Goldsmith s.n. (48a); Gomes & Sousa 4300 (62b); Goossens 4537 (1a); Gossweiler (1903-04) s.n. (40); (1903-04) s.n. (48a); (1916) s.n. (3b); (1919) s.n. (39); 504 (40); 930 (48a); 931 (40); 11698 (13); Graham Smith 570 (37); Greenway 873 (40); 1072 (40); 2569 (68); 4174 (68); 4178 (62b); 4190 (40); 4732 (68); 4742 (64); 6184 (41b); 6184 (48a); 6440 (6c); 6631 (13); 6659 (40); 9807 (48a); Greenway & Brenan 8295 (40); Gregory 193 (20b); 558 (20b); 612 (6d); Griffith in Moreau 385 (48a); 389 (48a); Groot cult. 320 (56); 368 (51c); 369 (51c); 371 (51a); 387 (26); 390 (40); 490 (51c); 496 (16); 635 (44b); 666 (17); 785 (51c); Grosvenor 383 (20a); 809 (56); Grosvenor & Renz 1066 (68); Guinea 2135 (3b); 2145 (63); Gutzwiller 879 (6e); 889 (64); 1645 (6d); 1697 (6e); 3046 (6d); 3548 (6e); 3563 (6e); 3579 (6e); Guy Babault 643 (63); 644 (63).

Haakman cult. s.n. (25); Haegen, de 112.75 (51a); 113.75 (51a); Hall 1705 (17); 2449 (63); GC 36621 (42); GC 36637 (52a); GC 39010 (48a); GC 39011 (16); GC 39020 (51c); GC 42314 (3b); GC 42518 (51b); Hall & Abbin GC 44643 (26); Hall & Enti GC 38111 (19); GC 38151 (17); GC 38163 (19); GC 38327 (3a); GC

39027 (63); GC 39030 (42); Hall & Hossain GC 38168 (51c); Hallé 925 (3b); 1686 (62b); 1729 (39); 2156 (49b); 2157 (63); 2159 (19); 2733 (1a); 2734 (42); 2738 (3b); 2843 (15); 2849 (3b); 2868 (51c); 2903 (17); 2903 (62b); 3078 (62b); 3079 (51c); 3101 (51c); 3244 (14); 3305 (46); 3353 (51c); 3380 (46); 3383 (42); 3400 (1a); 3403 (3b); 3404 (3b); 3743 (51c); 3845 (15); Hallé & Cairs 6058 (63); Hallé & Villiers 4329 (42); 4430 (1a); 4636 (13); 4727 (13); 4741 (6a); 4837 (13); 4849 (63); 4930 (26); 5210 (13); 5417 (26); 5419 (52a); 5423 (62b); 5456 (17); 5474 (3b); 5497 (3b); 5520 (42); Hambler 153 (63); 'Hammoniae' cult. Hort. 2302 (16); 2307 (16); Harley 1759 (26); 1760 (51a); 1761 (51b); 1762 (17); 1764 (42); 1765 (1a); 1795 (63); 1827 (42); 1833 (26); 1835 (51b); 1837 (48a); 1926 (4); 1932 (40); 1943 (48a); 1947 (42); 1950 (52b); 2025 (17); 2038 (3c); 2046 (19); 2066 (51c); 2083 (32b); 2130 (3c); 2168 (48a); 2204 (29); Harrington 546 (63); 547 (3b); Haswell 12 (17); Heidelberg Hort. cult. (1904)s.n. (48b); Hendrix 6935 (20a); Henriques 29 (24); Hepper 1430 (62b); 1431 (48a); 1507 (6a); 1798 (6a); 1918 (6c); 1923 (58); Hepper, Field & Mhoro 5462 (10); Herb. Reichenbach s.n. (51b); s.n. (51c); Herbigneaux 1 (17); Hilliard & Burtt 8451 (44a); Hirschberg, v. s.n. (40); Hitchen (1952) s.n. (56); Holland s.n. (17); (1900) s.n. (51c); 397 (4); Holmes 5 (6b); 10.1 (22); 10.2 (22); 12 (6b); 16 (64); 49 (41b); 49a (40); 254 (58); 350 (63); 365 (40); Honey 689 (48a); 730 (48a); Horne 1874 (13); Howard in King 149 (20b); Hulstaert 673 (3b); 1667 (51c); 1684 (51c); Humbert (1929) s.n. (63); 7697 (63); 8144 (5); 8309.3 (5); Humblot 378 (28); Hutchinson 4122 (64).

Imschoot, v. (1895) s.n. (39); (1896) s.n. (3b).

Jackson 271 (44a); Jacques-Felix s.n. (51b); 664 (62a); 1044 (17); 1785 (63); 2327 (6a); 2328 (61b); 2328a (54); 2999 (58); 5089 (58); Jacques-Georges 27521 (63); Jaeger 132 (6a); 232 (6a); 294 (6a); 568 (29); 871 (29); 1507 (62b); 4132 (6a); 4145 (29); 4161 (6a); 4229 (6a); 4232 (6a); 5552 (34a); 7169 (6a); 7189 (6a); 7192 (29); 7293 (29); 7385 (17); 8370 (6a); 8833 (58); Jans 1179 (3c); Jansen 6405 (58); Jespersen (1910) s.n. (48a); (1911) s.n. (44b); 1910 (42); Johansson 217 (52b); 414 (6a); 415 (32b); 417 (31); 426 (4); 435 (1b); 441 (17); 448 (59); 451 (59); 459 (29); 461 (63); 467 (42); 468 (3a); 477 (3b); 478 (43); 489 (43); 541 (19); 578 (17); 583 (3c); 585 (17); 586 (17); 596 (19); 617 (40); 631 (6a); 634 (51c); 637 (4); 640 (62b); 640 (63); 641 (1a); 647 (42); 648 (51c); 649 (40); 656 (40); 658 (4); 659 (17); 661 (52c); 662 (26); 667 (23); 668 (51a); 669 (27); 670 (51a); 671 (51a); 673 (51b); 675 (42); 675 (42); 677 (26); 678 (42); 679 (17); 683 (48a); 687 (13); 688 (26); 691 (13); 693 (27); 694 (23); 696 (34a); 699 (20b); 702 (13); 705 (4); 706 (34a); 710 (42); 711 (51b); 714 (48a); 714 (48a); 716 (29); 717 (32b); 718 (49b); 729 (4); 730 (52c); 744 (52c); 745 (29); 756 (17); 760 (63); 783 (3c); 811 (1b); 847 (1b); 855 (3b); 855 (51c); 857 (3b); 932 (6c); 934 (64); 946 (40); 949 (68); 1024 (10); 1061 (1b); Jones 1 (51c); 64 (29); 167 (6a); 168 (6a); 235 (50); Jones & Onochie FHI 18780 (63); Jongejan cult. 1 (49a); 6 (61a); 9 (13); 15 (57); 19 (40); 25 (51c); 26 (17); 29 (3c); 32 (51a); 51 (17); 54 (25); 55 (32b); 62 (3b); 63 (6a); 65 (17); 66 (19); 418 (17); 862 (17); 865 (19); 875 (46); Jordan 2085 (3a); 2086 (63).

Kalbreyer 140 (44a); Kassner 2380 (63); 2381 (63); 2959 (62b); 3096 (51a); 3240 (5); Keay FHI 21551 (48a); FHI 21552 (40); FHI 22668 (48a); FHI 22711 (57); FHI 22713 (51a); FHI 25398 (13); FHI 25426 (48a); FHI 25430 (26); FHI 25439 (20b); FHI 25473 (48a); FHI 25474 (51b); FHI 26720 (17); FHI 28074 (51c); FHI 28127 (51a); FHI 28128 (51a); FHI 28129 (39); FHI 28130 (48a); FHI 28132 (51a); FHI 28133 (57); FHI 28134 (51b); FHI 28231 (51a); FHI 28292 (51a); FHI 28293 (62b); FHI 28295 (51a); FHI 28296 (48a); FHI 37501 (44a); FHI 41840 (51a); FHI 42336 (13); Keay & Brenan FHI 22405 (17); Keay & Lightbody FHI 28374 (6a); Keay & Onochie FHI 21590 (57); Keay & Savory FHI 25242 (58); FHI 25245 (32a); FHI 25307 (52a); FHI 25309 (17); Kennedy 2730 (51a); Kerfoot s.n. (6e); Kew cult. Hort. s.n. (32b); (1892) s.n. (48a); (1899) s.n. (17); (1900) 175 (19); (1903) s.n. (1a); (1914) s.n. (17); (1914) s.n. (42); (1931) s.n. (3b); King 124 (35); 124b (35); 124c (35); 124d (35); 126b (40); 180b (48a); 180c (48a); Klaine 168 (40); Kokwero 3787 (6e); Koning, de 2802 (51c); 2813 (51c); 3122 (3a); 3123 (1a); 4013 (42); 4098 (1a); 5558 (3a); 6097 (48a); 6152 (49b); 6152b (40); 6167 (51a); 6214 (51a); 6360 (51c); 6361 (51c); 6405 (58); 6954 (3c); 6961 (17); 6996 (17); 6997 (17); Kornas 3168 (44a); 3169 (56); 3454 (62b).

Laan, v.d. 101 (13); 106 (3c); 170 (17); 241 (40); 299 (48a); 317 (17); 325 (51c); 327 (59); 328 (51c); 329 (51c); 342 (51c); 343 (51c); 353 (51c); 357 (40); 406 (3a); 453 (20a); 458 (25); 464 (26); 490 (58); 492 (51c); 496 (52a); 521 (17); 570 (15); 576 (51c); 581 (68); 593 (45); 599 (27); 600 (13); 604 (17); 607 (52a); 608 (17); 616 (62b); 617 (27); 644 (3b); 678 (63); 679 (13); 680 (13); 688 (62b); 689 (25); 700 (51c); 710 (17); 729 (13); 730 (3b); 737 (62b); 746 (3b); 752 (6e); 759 (26); 784 (51c); 788 (42); 797 (58); 805 (17); 818 (63);

819 (44a); 822 (17); 838 (54); 839 (62b); 840 (63); 844 (3b); 851 (13); 890 (62b); 895 (17); 899 (44a); 904 (39); 931 (1a); 939 (3b); 959 (40); 984 (62a); 1007 (52c); Lacomblez 25 (51b); Lacomblez in Vanderijst (1916) s.n. (51b); Laggiara s.n. (28); Lambinon 74/933 (5); Lange & Westwood 63 (51a); Langhe, de 98/57 (44a); Lankester 28 (6c); Latilo FHI 27332 (51a); FHI 32916 (17); FHI 41342 (51c); Latilo & Daramola FHI 28943 (58); Laurent (1893-1903) s.n. (42); (1897) s.n. (1a); (1897) s.n. (42); (1903) s.n. (42); (1903) s.n. (51c); (1903) s.n. (61a); (1905) s.n. (51c); 32 (1a); 124 (23); 135 (42); 136 (42); 178 (17); 190 (4); 232 (1a); 668 (51c); 685 (48a); 741 (42); 819 (3c); 820 (1a); 821 (48a); 883 (1a); 937 (41a); 1735 (25); 1740 (25); 1742 (42); 1747 (1a); 1748 (41a); 1750 (51c); 1766 (61a); 1771 (42); 1772 (36); 2034 (36); 2045 (17); 2055 (61a); 2056 (41a); 2059 (25); Lawton 2306 (44a); Leach 11276 (48a); 14981 (40); Leakey 7 (44a); Lebrun 385 (42); 1017 (61a); 1402 (36); 1403 (15); 1424 (3c); 4133 (51a); 4883 (5); 5210 (17); 5300 (51c); 5581 (51c); 6672 (42); 7878 (5); 9417 (20a); Lecomte s.n. (17); (43); Leedal 73 (22); 286 (44a); 446 (11); 484 (40); 581 (56); 589 (40); 883 (44a); 885 (40); 1031 (56); 1039 (40); 1340 (68); 1341 (66c); 1671 (20a); 1887 (64); 2635 (20a); 6113a (13); 6144 (13); 6305 (10); 6306 (20a); 6451 (44a); 7246 (66a); 7247 (65); 7252 (56 x 64); 7253 (13); Leeuwenberg 2488 (42); 2969b (34a); 3776 (44b); 4830 (17); 5355 (44b); 5357 (51c); 5427 (44b); 5435 (44b); 5438 (51b); 5466 (1c); 5532 (52a); 5596 (44b); 7133 (42); 7134 (62a); 9048 (62b); 9148 (62b); Leiden cult. Hort. 20283 (51a); 20289 (17); 21400 (19); 21472 (51a); 21693 (3a); 22296 (22); 22790 (17); 23602 (3b); 23603 (63); 23616 (42); 23617 (27); 23618 (62b); 23624 (3b); 23626 (68); 23628 (3c); 23630 (44a); 23633 (52a); 23638 (26); 23640 (62b); 23641 (42); 23644 (6a); 23645 (51a); 23646 (40); 23647 (42); 23648 (17); 23650 (6a); 23653 (58); 23654 (62b); 23657 (44a); 23660 (46); 23661 (51c); 23664 (51c); 23666 (4); 23669 (27); 23673 (42); 23675 (17); 23677 (3b); 23681 (40); 23686 (3b); 23687 (15); 23688 (48a); 23689 (17); 23693 (17); 23695 (51c); 23698 (3b); 23945 (3b); 23945 (52a); 23950 (32b); 23997 (32b); 24136 (42); 25043 (49b); 25065 (43); 25160 (49b); Lemaire 175 (22); Lemaire-Elias 153 (6b); 175 (22); 217 (56); Lendy s.n. (51a); Leonard 41 (39); 74 (68); 200 (42); 201 (1a); 1123 (51c); 1310 (17); 1418 (25); 1420 (63); 1420 (66a); 4503 (51c); 5105 (6d); 5109 (6e); 5186 (41a); Lescrauwaet 311 (1a); Letouzey 478 (6a); 2821 (42); 3819 (1c); 4762 (1a); 4763 (44b); 5617 (46); 8028 (13); 9194 (48a); 9216 (39); 9406 (44b); 9407 (17); 9417 (51c); 10029 (1a); 10777 (33); 11786 (58); 12338 (62a); 12885 (42); 13170 (51a); 13296 (33); 14317 (25); 14533 (3b); 14579 (63); 14652 (44b); 14661 (63); 14678 (44a); 14804 (51c); 14808 (51c); 14848 (51c); 15055 (17); 15169 (52a); 15200 (1a); 15259 (3b); Lewalle 1492 (20a); 6375 (5); Linder 334a (13); 669 (42); 669 (42); 827 (40); 938 (42); 938 (42); 1020 (3b); 1179 (51b); 1209 (51b); 1210 (51b); 1249 (61a); 1255 (51c); 1289 (51c); 1325 (42); 1347 (51c); 1393 (27); Lisowski 43449 (58); 65959 (63); 65962 (48a); 65963 (63); 65964 (48a); 66239 (20b); Lisowski, Malaisse & Symoens 11052/2 (62b); Lisowski, Malaisse & Symoens 8854 (48a); Loddiges s.n. (17); s.n. (19); s.n. (3a); s.n. (4); s.n. (40); s.n. (42); s.n. (51b); s.n. (51c); s.n. (52a); s.n. (62a); Lodewijk cult. 123 (4); 160 (17); 359 (51a); Louis 471 (3b); 2496 (26); 2858 (46); 3570 (17); 3610 (1a); 5051 (5); 6208 (42); 6219 (62a); 6302 (14); 6432 (62a); 6720 (48a); 7486 (3b); 7858 (42); 8251 (17); 8272 (14); 8275 (66a); 9089 (43); 9090 (17); 9093 (3b); 9784 (1a); 10572 (42); 10744 (48a); 10878 (3b); 11032 (1a); 12410 (1a); 12426 (59); 12570 (48a); 13375 (62b); 13537 (42); 13712 (66a); 13729 (48a); 13835 (3b); 13869 (41a); 14140 (42); 14155 (48a); 14398 (51c); 14399 (43); 14527 (1a); 14760 (3b); 15103 (25); 15640 (48a); 15647 (42); 15851 (39); 16371 (68); Louis II 12 (39); II 26 (48a); II 358 (23); Louis, Breteler & de Bruyn 320 (51c); 1209 (40); 1242 (13); 1316 (51c); 1316 (63); Lousteau (1901) s.n. (39); (1910) s.n. (3b); Loveridge 77 (51a); Lowe 3300 (58); 3335 (52a); Lucas 94 (6e); Lynes V 21 (44a).

Magogo & Glover 582 (40); Mahon (1899) s.n. (20b); Maitland 781 (33); 782 (33); Malaisse s.n. (64); 4448 (64); 4451 (22); 4833 (22); Mann s.n. (51c); 519 (17); 525 (3b); 526 (47); 527 (13); 642 (32a); 643 (6a); 648 (6d); 650 (6b); 783 (58); 1337 (6a); 1883 (18); 2111 (6a); 2121 (33); 2122 (63); 2124 (20a); 2126 (6b); Mantin (1895)s.n. (3b); Marmo 59 (51a); Martius 5 (48a); Mavi 625 (48a x 40); Mc.Clounie s.n. (44a); Mc.Loughlin 82 (56); Meel, v. 1359 (62b); 1705 (64); Meer, v. 209 (42); 211 (23); 1264 (3b); 1777 (63); 1821 (63); Meikle 522 (51c); 561 (51a); 562 (51a); 599 (62a); 600 (57); 601 (57); 602 (51a); 603 (51a); Meller s.n. (28); (1861) s.n. (44a); Melville 200 (40); 200 (42); 437 (13); 461 (6a); Mendonça 240 (40); Meyer s.n. (63); 57 (6d); 198 (6e); 8998 (48a); Miège (1965) s.n. (42); Migeod 376 (6a); Miles (1912) s.n. (40); (1913) s.n. (49a); 11 (43); 19 (16); 20 (51c); 21 (51a); 23 (3c); 31 (3a); 32 (40); Millen 189 (42); 189 (51c); Milne (1864) s.n. (57); Milne Redhead 501 (62b); 603 (41b); 945 (64); 996 (41b); 3074 (6b); 3183 (41b); 4297 (6a); 4349 (63); 4375a (64); Milne Redhead & Taylor 8199 (66c); 8842 (40); 8912 (6b); Moll 2842 (48a); Moloney s.n. (17); s.n. (4); s.n. (51c); (1891) s.n. (42); 219 (51a); Monod 11842 (24); 11971 (6a); 12130 (24); 12164 (24); 12249 (13); Monro 1988 (48a); Mooney 6926 (48a); Moore (1887) s.n. (17); Moreau 52/1

(40); 92/1 (62b); 6 (62b); 6 (64); 6a (64); 7 (68); 11 (20a); 12 (20a); 30 (6c); 58 (13); 65 (62b); 86 (41b); 87 (48a); 99 (13); 101 (6a); 101 (6c); 117 (48c); 117a (40); 152 (44a); 152a (44a); 168 (44a); 198 (40); 303 (44a); 309 (44a); 352 (44a); 357 (40); 384 (56); 385b (48a); 389 (48a); 414 (48a); 418 (48a); 435a (44a); 464 (67); 477 (11); 483 (40); 490 (6b); 508 (65); 572 (40); 606 (40); 699 (11); 713 (44a); 715 (48a); 746 (44a); 794 (66c); Moriarty 479 (44a); Morris 9 (62b); 9a (62b); 10 (11); 16 (68); 17 (40); 27 (28); 45 (48a); 46 (56); 53 (11); 57 (44a); 58 (56); 60 (64); 60a (64); 60b (64); 61 (20a); 61a (20a); 61b (20a); 61d (13); 88 (44a); 95 (13); 104 (40); 111 (62b); 131a (40); 134 (44a); 138 (11); 139 (13); 148 (20a); 164 (11); 166 (20a); 172 (56); 206 (66b); 214 (44a); 344 (11); 837 (62b); Morton (1953) s.n. (52b); A 3413 (17); GC s.n. (48a); GC s.n. (52b); GC 9413 (52b); GC 9414 (59); K 231 (63); SL 2636 (6a); SL 3329 (40); SL 3329b (42); SL 410 (6a); SL 411 (29); Morton & Gledhill SL 2979 (6a); SL 3141 (29); 3157 (34a); Morze 32 (41b); 34 (56); 35 (22); 75 (40); 135 (40); 136 (6a); 141 (44a); 142 (48a); Moss 19140 (48a); Mulder in Jongejan cult. 60 (5); Munnike (1985) s.n. (4); (1985) s.n. (61a); (1985) s.n. (62b); Musyoki & Hansen 1032 (40); Mutimushi 3480 (20b); Mylord (1954) s.n. (17); (1954) s.n. (26); (1954) s.n. (3c); (1954) s.n. (40); (1954) s.n. (43); (1954) s.n. (48a); (1954) s.n. (51a); (1954) s.n. (63).

Nash 12 (41b); 13 (62b); 90 (62b); Nash & King 178 (21); Ndi 51 (58); Nditapah SCA 190 (39); Ngoni 232 (28); 235 (68); Nichols s.n. (40); Nozeran (1955) s.n. (3c); (1955) s.n. (43); (1956) s.n. (40).

O'Brien cult. s.n. (31); (1891) s.n. (17); Oldeman 669 (51a); 700 (48a); 760 (51c); Olorunfemi FHI 25313 (63); Onochie FHI 32434 (51a); FHI 32436 (51a); Osmaston 1942 (12); Overlaet 848 (19).

Paris cult. Hort. s.n. (3b); (1924) s.n. (51a); (1954) s.n. (6a); (1986) s.n. (51b); Pauwels 5407 (48a); Pawek 346 (11); 2737 (22); 2939 (13); 3516 (20a); 5103 (22); 9668 (13); 9738 (6b); 10696 (68); 11154 (11); 12333 (22); 12417 (6b); 12598 (56); 13280 (6b); 13786 (6b); 13869 (22); 1773b (68); Pereira, Sarmento & Marques 1619 (40); Perez 716 (47); Perez Vera 106d (4); 109 (42); 121 (3b); 130 (48a); 133c (51a); 137 (23); 139b (43); 139c (43); 139d (43); 149 (40); 166 (46); 187 (3c); 201b (20b); 212 (17); 213a (52c); 213b (52c); 214 (58); 218 (62b); 219b (42); 221 (17); 222 (51c); 223 (3c); 231 (52a); 240 (52b); 241 (19); 248 (1b); 251 (48a); 254b (1c); 255 (42); 256 (26); 266 (51a); 267 (17); 268c (62a); 268d (62a); 270 (49b); 270b (49b); 292 (51c); 492 (51b); 493 (29); 529 (32b); 535b (40); 536 (48a); 542 (34a); 549 (27); 553 (49b); 554 (49a); 567 (52c); 570 (1a); 571 (1b); 578 (17); 590b (42); 601 (6a); 604 (46); 619 (44b); 623 (3b); 665 (3a); 709 (19); 710 (16); 711 (3a); 712 (1a); 713 (39); 714 (54); 715 (47); 720 (16); 734 (51a); 735 (48a); 753 (48a); 771 (3c); 813 (13); 814 (63); 857 (51c); 880 (63); Perrier 1908 (28); Petit Thouars, du s.n. (68); Pettersson 126 (40); 132 (62b); Pfennig s.n. (3b); Philcox, Leppard, Duri & Urayai 8957 (62b); 8964 (40); Phillips 1228 (44a); 3124 (44a); 3128 (6b); 3197 (22); 3313 (56); 4695 (22); Pierlot 1734 (6c); 1964 (6d); Piers s.n. (44a); Piers 36 (6d); Pittery 637 (1a); Pobeguin s.n. (3b); s.n. (62a); (1909) s.n. (48a); (1910) s.n. (52b); (1910) s.n. (6a); (1910-11) s.n. (58); (1913) s.n. (61a); 118 (39); 213b (4); 636 (48a); 696 (48a); 925 (48a); 929 (48a); 930 (48a); 1522 (63); 1974 (34a); 1975 (34a); 1976 (6a); 2184b (13); Polhill & Paulo 1530 (66c); 1684 (44a); 1840 (66c); Pope 108 (62b); Preis 108 (6c); Prempeh 29 (51a); Preuss 943 (63); 1215 (52a); 1217 (44b); 1225 (1c); 1241 (42); 1243 (17); 1299 (17); Purseglove 2624 (44a); 2680 (6e); Purves 12 (11); Putman 249 (51c); Pynaert 436 (51c); 447 (51c); 492 (4); 555 (51c); 566 (4); 707 (42); 1161 (4); 1321 (4); 1322 (42); 1451 (1a); 1456 (51c); 1466 (48a); 1572 (3c).

Quarre (1939) s.n. (48a); 4454 (64); 908 (64); Quintas 9 (40); 10 (13); 10 (49a); 1016 (51c).

Rawlins 763 (40); Raynal 9468 (58); 10486 (26); 10487 (52a); 10488 (48a); 13071 (6a); 18815 (13); 20485 (6d); Reekmans 10404 (64); Rehman 5857 (44a); Rendall s.n. (44a); Renvoize 1652 (68); Renvoize & Abdallah 1652 (68); Richards 1813 (62b); 3457 (63); 9302 (62b); 12670 (22); 13237 (48a); 15656b (48c); 16308 (62b); 16965 (6b); Richards in Methuen 415 (22); Roberty 13768 (4); 16450 (58); Robyns 672 (42); 672a (23); 4094 (5); 4094a (7); Rodgers 1129 (13); Rodgers & Hall 2634 (68); Rodgers & Vollesen 775 (62b); Rodgers, Hall & Mwasumbi WAR 2475 (68); Rose 1023 (51c); Rosevaer 91/36 (17); 53/37 (17); 56/37 (51c); 57/37 (13); 58/37 (51c); 59/37 (52a); 60/37 (57); Ross 230 (51c); 1870 (48a); Rudatis 1549 (44a); 1549 (61b); 1549 (6a); Rysselberghe (1903) s.n. (40); (1903) s.n. (46).

Sanders cult. s.n. (3a); Sanderson 898 (44a); Sanford 64/394 (17); 64/693 (51a); 65/1697 (51c); 65/1726 (43); 65/1727 (43); 66/137 (42); 66/261 (52a); 66/331 (42); 66/359 (17); 66/703 (51c); 66/79 (51c); 4002 (61b);

4035 (38); 4055 (32a); 4109 (6a); 4112 (6b); 4134 (48b); 4163 (6b); 4172 (6a); 4219 (51a); 4220 (51a); 4359 (29); 4740 (51c); 4835 (17); 4858 (17); 4944 (63); 5180 (44a); 5195 (26); 5239 (25); 5255 (62b); 5272 (43); 5273 (25); 5273 (51a); 5353 (40); 5354 (40); 5355 (44a); 5363 (62b); 5367 (44a); 5376 (25); 5456 (54); 5457 (6a); 5457 (6a); 5485 (13); 5561 (17); 5566 (25); 5575 (6a); 5576 (6c); 5655 (64); 5680 (42); 5682 (63); 5683 (13); 5686 (40); 5702 (51c); 5705 (42); 5706 (42); 5740 (51c); 5745 (52a); 5806 (3c); 5830 (13); 5831 (13); 5839 (13); 5861 (13); 5862 (51c); 5895 (4); 5978 (13); 5980 (20b); 6013 (6b); 6018 (6a); 6047 (52a); 6065 (52a); 6090 (43); 6140 (3b); 6186 (48a); 6197 (40); 6198 (40); 6219 (6a); 6255 (6a); 6267 (6a); Sangster 559 (26); Saunders s.n. (44a); Savory 159 (62b); Savory & Keay FHI 25098 (6a); FHI 25183 (48b); Saxer 46 (17); Scaetta 1617 (5); Schäfes 6884 (20a); Schayes 13 (48a); 1023 (48a); 1027 (40); 1028 (48a); 1097 (41b); 1098 (40); 1262 (22); 1308 (56); 1383 (62b); 1877 (20b); 1878 (20b); 1886 (58); Scheepers 745 (44a); Schlechter 12250 (62b); 12358 (44b); 12361 (17); 12369 (52a); 12373 (51c); 12377 (63); 12693 (48a); 12781 (48a); 12836 (6a); 12844 (58); 12896 (33); 15756 (13); 15759 (49b); 15760 (51a); 15784 (44a); 15791 (1c); Schlieben 671 (44a); 1622 (40); 3069 (44a); 3155 (65); 3284 (48a); 3284 (65); 3348 (13); 3578 (20a); Schmitz 2028 (48a); 2030 (48a); 3794 (62b); 4305 (48a); 4863 (48a); 5739 (40); 6177 (64); Schnell (1942) s.n. (51b); 1094 (29); 1224 (6a); 1824 (46); 2430 (29); 3845 (20b); Schuüff 5927 (44a); Scott Elliot s.n. (28); (1890) s.n. (56); (1892) s.n. (51c); 4023 (61a); 4915 (58); 5341 (40); 5341 (48a); 5778 (17); Seegeler 2665 (20a); Segerbäck 1129 (17); 1139 (42); 1310 (58); 1510 (51b); 1517 (52c); 1518 (4); 1519 (48a); 1520 (27); 1522 (42); 1523 (42); 1530 (52a); 1538 (19); 1557 (59); Senni 2434 (20a); Serand 13 (51a); Seret 165 (48a); 208 (48a); 295 (42); 335 (52b); 566 (46); 568 (40); 654 (51b); 809 (41a); 874 (48a); 999 (40); 1026 (51c); 1146 (46); 1210 (17); 1221 (42); Setten, v. 92 (51a); 113 (51c); 150 (51c); 211 (4); 230 (44a); 308 (58); 324 (51a); 358 (4); 388 (62b); 388 (63); 480 (17); 481 (44a); 494 (63); 495 (62b); 501 (20a); 525 (51a); 577 (3a); Sharland 1236 (17); 1815 (63); Sillitoe 142 (48a); Simon 11 (51c); Sita 817 (15); 2599 (4); 2848 (42); Smeathman s.n. (40); Smythe 114 (4); Snowden s.n. (27); 923 (20a); 940 (6a); Solheid 78 (51c); Sossweiler 11698 (62a); Spellenberg 376 (52a); StClair-Thompson 1091 (62b); Staff FHI 25272 (58); FHI 42079 (52a); Staff SCA 150b (58); Stauffer 334 (5); 1101 (48a); 1102 (5); Stewart 1193 (34b); 1194 (C1); Stolz K 53 (11); 453 (68); 454 (40); 638 (20a); 667 (40); 767 (11); 1023 (44a); 1411 (68); 1475 (28); 1784 (40); 1988 (20a); 2406 (68); Stolze & Scharff (1982) s.n. (9); Stopp 410 (3b); Strey 8295 (48a); Strid 2888 (48a); Surcouf 203 (40); Swarbrick SCA 473 (20a); SCA 475 (51a); Swynnerton 748 (40); 749 (44a); Symes 336 (6c); Symoens 5381 (63); 5393a (7); 6656 (48a); Synge 1453 (5).

Talbot (1911/12) s.n. (49a); 616 (63); 778 (1c); 779 (3b); 925 (63); 931 (6a); 932 (62b); 933 (27); 934 (17); 935 (23); 952 (13); 1268 (6a); 1576 (51c); 3288 (17); 3288 (51c); 3302 (52a); 3303 (51c); 3308 (1c); 3309 (3b); 3724 (1c); Tanner 240 (44a); 250 (44a); 254 (44a); Taylor (1887) s.n. (40); Testu, le s.n. (17); 668 (40); 2195 (51c); 2200 (39); 2246 (42); 2369 (52a); 3939 (63); 3939 (63); 4400 (48a); 4400 (48a); 5059 (39); 5177 (3b); 5178 (1c); 5183 (63); 5406 (52a); 5406a (52a); 5406b (52a); 5457 (62b); 5515 (53); 5527 (30); 5547 (30); 5562 (3b); 5563 (62b); 5595 (1c); 5707 (13); 5707 (13); 5720 (13 x 17); 5757 (53); 5760 (3b); 5768 (17); 5770 (48a); 5772 (36); 5776 (51c); 5777 (42); 5779 (1c); 5783 (51c); 5787 (3b); 5788 (49b); 5789 (63); 5796 (3b); 5865 (17); 5927 (63); 5929 (53); 6370 (42); 6370b (17); 6472 (13); 8947 (1a); 8949 (42); 8950 (52a); 8951 (3b); 9034 (44b); 9269 (46); Thomas 32 (13); 33 (62b); 125 (1c); 142 (23); 150 (20a); 152 (42); 161 (4); 189 (3b); 211 (51c); 467 (44b); 1663 (6a); 2741 (48a); Thonet 204 (42); Thorold CM 15 (51c); CM 16 (44b); CM 18 (52a); Thulin & Mhoro 967 (13); Tillett (1888) s.n. (48a); Tisserant in le Testu 212 (26); 239 (25); 383 (40); 416 (25); 427 (42); 465 (48a); 1133 (25); 1245 (62a); 1264 (42); 1551 (4); 1762 (48a); 1997 (25); 2628 (43); 2638 (3b); 2639 (40); 2659 (40); 2687 (40); 2747 (40); 2748 (48a); Torre & Paiva 10758 (62b); Trevor Lawrence (1890) s.n. (42); (1897/98) s.n. (17); (1900) s.n. (17); Trochain 8125 (40); Troupin 2458 (44b); 9713 (63); 9718 (6d); 11520 (63); 11526 (5); 15407 (63); 15480 (5); 15645 (6e); 15646 (6e); 15651 (63); 15652 (8); 15681 (22); 15683 (12); 15685 (5); 15693 (44a); 15695 (5); 15700 (6d); 15732 (7); 15734 (63); 15738 (5); 15766 (32b); 15768 (20a); 15771 (66a); 15779 (6e); 15782 (63); 15784 (6e); 15789 (63); 15790 (63); 15807 (44a); 15909 (20a); 15910 (12); 15911 (6a); 15913 (6e); 15951 (20a); 15952 (44a); Tweedie 47 (6a); 233 (6e); 293 (5); 492 (64); 613 (20a); 2593 (6c).

Vanderyst 22209 (13); Vasse 453 (48a); Vaughan 2130 (40); Veldhuizen, v. 628 (17); 749 (51b); Verdcourt 204 (68); 1705 (6e); 1711 (64); 1746 (68); Verdcourt, Greenway & Brass 201 (40); Vergnes s.n. (4); 203 (17); Verheyen & de Witte 3745 (20b); Verhoeff s.n. (39); Vermoesen 4 (1a); Versuchsanstalt Victoria s.n. (58); Vigne 4864 (19); 4869 (48a); 4870 (51a); Villiers 127 (51c); 166 (51c); 860 (39); 861 (52a); 872 (44b);

984 (3c); 985 (40); 988 (42); 989 (61b); 991 (49a); 1031 (1a); 1035 (62a); Villiers & Caratini 713 (16); Voorhoeve 436 (3c); Vroumsia Tchinaye 105 (46).

Wageningen cult. Hort. 00-1243 (39); 00-1244 (39); 00-1261 (51a); 00-1262 (51a); 00-1263 (51a); 00-1264 (51a); 00-1267 (51a); 00-155 (17); 00-431 (48a); 00-444 (40); 00-461 (4); 00-465 (59); 00-468 (51c); 00-469 (17); 00-479 (3a); 00-723 (51c); 00-726 (6a); 00-734 (17); 75-2-444 (58); 75-446 (26); 78-416 (17); 78-475 (3c); 78-526 (51c); 78-557 (51c); 78-565 (44a); 78-565 (44a); 78-567 (13); 78-743 (3a); 78-744 (17); 78-784 (42); 78-785 (51a); 78-788 (42); 78-793 (58); 79-240 (20a); 79-250 (44a); 79-272 (44a); 79-82 (62b); 79-82 (63); 79-83 (3b); 80-392 (17); 80-393 (17); 80-397 (25); 80-401 (17); 80-404 (17); 80-477a (62b); 80-483 (68); 80-501 (26); 80-510 (20a); 80-56 (42); 80-61 (45); 80-66 (51c); 80-82 (51c); 80-96 (51c); 80-98 (51c); 81-328 (48a); 81-17 (25); 81-27 (15); 81-34 (51c); 81-36 (51c); 81-7 (40); 83-202 (32b); 83-213 (6a); 83-214 (63); 83-217 (13); 83-219 (54); 83-222 (27); 83-231 (3c); 83-233 (13); 83-234 (63); 83-236 (3b); 83-239 (3b); 83-242 (13); 83-304 (3b); 83-314 (3b); 83-322 (27); 83-323 (3b); 83-325 (17); 83-332 (63); 83-335 (13); 83-337 (52c); 83-339 (62b); 83-340 (3b); 83-351 (62b); 83-354 (6a); 83-362 (51c); 83-367 (3b); 83-394 (46); 83-402 (17); 83-414 (3b); 83-418 (52a); 83-419b (52a); 83-692 (17); 84-2 (6e); 85-287b (3b); 85-312 (39); Waterman & McKeay 814 (13); Weigel in Zenker 56 (1c); Wells & Edwards 106 (48a); Welwitsch 689 (19); 697 (3c); Westwood 3 (59); 48 (48a); 62 (52b); 73 (1c); 134 (19); 156 (40); 183 (48a); 209 (48a); 578 (40); 596 (44a); 646 (40); 654 (22); Whyte (1904) s.n. (44b); Wild 2264 (40); 2804 (20a); 2808 (56); 2813 (62b); 3537 (48a); 3543 (44a); 5256 (68); Wild & Chase 5544 (44a); 5545 (48a); 5546 (40); Wilde, de, J. 356 (51b); 559 (19); 688 (47); 805 (48a); 980 (49a); 1012 (40); 1013 (40); 1025 (42); 3292 (51c); 7667a (3c); 8132 (1a); 8475 (3b); Wilde, de, J., Arends & Groenendijk 100 (51c); 102 (51c); 129 (45); Wilde, de, J., Arends, Louis, Bouman & Karper 124 (13); 183 (2); 195 (13); 774 (3b); 901 (17); 938 (52a); Wilde, de, J., de Wilde, W., & de Wilde-Duyfjes 4276 (6a); Wilde, de, W., & de Wilde-Duyfjes 1367 (62b); 1691 (58); 2121 (51c); 2137 (44b); 2259 (1c); 2749 (1c); 3732 (51b); 8912 (48a); Williamson 15 (48a); 20 (44a); 21 (66a); 22 (20b); 23 (62b); 305 (44a); 835 (6b); 839 (6b); 1397a (56); 1398 (44a); 1822 (63); 1823 (63); 1824 (40); 1860 (64); 2153 (44a); Williamson & Gassner 2406 (13); 2416 (6a); 2428 (6a); 2430 (6a); 2433 (63); 2445 (40); 2449 (6a); 2453 (13); 2455 (56); Williamson & Simon 442 (13); 1822 (63); 1824 (40); 1843 (63); 1844 (6b); Wimbush & King 129 (48b); 150 (6d); Winkler 14a (1c); Wit, de (1960) s.n. (31); Witte, de 2548 (64); 3744 (62b); 8030 (51c); Wright 58/18 (59); Wrigley 31 (51c); 532 (62b); 543 (17); Wrigley & Melville 200 (42); Wubben cult. s.n. (17); s.n. (26); s.n. (3a); s.n. (48a); s.n. (51c); s.n. (52b); s.n. (62b); s.n. (63); s.n. (6a); s.n. (6c).

Yeoward (1939) s.n. (19).

Zenker (1895) s.n. (1c); 45 (1c); 254 (44b); 524 (44b); 832 (1c); 1558 (42); 2267 (44b); 2452 (1c); 2454 (44b); 2690 (1c); 2894 (52a); 2895 (44b); 3336 (62b); 3902 (63); 3903 (17); 3903 (51c); 3926 (17); 4735 (6a); 4736 (61b); Zimmermann s.n. (68); (1916) s.n. (20a); 10 (13); Zimmermann in Herb. Amani (1917) s.n. (40).

LEGENDS OF THE FIGURES

Fig. 1. *Bulbophyllum schinzianum* Krzl. var. *schinzianum.*
a - plant; b - flower; c - flower analysis, from left to right: median sepal, petal, lateral sepal, lip; d, e - lip, left: adaxially, right: abaxially; f–h - lip, adaxially; i - column and lip, lateral view; j - anther, left: adaxially, right: abaxially; k - pollinia, left: single pair, right: both pairs. a–d, i - Baldwin 10100; e - v.d. Burg 1241; f - Pittery 637; g - De Wilde 8132; h - Laurent 232; j, k - Edinburgh cult. s.n.

Fig. 2. *Bulbophyllum schinzianum* Krzl. var. *irigaleae* (Cribb & Perez-Vera) J.J. Vermeulen.
a - plant; b - flower; c - flower analysis, from left to right: median sepal, petal, lateral sepal, lip; d - lip, left: adaxially, right: abaxially; e - column and lip, lateral view. a–e - Perez-Vera 571.

Fig. 3. *Bulbophyllum schinzianum* Krzl. var. *phaeopogon* (Schltr) J.J. Vermeulen.
a - plant; b - flower; c - flower analysis, from left to right: median sepal, petal, lateral sepal, lip; d - lip, left: adaxially, right: abaxially; e - column and lip, lateral view; f - pollinia, left: single pair, right: both pairs (stipes missing!); g - anther, left: abaxially, right: adaxially. a–e - Le Testu 5178; f–g - Chew 12.

Fig. 4. *Bulbophyllum subligaculiferum* J.J. Vermeulen.
a - plant; b - flower analysis, from left to right: median sepal, petal, lateral sepal, lip; c - lip, left: adaxially, right: abaxially; d - column and lip, lateral view; e - longitudinal section through base of lip; f - anther, above: adaxially, below: abaxially; g - pollinia, left: single pair, right: both pairs. a–g - De Wilde, Arends, Louis, Bouman & Karper 183.

Fig. 5. *Bulbophyllum saltatorium* Lindl. var. *saltatorium.*
a - plant; b - flower; c - flower analysis, from left to right: median sepal, petal, lateral sepal, lip; d - lip, left: adaxially, right: abaxially; e–f - column and lip, lateral view. a–f - Sanders cult. s.n.

Fig. 6. *Bulbophyllum saltatorium* Lindl. var. *albociliatum* (Finet) J.J. Vermeulen, entity a.
a - plant; b - flower; c - flower analysis, from left to right: median sepal, petal, lateral sepal, lip; d - lip, left: adaxially, right: abaxially; e - column and lip, lateral view; f - anther, left: adaxially, right: abaxially; g - pollinia, left: single pair, with detached stipes, right: both pairs. a–e - Jongejan cult. 62; f–g - Dekker 434.

Fig. 7. *Bulbophyllum saltatorium* Lindl. var. *albociliatum* (Finet) J.J. Vermeulen, entity b.
a - Infloresc ense; b - plant; c - flower analysis, from left to right: median sepal, petal, lateral sepal, lip; d - lip, left: adaxially, right: abaxially; e - column and lip, lateral view. a - Glasnevin cult. s.n.; b - Kew cult. s.n.; c–e - Pfennig s.n.

Fig. 8. *Bulbophyllum saltatorium* Lindl. var.. *albociliatum* (Finet) J.J. Vermeulen.
a–b plant; c–d flower analysis, from left to right: median sepal, petal, lateral sepal, lip; e–f lip, left: adaxially, right: abaxially; g–h column, lateral view. a, c, e, g - Le Testu 5787; b, d, f, h - Le Testu 5562.

Fig. 9. *Bulbophyllum saltatorium* Lindl. var. *calamarium* (Lindl.) J.J. Vermeulen.
a - plant; b - flower; c - flower analysis, from left to right: median sepal, petal, lateral sepal, lip; d - lip, left: adaxially, right: abaxially; e - column and lip, lateral view; f - anther, adaxially; g - pollinia, left: single pair, right: both pairs. a–e - Jongejan cult. 29; f–g - De Wilde 7667a.

Fig. 10. *Bulbophyllum barbigerum* Lindl.
a - plant; b - flower; c - flower analysis, from left to right: median sepal, petal, lateral sepal, lip; d - lip, adaxially; e - column and lip, lateral view; f - anther, left: adaxially, right: abaxially; g - pollinia, left: single pair, right: both pairs. a–e - Lodewijk cult. 123; f–g - Wageningen cult. Hort. 00-461 = Leiden cult. Hort. 23666.

Fig. 11. *Bulbophyllum vulcanicum* Krzl.
a - plant; b - flower; c - flower analysis, from left to right: median sepal, petal, lateral sepal, lip; d - lip, left: adaxially, right: abaxially; e - column and lip, lateral view; f - anther, left: adaxially, right: abaxially; g - pollinia, left: single pair, right: both pairs. a–g - Mulder in Jongejan cult. 60.

Fig. 12. *Bulbophyllum cochleatum* Lindl. var. *cochleatum*, entity a.
a - plant; b - flower; c - flower analysis, from left to right: median sepal, petal, lateral sepal, lip; d - lip, left: adaxially, right: abaxially; e - column and lip, lateral view; f - anther, left: adaxially, right: abaxially; g - pollinia, left: single pair, right: both pairs. a–e - Wubben cult. s.n.; f–g - Jongejan cult. 63.

Fig. 13. *Bulbophyllum cochleatum* Lindl. var. *cochleatum*, entity b.
a - plant; b - flower; c - flower analysis, from left to right: median sepal, petal, lateral sepal, lip; d - lip, left: adaxially, right: abaxially; e - column and lip, lateral view. a - Mann 1337; b–e - Melville 461,

Fig. 14. *Bulbophyllum cochleatum* Lindl. var. *gravidum* (Lindl.) J.J. Vermeulen, entity c. (a, f, g) and d. (b–e, h).
a–b - plant (back pseudobulb of plant in fig. b has lost one leaf); c - flower; d - flower analysis, from left to right: median sepal, petal, lateral sepal, lip; e - lip, left: adaxially, right: abaxially; f - lip, adaxially; g–h - column and lip, lateral view. a - Mann 650; b–e, h - Milne Redhead & Taylor 8912; f–g Sanford 4163.

Fig. 15. *Bulbophyllum cochleatum* Lindl. var. *bequaertii* (De Wild.) J.J. Vermeulen.
a - plant; b - flower; c - flower analysis, from left to right: median sepal, petal, lateral sepal, lip; d - lip, left: adaxially, right: abaxially; e - column and lip, lateral view. a–e - Wubben cult. s.n.

Fig. 16. *Bulbophyllum cochleatum* Lindl. var. *tenuicaule* (Lindl.) J.J. Vermeulen.
a - plant; b - flower; c - flower analysis, from left to right: median sepal, petal, lateral sepal, lip; d - lip, left: adaxially, right: abaxially; e - column and lip, lateral view. a–e - Wimbush & King 150.

Fig. 17. *Bulbophyllum cochleatum* Lindl. var. *brachyanthum* (Lindl.) J.J. Vermeulen.
a - plant; b - flower; c - flower analysis, from left to right: median sepal, petal, lateral sepal, lip; d - lip, left: adaxially, right: abaxially; e - column and lip, lateral view. a - Gardner 2548; b–e - Meyer 198.

Fig. 18. *Bulbophyllum burttii* Summerh.
a - plant; b - flower; c - flower analysis, from left to right: median sepal, petal, lateral sepal, lip; d - lip, left: adaxially, right: abaxially; e - column and lip, lateral view; f - anther, left: adaxially, right: abaxially; g - pollinia, left: single pair, right: both pairs. a - Burtt 3164; b–e - Germain 2996; f–g - Symoens 5393a.

Fig. 19. *Bulbophyllum kivuense* J.J. Vermeulen.
a - plant; b - flower; c - flower analysis, from left to right: median sepal, petal, lateral sepal, lip; d - lip, left: adaxially, right: abaxially; e - column and lip, lateral view; f - anther, left: adaxially, right: abaxially; g - pollinia, left: single pair, right: both pairs. a–g - Troupin 15652.

Fig. 20. *Bulbophyllum inornatum* J.J. Vermeulen.
a - plant; b - flower; c - flower analysis, from left to right: median sepal, petal, lateral sepal, lip; d - lip, left: adaxially, right: abaxially; e - column and lip, lateral view; f - anther, left: adaxially, right: abaxially; g - pollinia, left: single pair, right: both pairs. a–g - Stolze & Scharff s.n.

Fig. 21. *Bulbophyllum bavonis* J.J. Vermeulen.
a - plant; b - flower; c - flower analysis, from left to right: median sepal, petal, lateral sepal, lip; d - lip, left: adaxially, right: abaxially; e - column and lip, lateral view; f - anther, left: adaxially, right: abaxially; g - pollinia, left: single pair, right: both pairs. a–g - Dowsett Lemaire 674.

Fig. 22. *Bulbophyllum stolzii* Schltr.
a - plant; b - flower; c - flower analysis, from left to right: median sepal, petal, lateral sepal, lip; d - lip, left: adaxially, right: abaxially; e - column and lip, lateral view; f - anther, left: adaxially, right: abaxially; g - pollinia, left: single pair, right: both pairs. a–b, f - Stolz 767; c–e, g - Cribb & Grey Wilson 10818.

Fig. 23. *Bulbophyllum prorepens* Summerh.
a - plant; b - flower; c - flower analysis, from left to right: median sepal, petal, lateral sepal, lip; d - lip, left: adaxially, right: abaxially; e - column and lip, lateral view; f - anther, left: adaxially, right: abaxially; g - pollinia, left: both pairs, right: single pair. a–g - Burtt 3152.

Fig. 24. *Bulbophyllum intertextum* Lindl.
a–b - plant; c - flower; d - flower analysis, from left to right: median sepal, petal, lateral sepal, lip; e - lip, left: adaxially, right: abaxially; f - column and lip, lateral view; g - column, frontal view; h - anther, left: adaxially, right: abaxially; i - pollinia, left: a single, right: both. a, c–i - Jongejan cult. 6; b - Waterman & Mc Keay 814.

Fig. 25. *Bulbophyllum vanum* J.J. Vermeulen.
a - plant; b - top part inflorescence; c - flower analysis, from left to right: median sepal, petal, lateral sepal, lip; d - lip, left: adaxially, right: abaxially; e - column and lip, lateral view; f - anther, left: adaxially, right: abaxially; g - pollinia, left: single pair, right: both pairs. a–g - Breteler & De Wilde 642.

Fig. 26. *Bulbophyllum capituliflorum* Rolfe.
a–b - plant; c - flower; d - flower analysis, from left to right: median sepal, petal, lateral sepal, lip; e - lip, left: adaxially, right: abaxially; f - column and lip, lateral view; g - anther, left: adaxially, right: abaxially; h - pollinia, left: single pair, right: both pairs. a, d, g, h - Wageningen cult. Hort. 81-027 = Leiden cult. Hort. 23687; b, c, e, f - Glasnevin cult. s.n.

Fig. 27. *Bulbophyllum pipio* Rchb. f.
a - plant; b - flower; c - flower analysis, from left to right: median sepal, petal, lateral sepal, lip; d - lip, left: adaxially, right: abaxially; e - column and lip, lateral view; f - anther, left: adaxially, right: abaxially; g - pollinia, left: single pair, right: both pairs. a–e - Groot cult. 496; f, g - Adames 100.

Fig. 28. *Bulbophyllum pumilum* (Sw.) Lindl., entity a.
a - plant; b - flower; c - flower analysis, from left to right: median sepal, petal, lateral sepal, lip; d - lip, left: adaxially, right: abaxially; e - column and lip, lateral view; f - anther, left: adaxially, right: abaxially; g - pollinia, left: single pair, right: both pairs. a–e - Jongejan cult. 65; f, g - Jongejan cult. 26.

Fig. 29. *Bulbophyllum pumilum* (Sw.) Lindl., entity b. (a–e) and c. (f–j).
a - plant; b - flower; c - flower analysis, from left to right: median sepal, petal, lateral sepal, lip; d - lip, left: adaxially, right: abaxially; e - column and lip, lateral view; f - plant; g - flower; h - flower analysis, from left to right: median sepal, petal, lateral sepal, lip; i - lip, left: adaxially, right: abaxially; j - column and lip, lateral vieuw. a–e - Jongejan cult. 51; f–j - Lodewijk cult. 160.

Fig. 30. *Bulbophyllum pumilum* (Sw.) Lindl., entity d. (a–e) and e. (f–j).
a - plant; b - flower; c - flower analysis, from left to right: longitudinal section through median sepal, median sepal, petal, lateral sepal, lip; d - lip, left: adaxially, right: abaxially; e - column and lip, lateral view; f - plant; g - flower; h - flower analysis, from left to right: median sepal, petal, lateral sepal, lip; i - lip, left: adaxially, right: abaxially; j - column and lip, lateral view. a–e - Groot cult. 666; f–j - Keay & Savory FHI 25309.

Fig. 31. *Bulbophyllum coriscense* Rchb. f.
a - plant; b - flower; c - flower analysis, from left to right: median sepal, petal, lateral sepal, lip; d - lip, left: adaxially, right: abaxially; e - column and lip, lateral view; f - anther, left: adaxially, right: abaxially; g - pollinia, left: a single, right: both. a–g - Mann 1883.

Fig. 32. *Bulbophyllum cocoinum* Lindl.
a - plant; b - flower; c - flower analysis, from left to right: median sepal, petal, lateral sepal, lip; d - lip, left: adaxially, right: abaxially; e - column and lip, lateral view; f - anther, above: adaxially, below: abaxially; g - pollinia, above: both, below: a single. a–e - Jongejan cult. 66; f–g - Jongejan cult. 865.

Fig. 33. *Bulbophyllum josephii* (Kuntze) Summerh. var. *josephii*; E. African specimen (a –e, i, j) and Cameroun specimen (f–h).
a - plant; b - flower; c - flower analysis, from left to right: median sepal, petal, lateral sepal, lip; d - lip, left: adaxially, right: abaxially; e - column and lip, lateral view; f - lip, left: adaxially, right: abaxially; g - column and lip, lateral view; h - petal; i - anther, left: adaxially, right: abaxially; j - pollinia, left: single pair, right: both pairs. a–c - Stolz 638; d, e - Cribb, Grey Wilson & Mwasumbi 11315; f–h - Mann 2124; i, j - Schlieben 3578.

Fig. 34. *Bulbophyllum josephii* (Kuntze) Summerh. var. *mahonii* (Rolfe) J.J. Vermeulen.
a - plant; b - flower; c - flower analysis, from left to right: median sepal, petal, lateral sepal, lip; d - lip, left: adaxially, right: abaxially; e - column and lip, lateral view. a - Mahon s.n.; b–e - Williamson 22.

Fig. 35. *Bulbophyllum josephii* (Kuntze) Summerh. var. *mahonii* (Rolfe) J.J. Vermeulen; W. African specimen (a, c, e, f) and Cameroun specimen (b, d, g, h).
a, b - plant; c, d - flower analysis, from left to right: median sepal, petal, lateral sepal, lip; e, g - lip, left: adaxially, right: abaxially; f, h - column and lip, lateral view. a, c, e, f - Adams 640; b - Keay FHI 25439; d, g, h - Gregory 558.

Fig. 36. *Bulbophyllum calvum* Summerh.
a - plant; b - flower; c - flower analysis, from left to right: median sepal, petal, lateral sepal, lip; d - lip, left: adaxially, right: abaxially; e - column and lip, lateral view. a–e - Nash & King 178.

Fig. 37. *Bulbophyllum expallidum* J.J. Vermeulen.
a - plant; b - flower; c - flower analysis, from left to right: median sepal, petal, lateral sepal, lip; d - lip, left: adaxially, right: abaxially; e - column and lip, lateral view; f - anther, above: adaxially, below: abaxially; g - pollinia, above: both, below: a single. a–e - Holmes 10.2; f–g - Leiden cult. Hort. 22296.

Fig. 38. *Bulbophyllum ivorense* Cribb & Perez-Vera.
a - plant; b - flower; c - flower analysis, from left to right: median sepal, petal, lateral

sepal, lip; d - lip, left: adaxially, right: abaxially; e - column and lip, lateral view; f - anther, left: adaxially, right: abaxially; g - pollinia, left: single pair, right: both pairs. a - Perez-Vera 137; b–g - Adames 588.

Fig. 39. *Bulbophyllum mediocre* Summerh.
a - plant; b - flower; c - flower analysis, from left to right: median sepal, petal, lateral sepal, lip; d - lip, left: adaxially, right: abaxially; e - column and lip, lateral view; f - anther, left: adaxially, right: abaxially; g - pollinia, left: single pair, right: both pairs. a–e - Monod 11842; f–g - Monod 12130.

Fig. 40. *Bulbophyllum schimperanum* Krzl.
a - plant; b - flower; c - flower analysis, from left to right: median sepal, petal, lateral sepal, lip; d - lip, left: adaxially, right: abaxially; e - column and lip, lateral view; f - anther, left: adaxially, right: abaxially; g - pollinia, left: single pair, right: both pairs. a–g - Jongejan cult. 51.

Fig. 41. *Bulbophyllum oxychilum* Schltr.
a - plant; b - part of the inflorescence; c - flower analysis, from left to right: median sepal, petal, lateral sepal, lip; d - lip, left: adaxially, right: abaxially; e - column and lip, lateral view; f - anther, left: adaxially, right: abaxially; g - pollinia, left: single pair, right: both pairs. a–e - Groot cult. 387; f, g - Adames 664.

Fig. 42. *Bulbophyllum nigritianum* Rendle.
a - plant; b - flower; c - flower analysis, from left to right: median sepal, petal, lateral sepal, lip; d - lip, left: adaxially, right: abaxially; e - column and lip, lateral view; f - anther, left: adaxially, right: abaxially; g - pollinia, left: single pair, right: both pairs. a–e - Deighton 4946; f, g - Bowden 135-23.

Fig. 43. *Bulbophyllum humblotii* Rolfe.
a - plant; b - flower; c - flower analysis, from left to right: median sepal, petal, lateral sepal, lip; d - lip, left: adaxially, right: abaxially; e - column and lip, lateral view; f - anther, left: adaxially, right: abaxially; g - pollinia, left: single pair, right: both pairs. a–e - Ngoni 232; f, g - Ball 316;

Fig. 44. *Bulbophyllum scariosum* Summerh.
a - plant; b - flower, right: floral bract; c - flower analysis, from left to right: median sepal, petal, lateral sepal, lip; d - lip, left: adaxially, right: abaxially; e - column and lip, lateral view; f - anther, left: adaxially, right: abaxially; g - pollinia, left: single pair, right: both pairs. a–e - Adames 801; f, g - Bos 2390.

Fig. 45. *Bulbophyllum pandanetorum* Summerh.
a - plant; b - flower with floral bract; c - flower analysis, from left to right: median sepal, petal, lateral sepal, lip; d - lip, left: adaxially, right: abaxially; e - column and lip, lateral view; a, d - Le Testu 5527; b, c, e - Le Testu 5547.

Fig. 46. *Bulbophyllum denticulatum* Rolfe.
a - plant; b - flower; c - flower analysis, from left to right: median sepal, petal, lateral sepal, lip; d - lip, left: adaxially, right: abaxially; e - column and lip, lateral view; f - anther, left: adaxially, right: abaxially; g - pollinia, left: single pair, right: both pairs. a–e - O'Brien cult. s.n.; f, g - De Wit s.n.

Fig. 47. *Bulbophyllum comatum* Lindl. var. *comatum*.
a - plant; b - flower; c - flower analysis, from left to right: median sepal, petal, lateral sepal, lip, floral bract; d - lip, left: adaxially, right: abaxially; e - column and lip, lateral view; f - column, frontal view; g - petal. a–g - Keay & Savory FHI 25245.

Fig. 48. *Bulbophyllum comatum* Lindl. var. *inflatum* (Rolfe) J.J. Vermeulen.
a - plant; b - flower; c - flower analysis, from left to right: median sepal, petal, lateral sepal, lip, floral bract; d - lip, left: adaxially, right: abaxially; e - column and lip, lateral view; f - anther, left: adaxially, right: abaxially; g - pollinia, left: single pair, right: both pairs. a–e - Jongejan cult. 55; f, g - Geerling & Bokdam 1791.

Fig. 49. *Bulbophyllum bifarium* Hook. f.
a - plant; b - part of inflorescence; c - flower analysis, from left to right: median sepal, petal, lateral sepal, lip; d - lip, above: adaxially, below: abaxially; e - column and lip, lateral view; f - anther, left: adaxially, right: abaxially; g - pollinia, left: single pair viewed from different directions, right: both pairs. a–g - Letouzey 13296.

Fig. 50. *Bulbophyllum bidenticulatum* J.J. Vermeulen ssp. *bidenticulatum*
a - plant; b - part of inflorescence; c - flower analysis, from left to right: median sepal, petal, lateral sepal, lip; d - lip, left: adaxially, right: abaxially; e - column and lip, lateral view; f - anther, above: adaxially, below: abaxially; a–f - Morton & Gledhill S.L. 3157.

Fig. 51. *Bulbophyllum bidenticulatum* J.J. Vermeulen ssp. *joyceae* J.J. Vermeulen.
a - plant; b - part of inflorescence; c - flower analysis, from left to right: median sepal, petal, lateral sepal, lip; d - lip, left: adaxially, right: abaxially; e - column and lip, lateral view; f - anther, left: adaxially, right: abaxially; g - pollinia, left: single pair, right: both pairs. a–g - Stewart 1193.

Fig. 52. *Bulbophyllum nigericum* Summerh.
a - plant; b - part of inflorescence; c - flower analysis, from left to right: median sepal, petal, lateral sepal, lip; d - lip, left: adaxially, right: abaxially; e - column and lip, lateral view. a–e - King 124.

Fig. 53. *Bulbophyllum renkinianum* (Laurent) De Wild.
a - plant; b - flower; c - flower analysis, from left to right: median sepal, petal, lateral sepal, lip; d - lip, left: adaxially, right: abaxially; e - column and lip, lateral view; f - anther, above: adaxially, below: abaxially; g - pollinia, both pairs. a–g - Le Testu 5772.

Fig. 54. *Bulbophyllum dolabriforme* J.J. Vermeulen.
a - plant; b - flower; c - flower analysis, from left to right: median sepal, petal, lateral sepal, lip; d - lip, left: adaxially, right: abaxially; e - column and lip, lateral view; f - anther, left: adaxially, right: abaxially; g - pollinia, left: single pair, right: both pairs. a–g - Graham & Smith 570.

Fig. 55. *Bulbophyllum curvimentatum* J.J. Vermeulen.
a - plant; b - flower analysis, from left to right: median sepal, petal, lateral sepal, lip; c - lip, left: adaxially, right: abaxially; d - column and lip, lateral view; e - anther, left: adaxially, right: abaxially. a–e - Sanford 4035.

Fig. 56. *Bulbophyllum colubrinum* (Rchb. f.) Rchb. f.
a - plant; b - flower; c - flower analysis, from left to right: median sepal, petal, lateral sepal, lip; d - lip, left: adaxially, right: abaxially; e - column and lip, lateral view; f - anther, left: adaxially, right: abaxially; g - pollinia, left: single pair, right: both pairs. a - Adames 128; b–g - van Imschoot s.n.

Fig. 57. *Bulbophyllum maximum* (Lindl.) Rchb. f., entity a.
a - plant; b - flower; c - flower analysis, from left to right: median sepal, petal, lateral sepal, lip; d - lip, left: adaxially, right: abaxially; e - column and lip, lateral view. a - De Koning s.n.; b–e - Bowling G.C. 36615.

Fig. 58. *Bulbophyllum maximum* (Lindl.) Rchb. f., entity b.
a - plant; b - inflorescence; c - flower; d - flower analysis, from left to right: median sepal, petal, lateral sepal, lip; e - lip, left: adaxially, right: abaxially; f - column and lip, lateral view; g - anther, left: adaxially, right: abaxially; h - pollinia, left: single pair, right: both pairs. a, c–g - Jongejan cult. 19; b - Groot cult. 390.

Fig. 59. *Bulbophyllum injoloense* De Wild. ssp. *injoloense*.
a - plant; b - flower analysis, from left to right: median sepal, petal, lateral sepal, lip; c - lip, left: adaxially, right: abaxially; d - column and lip, lateral view. a–d - Laurent 937.

Fig. 60. *Bulbophyllum injoloense* De Wild. ssp. *pseudoxypterum* (J.J. Vermeulen) J.J. Vermeulen.
a - plant; b - flower; c - flower analysis, from left to right: median sepal, petal, lateral sepal, lip; d - lip, left: adaxially, right: abaxially; e - column and lip, lateral view; f - anther, left: adaxially, right: abaxially. a–e - Moreau 86; f - Nash 12.

Fig. 61. *Bulbophyllum imbricatum* Lindl.
a, b - plant; c - flower; d - flower analysis, from left to right: median sepal, petal, lateral sepal, lip; e - lip, left: adaxially, right: abaxially; f - column and lip, lateral view; g - anther, left: adaxially, right: abaxially; h - pollinia, left: single pair, right: both pairs. a - Sanford 5705; b, c - Adames 685; d–f - Segerbaeck 1139; g, h - Le Testu 8949.

Fig. 62. *Bulbophyllum magnibracteatum* Summerh.
a, b - plant; c - flower; d - flower analysis, from left to right: median sepal, petal, lateral sepal, lip; e - lip, left: adaxially, right: abaxially; f - column and lip, lateral view; g - anther, left: adaxially, right: abaxially; a, c–f - Perez-Vera 139b; b - Perez-Vera 139d; g - Mylord s.n.

Fig. 63. *Bulbophyllum sandersonii* (Hook. f.) Rchb. f. ssp. *sandersonii.*
a - plant; b - inflorescence; c - flower; d - flower analysis, from left to right: median sepal, petal, lateral sepal, lip; e - lip, left: adaxially, right: abaxially; f - column and lip, lateral view; g - anther, left: adaxially, right: abaxially; h - pollinia, left: single pair, right: both pairs. a - Polhill & Paulo 1684; b–f - Gassner 16; g, h - Wubben cult. s.n.

Fig. 64. *Bulbophyllum sandersonii* (Hook. f.) Rchb. f. ssp. *stenopetalum* (Krzl.) J.J. Vermeulen.
a - plant; b - flower; c - flower analysis, from left to right: median sepal, petal, lateral sepal, lip; d - lip, left: adaxially, right: abaxially; e - column and lip, lateral view; f - anther, left: adaxially, right: abaxially; g - pollinia, left: single pair, right: both pairs, with detached stipes. a–g - Groot cult. 635.

Fig. 65. *Bulbophyllum lizae* J.J. Vermeulen.
a - plant; b - flower; c - flower analysis, from left to right: median sepal, petal, lateral sepal, lip; d - lip, left: adaxially, right: abaxially; e - column and lip, lateral view; f - anther, left: adaxially, right: abaxially; g - pollinia, left: both pairs, right: single pair. a–g - De Wilde, Arends & Groenendijk 129.

Fig. 66. *Bulbophyllum purpureorhachis* (De Wild.) Schltr.
a - plant; b - flower; c - flower analysis, from left to right: median sepal, petal, lateral sepal, lip; d - lip, left: adaxially, right: abaxially; e - column and lip, lateral view; f - anther, left: adaxially, right: abaxially; g - pollinia, left: single pair, right: both pairs. a–e - Letourey 5617; f, g - Le Testu 9269.

Fig. 67. *Bulbophyllum falcipetalum* Lindl.
a - plant; b - flower; c - flower analysis, from left to right: median sepal, petal, lateral sepal, lip, floral bract; d - lip, left: adaxially, right: abaxially; e - column and lip, lateral view; f - column, frontal view; g - anther, left: adaxially, right: abaxially; h - pollinia, left: single pair, right: both pairs. a–f - Perez 716; g, h - De Wilde 688.

Fig. 68. *Bulbophyllum scaberulum* (Rolfe) Bolus var. *scaberulum*, entity a. (f), entity b. (a), entity c. (b–e, g, h).
a, b - plant; c - flower; d - flower analysis, from left to right: median sepal, petal, lateral sepal, lip; e - lip, above: adaxially, below: abaxially; f - column and lip, lateral view; g - anther, left: adaxially, right: abaxially; h - pollinia, left: single pair, right: both pairs. a - Pynaert 1466; b - Dekker 431; c–e, g, h - De Koning 6097. f - Wubben cult. s.n.

Fig. 69. *Bulbophyllum scaberulum* (Rolfe) Bolus var. *fuerstenbergianum* (De Wild.) J.J. Vermeulen.
a - plant; b - flower; c - flower analysis, from left to right: median sepal, petal, lateral sepal, lip; d - lip, left: adaxially, right: abaxially; e - column and lip, lateral view; f - column, frontal view. a - Savory & Keay FHI 25183; b–f - Wimbush & King 129.

Fig. 70. *Bulbophyllum scaberulum* (Rolfe) Bolus var. *crotalicaudatum* J.J. Vermeulen.
a - plant; b - inflorescence; c - flower; d - flower analysis, from left to right: median sepal, petal, lateral sepal, lip; e - lip, left: adaxially, right: abaxially; f - column and lip, lateral view. a–f - Richards 15656b.

Fig. 71. *Bulbophyllum resupinatum* Ridl. var. *resupinatum*.
a - plant; b - part of the inflorescence; c - flower analysis, from left to right: median sepal, petal, lateral sepal, lip; d - lip, left: adaxially, right: abaxially; e - column and lip, lateral view. a–e - Jongejan cult. 1.

Fig. 72. *Bulbophyllum resupinatum* Ridl. var. *filiforme* (Krzl.) J.J. Vermeulen, entity a.
a - plant; b - flower; c - flower analysis, from left to right: median sepal, petal, lateral sepal, lip; d - lip, left: adaxially, right: abaxially; e - column and lip, lateral view. a–e - Perez-Vera 270.

Fig. 73. *Bulbophyllum resupinatum* Ridl. var. *filiforme* (Krzl.) J.J. Vermeulen, entity b.
a, b - plant; c - flower; d - top of the inflorescence; e, f - flower analysis, from left to right: median sepal, petal, lateral sepal, lip; g - lip, left: adaxially, right: abaxially; h - column and lip, lateral view; i - anther, left: adaxially, right: abaxially; j - pollinia, left: single pair, right: both pairs, with a detached stipes. a, c, e, g, h - Perez-Vera 553; b, d, f - Bequaert 6705; i, j - De Koning 6152.

Fig. 74. *Bulbophyllum parvum* Summerh.
a, b - plant; c - flower analysis, from left to right: median sepal, petal, lateral sepal, lip, floral bract; d - lip, left: adaxially, right: abaxially; e - column and lip, lateral view. a–e - Jones 235.

Fig. 75. *Bulbophyllum falcatum* (Lindl.) Rchb. f. var. *falcatum*.
a - plant; b - flower; c - flower analysis, from left to right: median sepal, petal, lateral sepal, lip; d - lip, left: adaxially, right: abaxially; e - column and lip, lateral view; f - anther, left: adaxially, right: abaxially; g - pollinia, left: single pair, right: both pairs. a–e - Groot cult. 371; f, g - Leiden cult. Hort. 20283.

Fig. 76. *Bulbophyllum falcatum* (Lindl.) Rchb. f. var. *bufo* (Lindl.) J.J. Vermeulen.
a - plant; b - flower; c - flower analysis, from left to right: median sepal, petal, lateral sepal, lip; d - lip, left: adaxially, right: abaxially; e - column and lip, lateral view; a–e - Linder 1209.

Fig. 77. *Bulbophyllum falcatum* (Lindl.) Rchb. f. var. *velutinum* (Lindl.) J.J. Vermeulen, entity a. (a, c, f–j), entity b. (b, d, e).
a, b, c - plant; d - part of the inflorescence, lateral view; e - top of the inflorescence, frontal view; f - flower; g - flower analysis, from left to right: median sepal, petal, lateral sepal, lip; h - lip, left: adaxially, right: abaxially; i - column and lip, lateral view; j - top column, frontal view; a, f - Groot cult. 490; b, d, e, - Jongejan cult. 25; c, g–j - Groot cult. 369.

Fig. 78. *Bulbophyllum calyptratum* Krzl. var. *calyptratum.*
a, b - plant; c - flower; d - flower analysis, from left to right: median sepal, petal, lateral sepal, lip; e - lip, left: adaxially, right: abaxially; f - column and lip, lateral view; g - column h - anther, left: adaxially, right: abaxially; i - pollinia, left: single pair, right: both pairs. a - Staff FHI 42079; b - Deighton 5026; c–i - Segerbaeck 1182.

Fig. 79. *Bulbophyllum calyptratum* Krzl. var. *graminifolium* (Summerh.) J.J. Vermeulen.
a - plant; b - flower, left: frontal view, right: lateral view; c - flower analysis, from left to right: median sepal, petal, lateral sepal, lip; d - lip, left: adaxially, right: abaxially; e - column and lip, lateral view. a–e - Wubben cult. s.n.

Fig. 80. *Bulbophyllum calyptratum* Krzl. var. *lucifugum* (Summerh.) J.J. Vermeulen.
a - plant; b - flower, frontal view; c - flower, lateral view; d - flower analysis, from left to right: median sepal, petal, lateral sepal, lip; e - lip, left: adaxially, right: abaxially; f - column and lip, lateral view. a–f - Deighton 4377.

Fig. 81. *Bulbophyllum carnosilabium* Summerh.
a - plant (right inflorescence also basal, not on top of pseudobulb); b - flower, lateral view; c - flower analysis, from left to right: median sepal, petal, lateral sepal, lip; d - lip, left: adaxially, right: abaxially; e - column and lip, lateral view; f - anther, left: adaxially, right: abaxially; g - pollinia, left: single pair, right: both pairs. a - Le Testu 5929; b–g - Le Testu 6473.

Fig. 82. *Bulbophyllum carnosisepalum* J.J. Vermeulen.
a - plant; b - top inflorescence; c - flower analysis, from left to right: median sepal, section through median sepal, petal, lateral sepal, section through lateral sepal, lip; d - lip, left: adaxially, right: abaxially; e - column and lip, lateral view; f - anther, left: adaxially, right: abaxially; g - pollinia, both pairs. a–f - Bruneel 465; g - Sanford 5456.

Fig. 83. *Bulbophyllum ballii* Cribb.
a - plant, flowering; b - plant, fruiting; c - flower; d - flower analysis, from left to right: median sepal, petal, lateral sepal, lip; e - lip, left: adaxially, right: abaxially; f - column and lip, lateral view; g - anther, left: adaxially, right: abaxially. a, c–g - Ball 258; b - Ball 1323.

Fig. 84. *Bulbophyllum elliotii* Rolfe.
a - plant; b - inflorescence; c - flower analysis, from left to right: median sepal, petal,

lateral sepal, lip; d - lip, left: adaxially, right: abaxially; e - column and lip, lateral view; f - variation in the shape of the petals; g - anther, left: adaxially, right: abaxially; h - pollinia, left: single pair, right: both pairs. a–e, g–h - Groot cult. 320; f - from left to right: Williamson 1397a - Brummitt & Pawek 11097 - Scott Elliot (1890) s.n. - Grosvenor 809 - Lemaire Elias 217.

Fig. 85. *Bulbophyllum porphyrostachys* Summerh.
a - plant; b - flower; c - flower analysis, from left to right: median sepal, petal, lateral sepal, lip; d - lip, adaxially; e - lip, abaxially; f - column and lip, lateral view; g - anther, left: adaxially, right: abaxially; h - pollinia, left: single pair, right: both pairs. a–h - Jongejan cult. 15.

Fig. 86. *Bulbophyllum lupulinum* Lindl.
a - plant; b - flower; c - flower analysis, from left to right: median sepal, petal, lateral sepal, lip; d - lip, above: adaxially, below: abaxially; e - column and lip, lateral view; f - anther, left: adaxially, right: abaxially; g - pollinia, above: both pairs, below: single pair. a–e - Hepper 1923; f–g - Ndi 51.

Fig. 87. *Bulbophyllum tetragonum* Lindl.
a - plant; b - inflorescence; c - flower; d - flower analysis, from left to right: median sepal, petal, lateral sepal, lip; e - lip, left: adaxially, right: abaxially; f - column and lip, lateral view; g - anther, left: adaxially, right: abaxially; h - pollinia, left: single pair, right: both pairs. a–h - Wright 58/18.

Fig. 88. *Bulbophyllum horridulum* J.J. Vermeulen.
a - plant; b - flower; c - flower analysis, from left to right: median sepal, petal, section through petal, lateral sepal, id. in lateral view, lip; d - lip, left: adaxially, right: abaxially; e - column and lip, lateral view; f - anther, left: adaxially, right: abaxially; g - pollinia, left: both pairs, right: single pair. a–g - Christiaensen 2458.

Fig. 89. *Bulbophyllum acutebracteatum* De Wild. var. *acutebracteatum*.
a - plant; b - inflorescence; c - flower analysis, from left to right: median sepal, petal, lateral sepal, lip; d - lip, left: adaxially, right: abaxially; e - column and lip, lateral view; f - anther, left: adaxially, right: abaxially; g - pollinia, left: single pair, right: both pairs. a–g - Jongejan cult. 6.

Fig. 90. *Bulbophyllum acutebracteatum* De Wild. var. *rubrobrunneopapillosum* (De Wild.) J.J. Vermeulen.
a - plant; b - part of inflorescence; c - flower analysis, from left to right: median sepal, petal, lateral sepal, lip; d - floral bract; e - lip, left: adaxially, right: abaxially; f - column and lip, lateral view. a–f - Bequaert 6706.

Fig. 91. *Bulbophyllum fuscum* Lindl. var. *fuscum*.
a - plant; b - part of inflorescence; c - flower analysis, from left to right: median sepal, petal, lateral sepal, lip, floral bract; d - lip, left: adaxially, right: abaxially; e, f - column and lip, lateral view. a–e - Leeuwenberg 7134; f - Deighton 5244.

Fig. 92. *Bulbophyllum fuscum* Lindl. var. *melinostachyum* (Schltr) J.J. Vermeulen.
a - plant; b - part of inflorescence; c - flower analysis, from left to right: median sepal, petal, lateral sepal, lip; d - lip, left: adaxially, right: abaxially; e - column and lip, lateral view; f - anther, left: adaxially, right: abaxially; g - pollinia, above: both pairs, below: single pair. a–e - Wubben cult. s.n.; f, g - Galt 15.

Fig. 93. *Bulbophyllum oreonastes* Rchb. f.
a, b - plant; c - part of inflorescence; d - flower analysis, from left to right: median sepal, petal, lateral sepal, lip, floral bract; e - lip, left: adaxially, right: abaxially; f - column and lip, lateral view; g - anther, left: adaxially, right: abaxially; h - pollinia, left: single pair, right: both pairs. a, c - Wubben cult. s.n.; b - Milne Redhead 4349; d–f - Harrington 546; g–h - Johansson 760.

Fig. 94. *Bulbophyllum encephalodes* Summerh.
a - plant; b - part of inflorescence; c - flower analysis, from left to right: median sepal, petal, lateral sepal, lip; d - lip, left: adaxially, right: abaxially; e - column and lip, lateral view; f - anther, left: adaxially, right: abaxially; g - pollinia, left: single pair, right: both pairs. a–e - Cribb & Grey Wilson 10255; f, g - Bain s.n.

Fig. 95. *Bulbophyllum concatenatum* Cribb & Taylor.
a - plant; b - flower; c - flower analysis, from left to right: median sepal, petal, lateral sepal, lip; d - floral bract; e - lip, left: adaxially, right: abaxially; f - column and lip, lateral view; g - anther, adaxially; h - pollinia, both pairs. a - Drummond & Hemsley 1690; b–h - Schlieben 3155.

Fig. 96. *Bulbophyllum unifoliatum* De Wild. ssp. *unifoliatum*.
a - plant; b - part of inflorescence; c - flower analysis, from left to right: median sepal, petal, lateral sepal, lip, floral bract; d - lip, left: adaxially, right: abaxially; e - lip, adaxially; f - column and lip, lateral view; g - anther, left: adaxially, right: abaxially; h - pollinia, above: both pairs, below: single pair. a–d, f - Bequaert 6616; e, g–h - Louis 13712.

Fig. 97. *Bulbophyllum unifoliatum* De Wild. ssp. *infracarinatum* (Williamson) J.J. Vermeulen.
a - plant; b - part of inflorescence; c - flower analysis, from left to right: median sepal, petal, lateral sepal, lip; d - floral bract; e - lip, left: adaxially, right: abaxially; f - column and lip, lateral view; g - anther, left: adaxially, right: abaxially; h - pollinia, above: both pairs, below: single pair. a–h - Ball 986.

Fig. 98. *Bulbophyllum unifoliatum* De Wild. ssp. *flectens* (Cribb & Taylor) J.J. Vermeulen.
a - plant; b - flower; c - flower analysis, from left to right: median sepal, petal, lateral sepal, lip; d - floral bract; e - lip, left: adaxially, right: abaxially; f - column and lip, lateral view; g - anther, left: adaxially, right: abaxially; h - pollinia, above: both pairs, below: single pair. a–h - Cribb, Grey Wilson & Mwasumbi 11446.

Fig. 99. *Bulbophyllum gilgianum* Krzl.
a - plant; b - inflorescence; c - flower analysis, from left to right: section through median sepal, median sepal, petal, section through lateral sepal, lateral sepal, lip, floral bract; d - lip, left: adaxially, right: abaxially; e - column and lip, lateral view; f - column, front view; g - anther, left: adaxially, right: abaxially; h - pollinia, left: single pair, right: both pairs. a–g - Moreau 464; h - Cribb & Grey Wilson 10420.

Fig. 100. *Bulbophyllum longiflorum* Thouars
a - plant; b - flower; c - flower analysis, from left to right: median sepal, petal, lateral sepal, lip; d - lip, left: adaxially, right: abaxially; e - column and lip, lateral view; f - anther, left: abaxially, right: adaxially; g - pollinia, left: single pair, right: both pairs. a–g - Wageningen cult. Hort. 80-483 = Leiden cult. Hort. 23626.

Fig. 101. *Chaseella pseudohydra* Summerh.
a - plant; b - pseudobulb and leaves; c - top part inflorescence; d - flower analysis, from left to right: median sepal, petal, lateral sepal, lip; e - lip, left: adaxially, right: abaxially; f - column and lip, lateral view; g - anther, left: adaxially, right: abaxially; h - pollinia, left: single, right: both. a–h - Chase SRGH 22738.

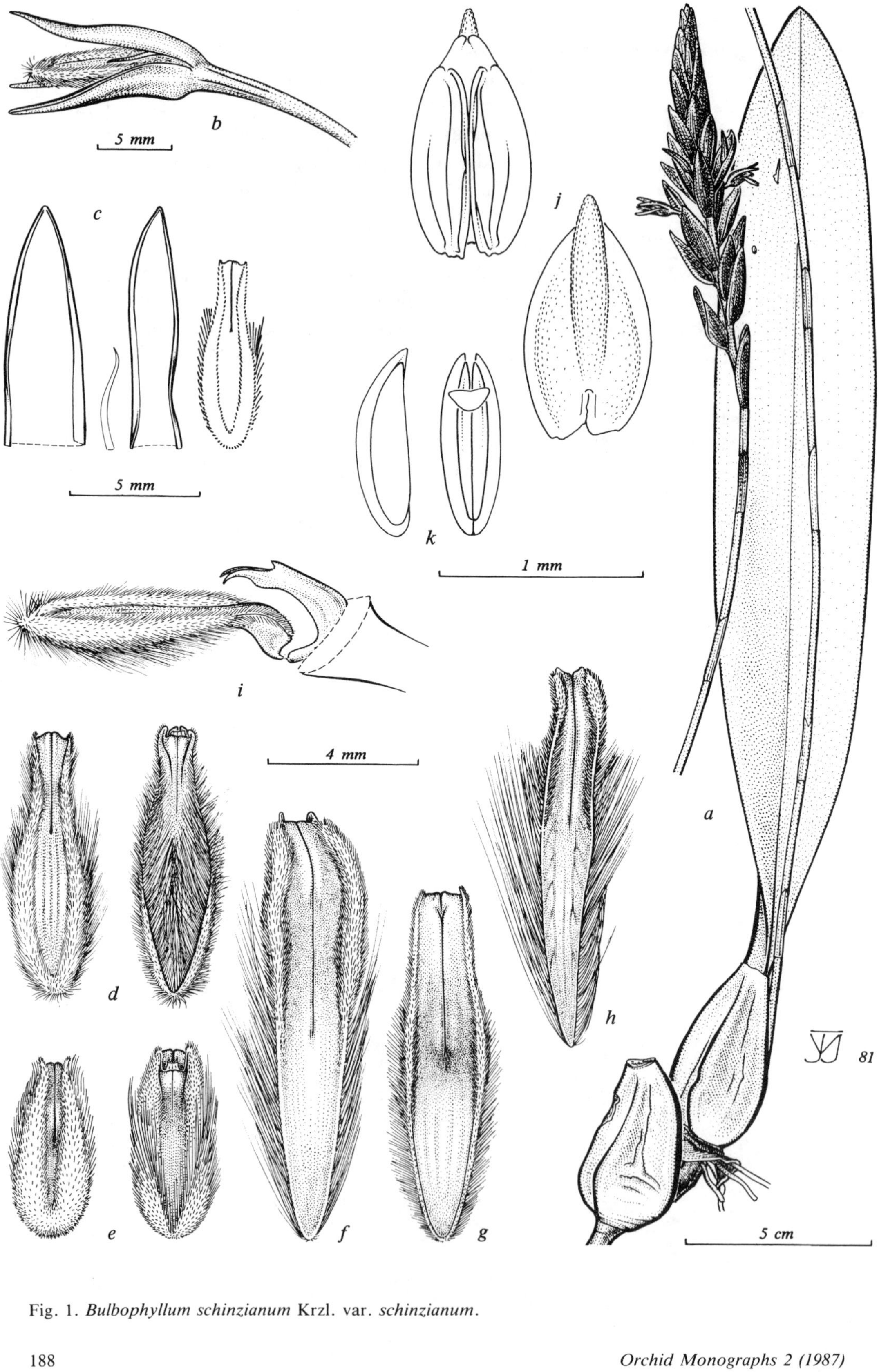

Fig. 1. *Bulbophyllum schinzianum* Krzl. var. *schinzianum*.

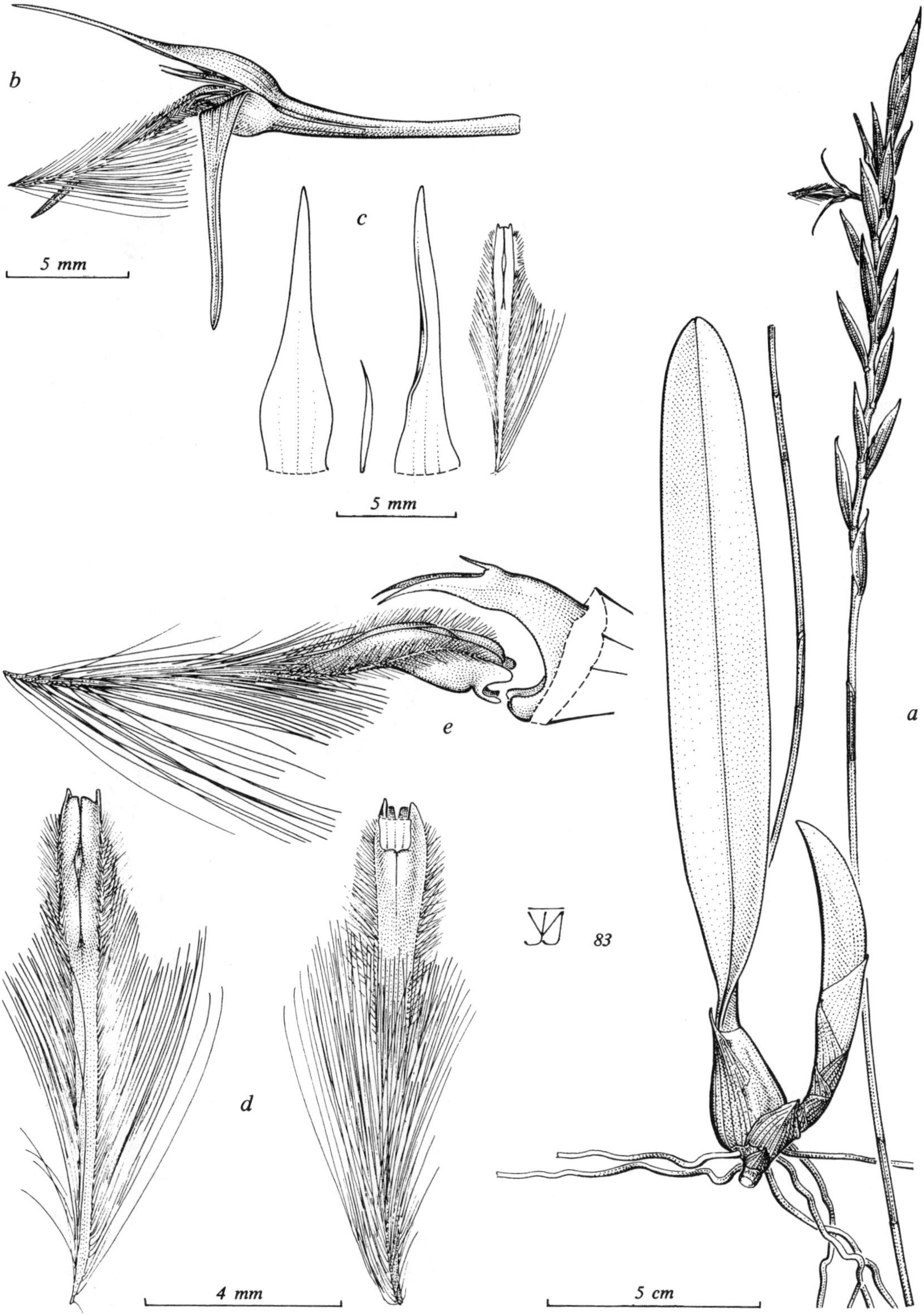

Fig. 2. *Bulbophyllum schinzianum* Krzl. var. *irigaleae* (Cribb & Perez-Vera) J.J. Vermeulen.

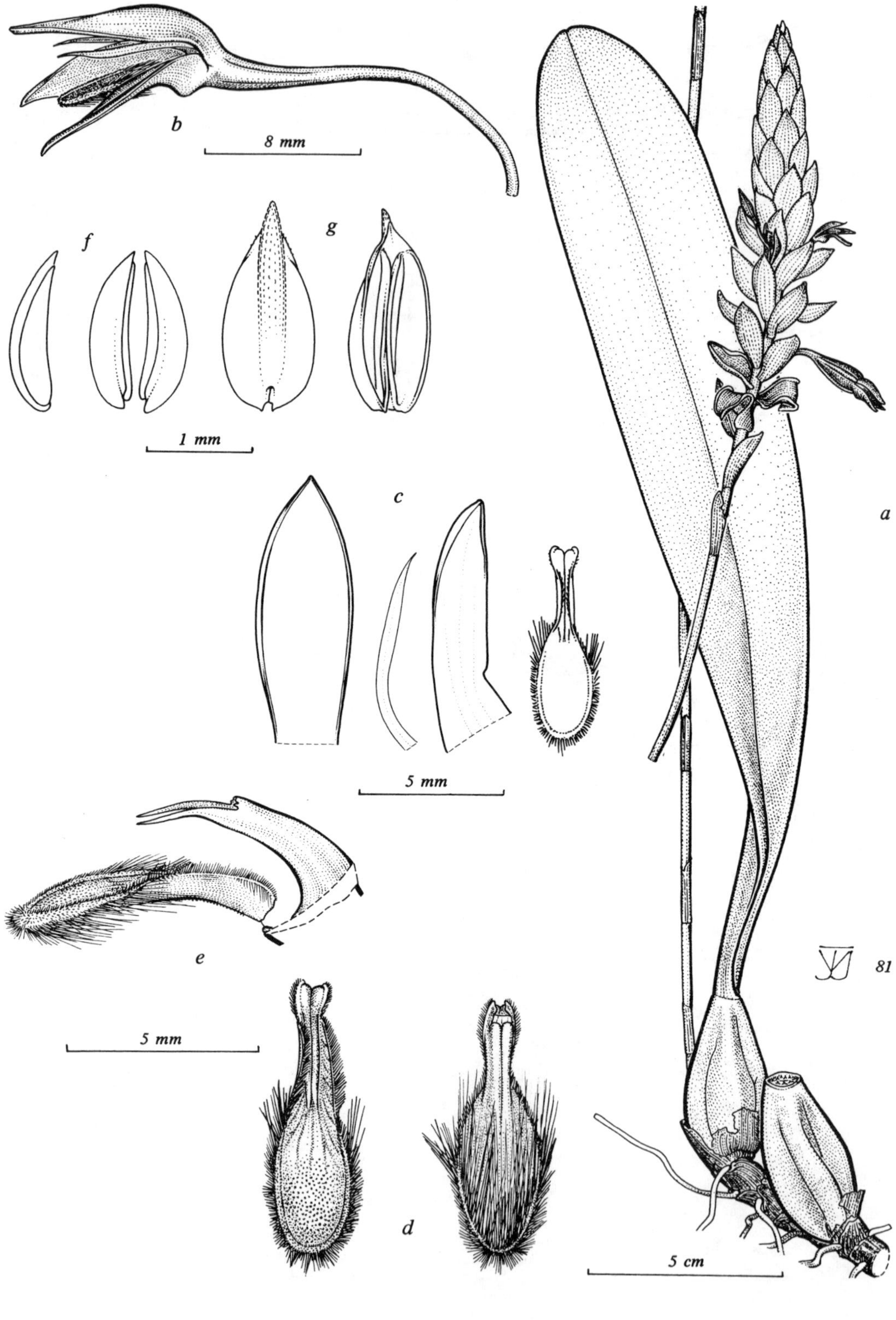

Fig. 3. *Bulbophyllum schinzianum* Krzl. var. *phaeopogon* (Schltr) J.J. Vermeulen.

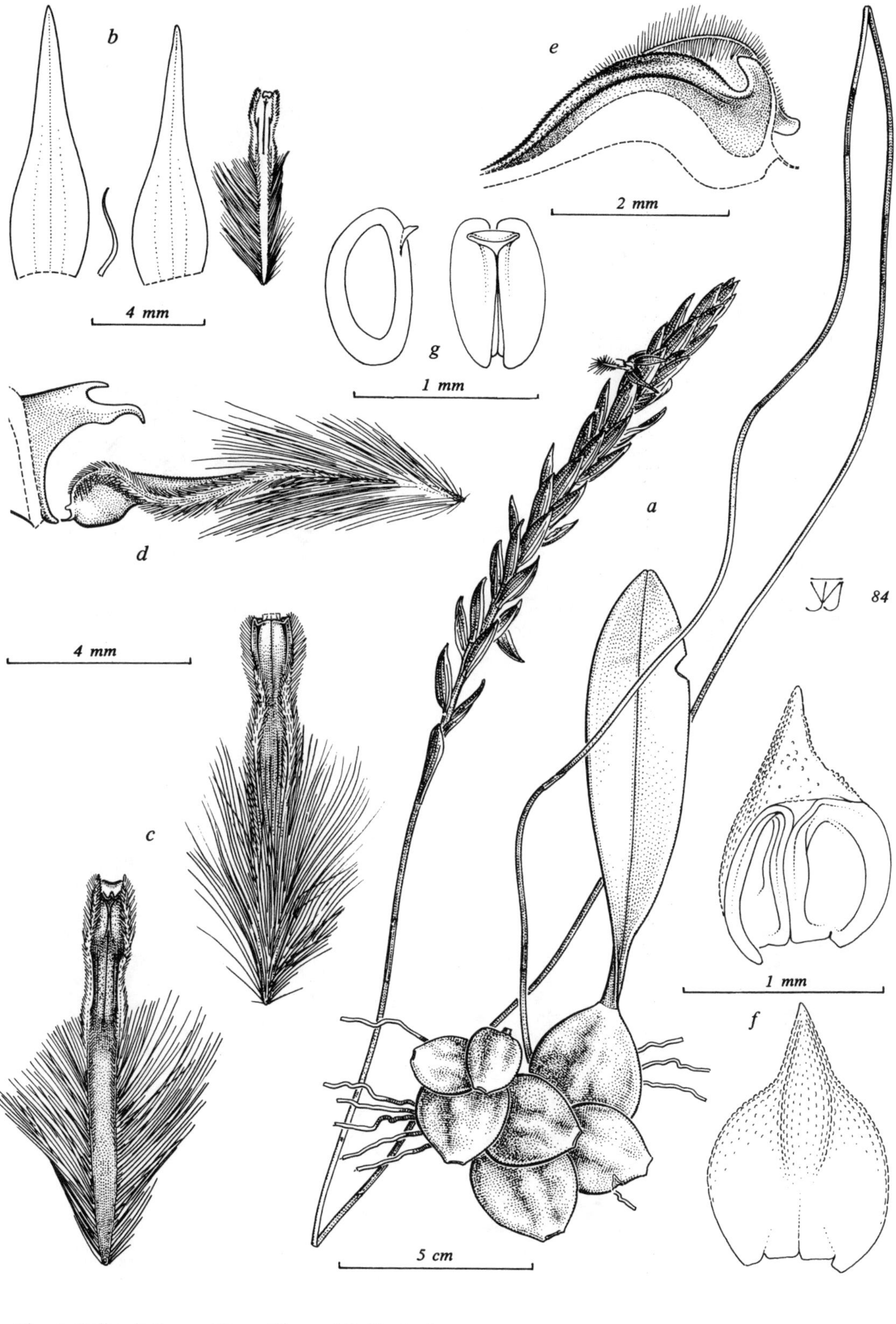

Fig. 4. *Bulbophyllum subligaculiferum* J.J. Vermeulen.

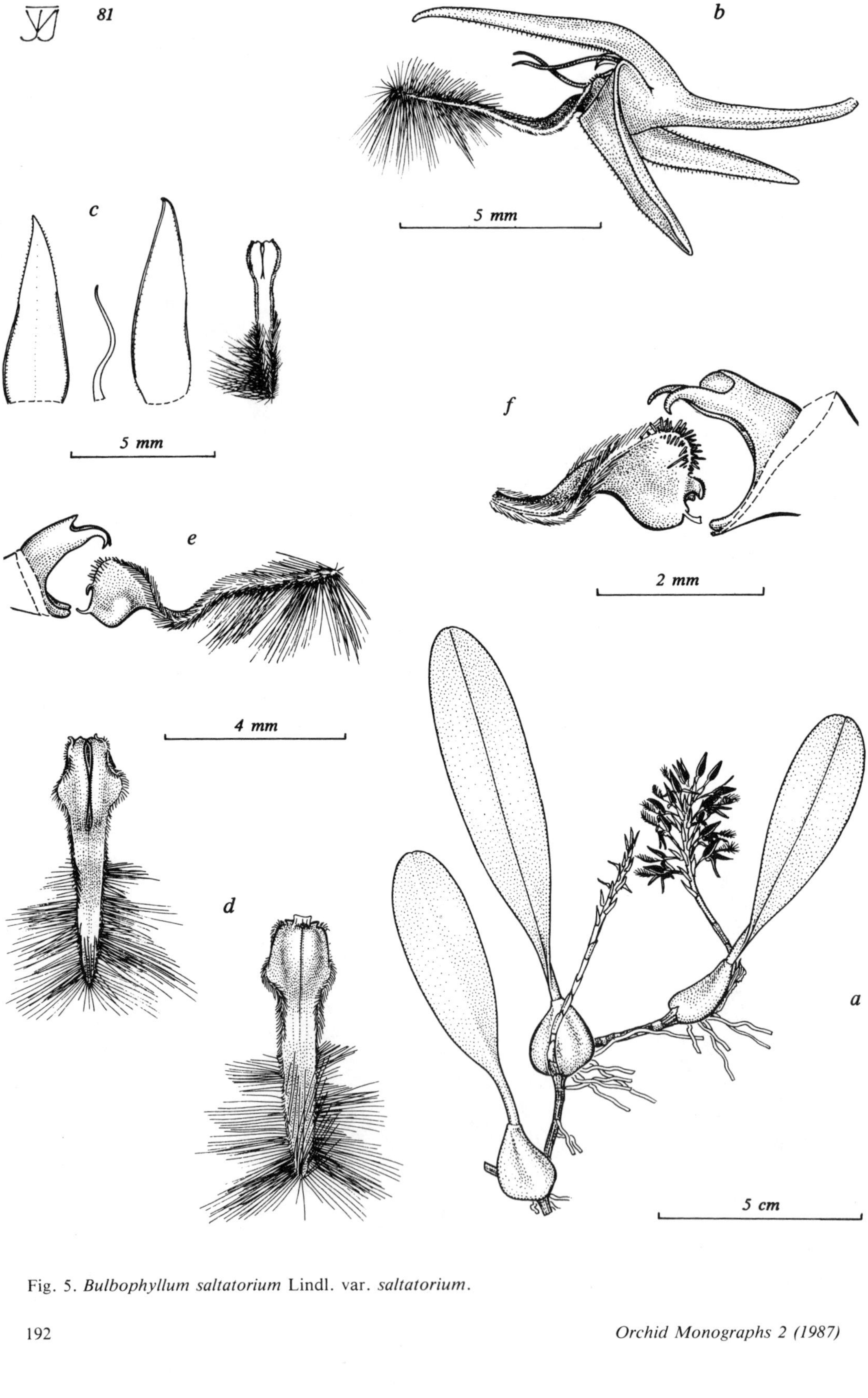

Fig. 5. *Bulbophyllum saltatorium* Lindl. var. *saltatorium*.

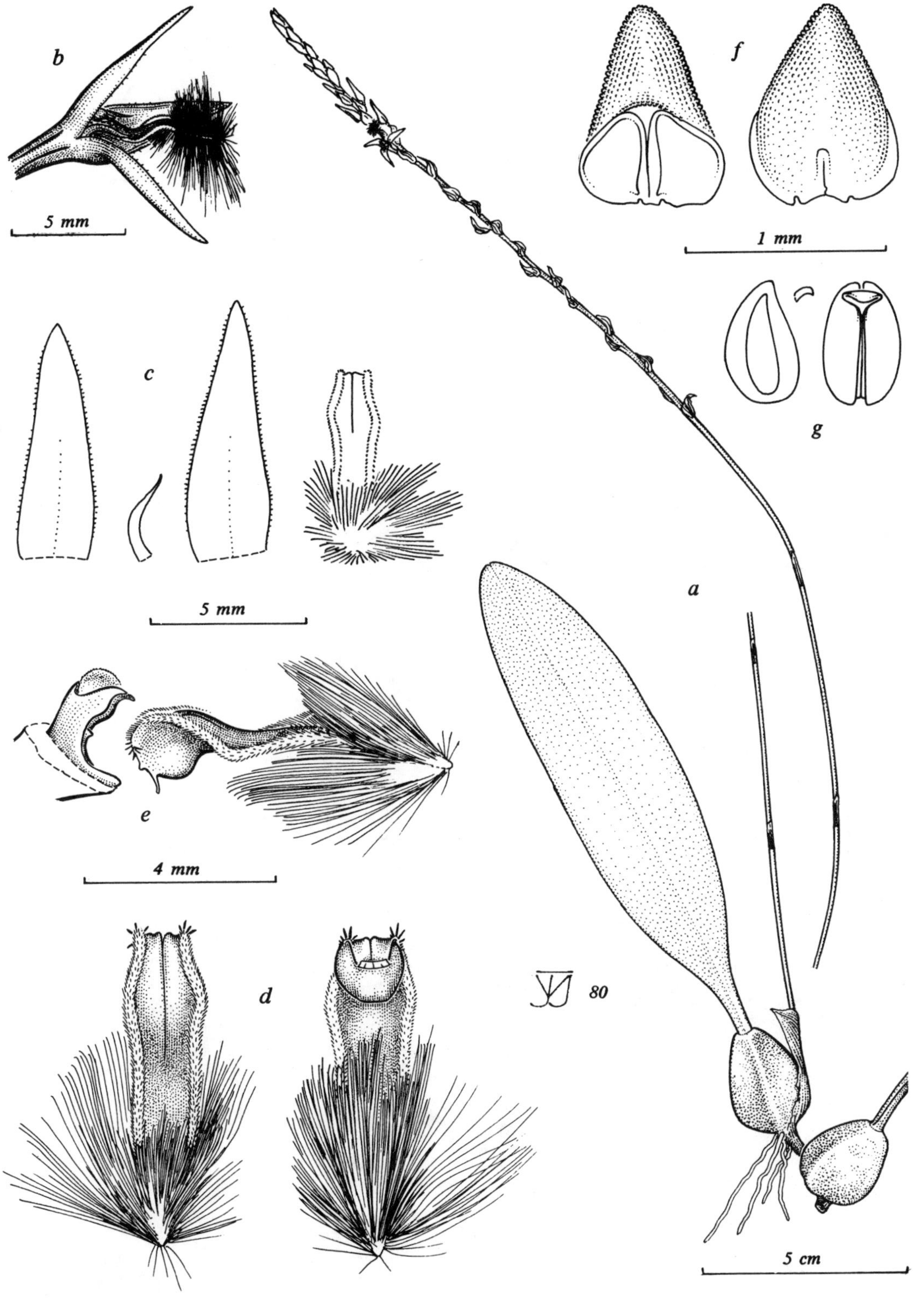

Fig. 6. *Bulbophyllum saltatorium* Lindl. var. *albociliatum* (Finet) J.J. Vermeulen, entity a.

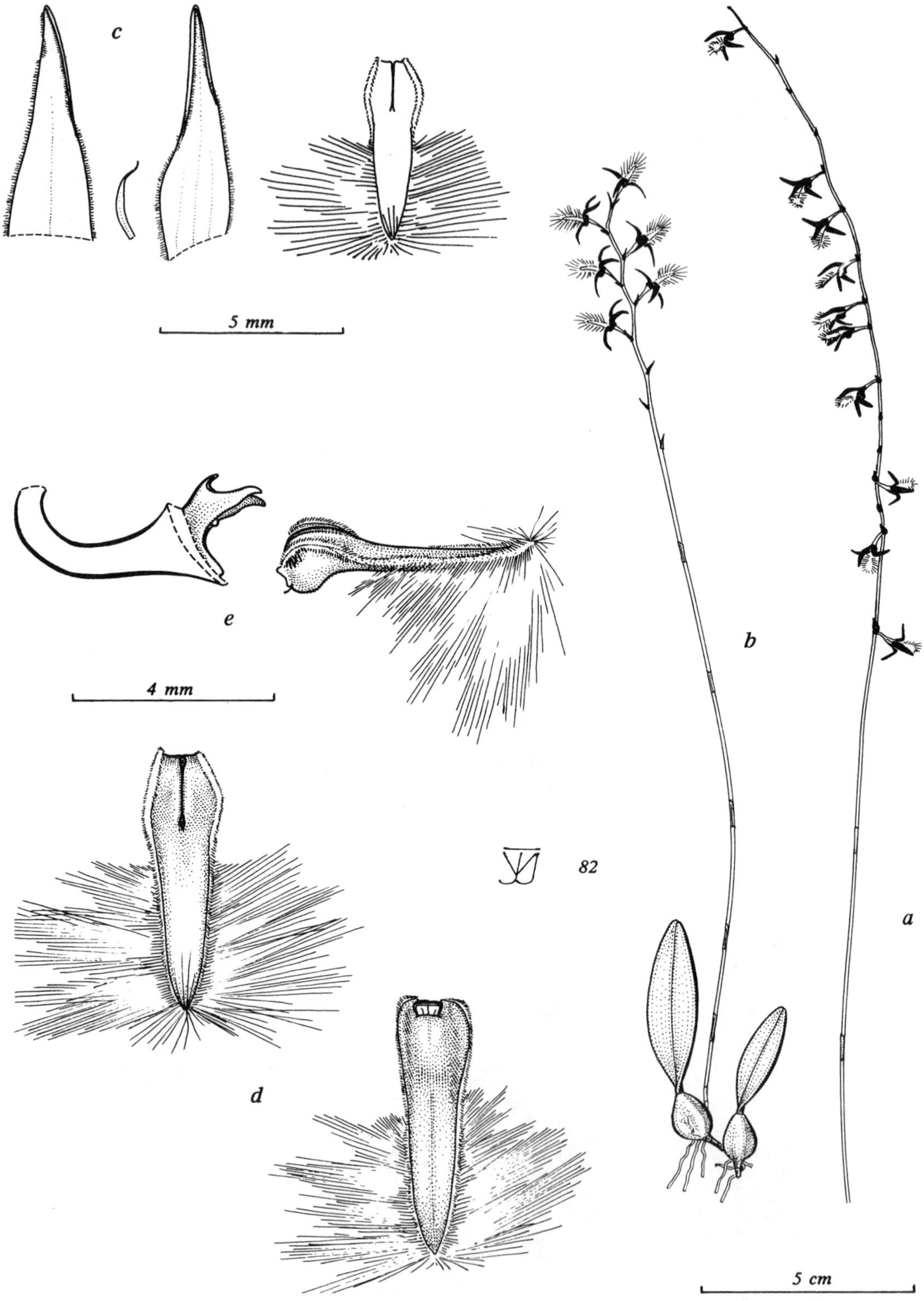

Fig. 7. *Bulbophyllum saltatorium* Lindl. var. *albociliatum* (Finet) J.J. Vermeulen, entity b.

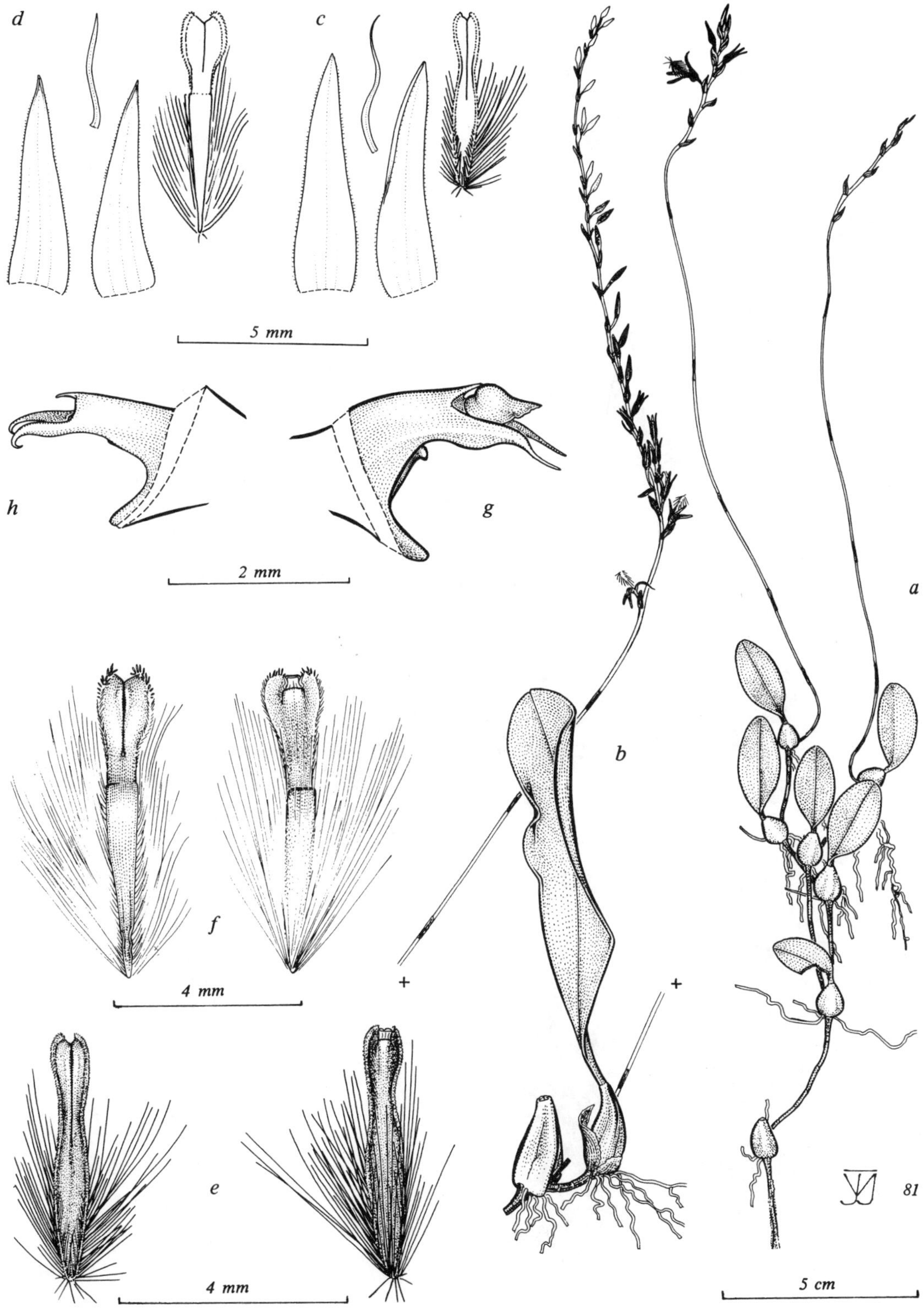

Fig. 8. *Bulbophyllum saltatorium* Lindl. var. *albociliatum* (Finet) J.J. Vermeulen.

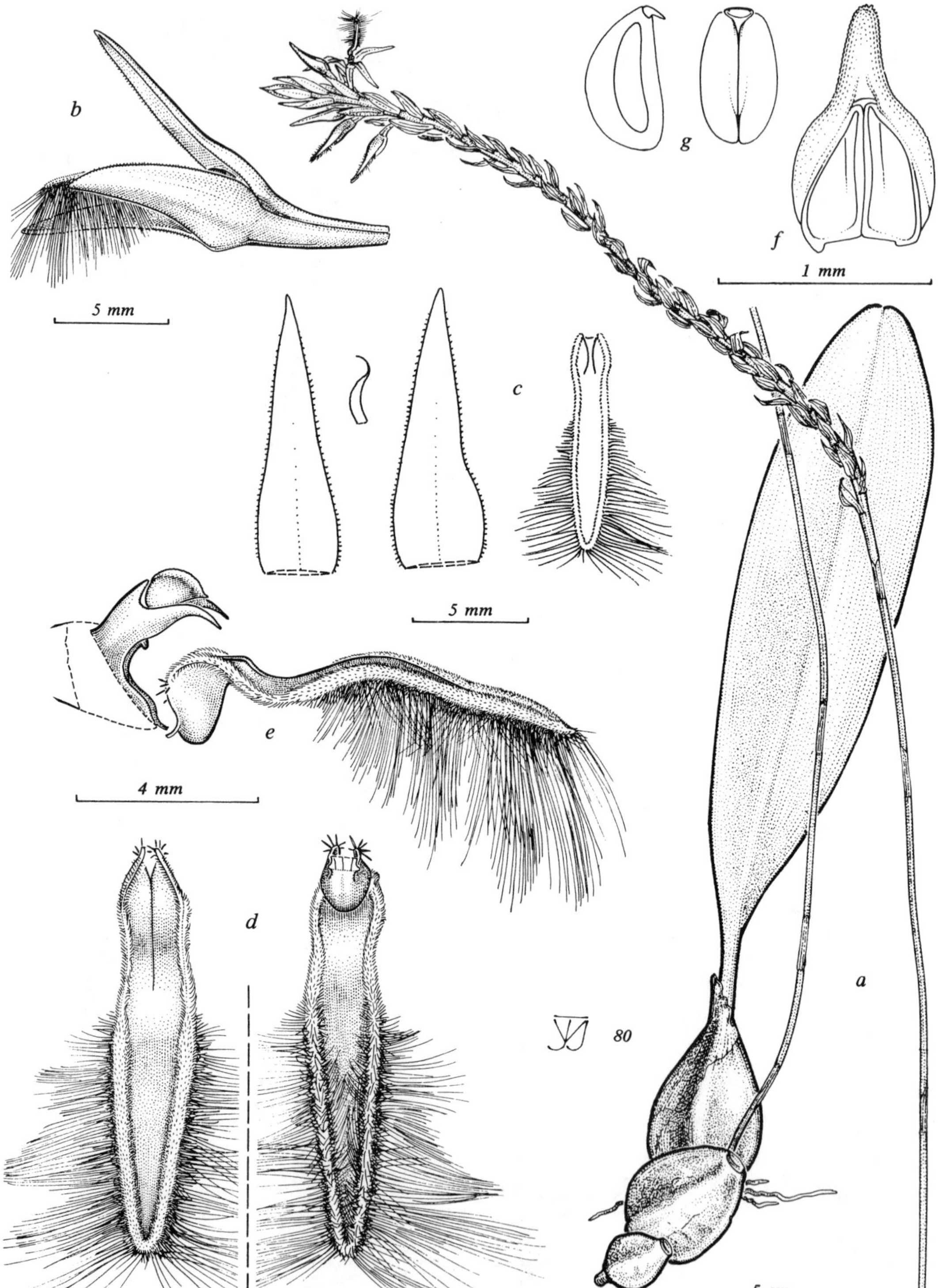

Fig. 9. *Bulbophyllum saltatorium* Lindl. var. *calamarium* (Lindl.) J.J. Vermeulen.

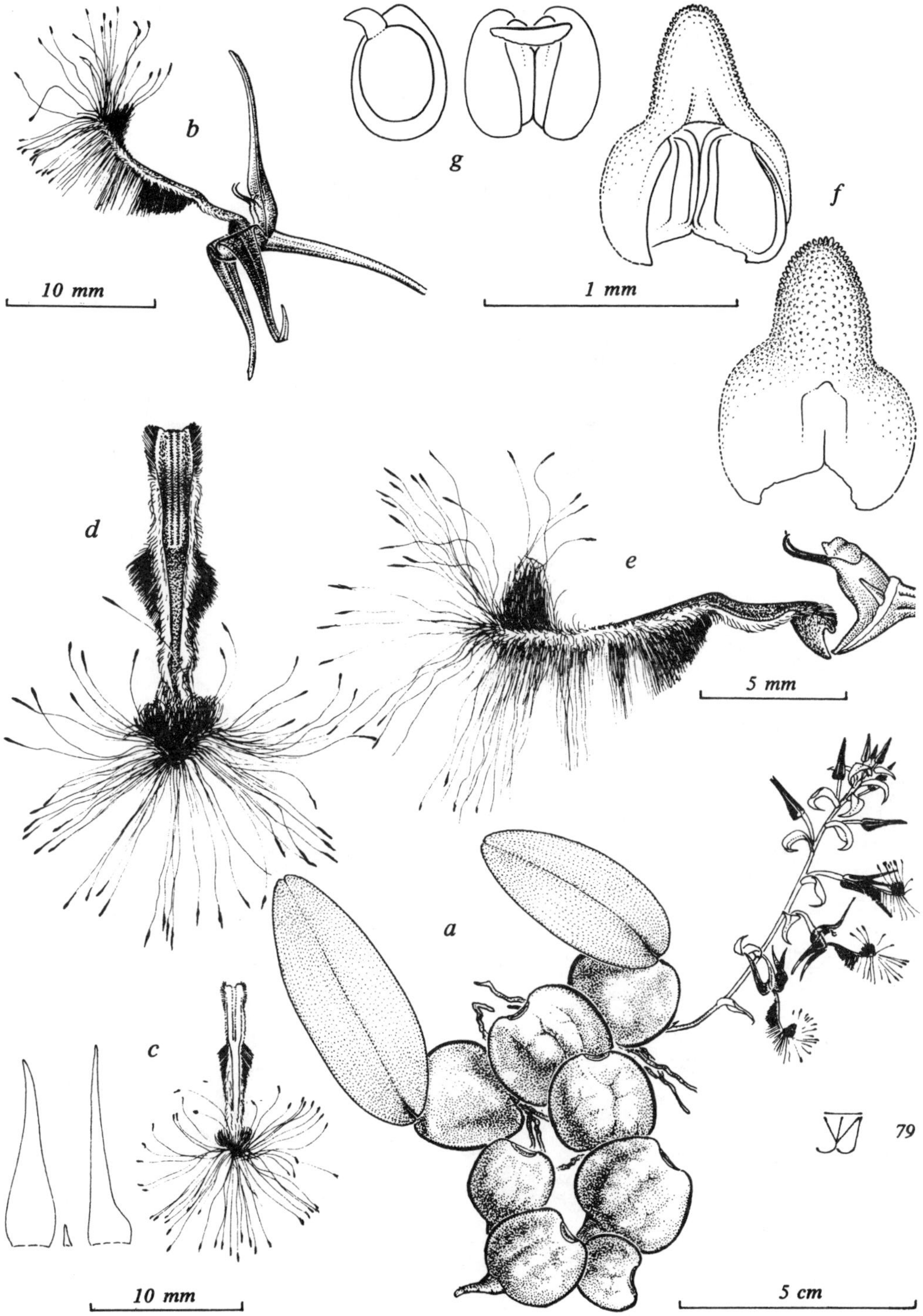

Fig. 10. *Bulbophyllum barbigerum* Lindl.

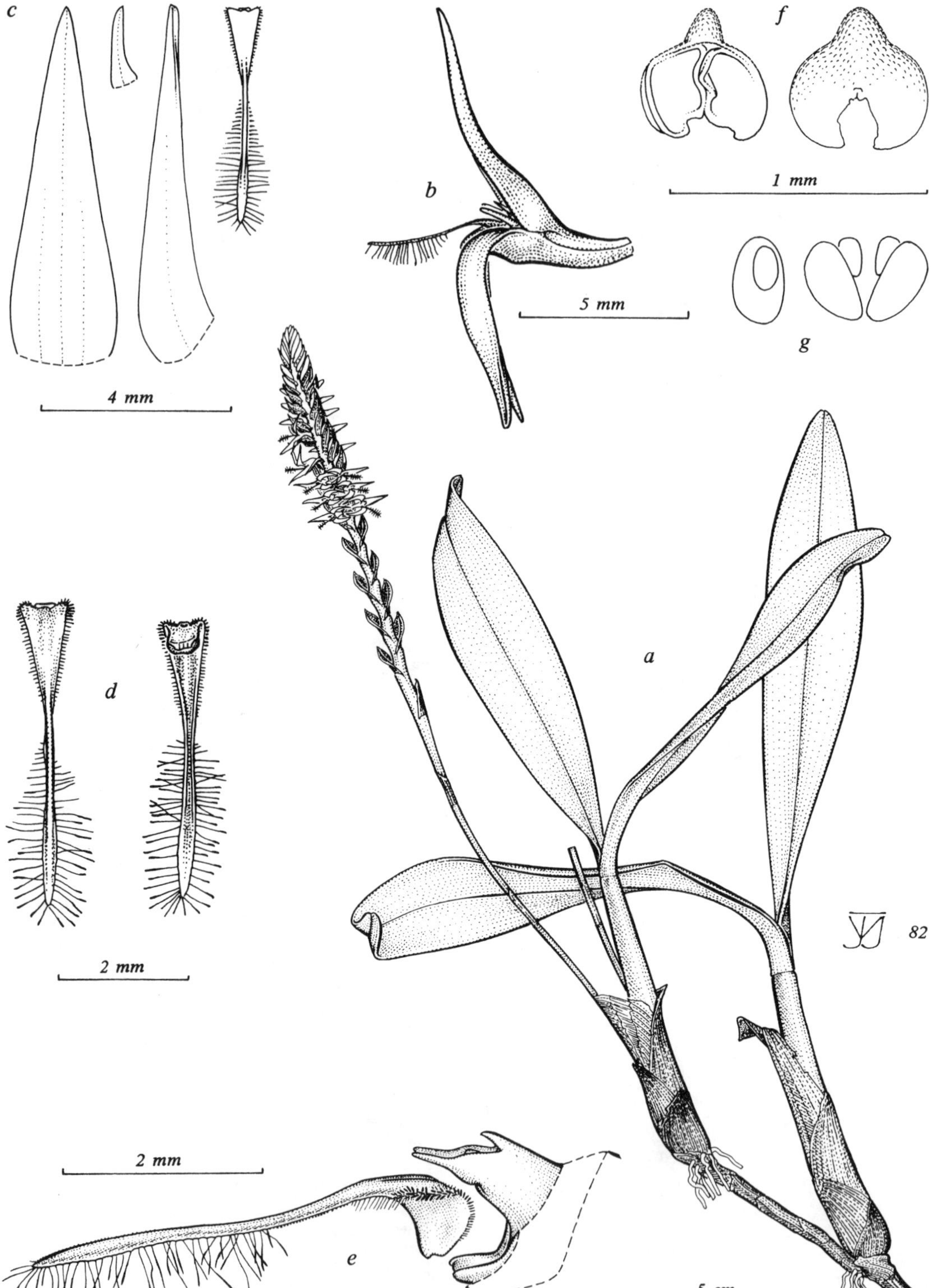

Fig. 11. *Bulbophyllum vulcanicum* Krzl.

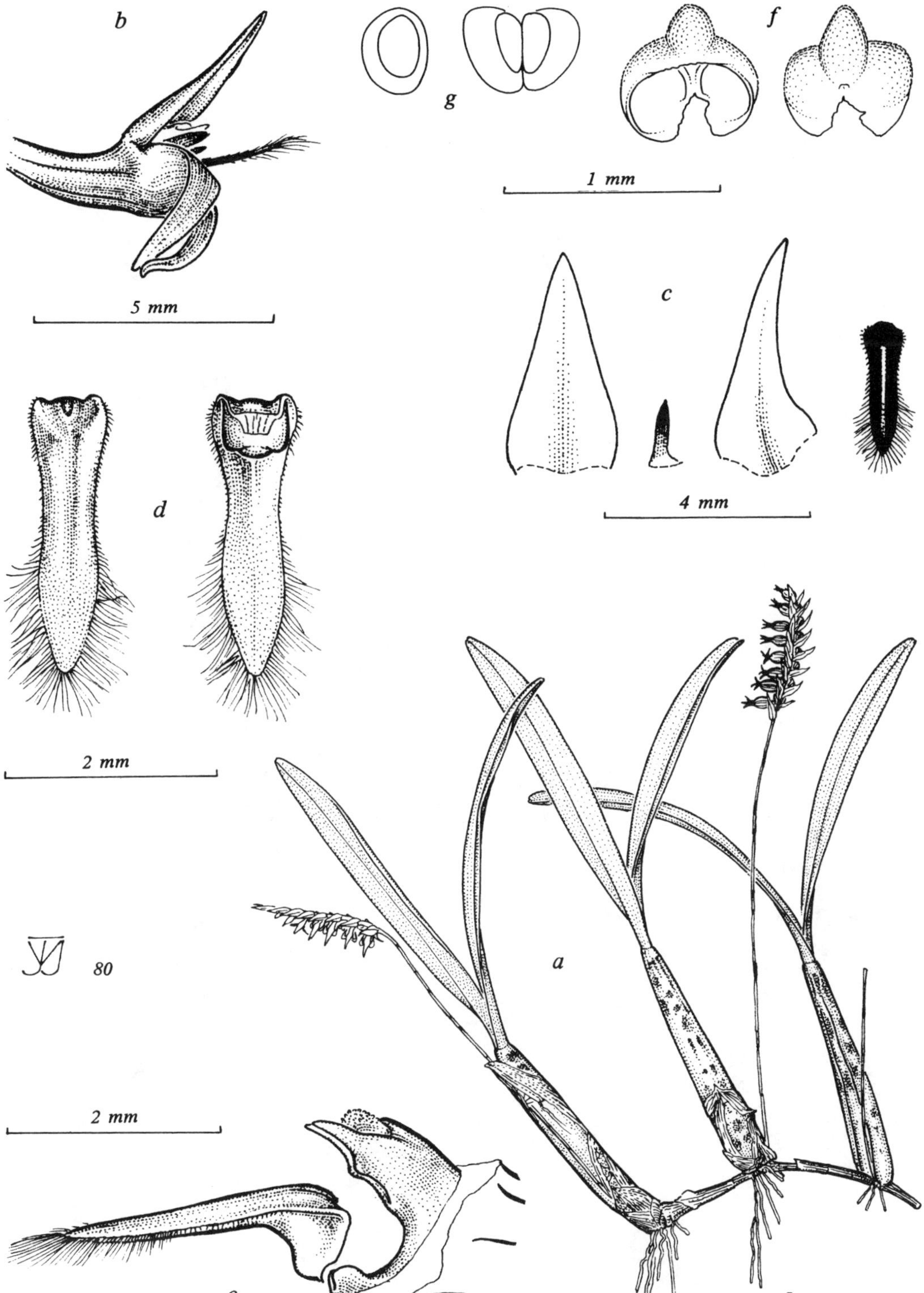

Fig. 12. *Bulbophyllum cochleatum* Lindl. var. *cochleatum*, entity a.

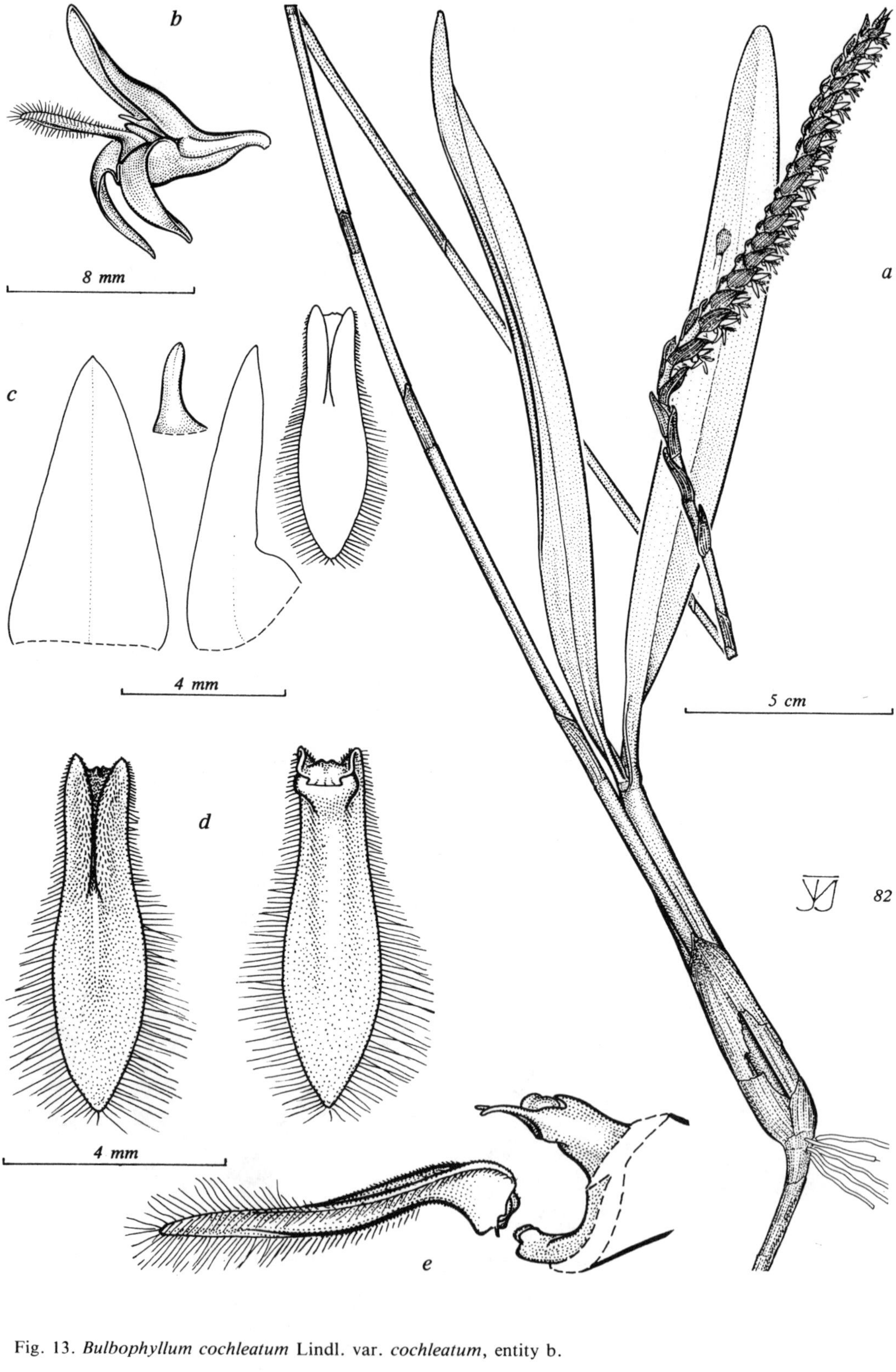

Fig. 13. *Bulbophyllum cochleatum* Lindl. var. *cochleatum*, entity b.

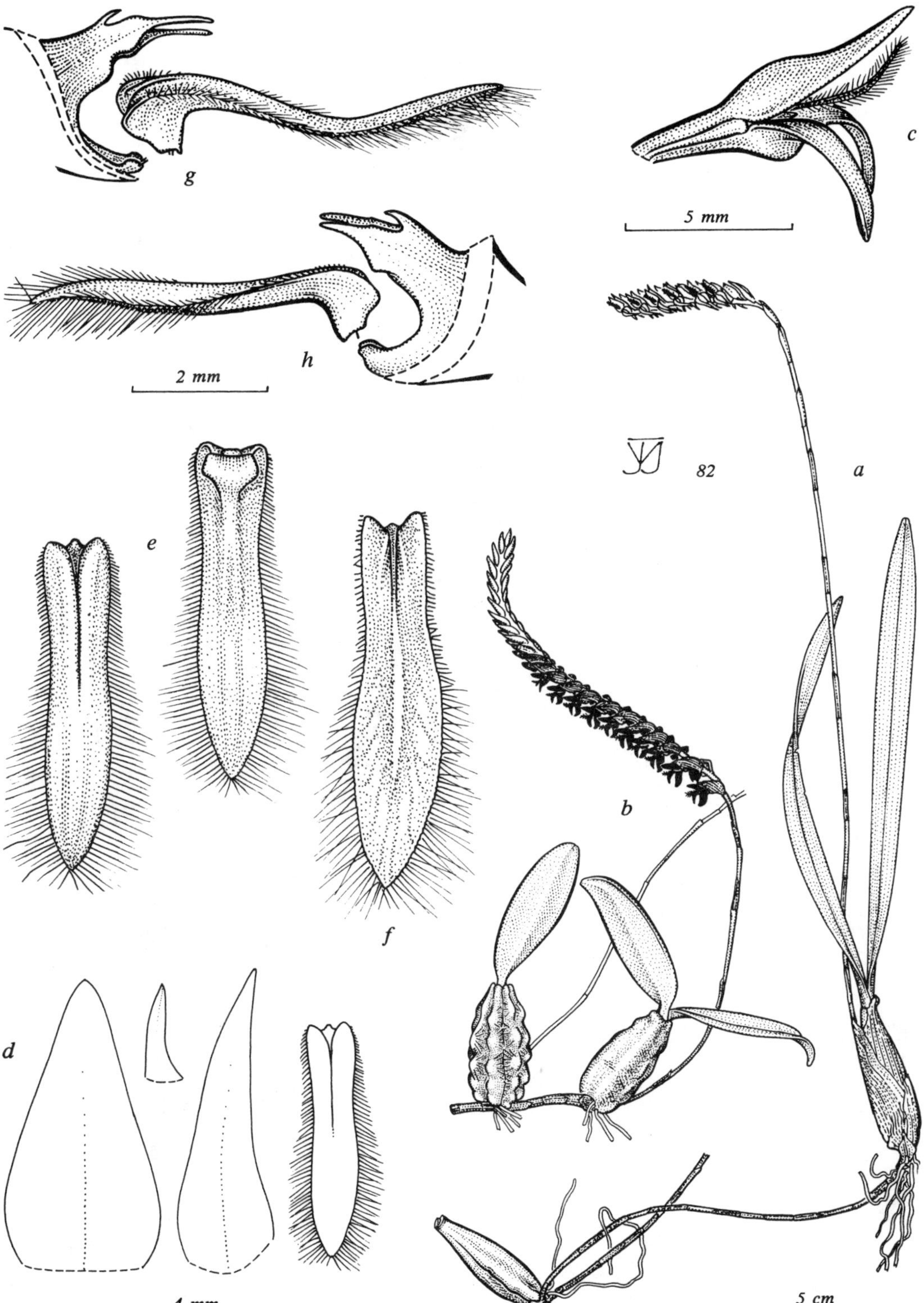

Fig. 14. *Bulbophyllum cochleatum* Lindl. var. *gravidum* (Lindl.) J.J. Vermeulen, entity c. (a, f, g) and d. (b–e, h).

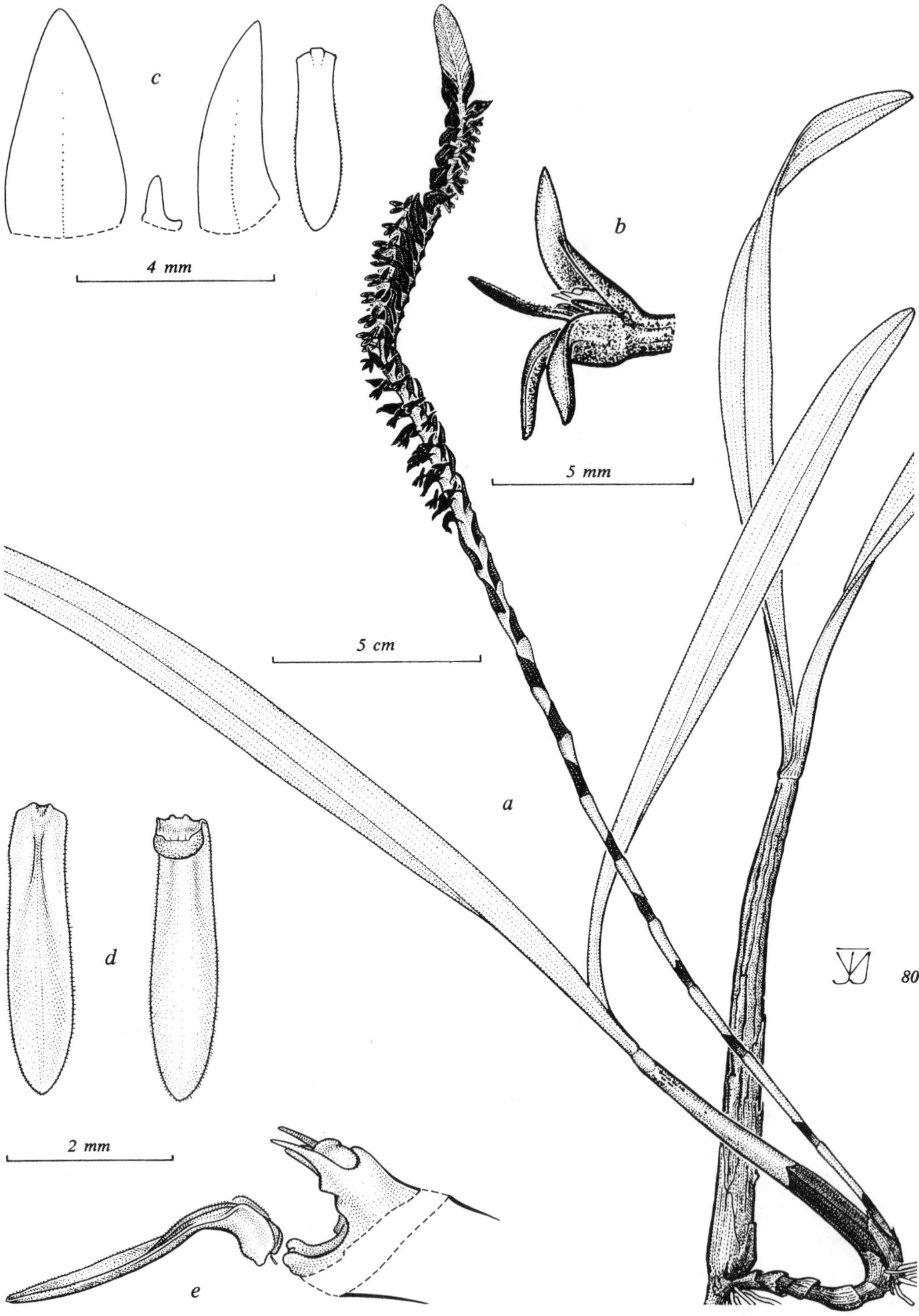

Fig. 15. *Bulbophyllum cochleatum* Lindl. var. *bequaertii* (De Wild.) J.J. Vermeulen.

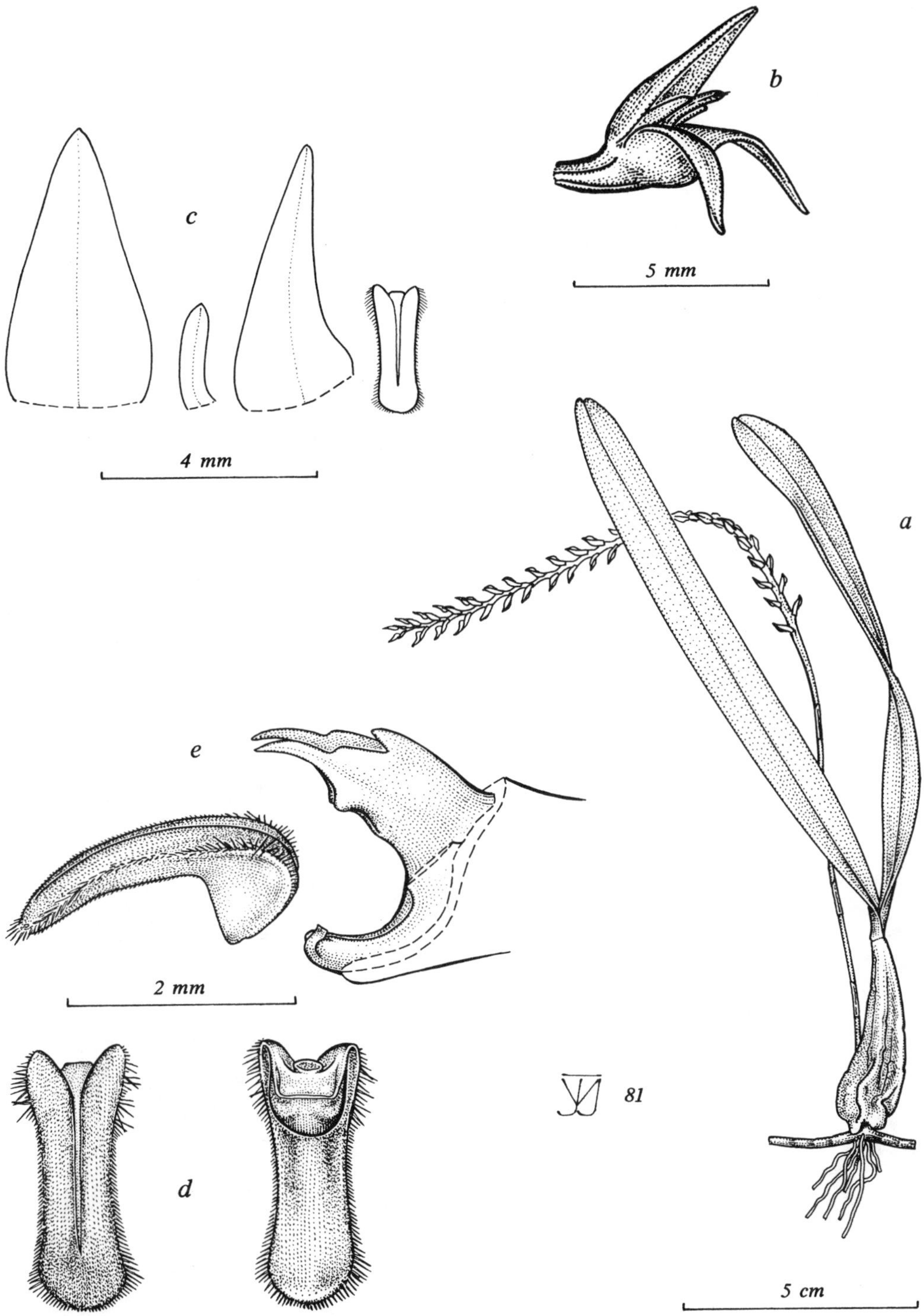

Fig. 16. *Bulbophyllum cochleatum* Lindl. var. *tenuicaule* (Lindl.) J.J. Vermeulen.

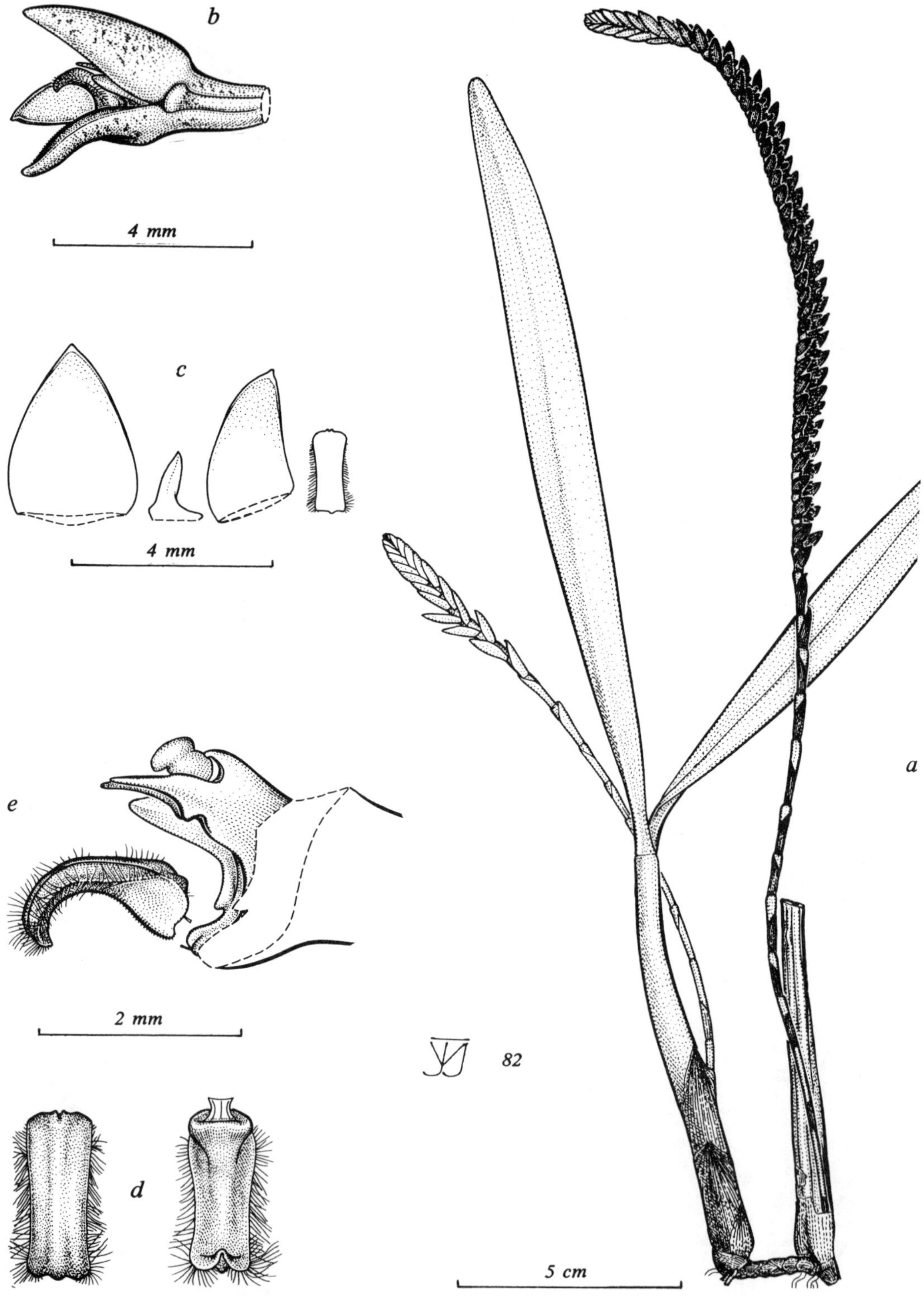

Fig. 17. *Bulbophyllum cochleatum* Lindl. var. *brachyanthum* (Lindl.) J.J. Vermeulen.

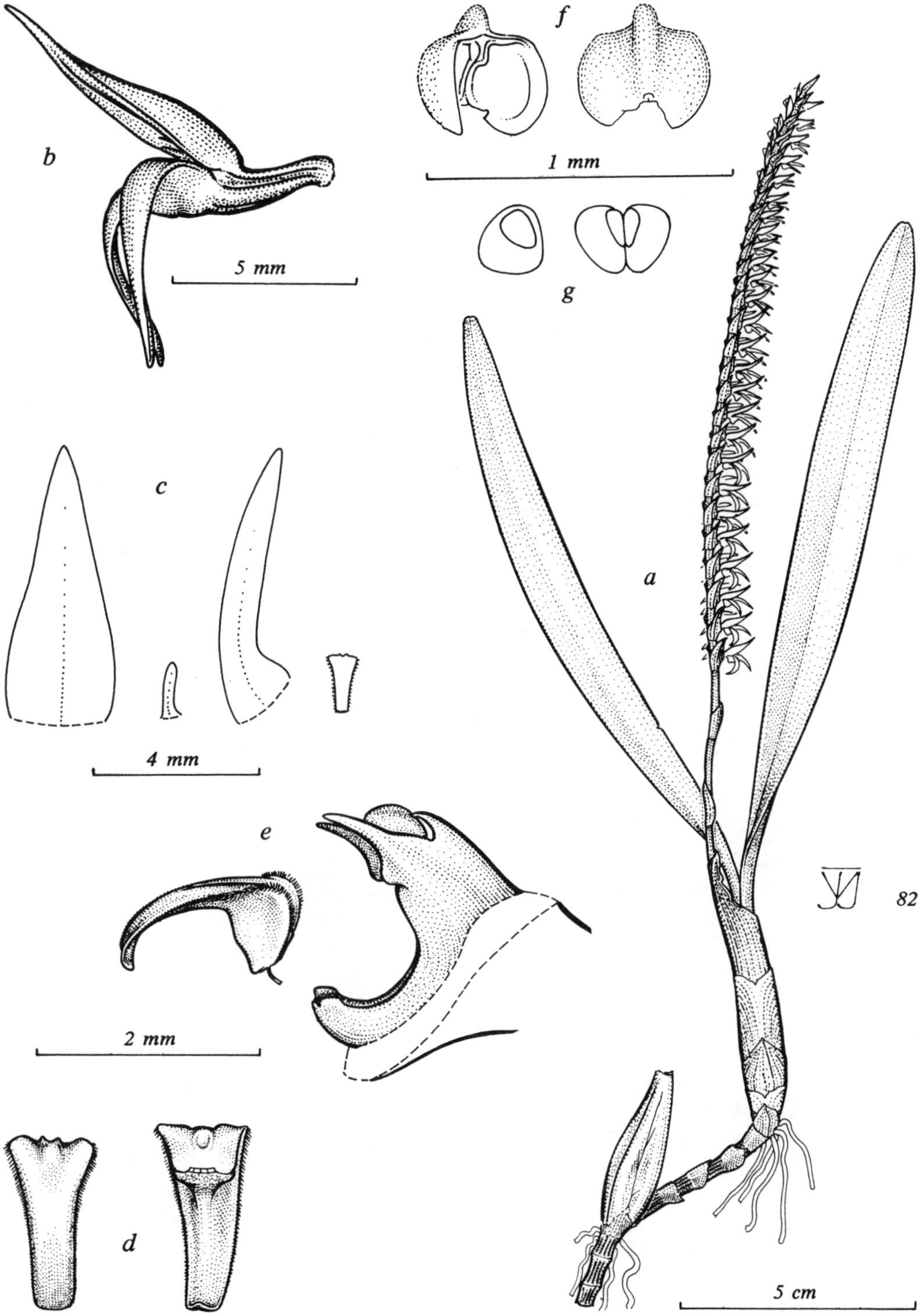

Fig. 18. *Bulbophyllum burttii* Summerh.

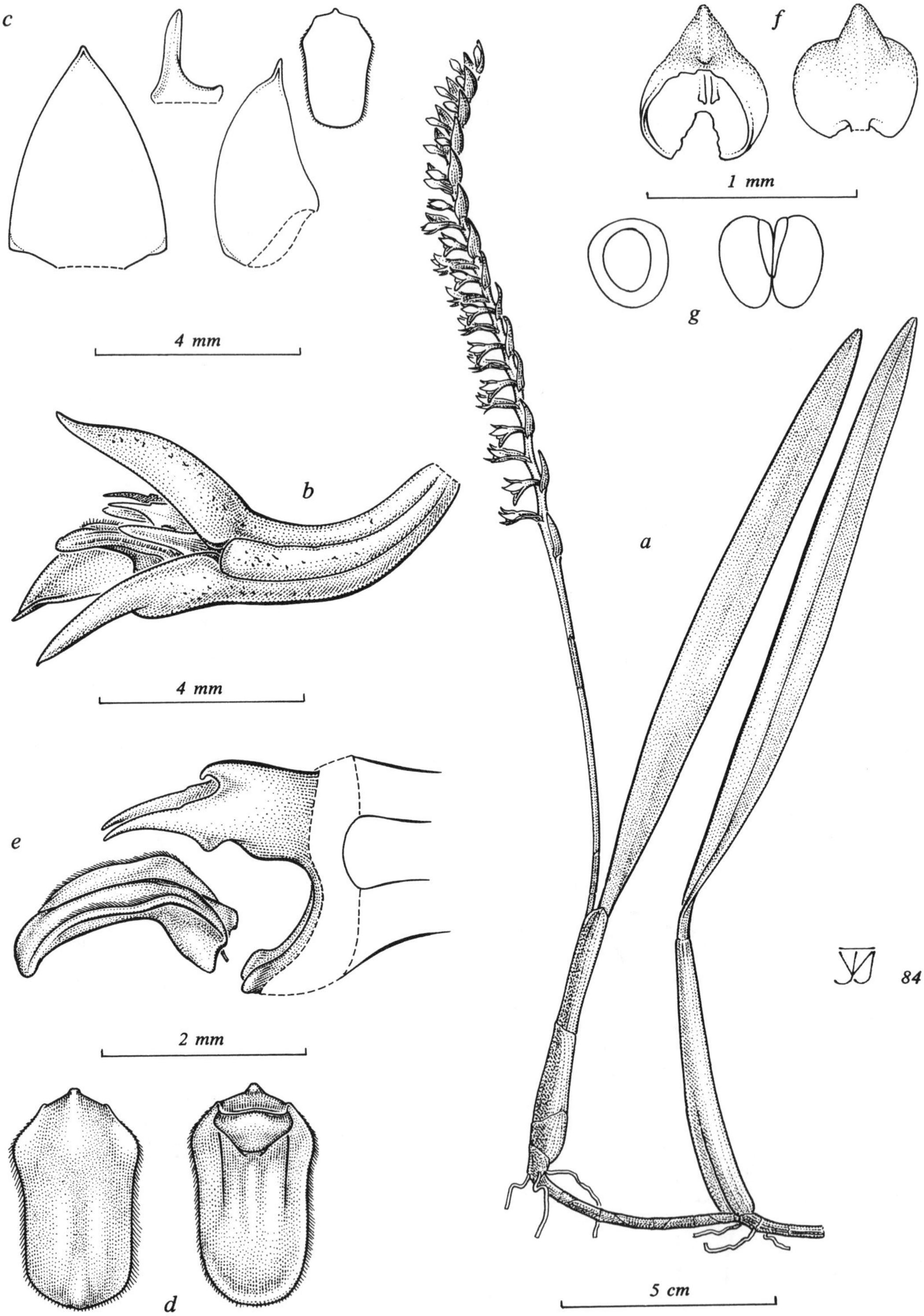

Fig. 19. *Bulbophyllum kivuense* J.J. Vermeulen.

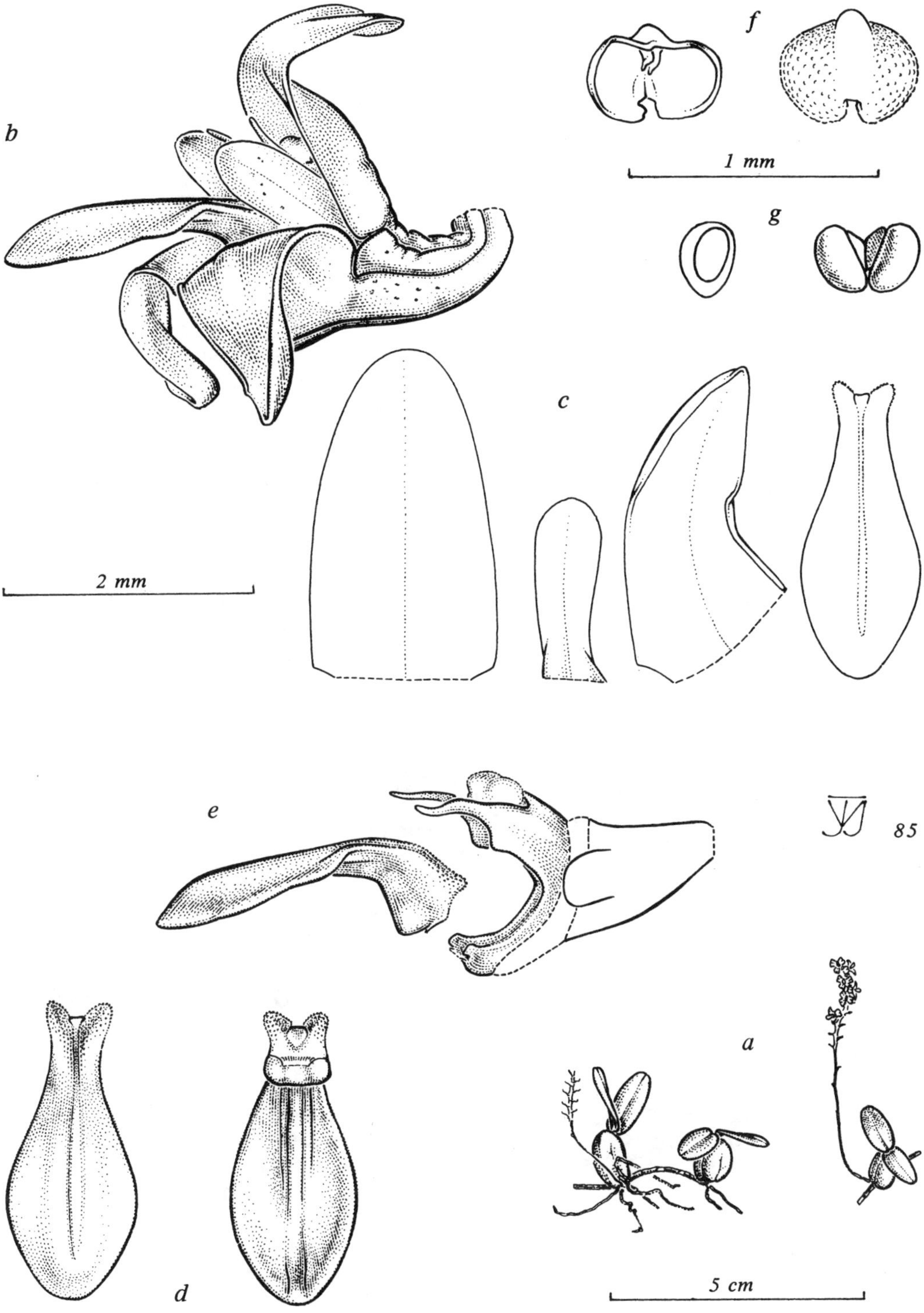

Fig. 20. *Bulbophyllum inornatum* J.J. Vermeulen.

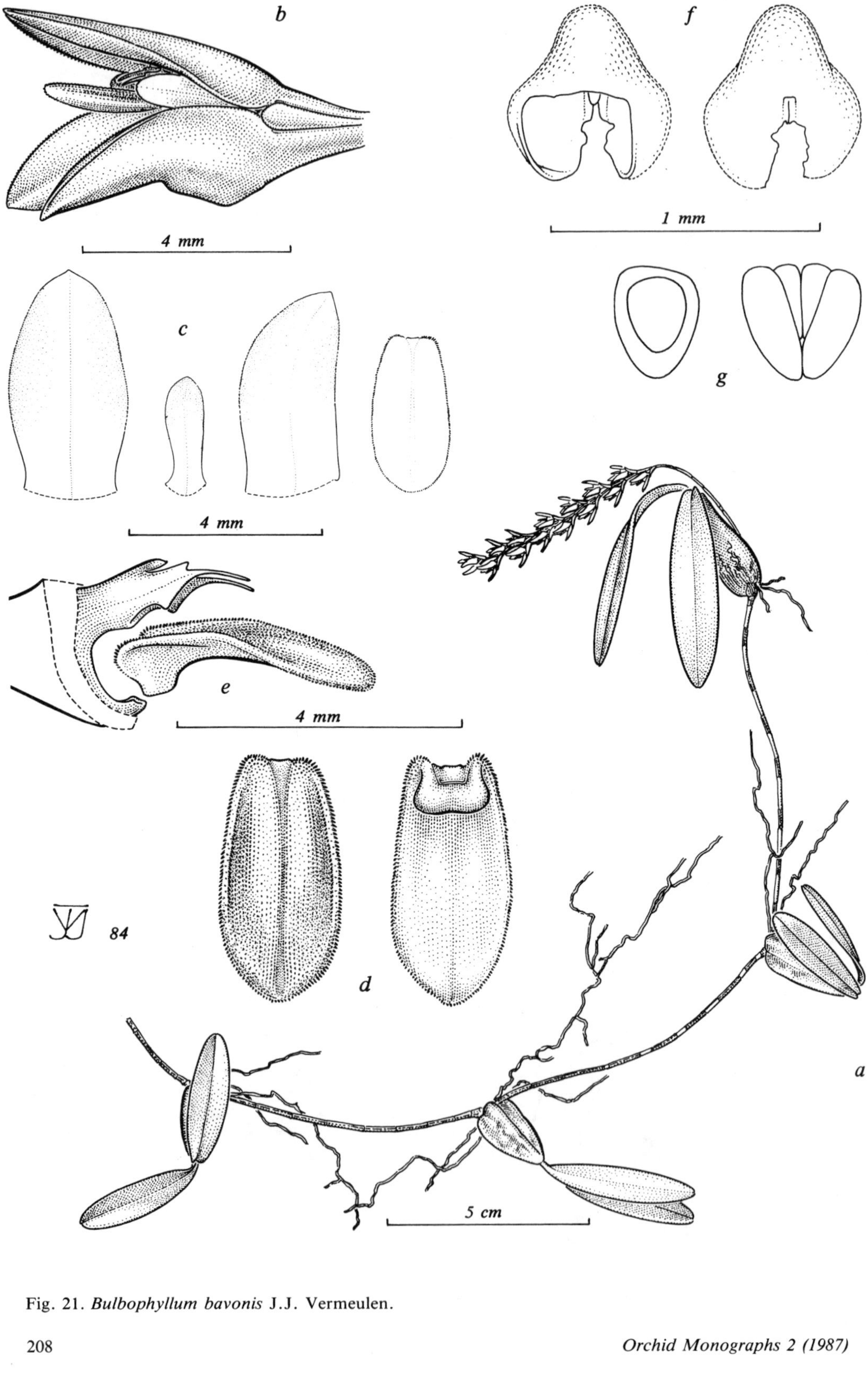

Fig. 21. *Bulbophyllum bavonis* J.J. Vermeulen.

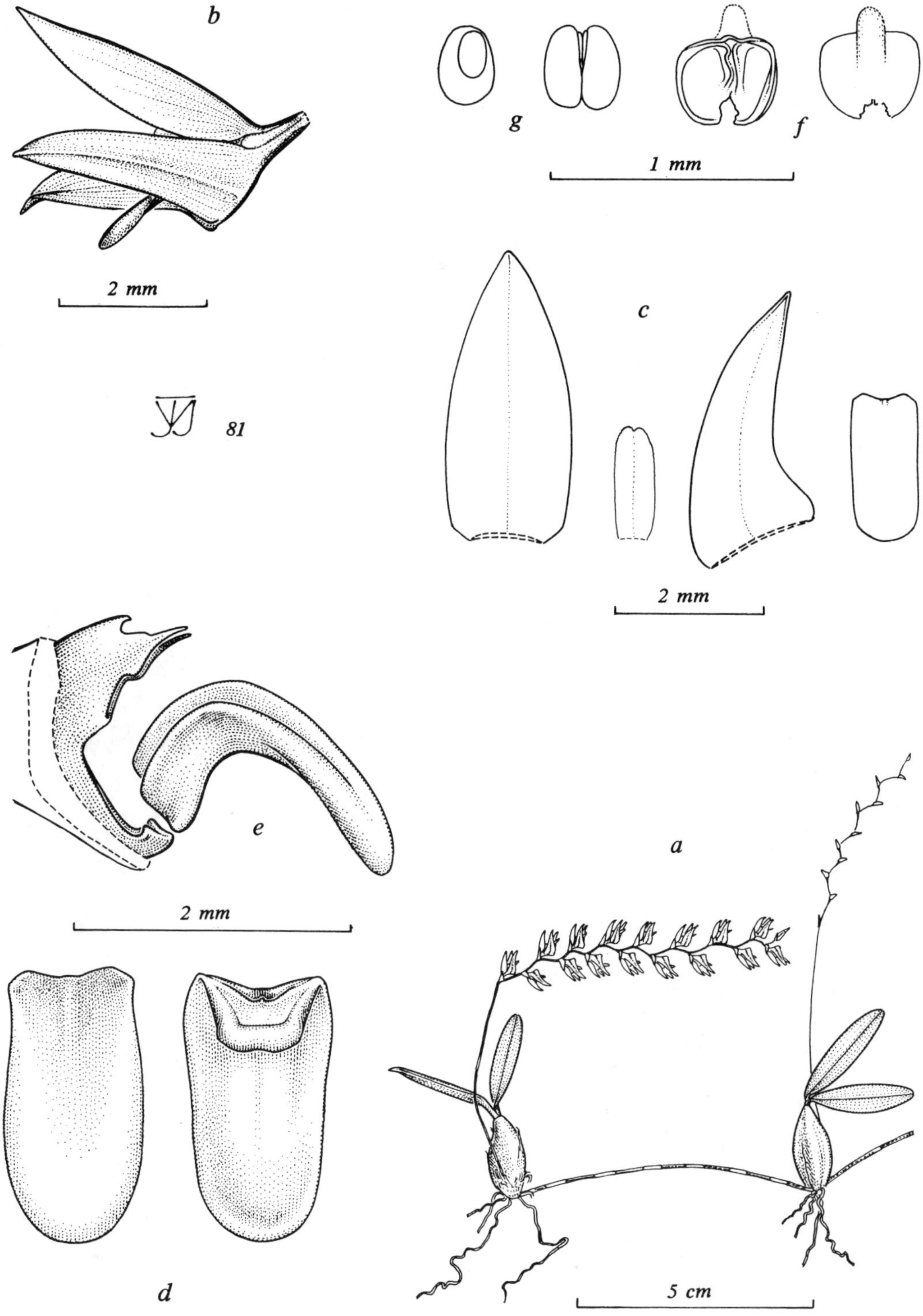

Fig. 22. *Bulbophyllum stolzii* Schltr.

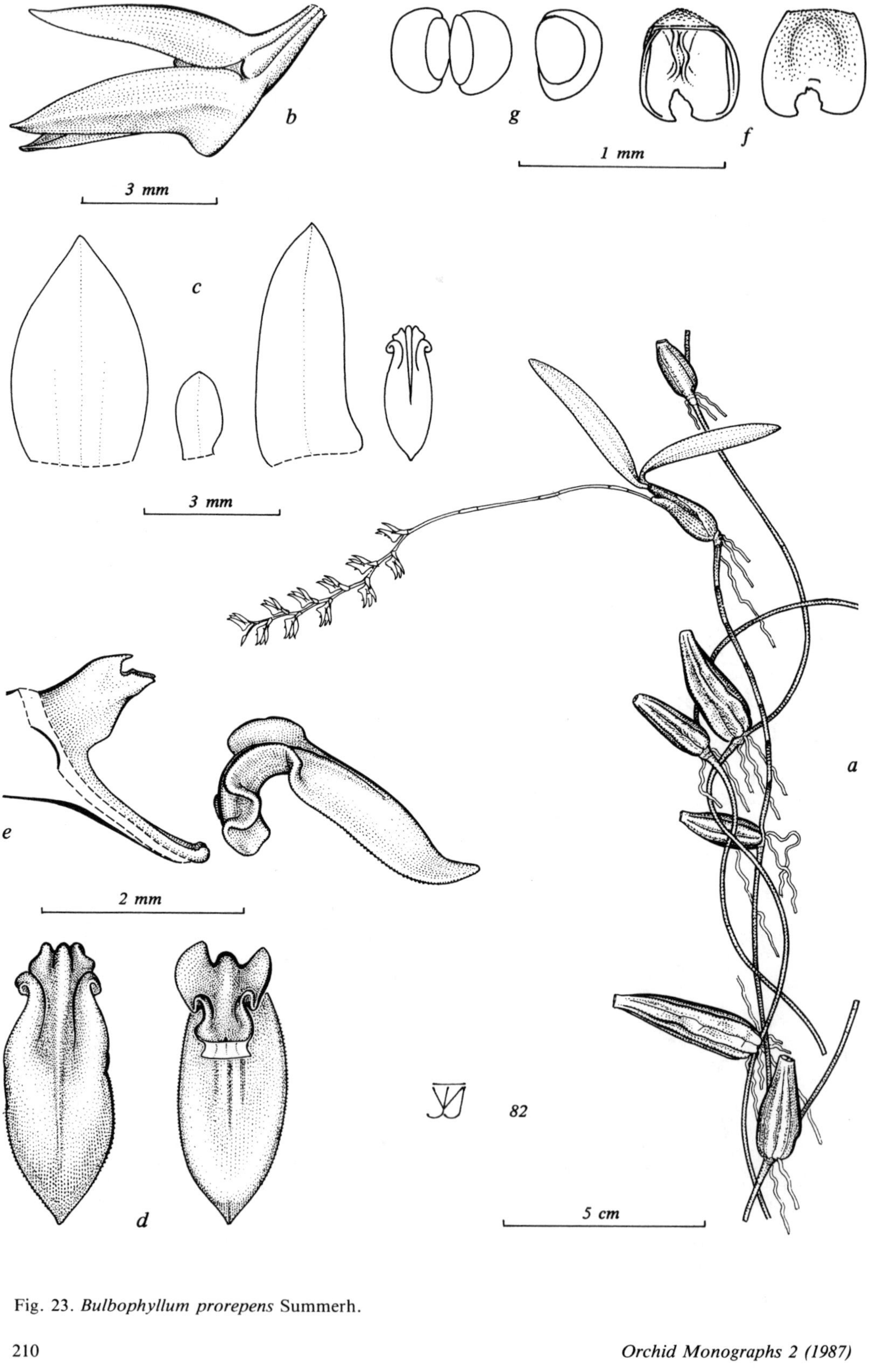

Fig. 23. *Bulbophyllum prorepens* Summerh.

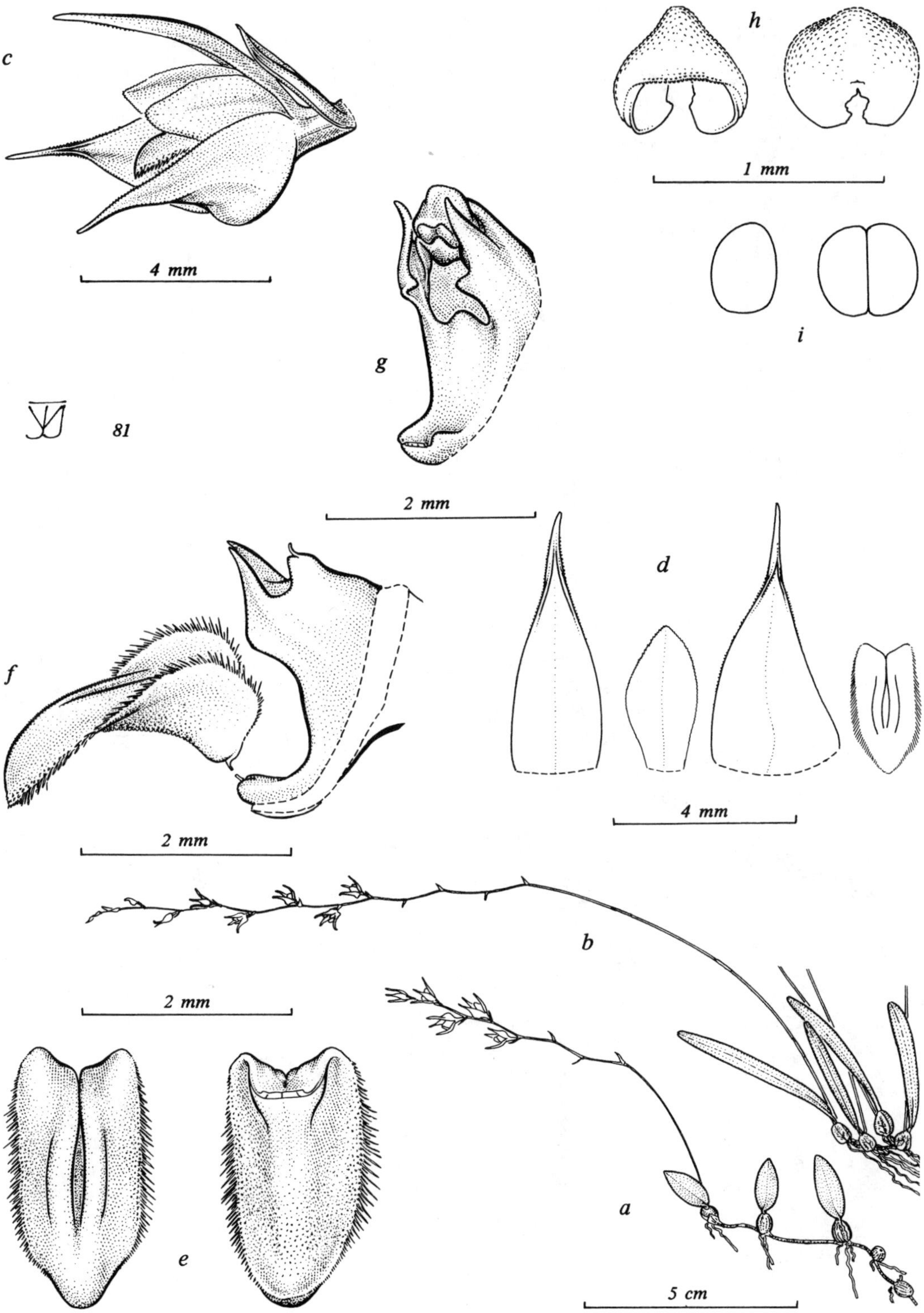

Fig. 24. *Bulbophyllum intertextum* Lindl.

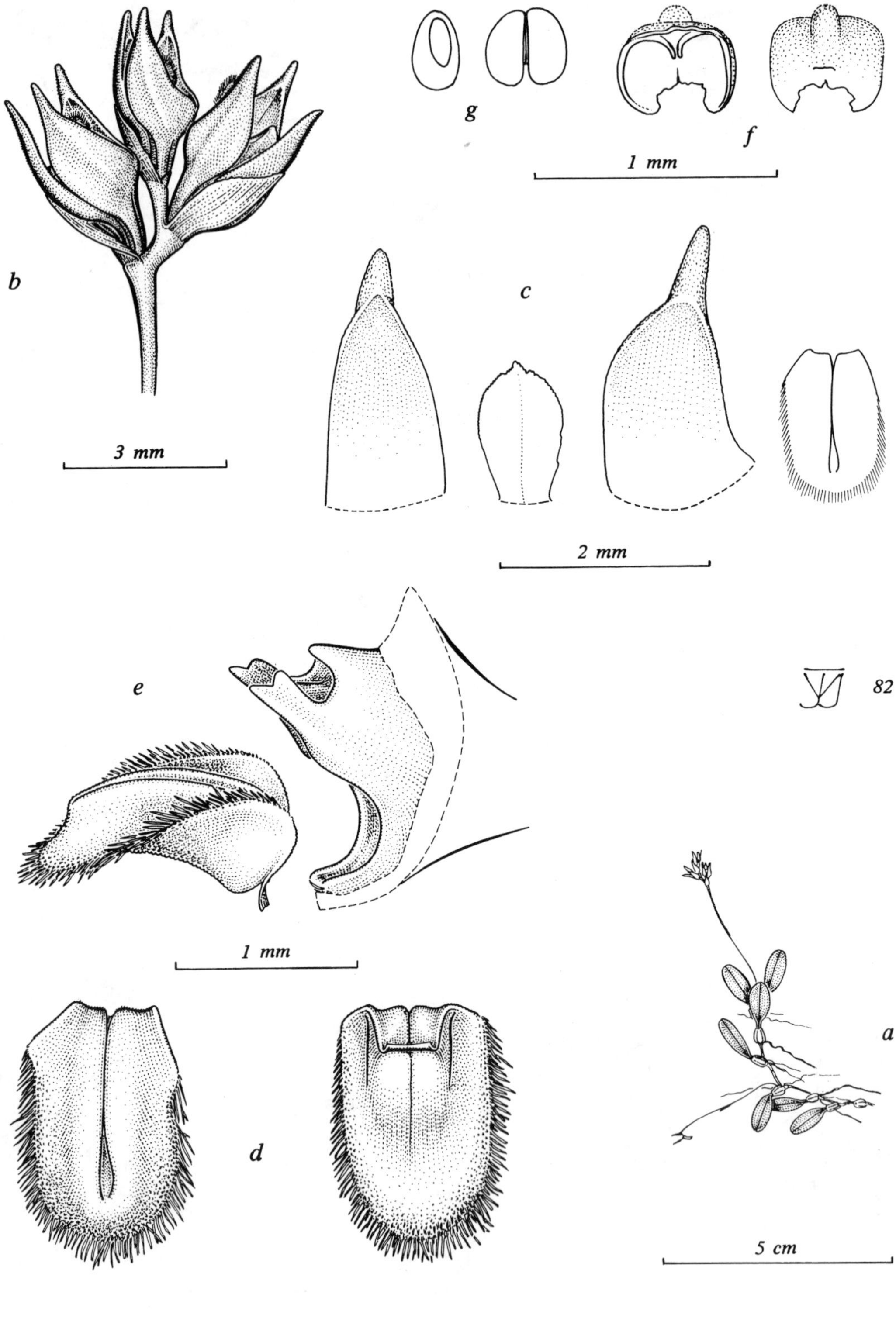

Fig. 25. *Bulbophyllum vanum* J.J. Vermeulen.

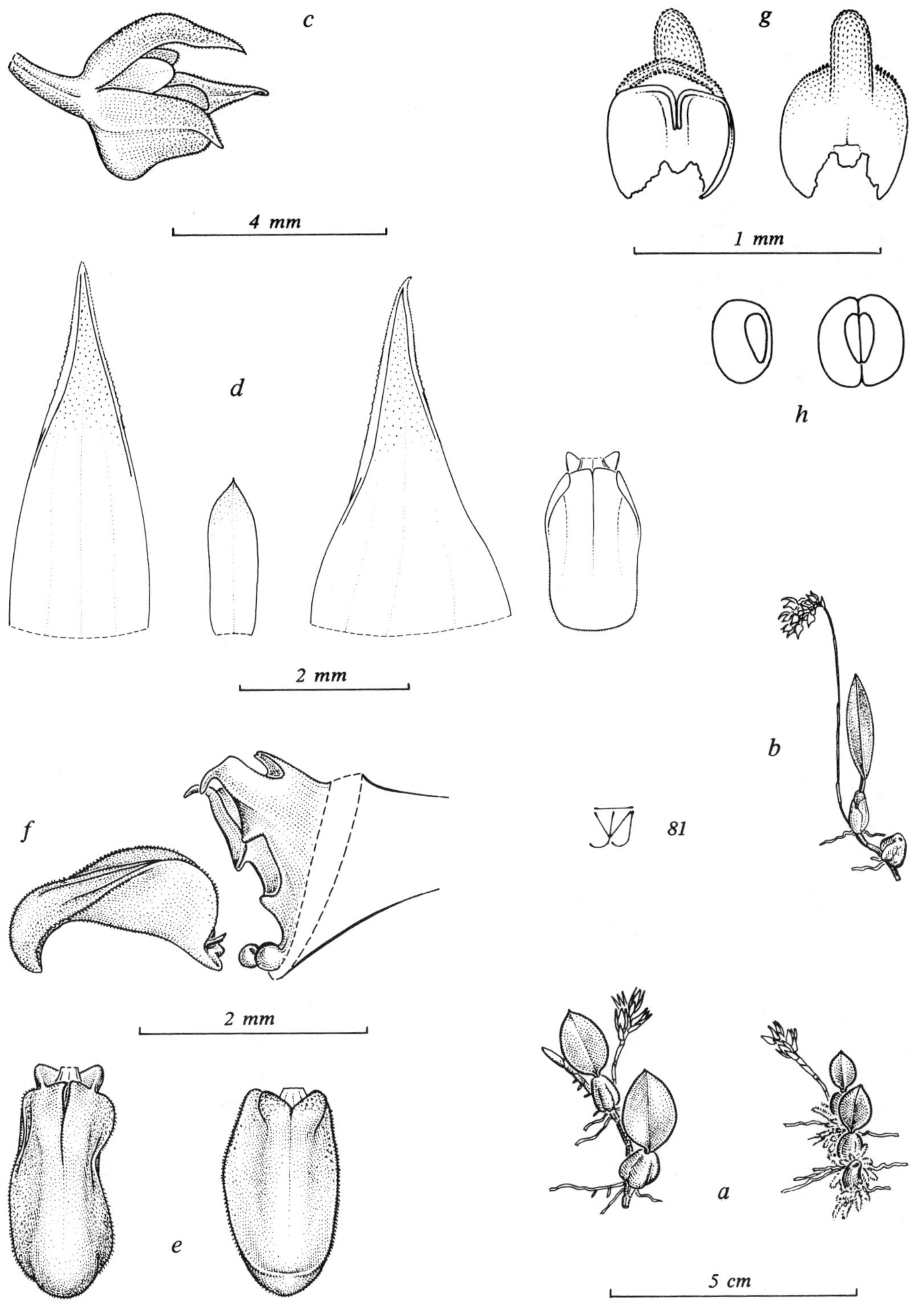

Fig. 26. *Bulbophyllum capituliflorum* Rolfe.

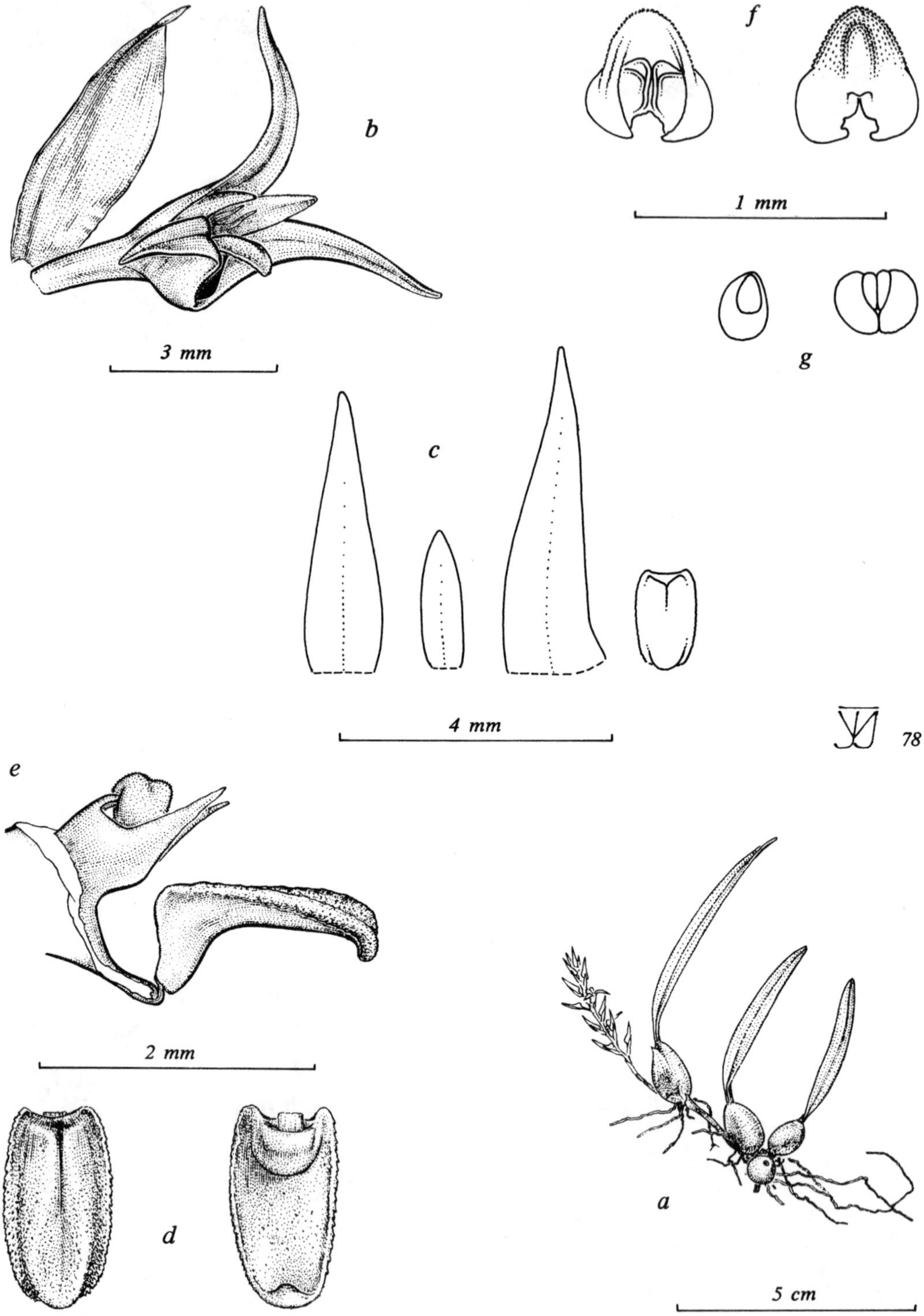

Fig. 27. *Bulbophyllum pipio* Rchb. f.

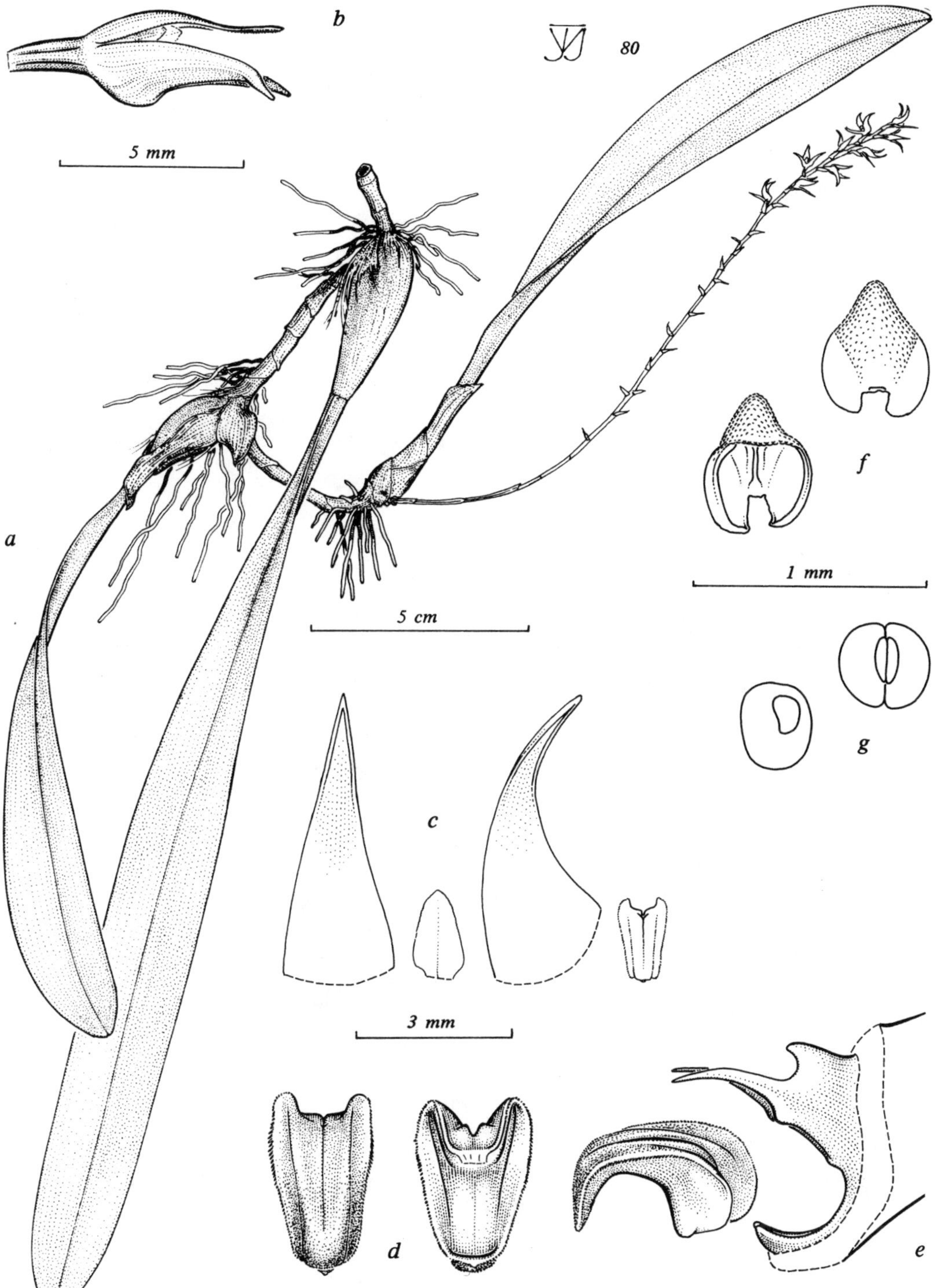

Fig. 28. *Bulbophyllum pumilum* (Sw.) Lindl.; entity a.

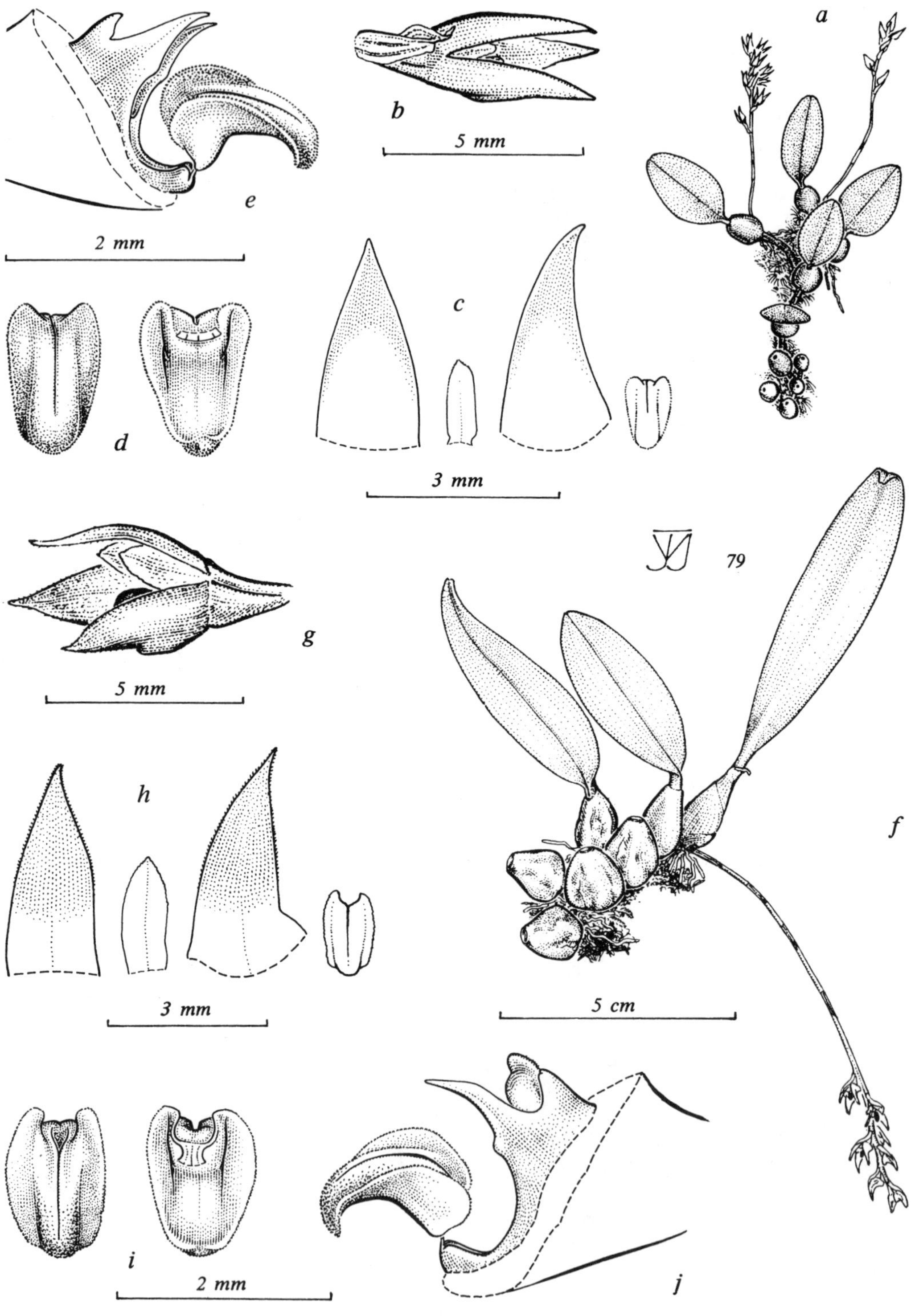

Fig. 29. *Bulbophyllum pumilum* (Sw.) Lindl., entity b. (a–e) and c. (f–j).

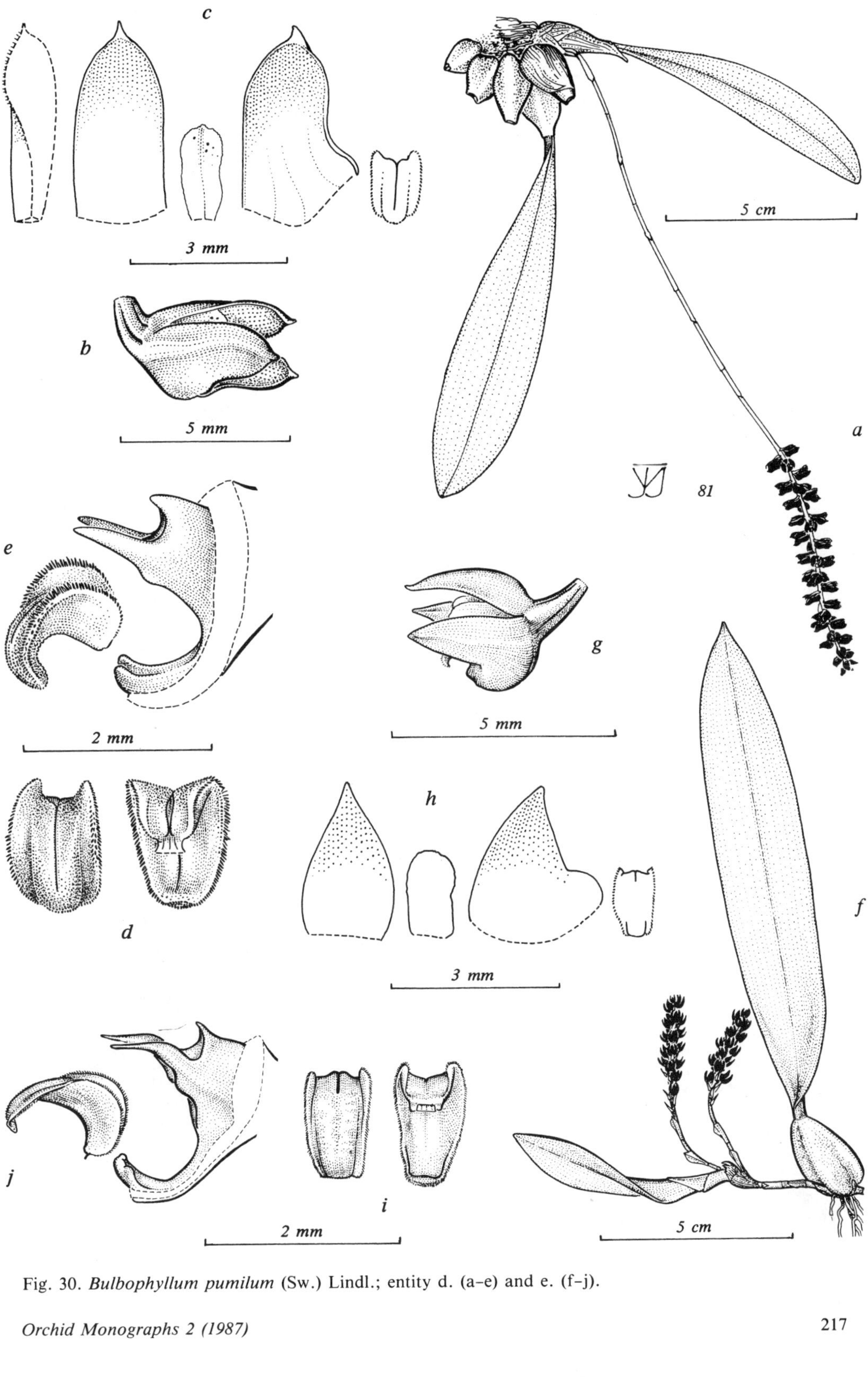

Fig. 30. *Bulbophyllum pumilum* (Sw.) Lindl.; entity d. (a–e) and e. (f–j).

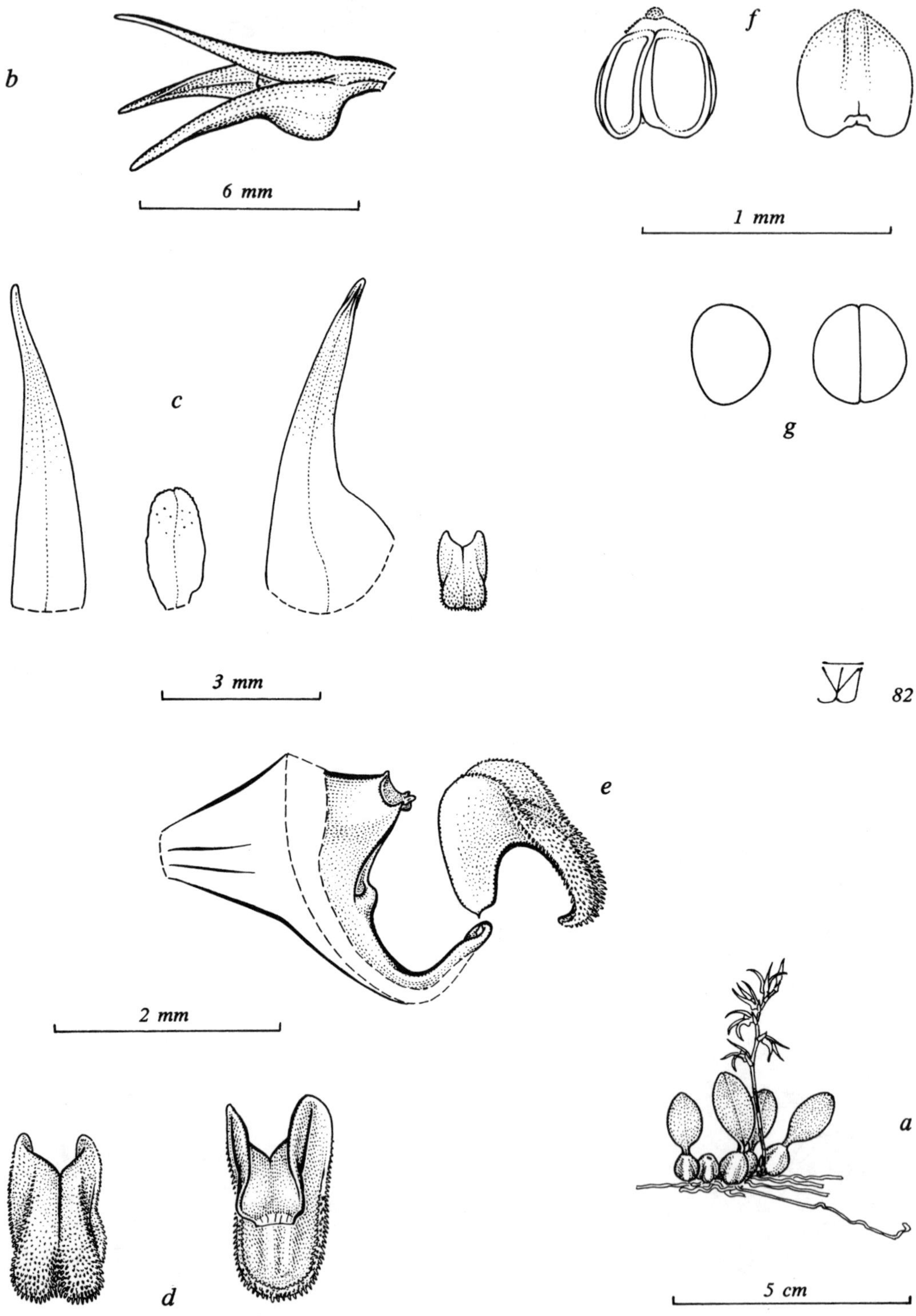

Fig. 31. *Bulbophyllum coriscense* Rchb. f.

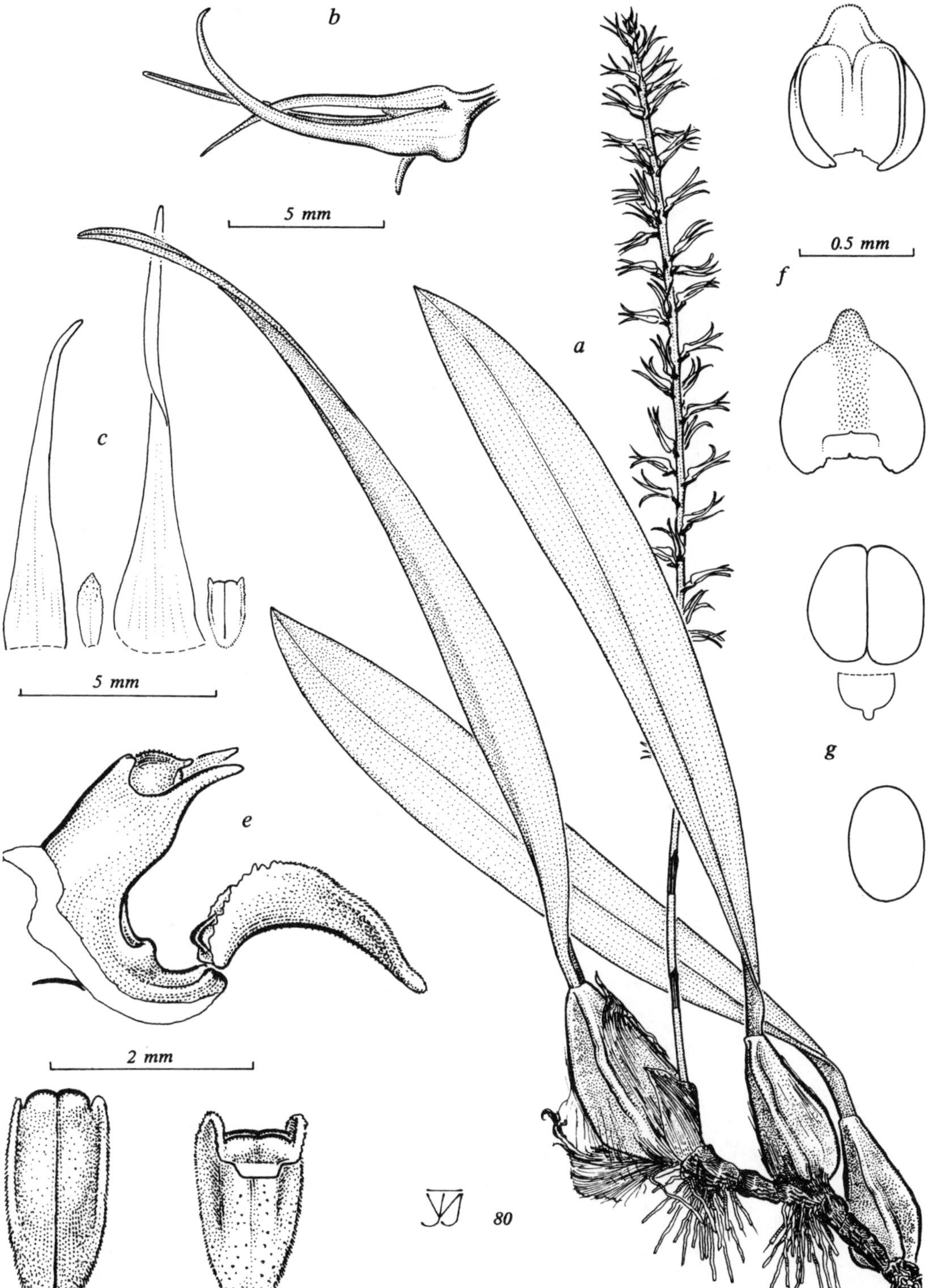

Fig. 32. *Bulbophyllum cocoinum* Lindl.

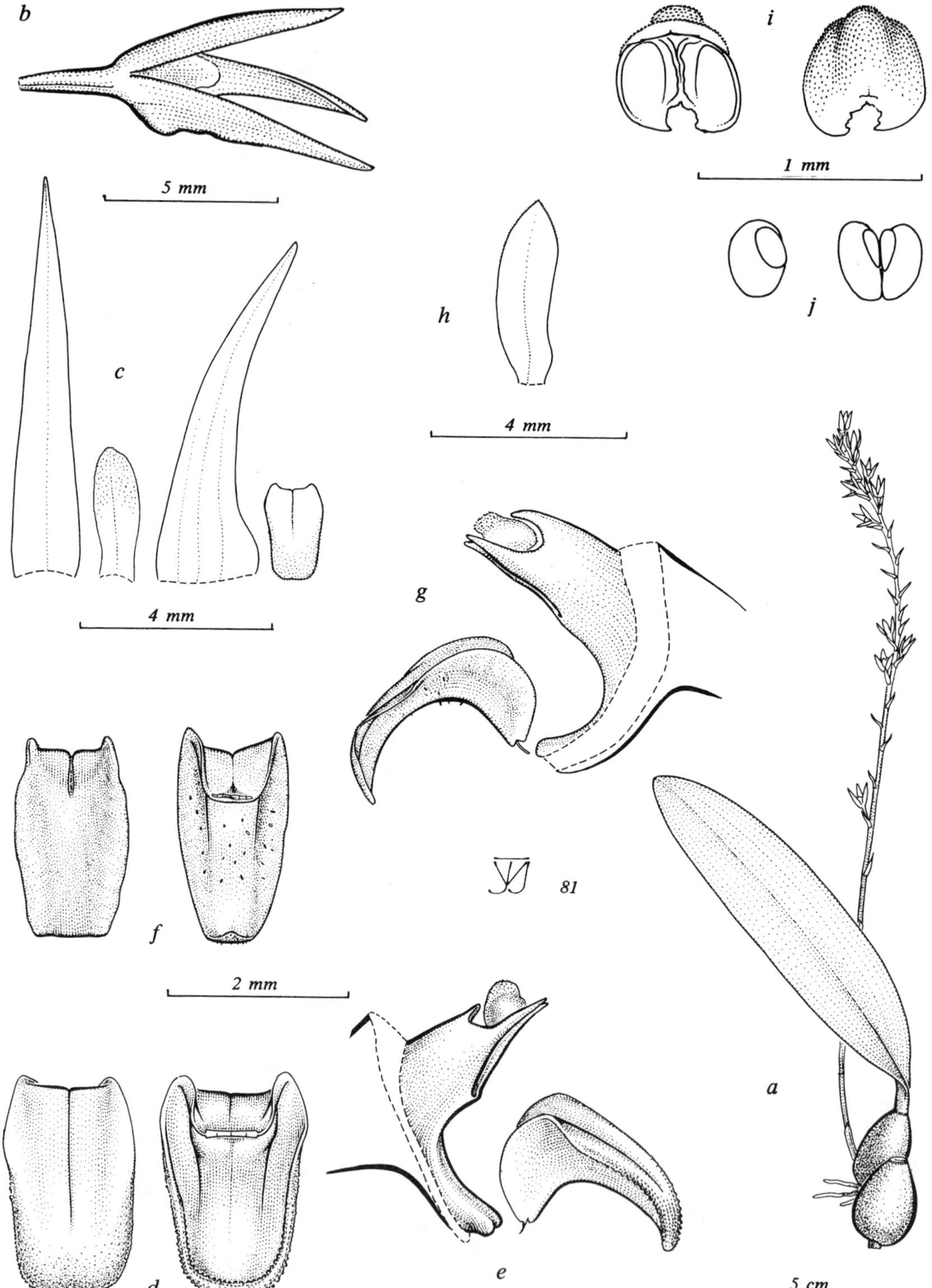

Fig. 33. *Bulbophyllum josephii* (Kuntze) Summerh. var. *josephii*; E. African specimen (a–e, i, j) and Cameroun specimen (f–h).

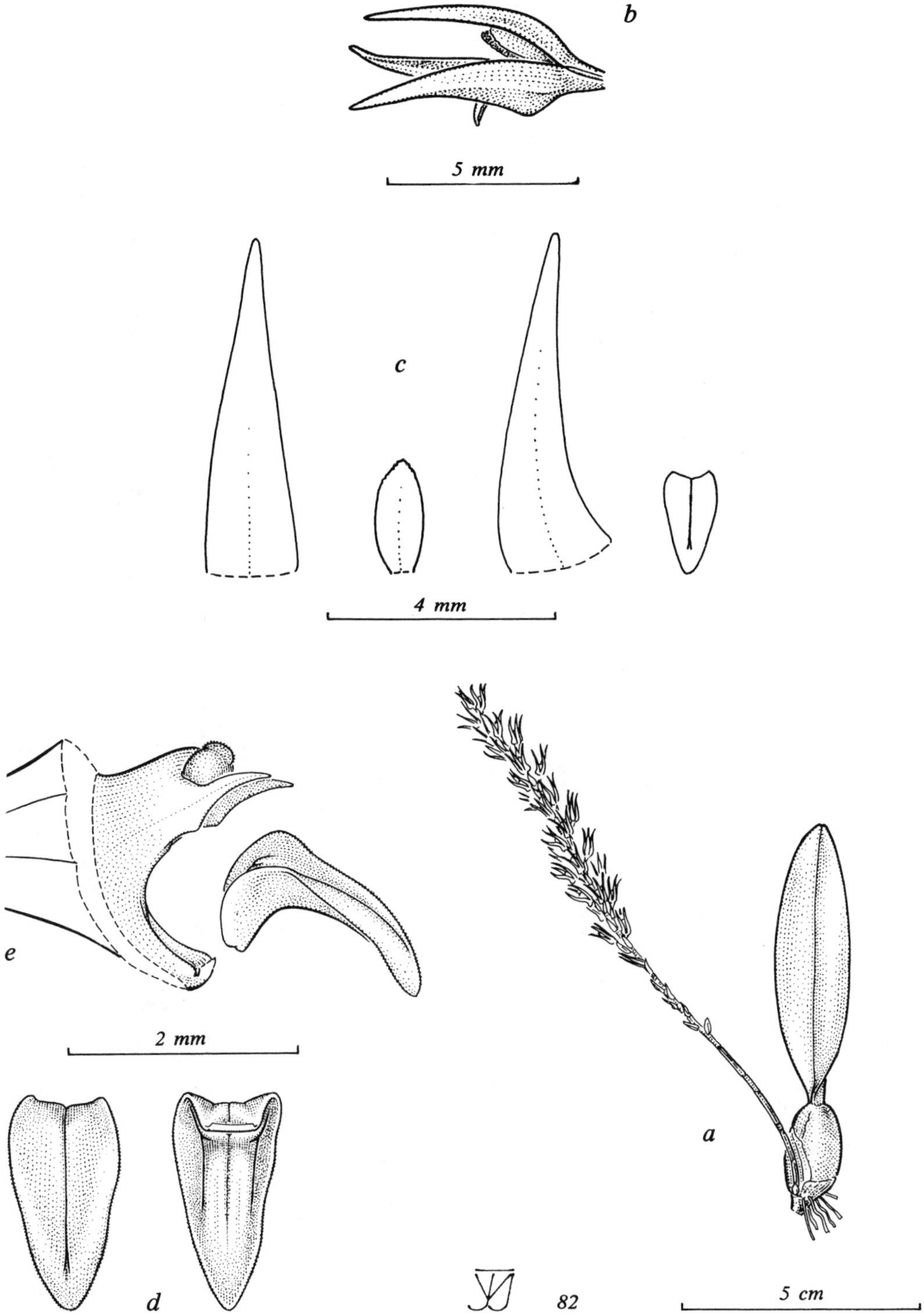

Fig. 34. *Bulbophyllum josephii* (Kuntze) Summerh. var. *mahonii* (Rolfe) J.J. Vermeulen.

Fig. 35. *Bulbophyllum josephii* (Kuntze) Summerh. var. *mahonii* (Rolfe) J.J. Vermeulen; W. African specimen (a, c, e, f) and Cameroun specimen (b, d, g, h).

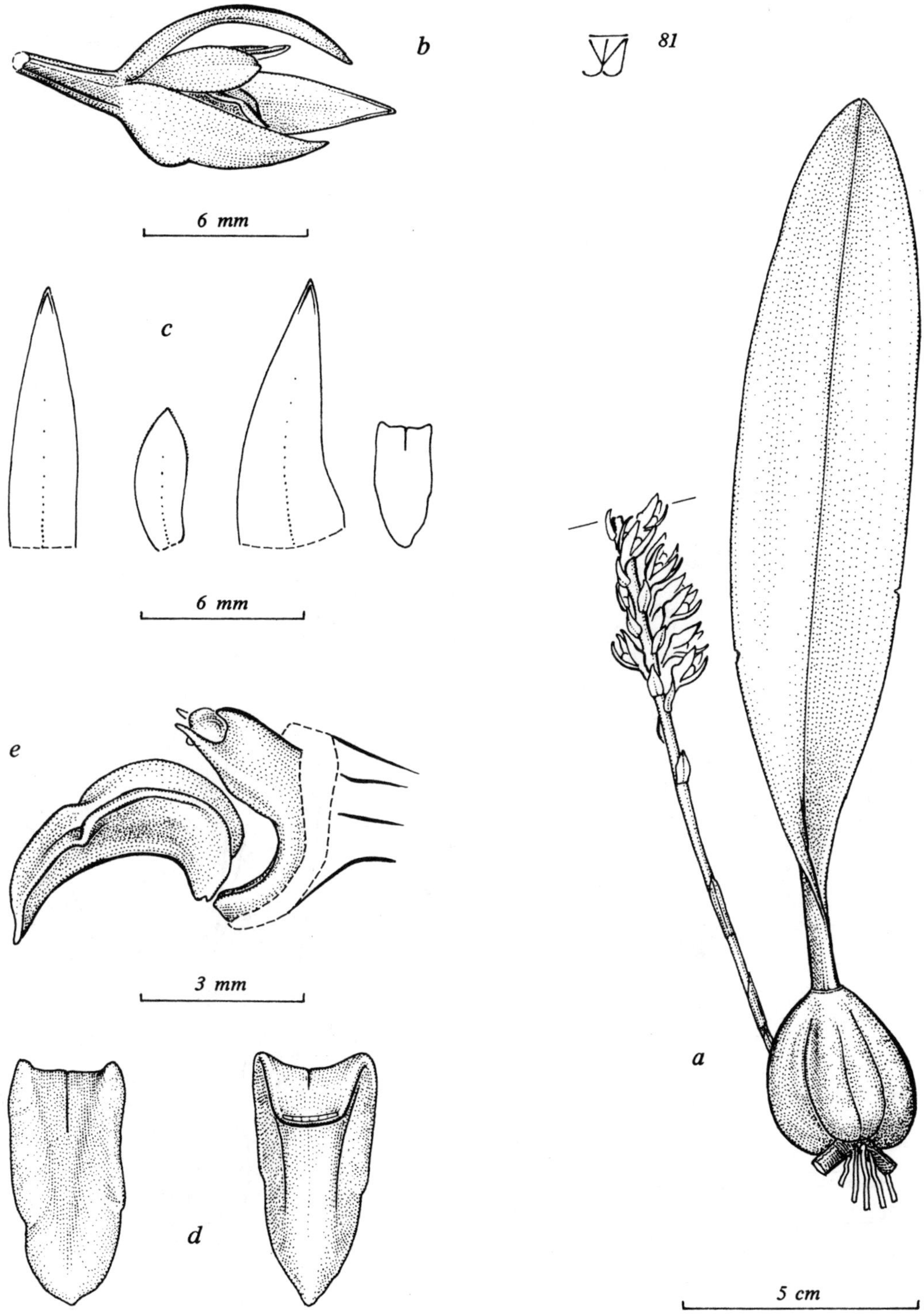

Fig. 36. *Bulbophyllum calvum* Summerh.

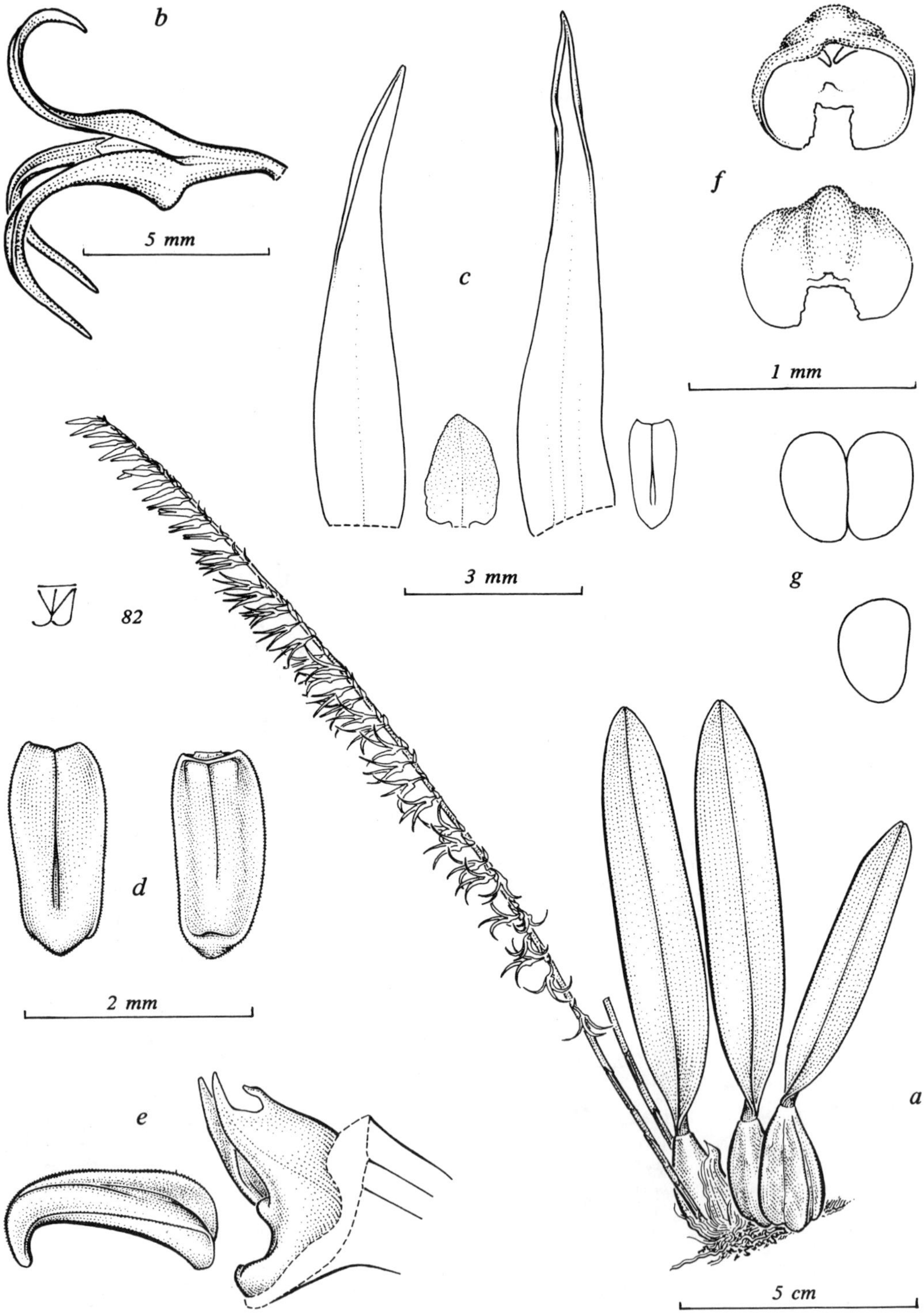

Fig. 37. *Bulbophyllum expallidum* J.J. Vermeulen.

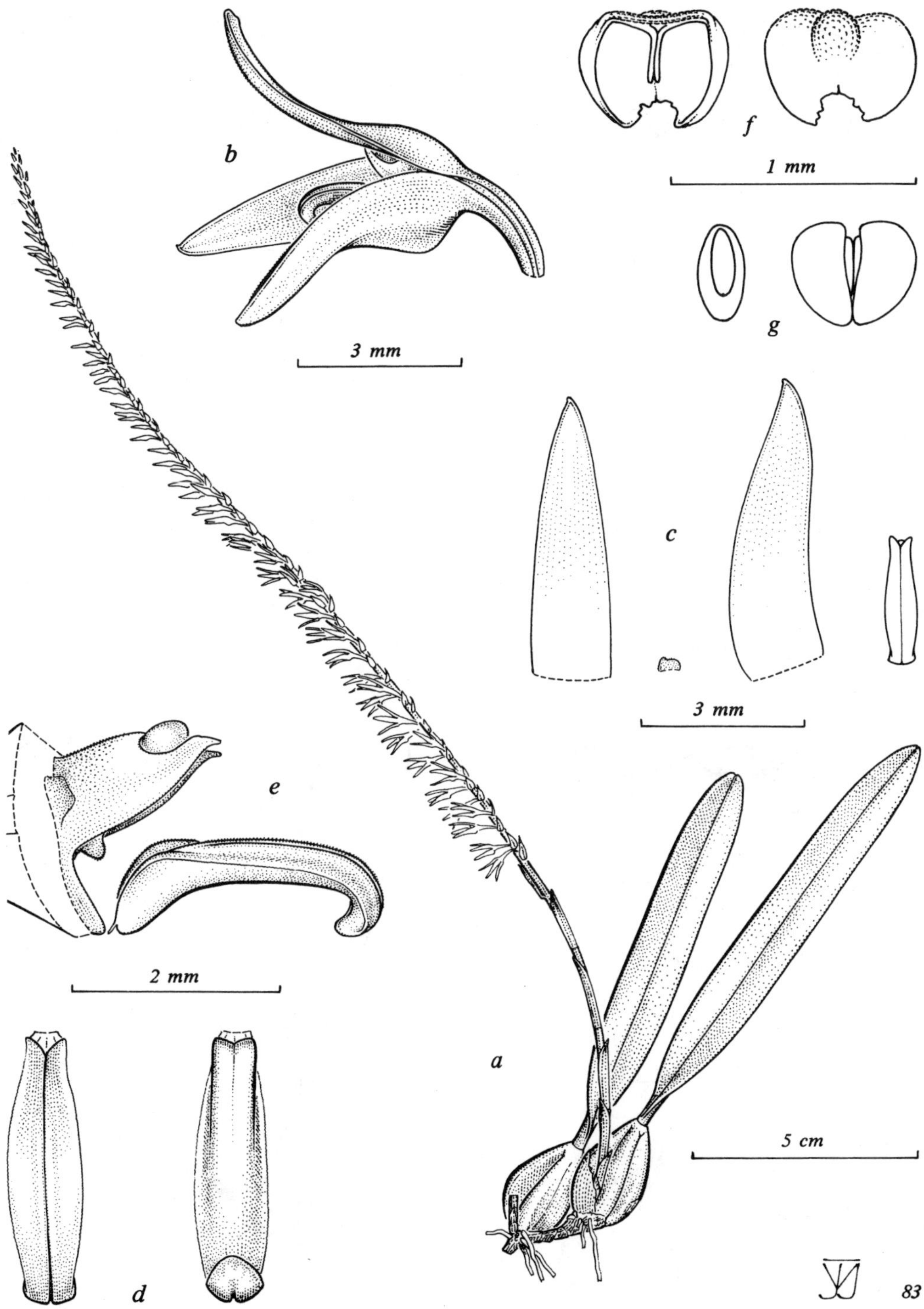

Fig. 38. *Bulbophyllum ivorense* Cribb & Perez-Vera.

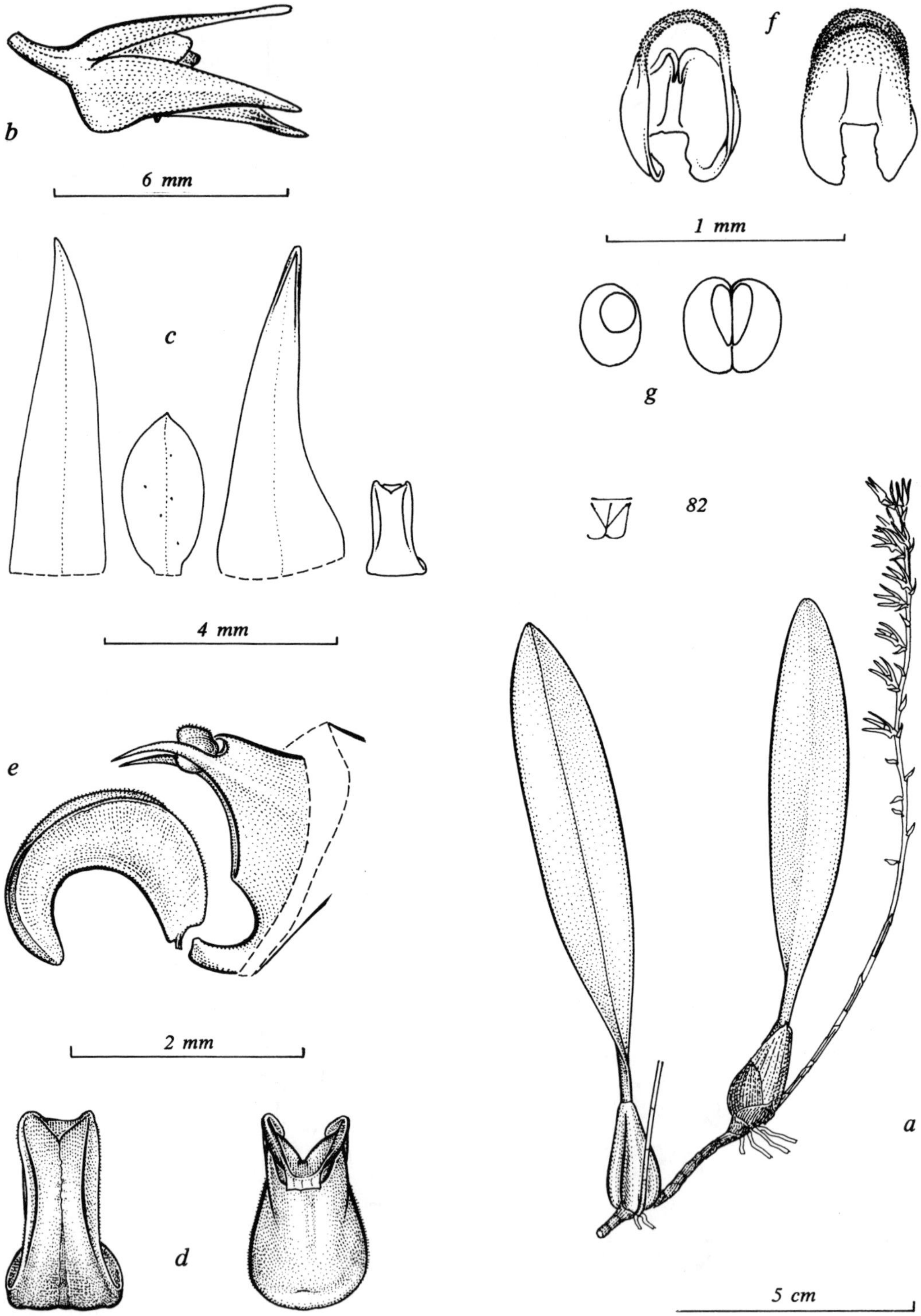

Fig. 39. *Bulbophyllum mediocre* Summerh.

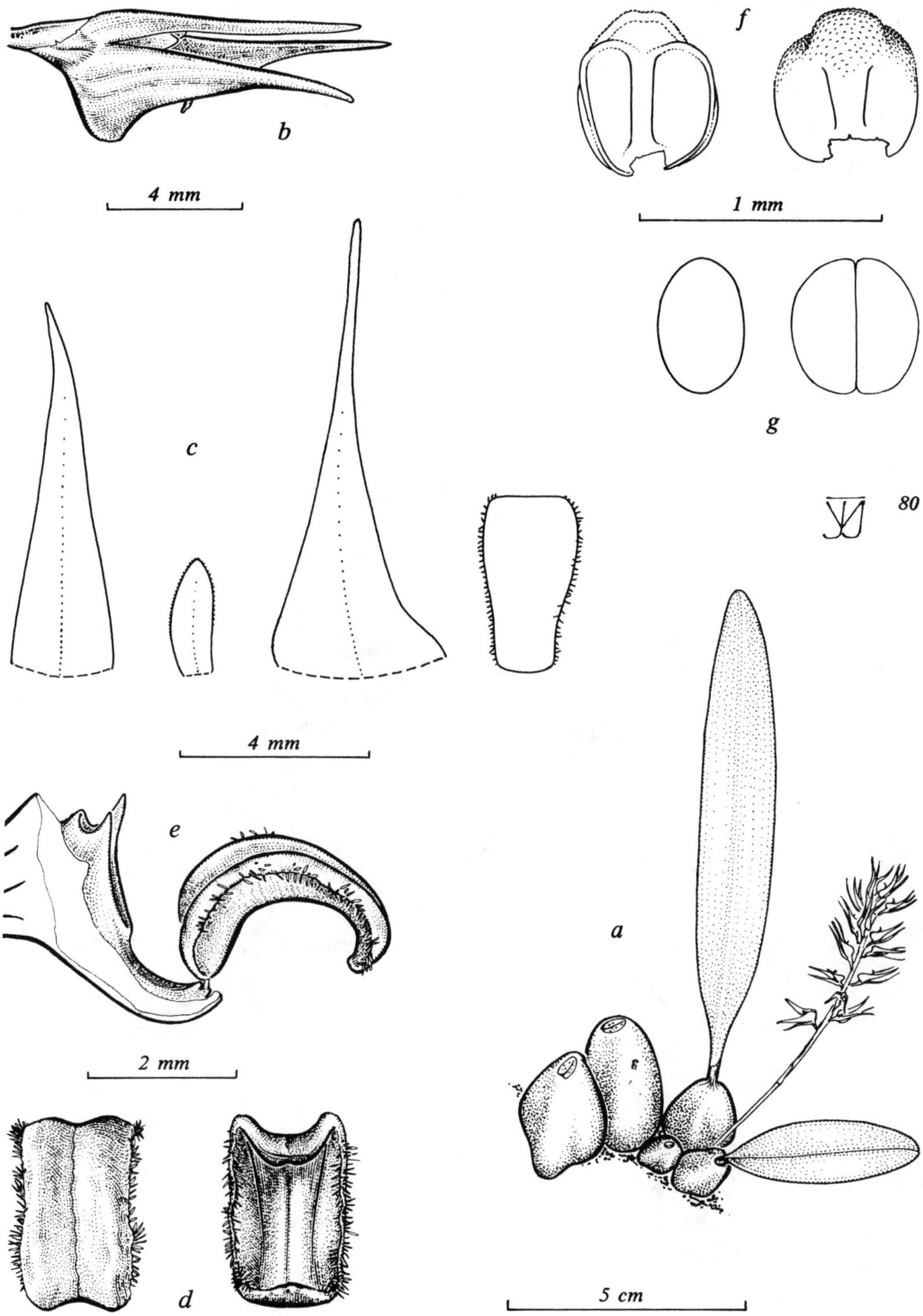

Fig. 40. *Bulbophyllum schimperanum* Krzl.

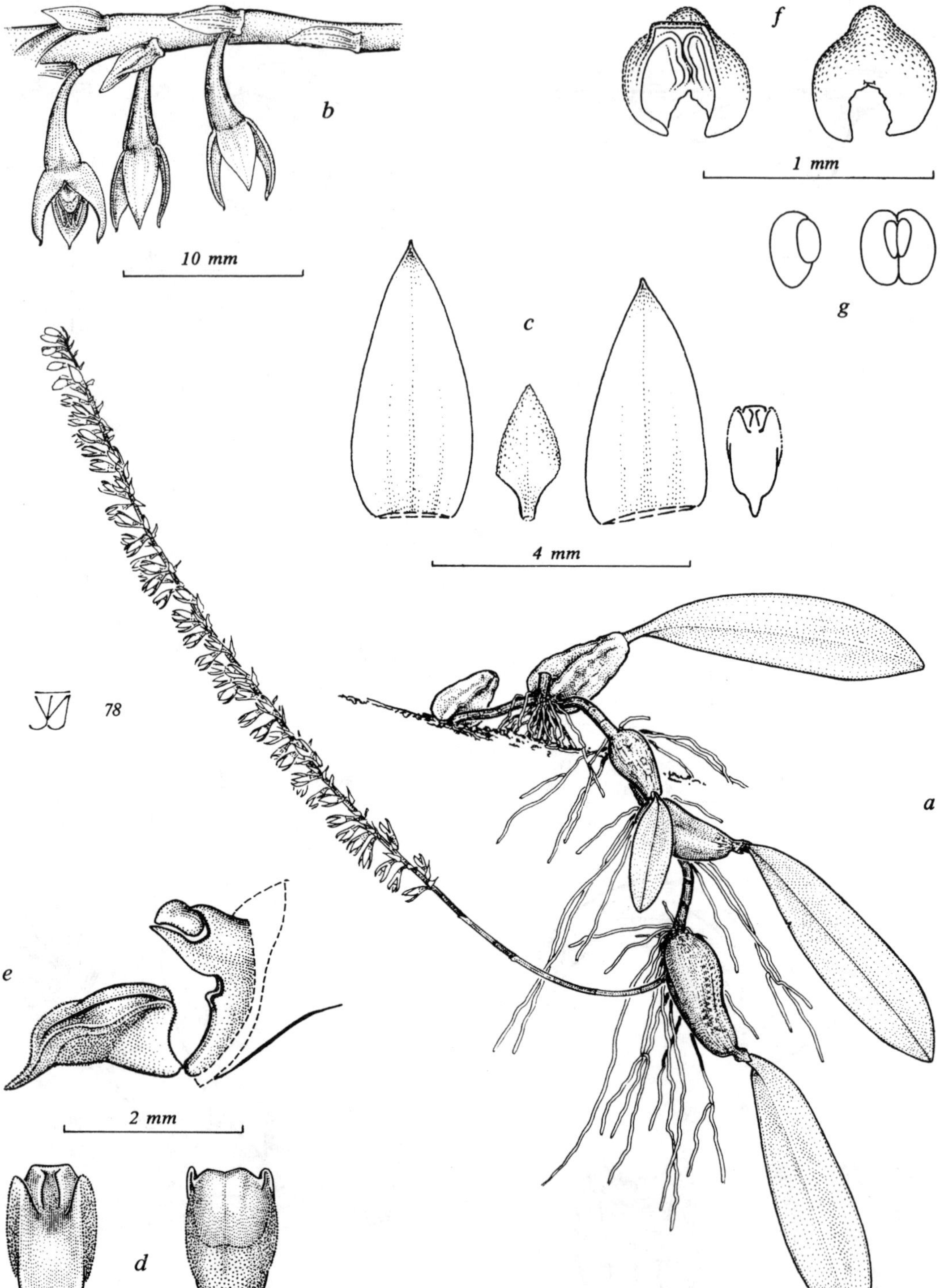

Fig. 41. *Bulbophyllum oxychilum* Schltr.

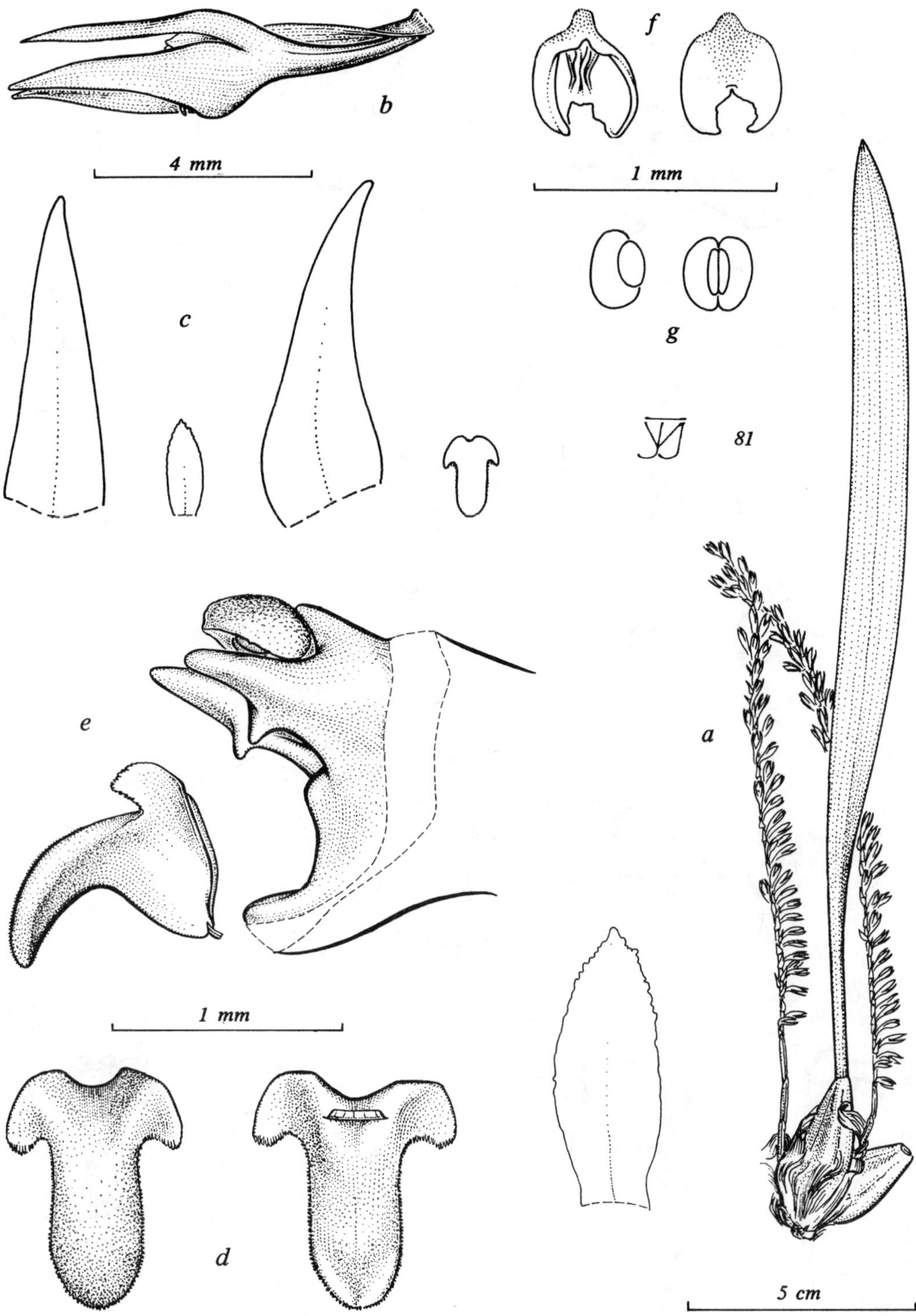

Fig. 42. *Bulbophyllum nigritianum* Rendle.

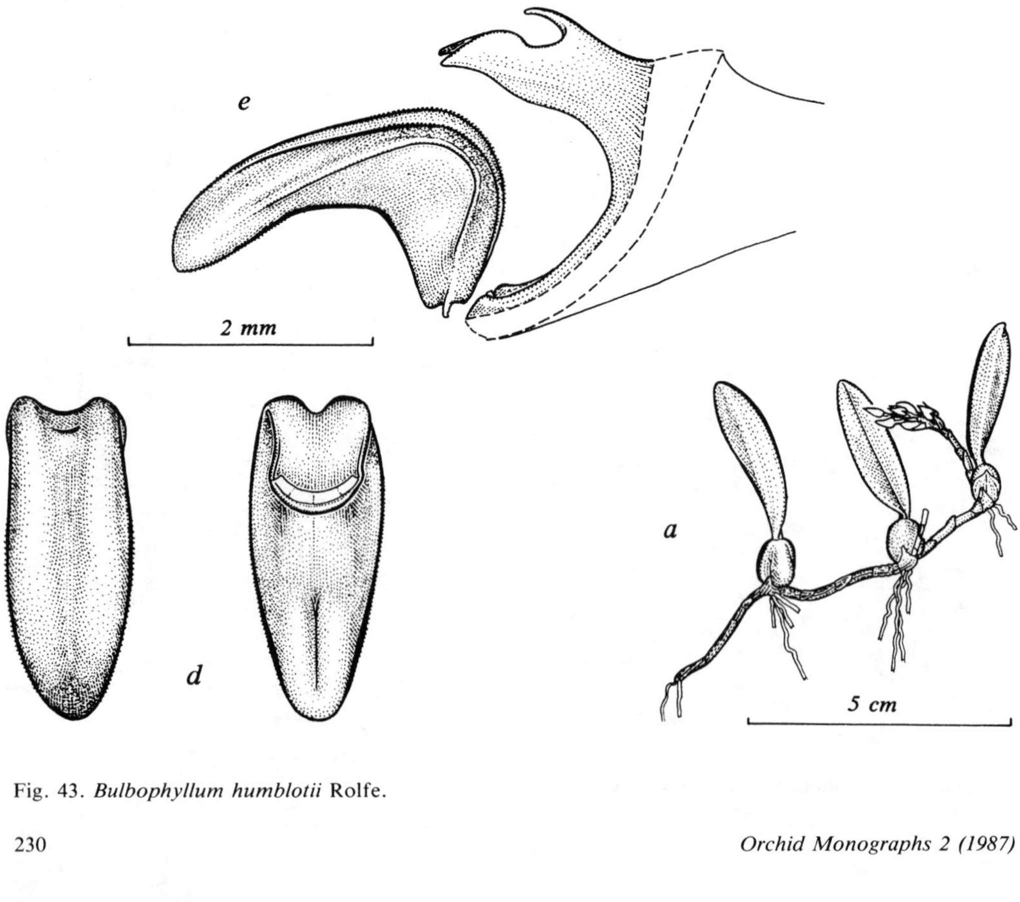

Fig. 43. *Bulbophyllum humblotii* Rolfe.

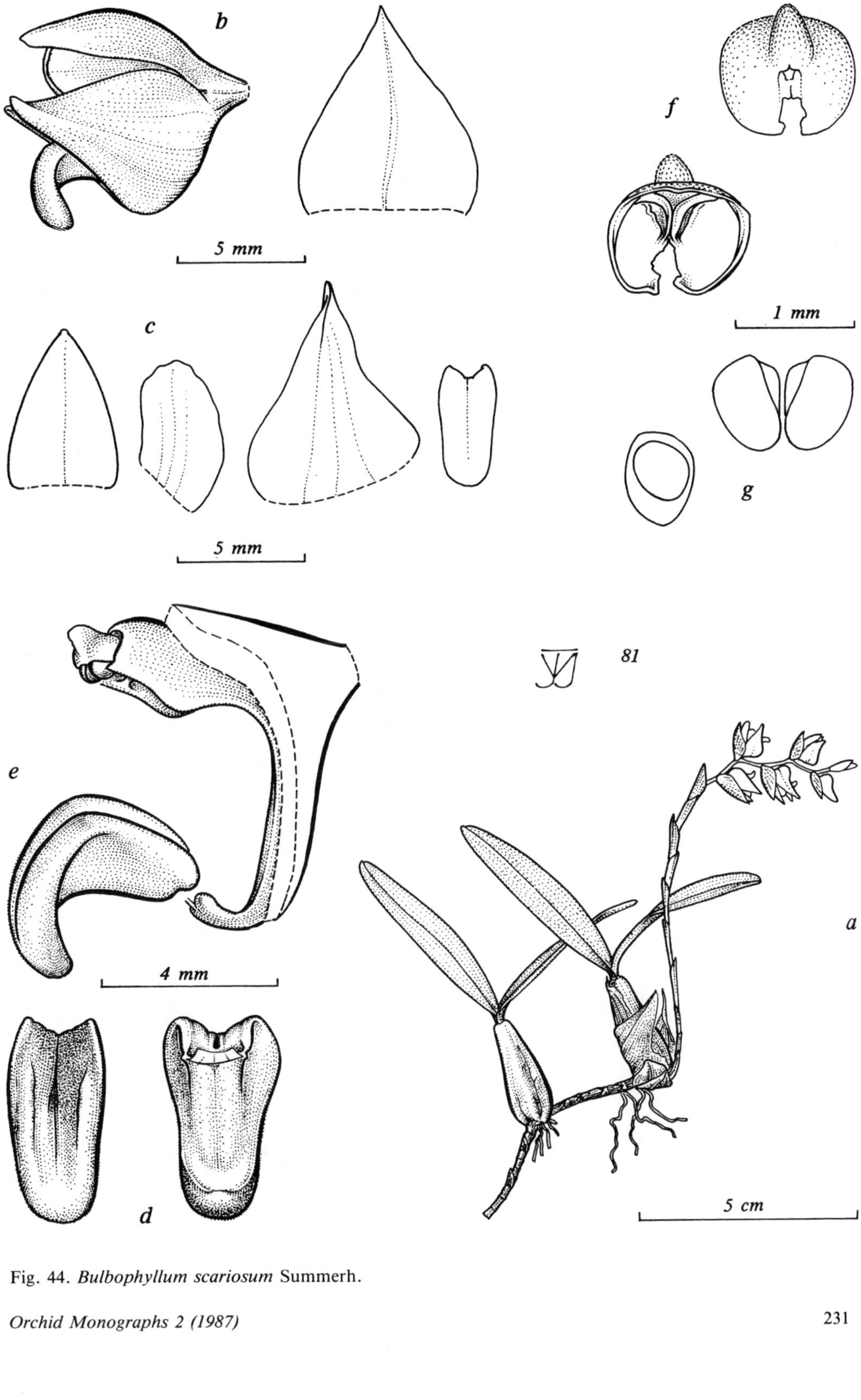

Fig. 44. *Bulbophyllum scariosum* Summerh.

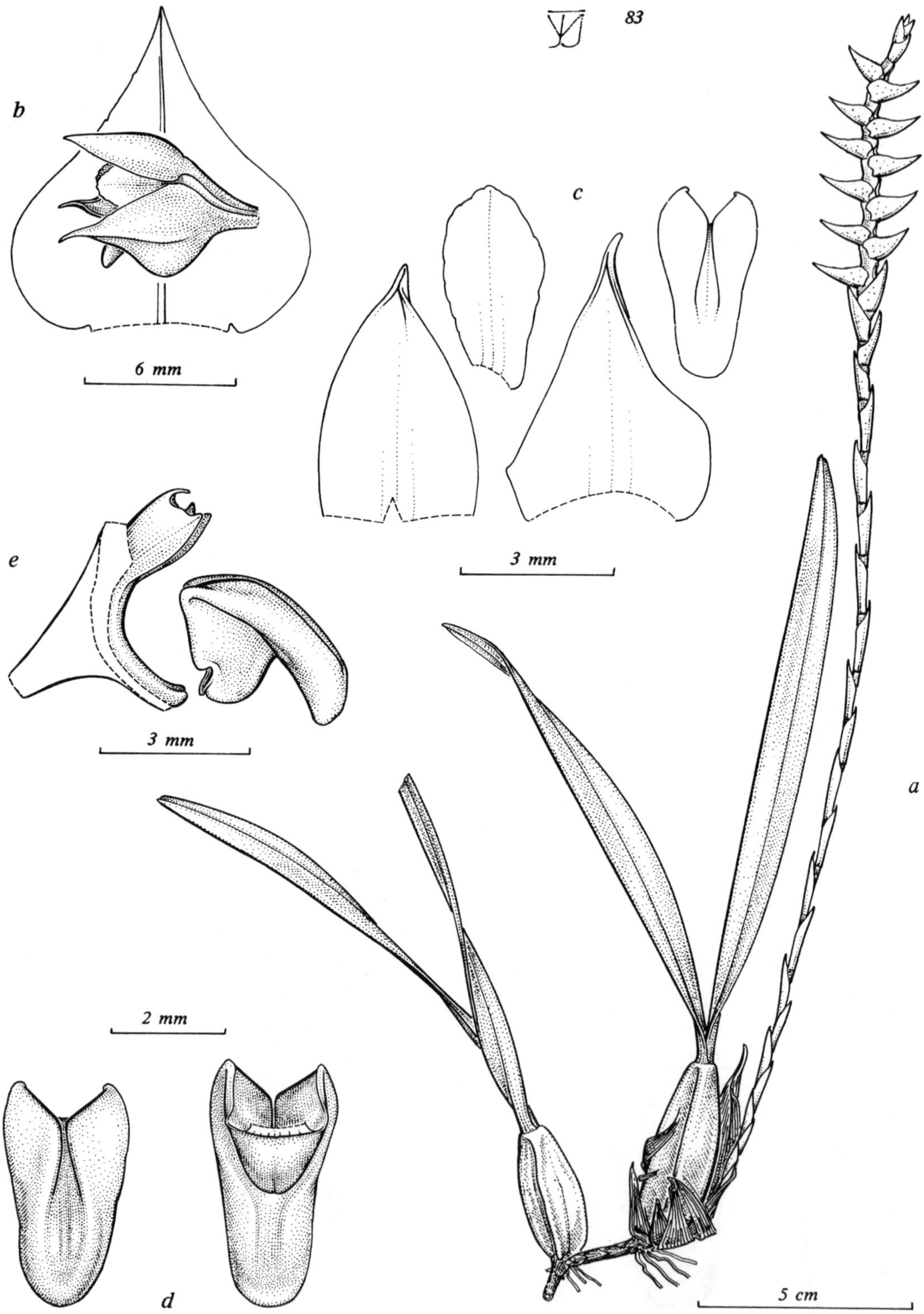

Fig. 45. *Bulbophyllum pandanetorum* Summerh.

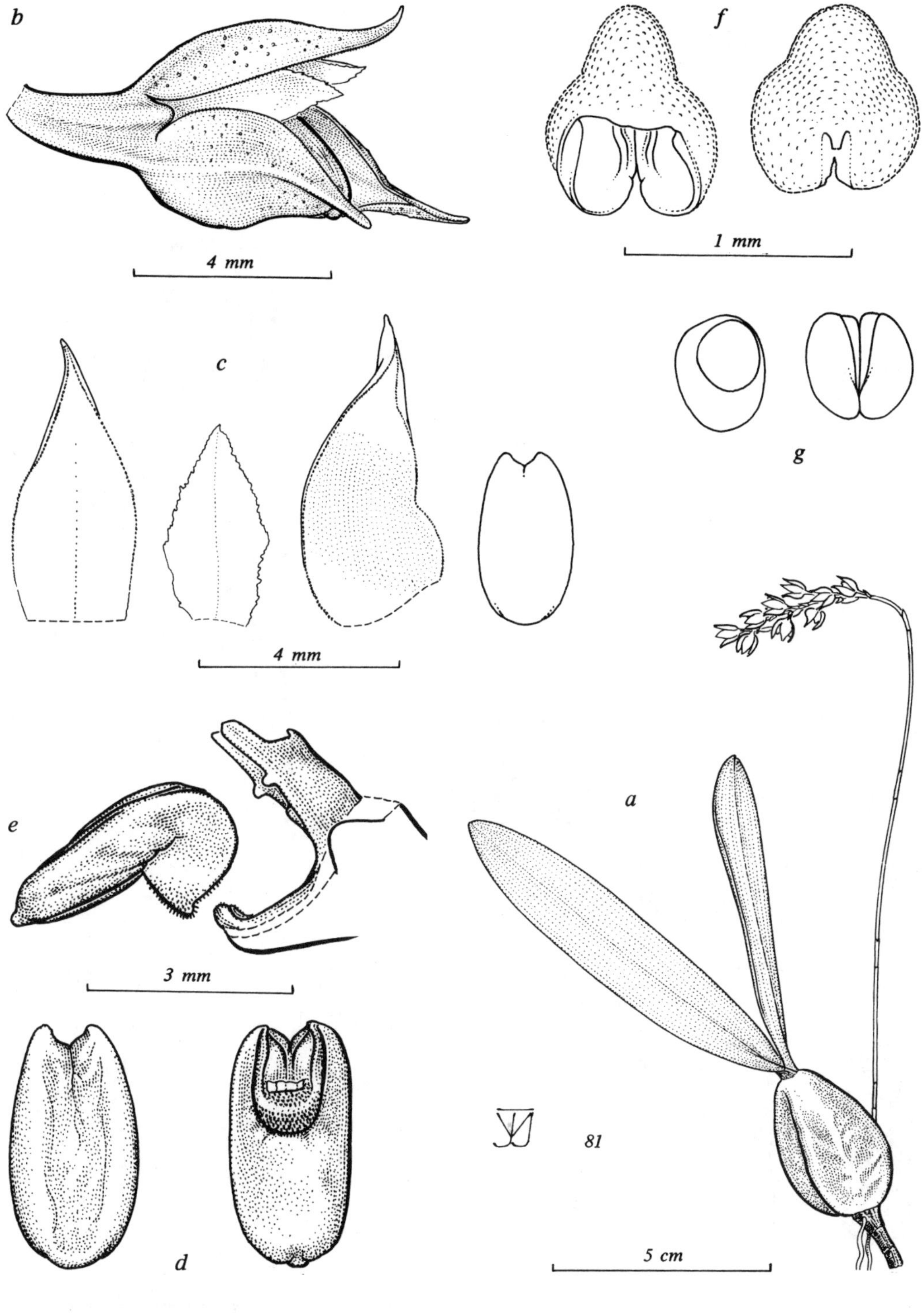

Fig. 46. *Bulbophyllum denticulatum* Rolfe.

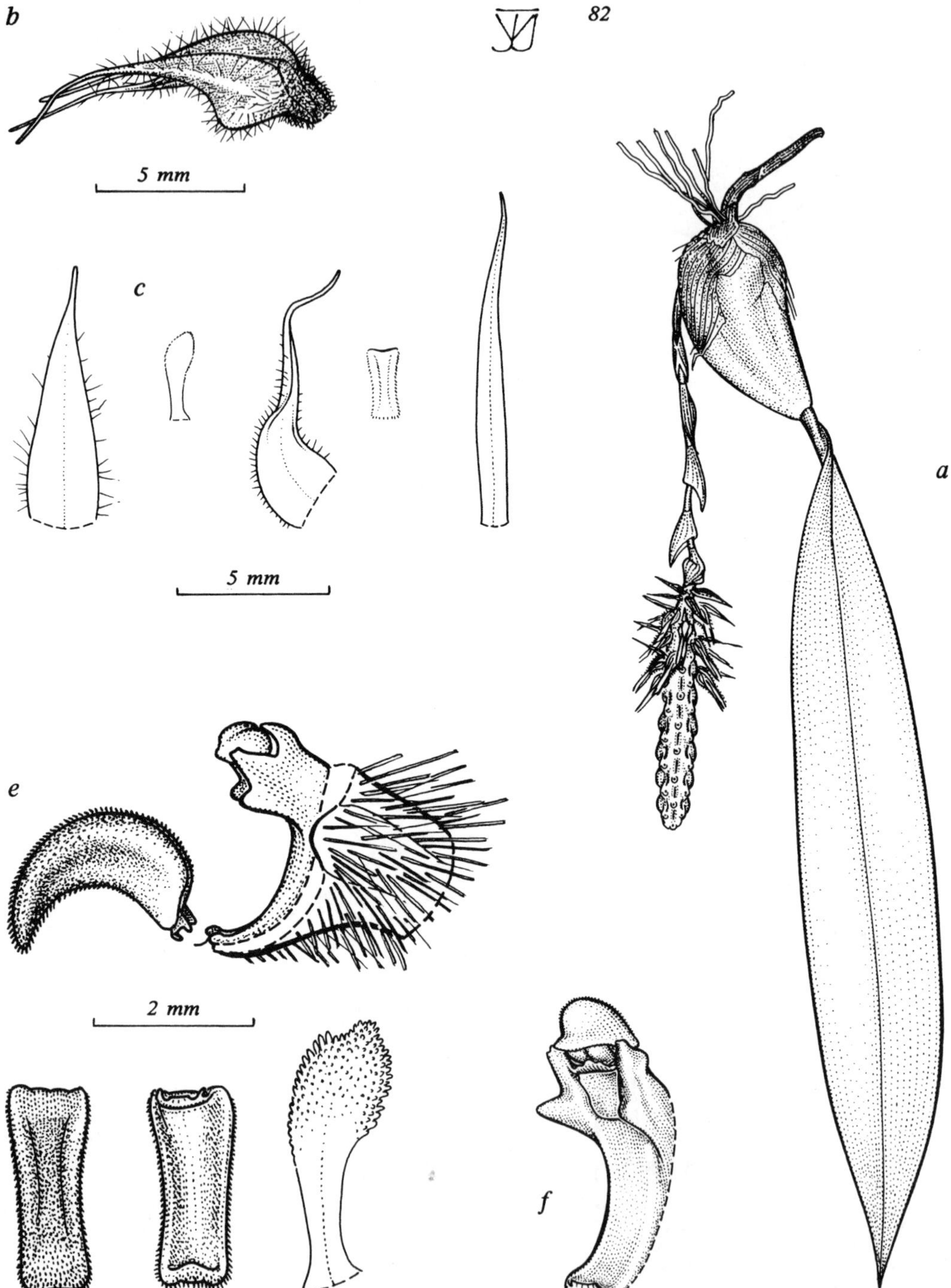

Fig. 47. *Bulbophyllum comatum* Lindl. var. *comatum*.

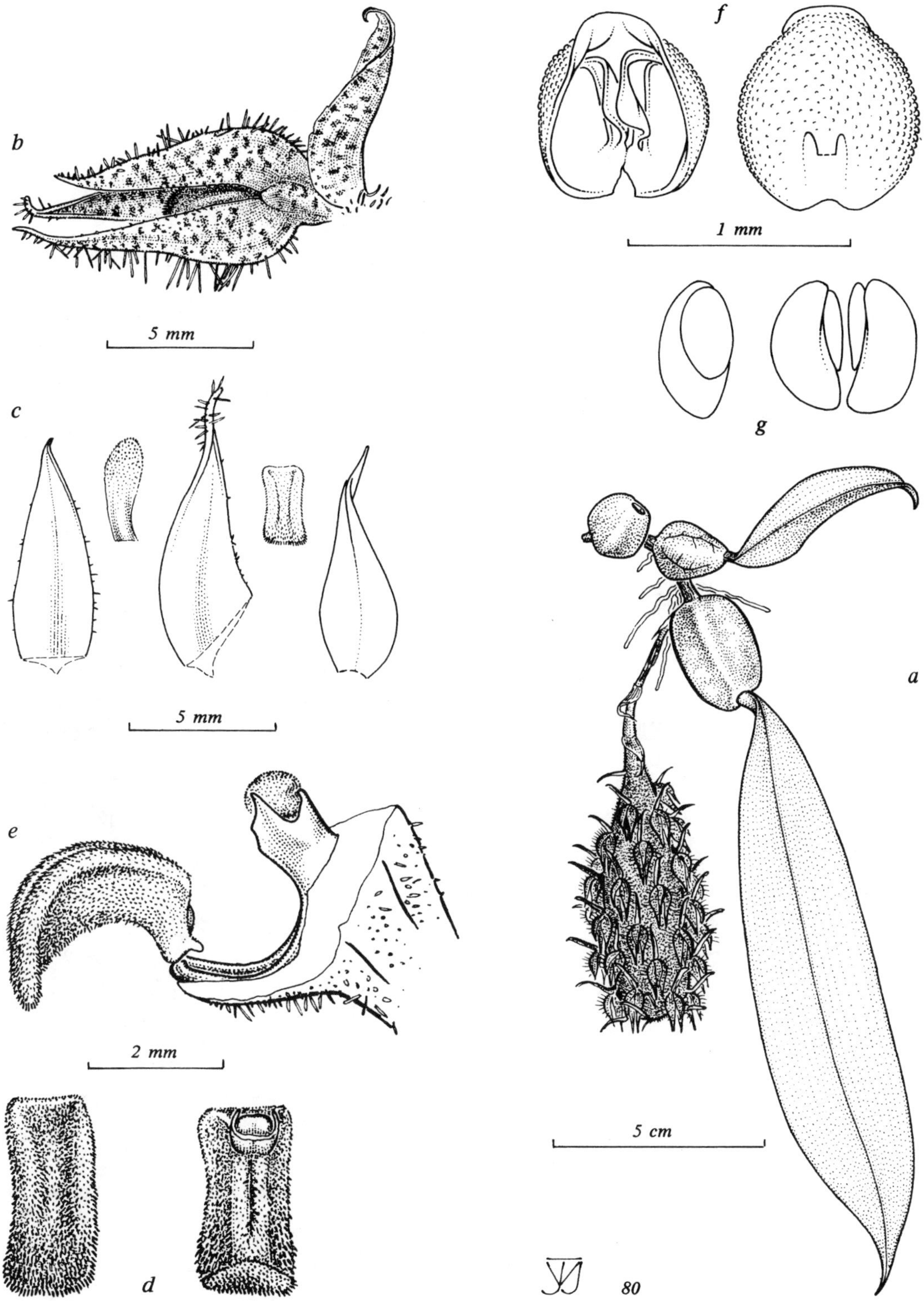

Fig. 48. *Bulbophyllum comatum* Lindl. var. *inflatum* (Rolfe) J.J. Vermeulen.

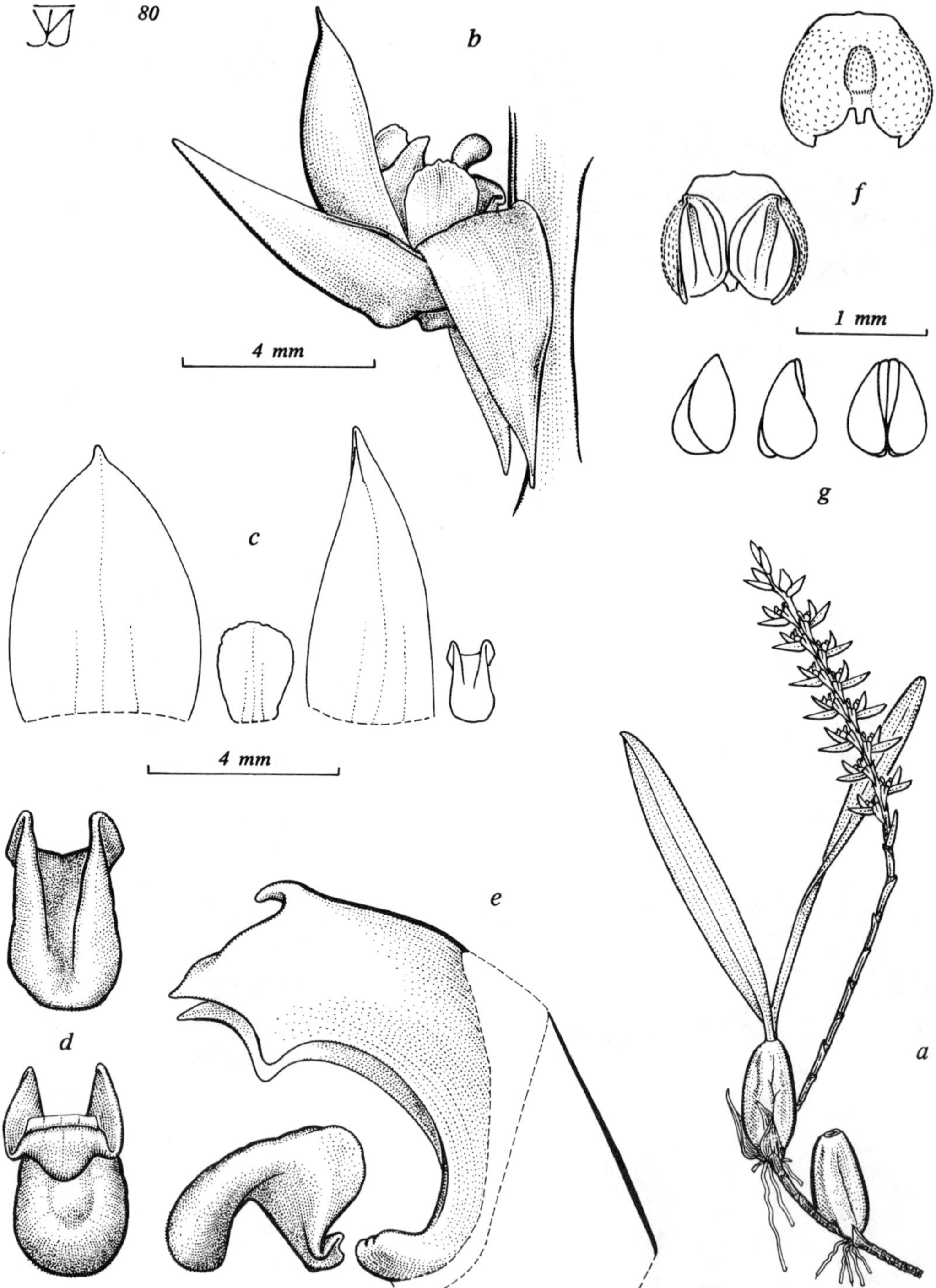

Fig. 49. *Bulbophyllum bifarium* Hook. f.

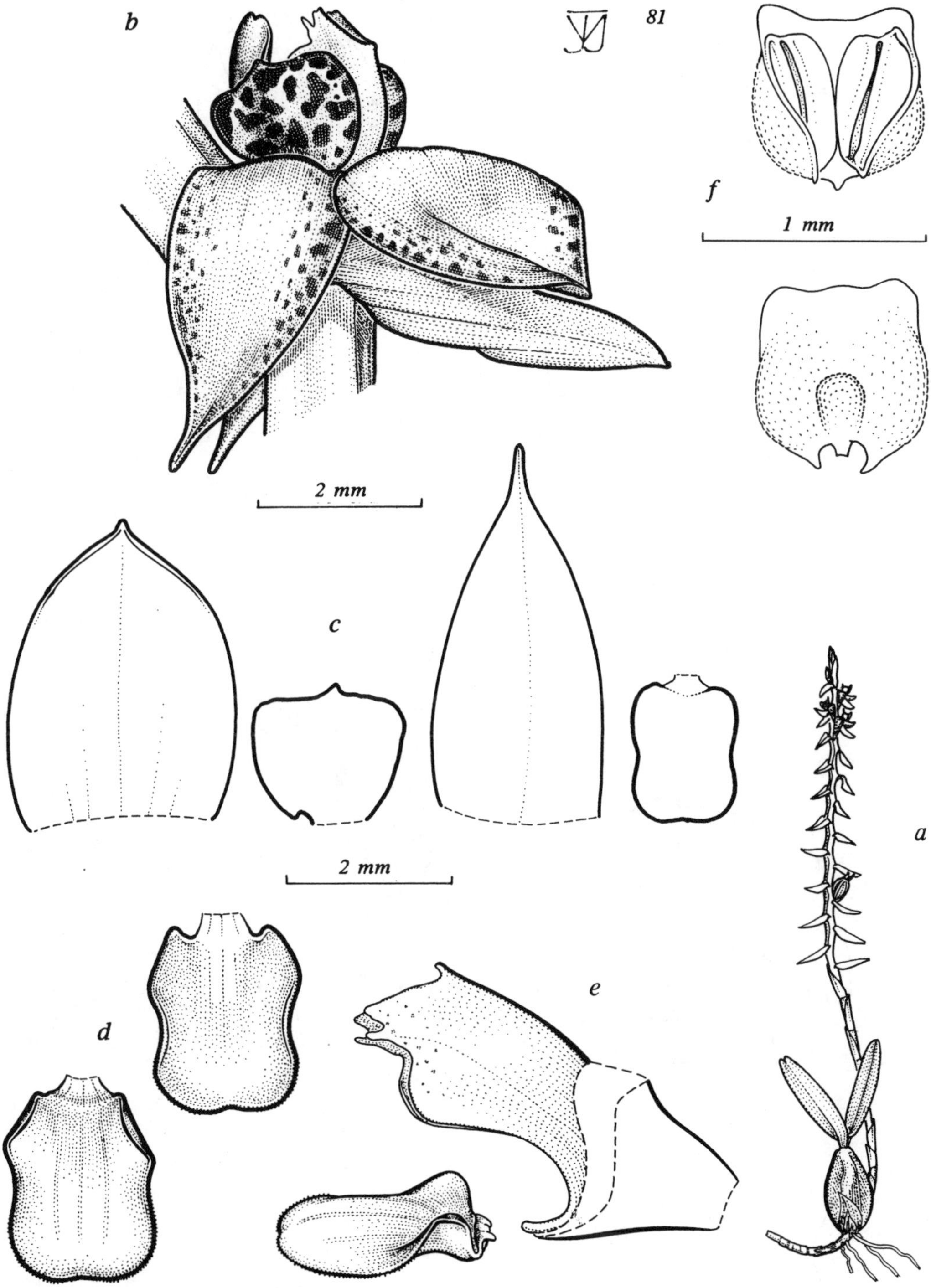

Fig. 50. *Bulbophyllum bidenticulatum* J.J. Vermeulen ssp. *bidenticulatum*

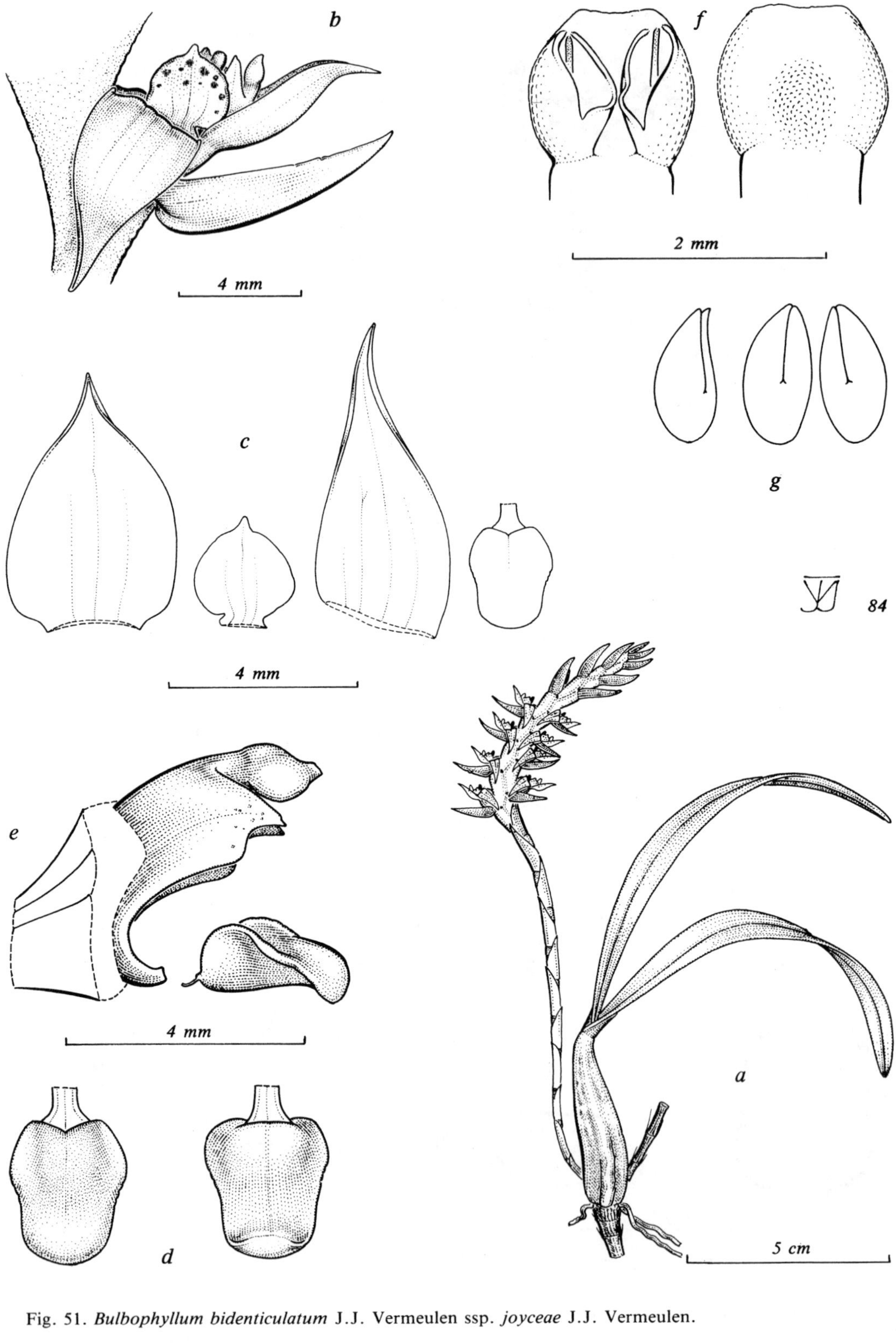

Fig. 51. *Bulbophyllum bidenticulatum* J.J. Vermeulen ssp. *joyceae* J.J. Vermeulen.

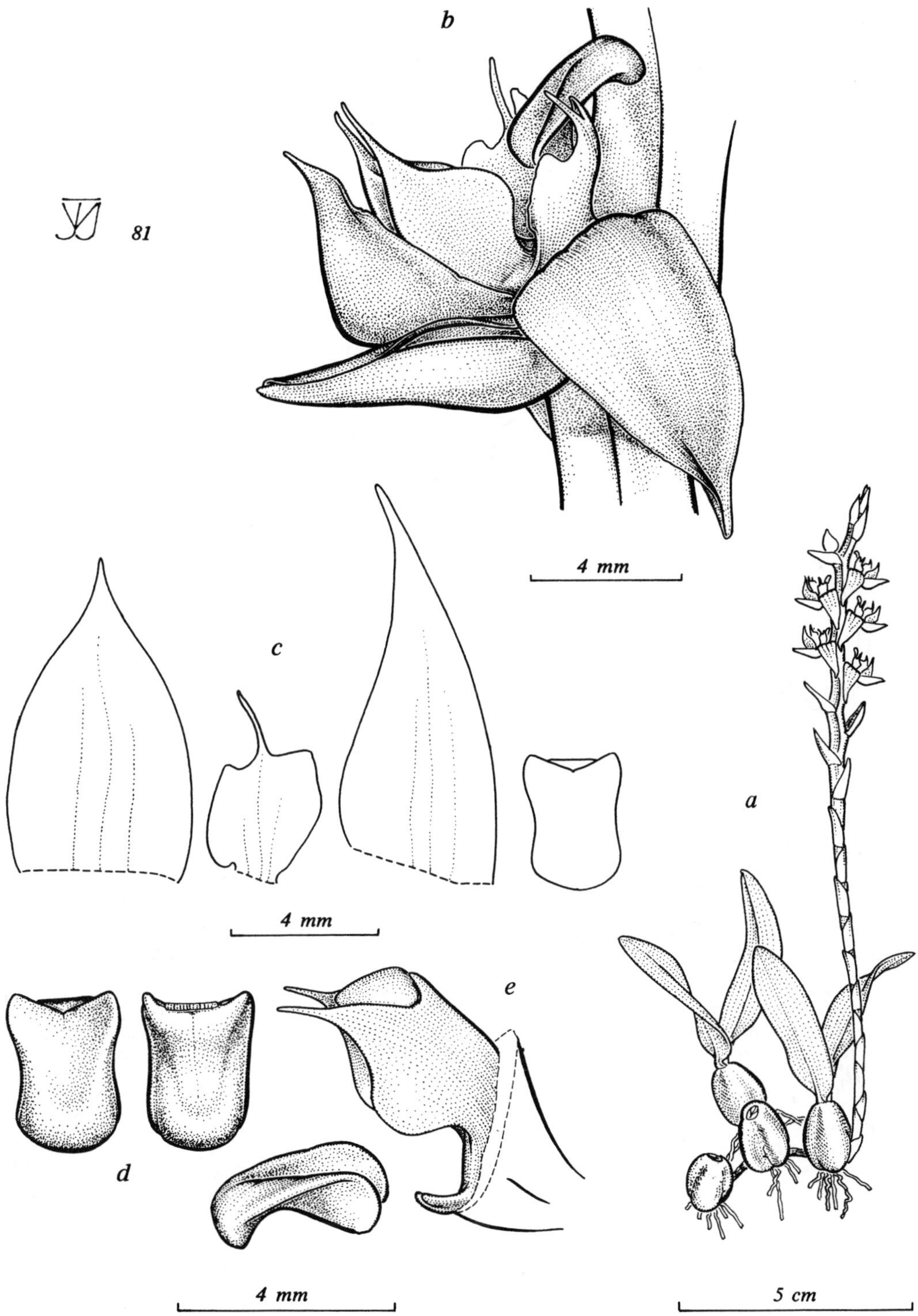

Fig. 52. *Bulbophyllum nigericum* Summerh.

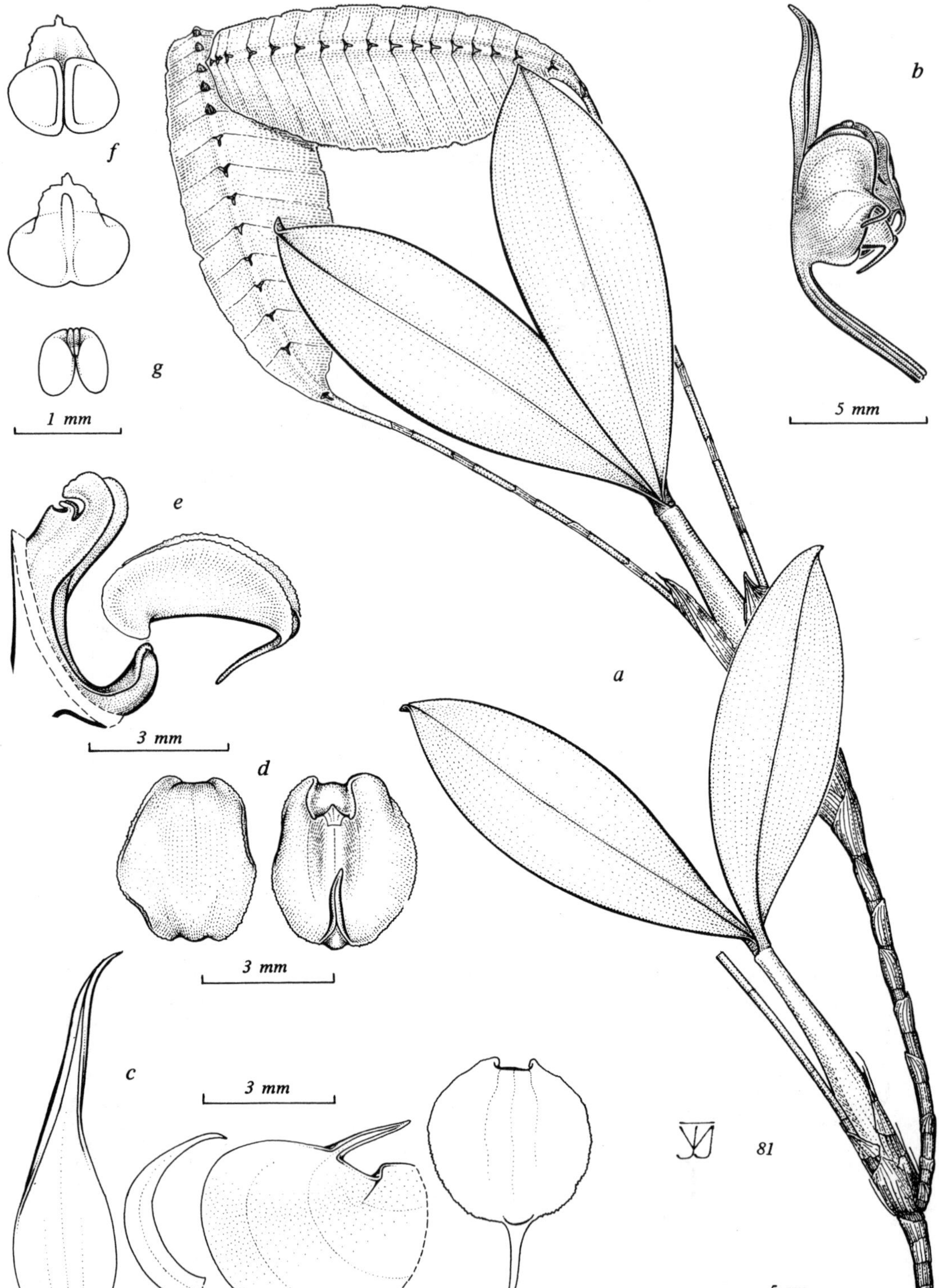

Fig. 53. *Bulbophyllum renkinianum* (Laurent) De Wild.

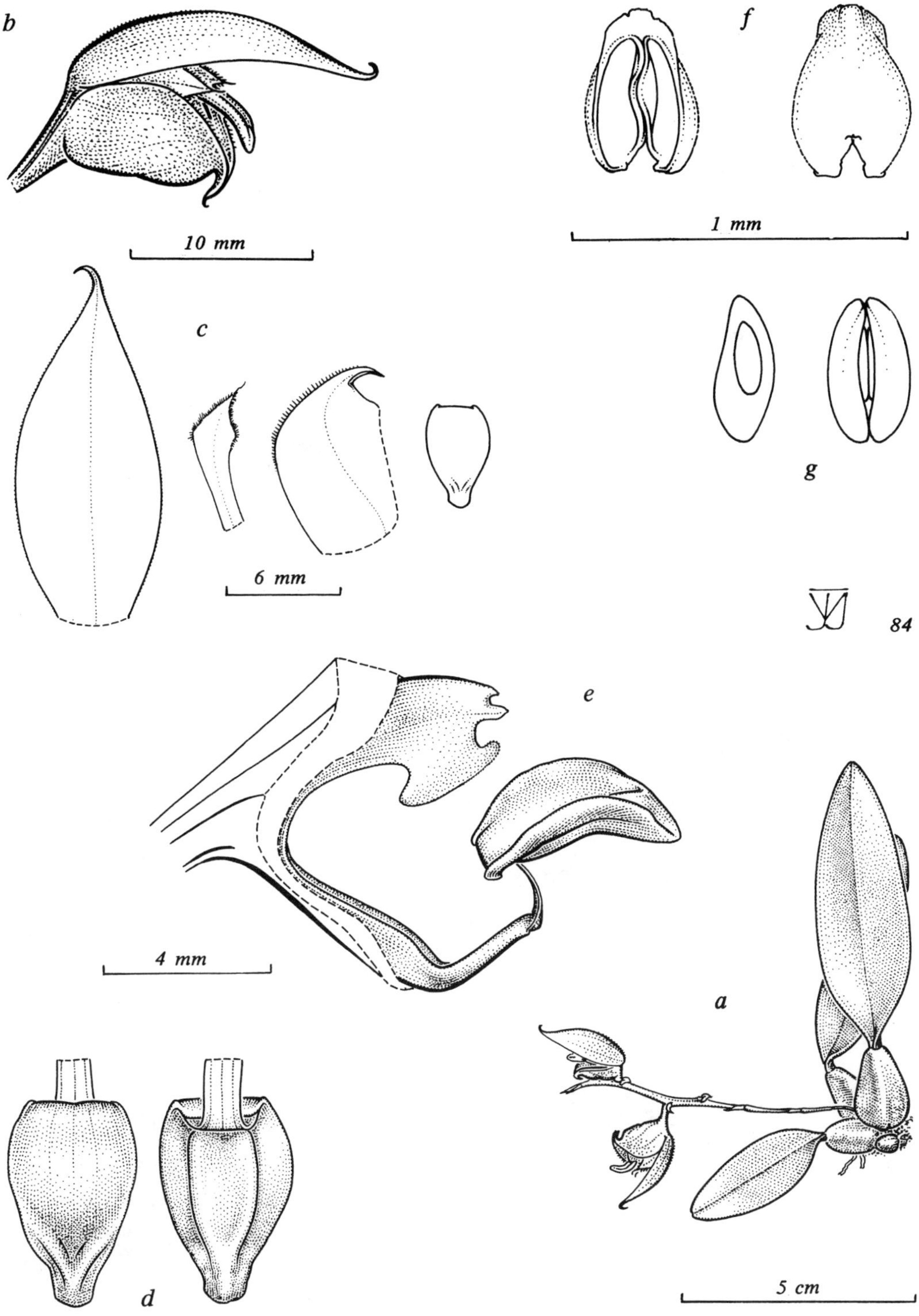

Fig. 54. *Bulbophyllum dolabriforme* J.J. Vermeulen.

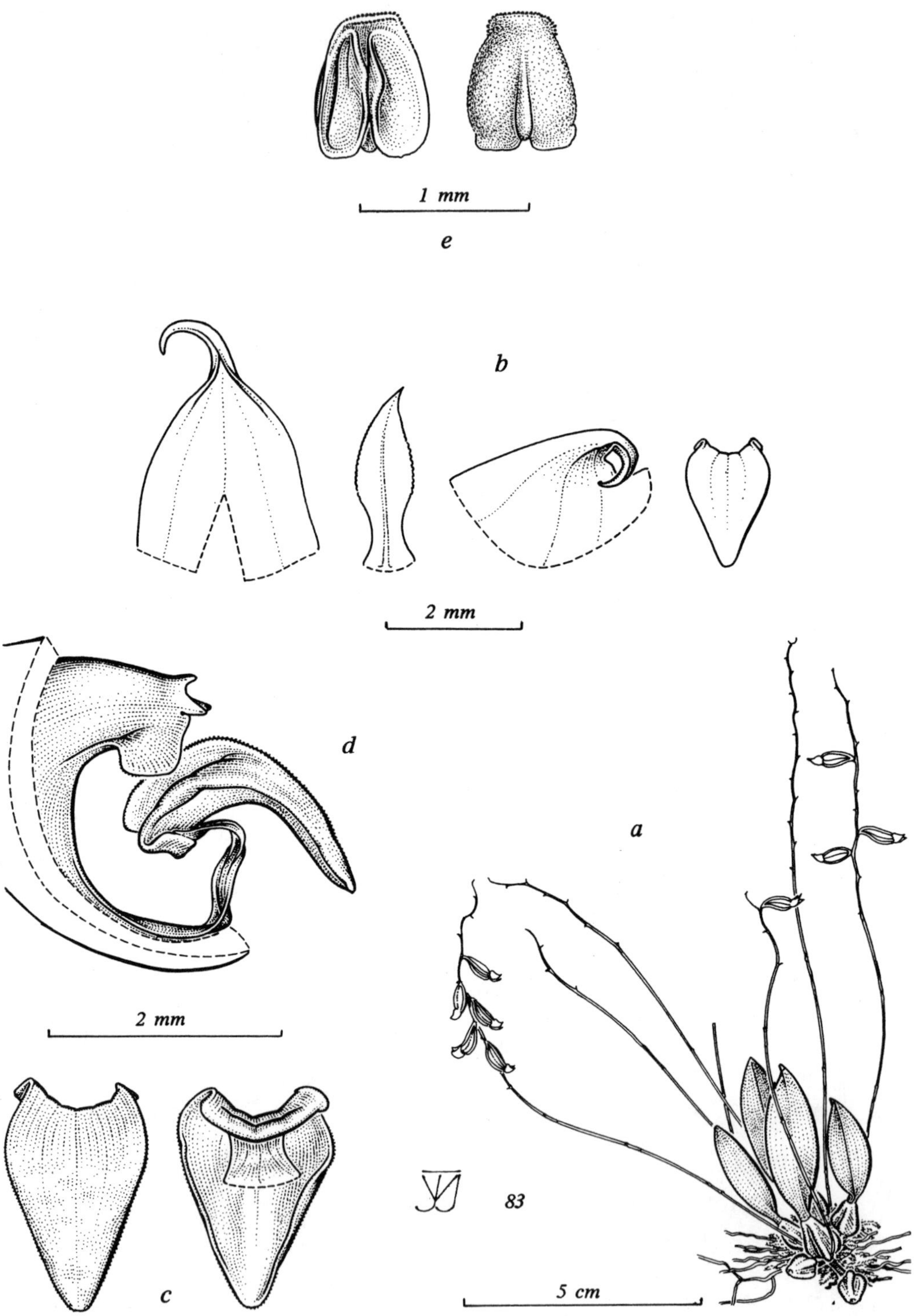

Fig. 55. *Bulbophyllum curvimentatum* J.J. Vermeulen.

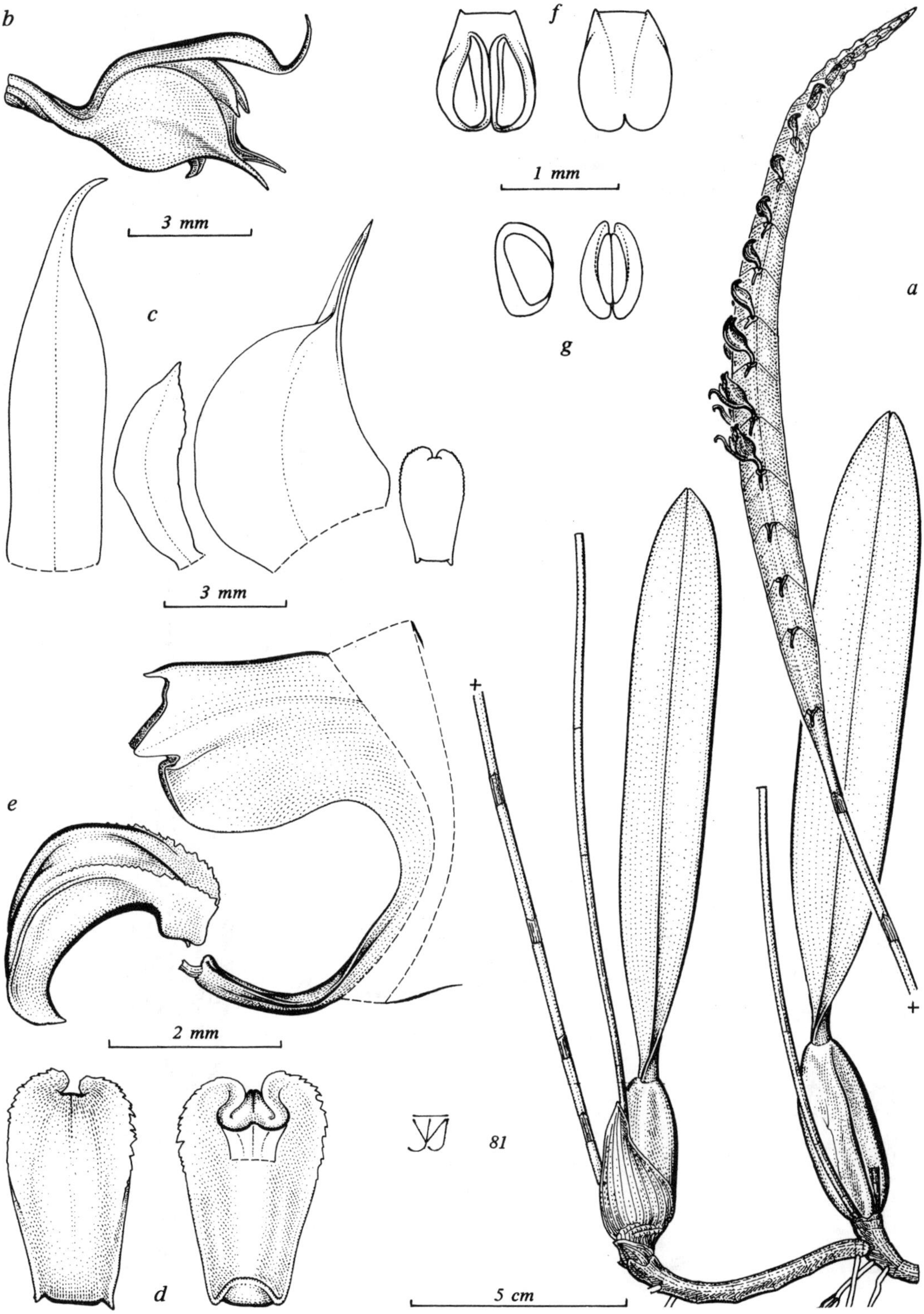

Fig. 56. *Bulbophyllum colubrinum* (Rchb. f.) Rchb. f.

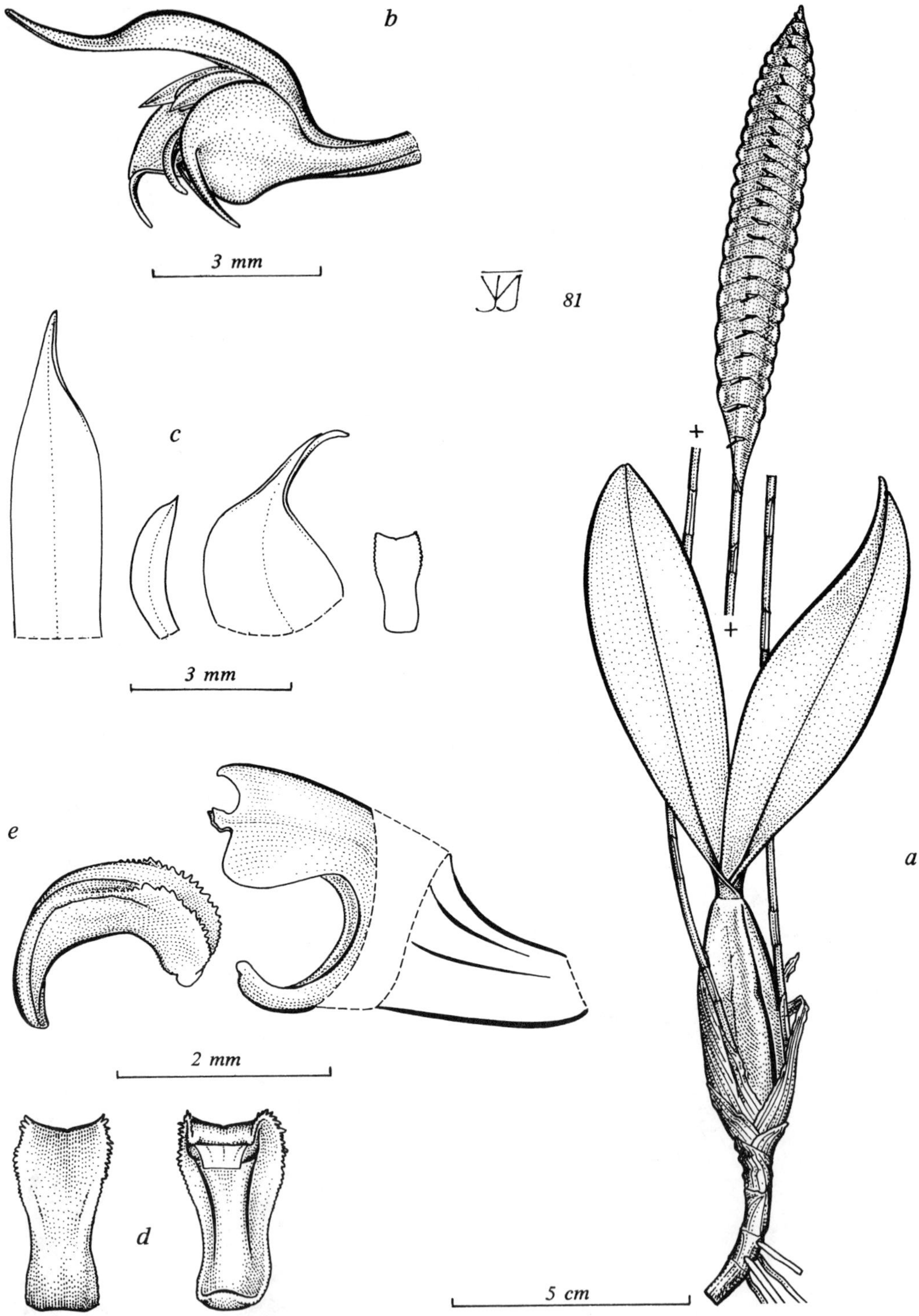

Fig. 57. *Bulbophyllum maximum* (Lindl.) Rchb. f.; entity a.

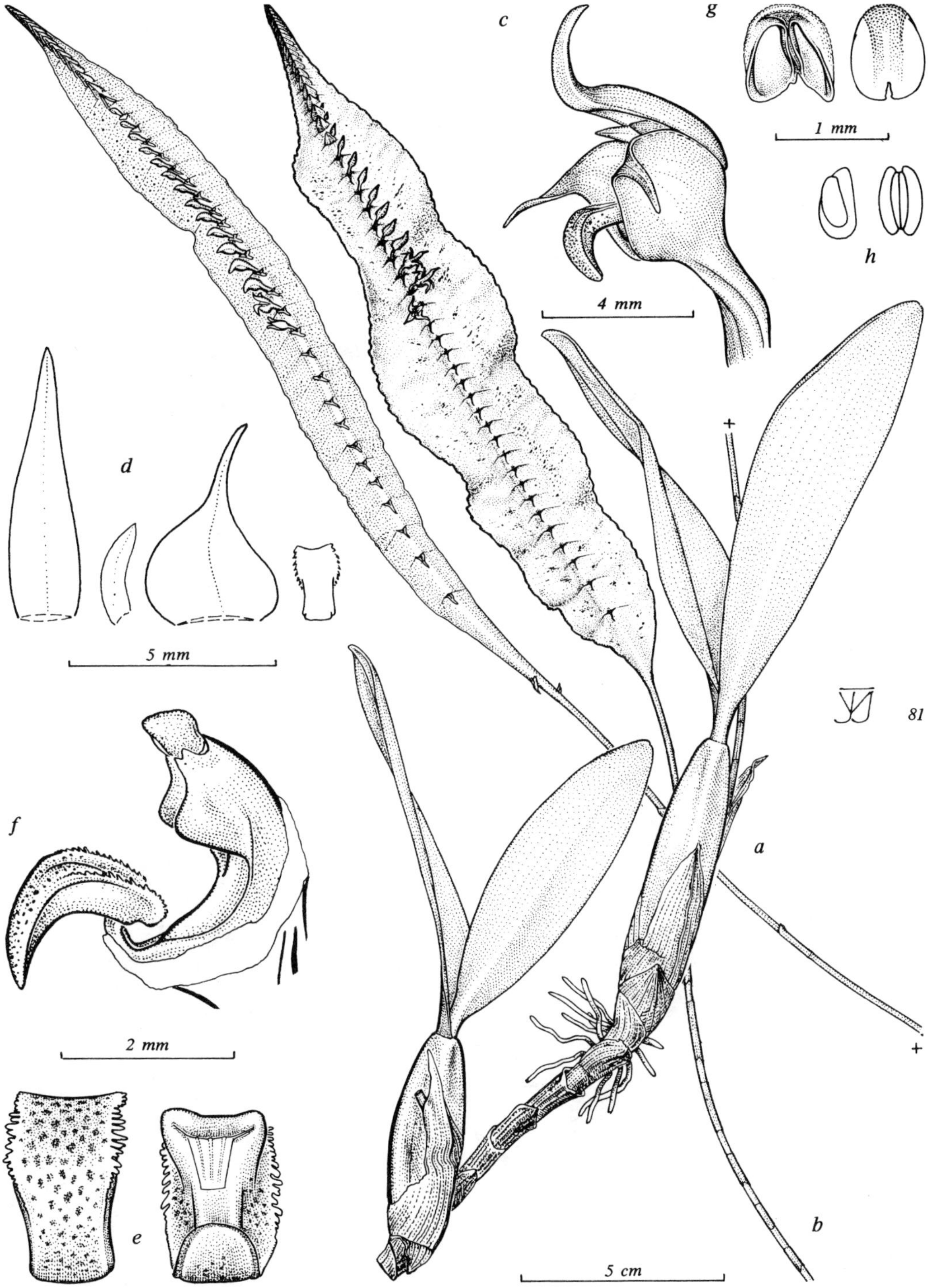

Fig. 58. *Bulbophyllum maximum* (Lindl.) Rchb. f.; entity b.

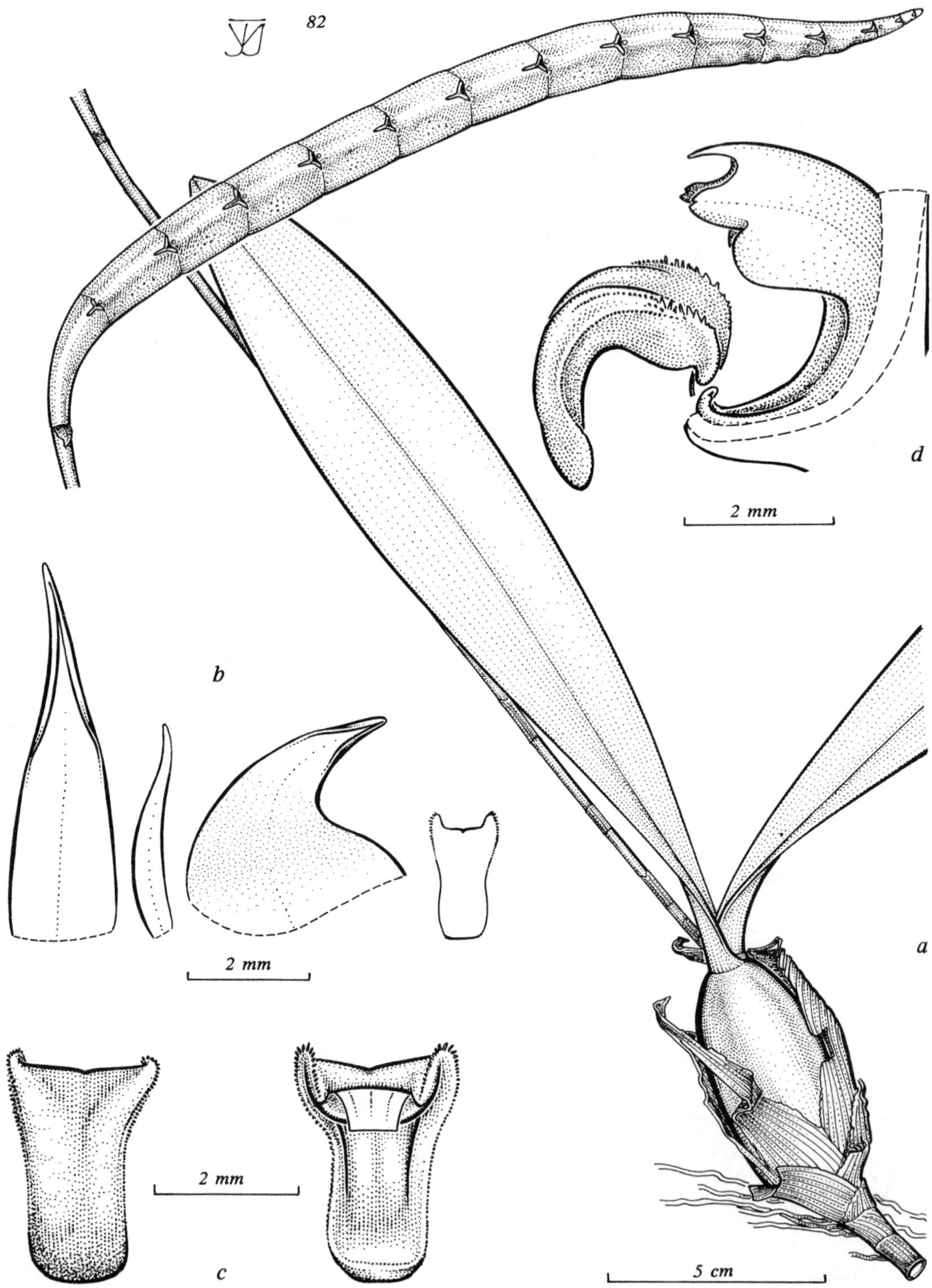

Fig. 59. *Bulbophyllum injoloense* De Wild. ssp. *injoloense*.

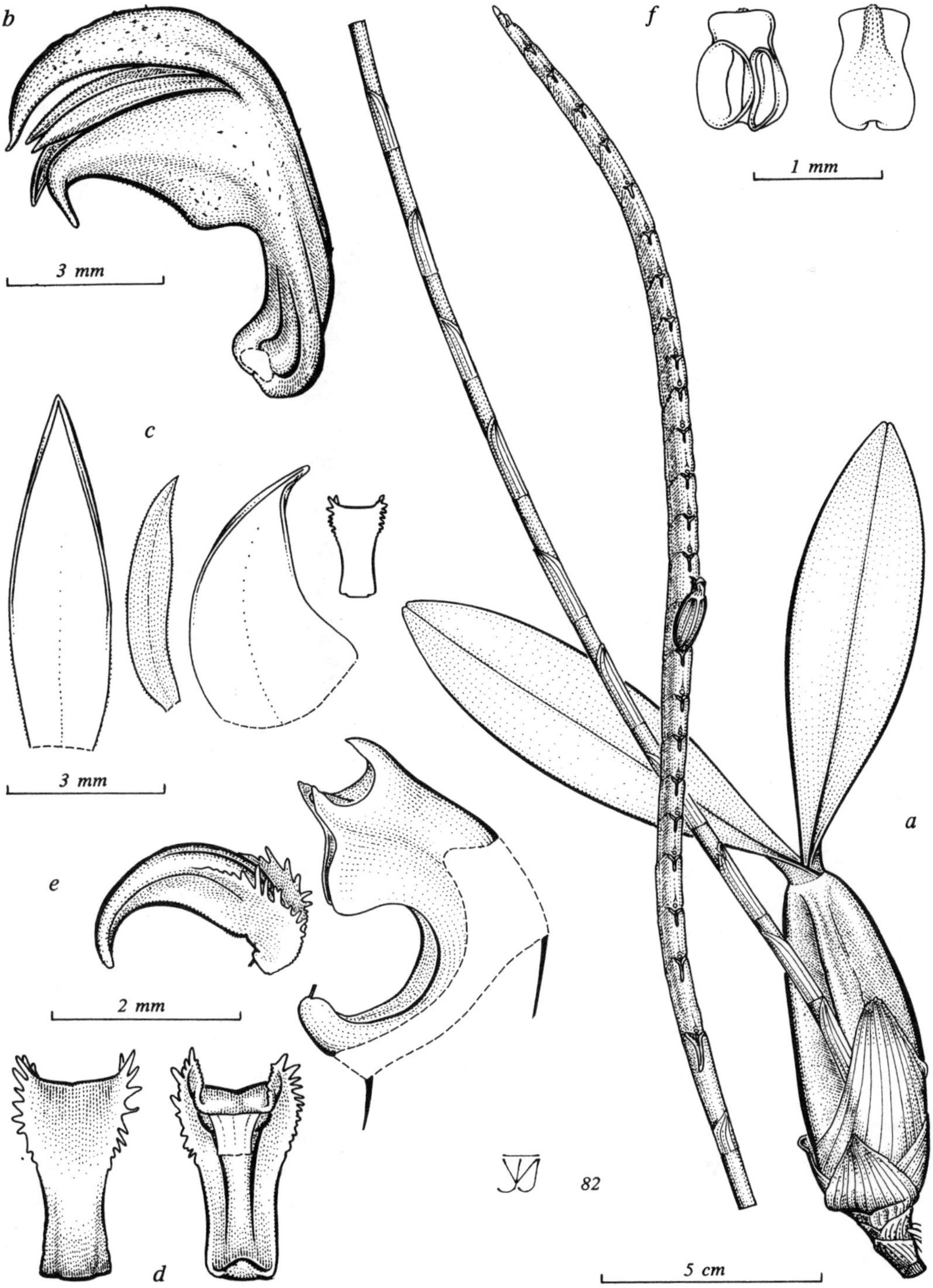

Fig. 60. *Bulbophyllum injoloense* De Wild. ssp. *pseudoxypterum* (J.J. Vermeulen) J.J. Vermeulen.

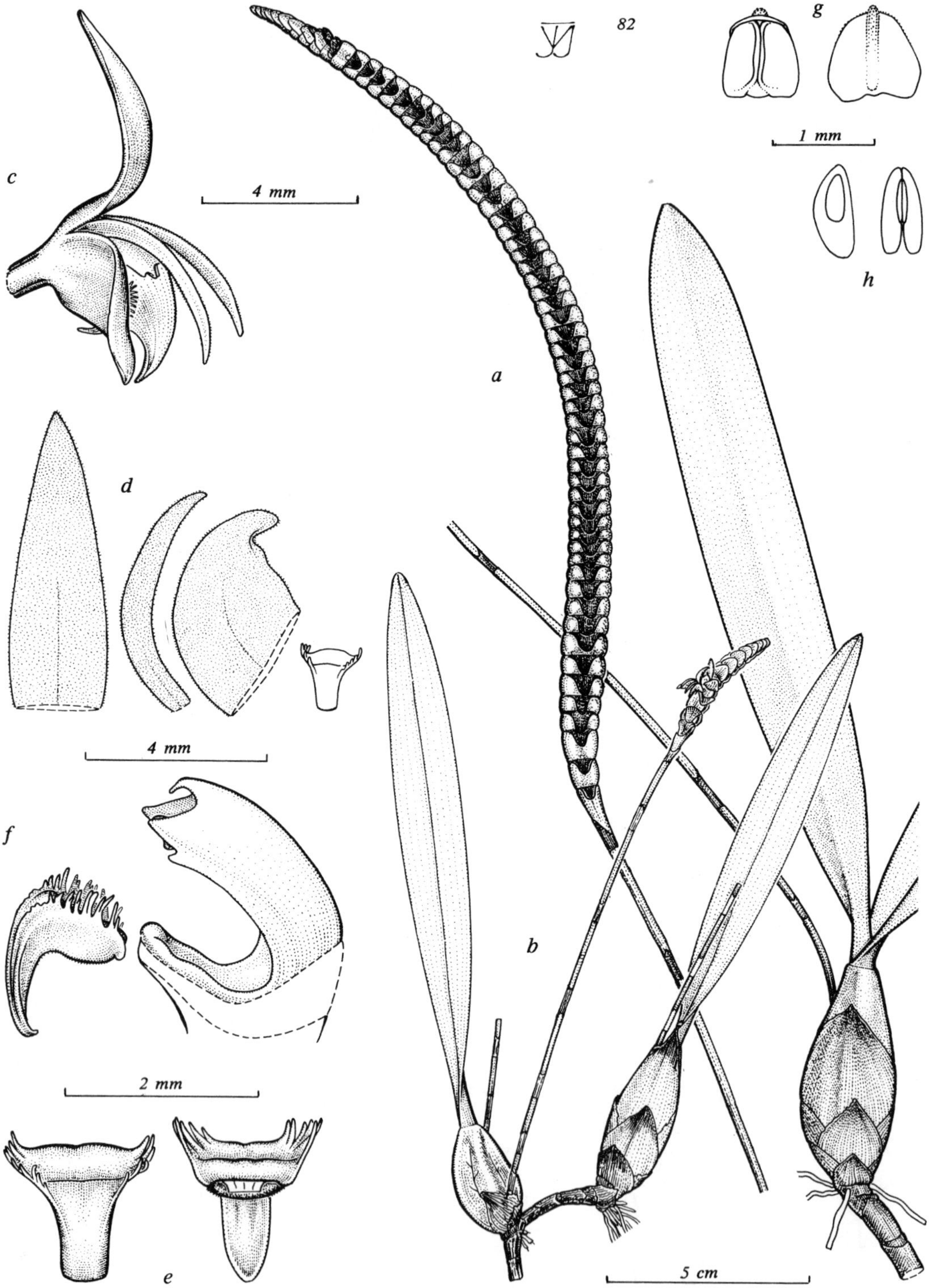

Fig. 61. *Bulbophyllum imbricatum* Lindl.

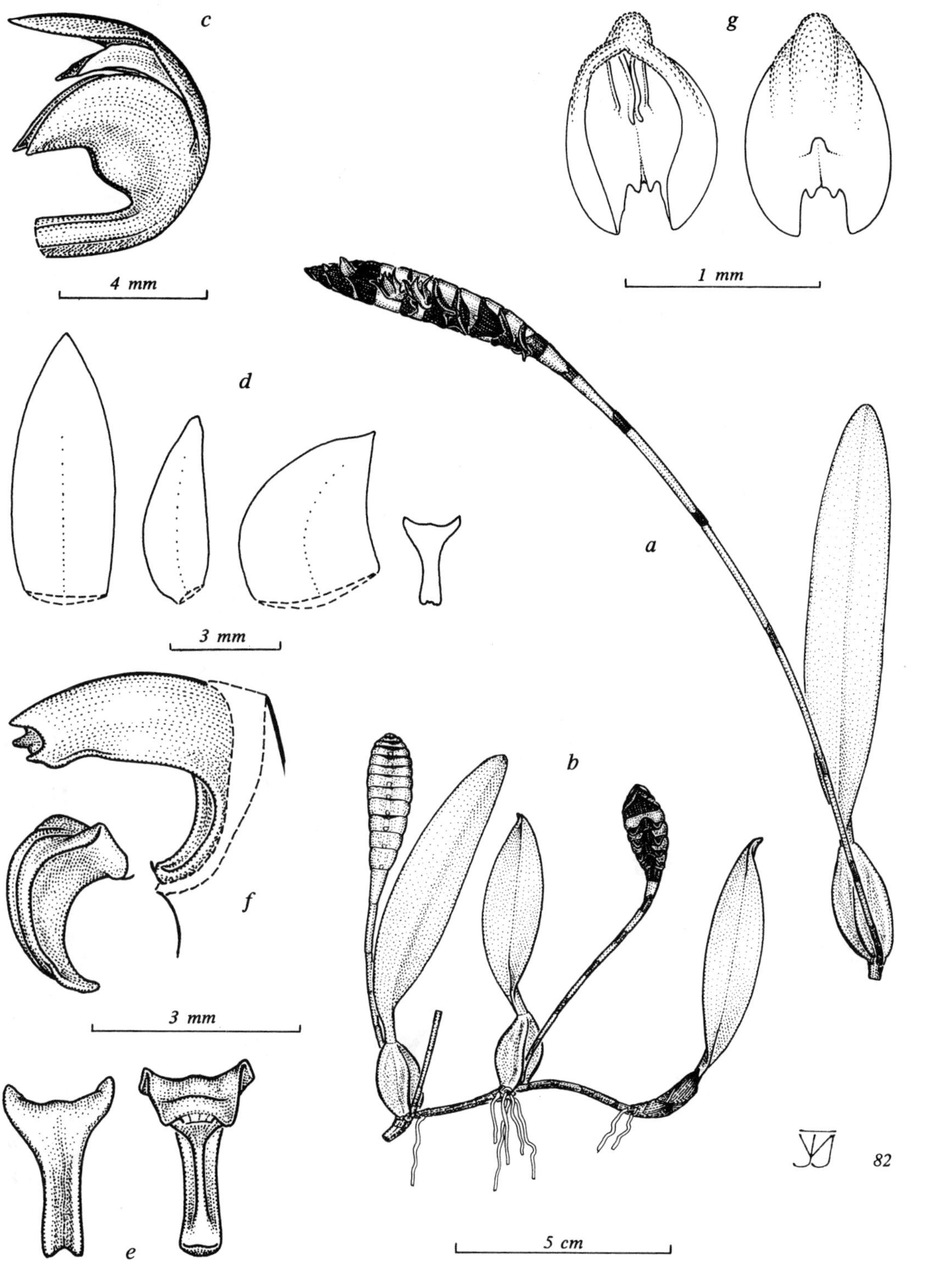

Fig. 62. *Bulbophyllum magnibracteatum* Summerh.

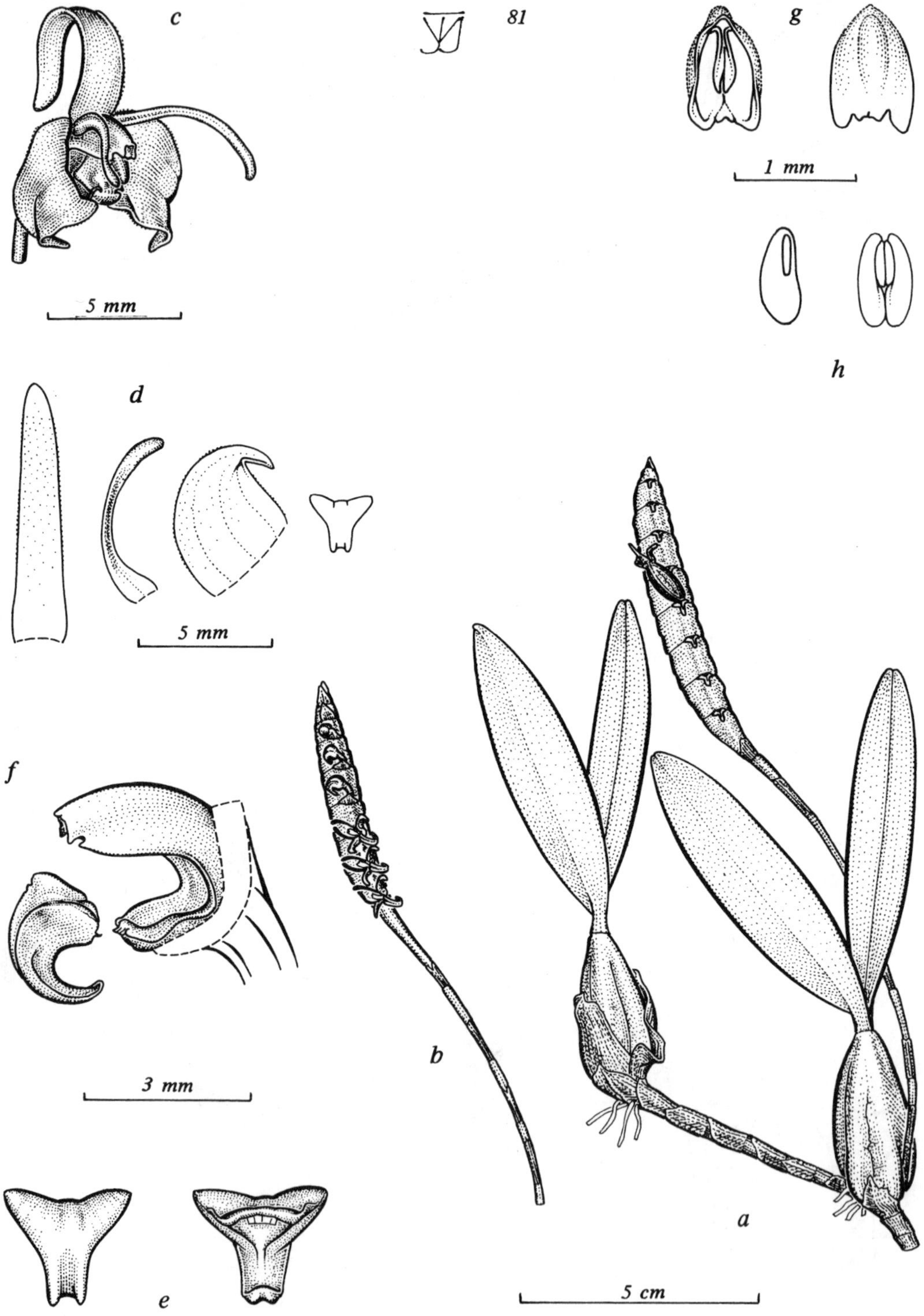

Fig. 63. *Bulbophyllum sandersonii* (Hook. f.) Rchb. f. ssp. *sandersonii*.

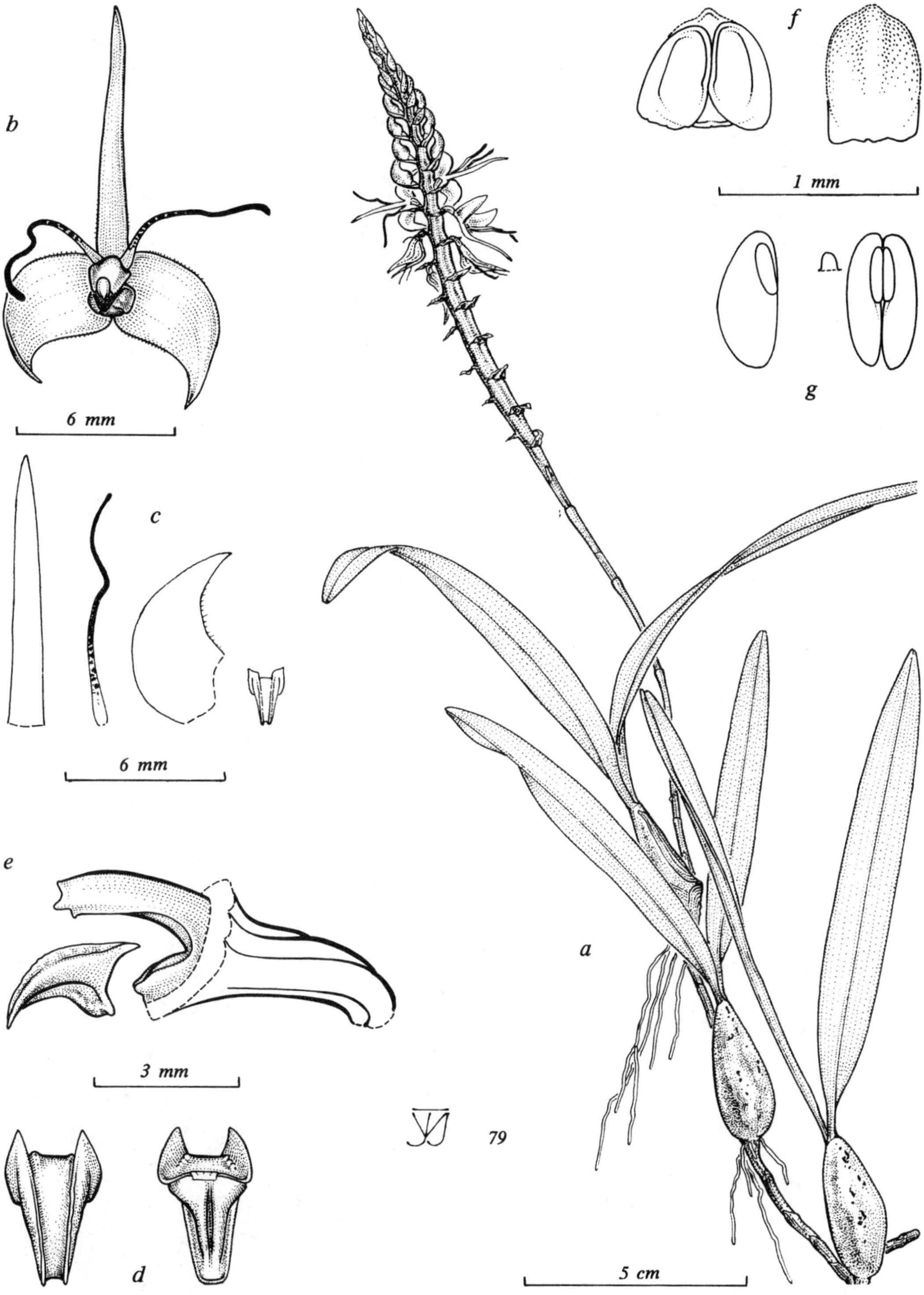

Fig. 64. *Bulbophyllum sandersonii* (Hook. f.) Rchb. f. ssp. *stenopetalum* (Krzl.) J.J. Vermeulen.

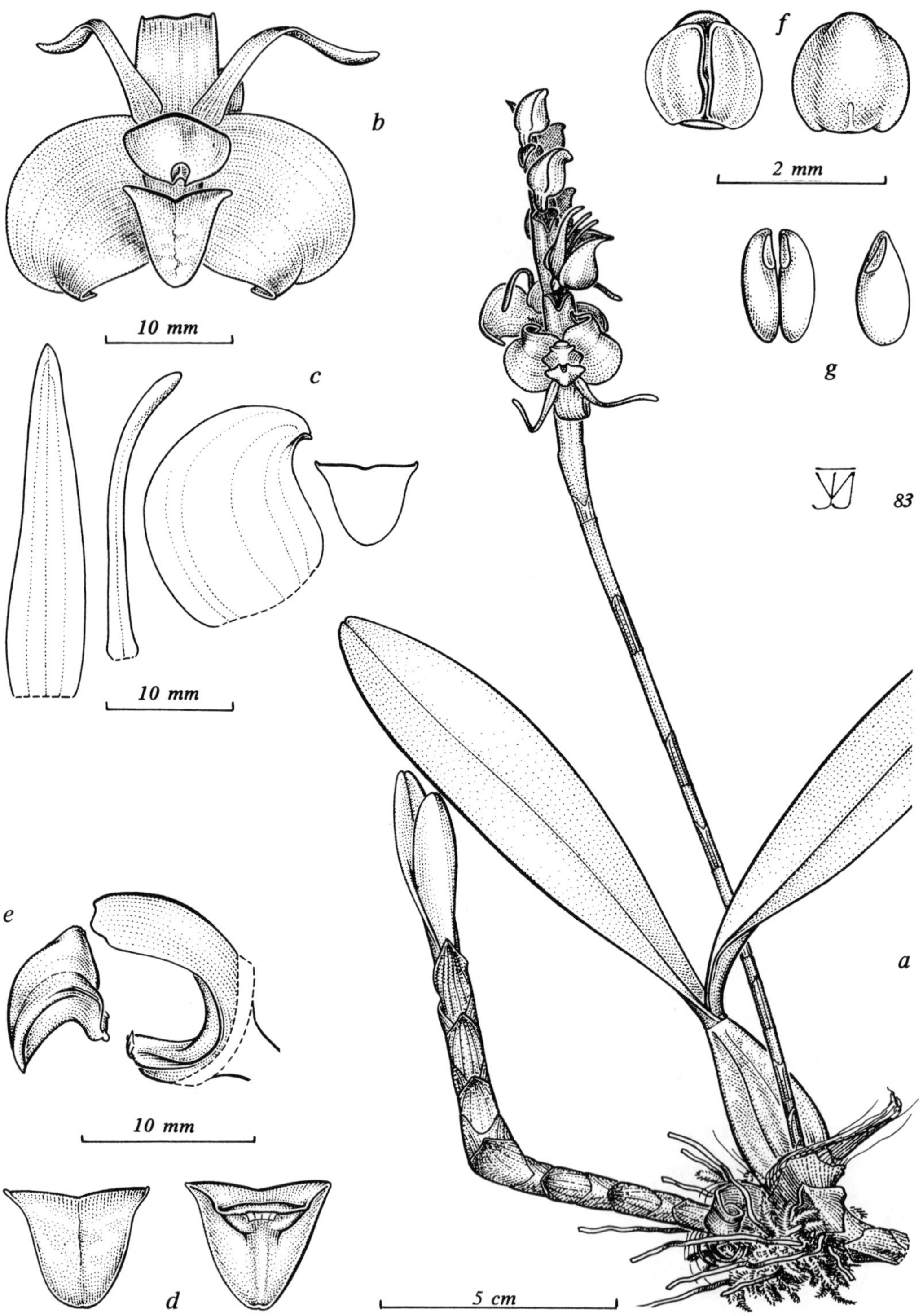

Fig. 65. *Bulbophyllum lizae* J.J. Vermeulen.

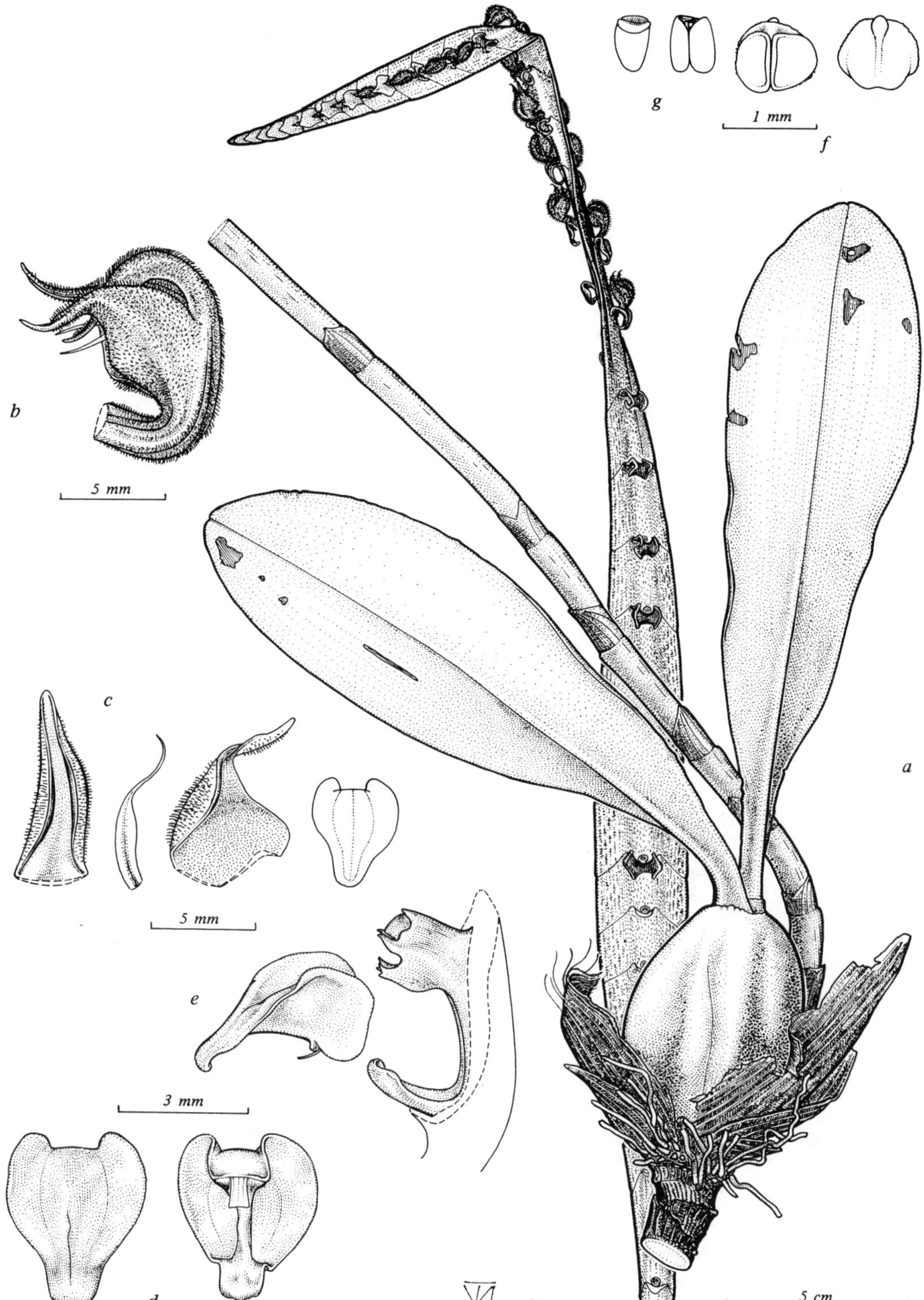

Fig. 66. *Bulbophyllum purpureorhachis* (De Wild.) Schltr.

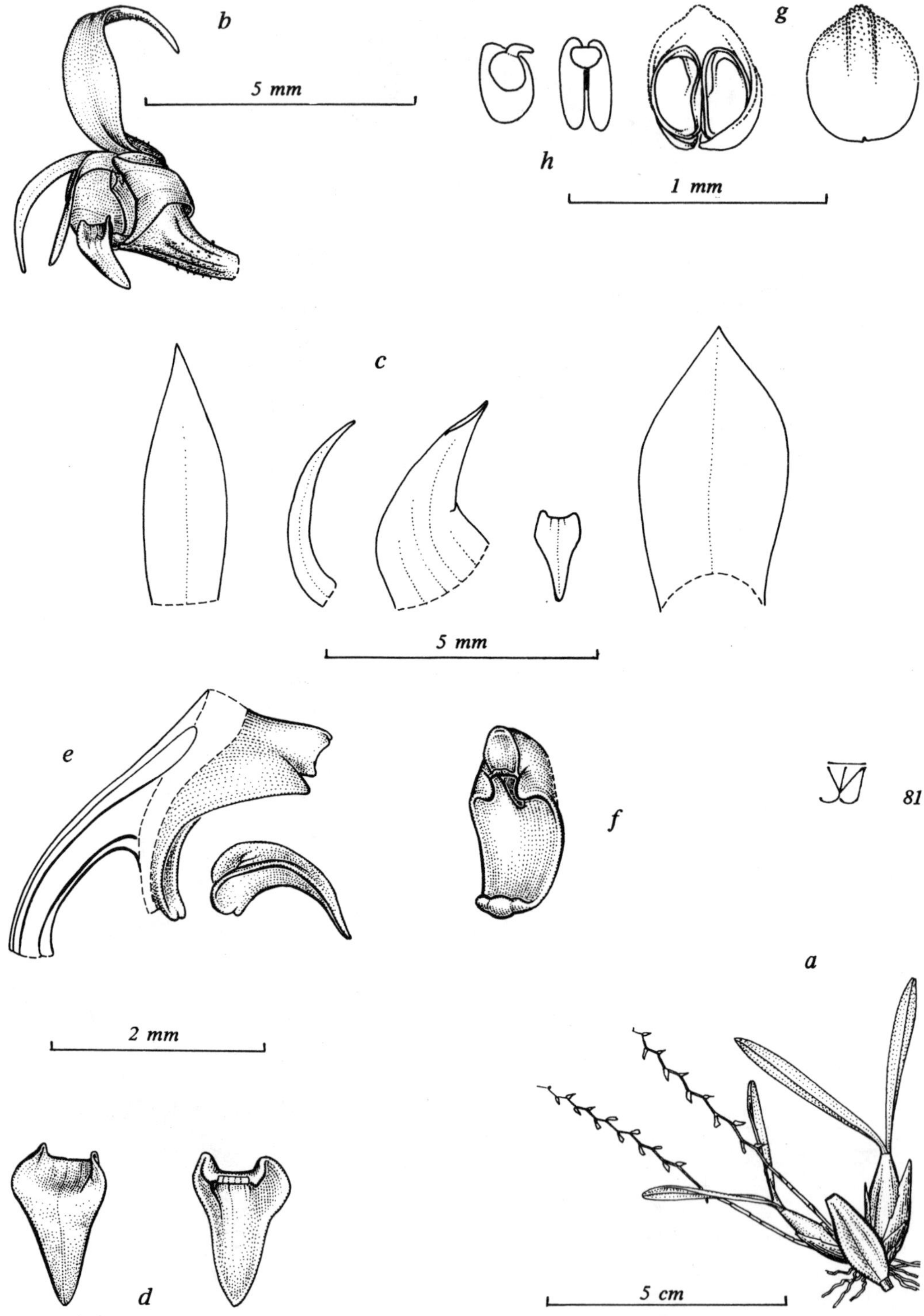

Fig. 67. *Bulbophyllum falcipetalum* Lindl.

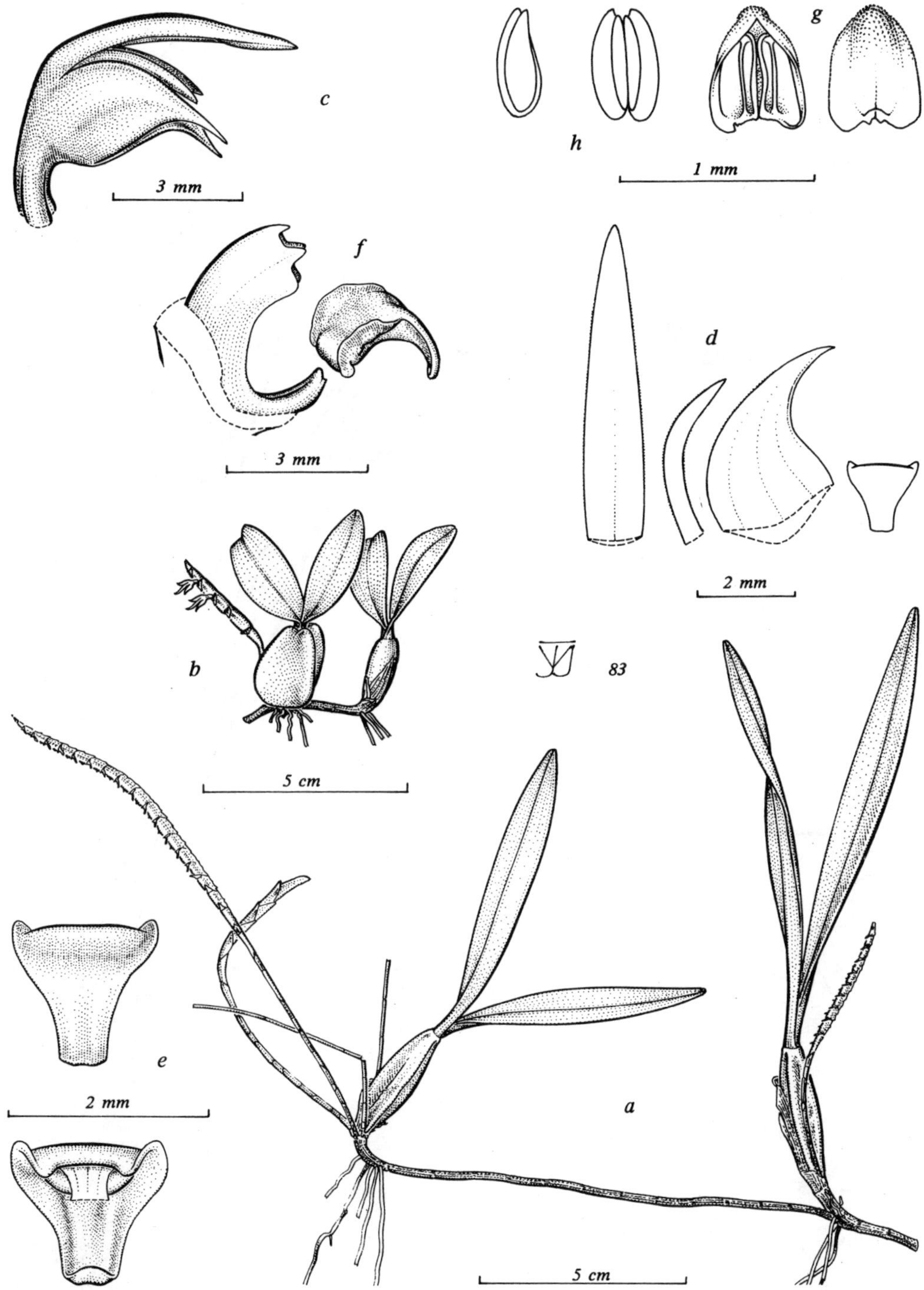

Fig. 68. *Bulbophyllum scaberulum* (Rolfe) Bolus var. *scaberulum*, entity a. (f), entity b. (a), entity c. (b–e, g, h).

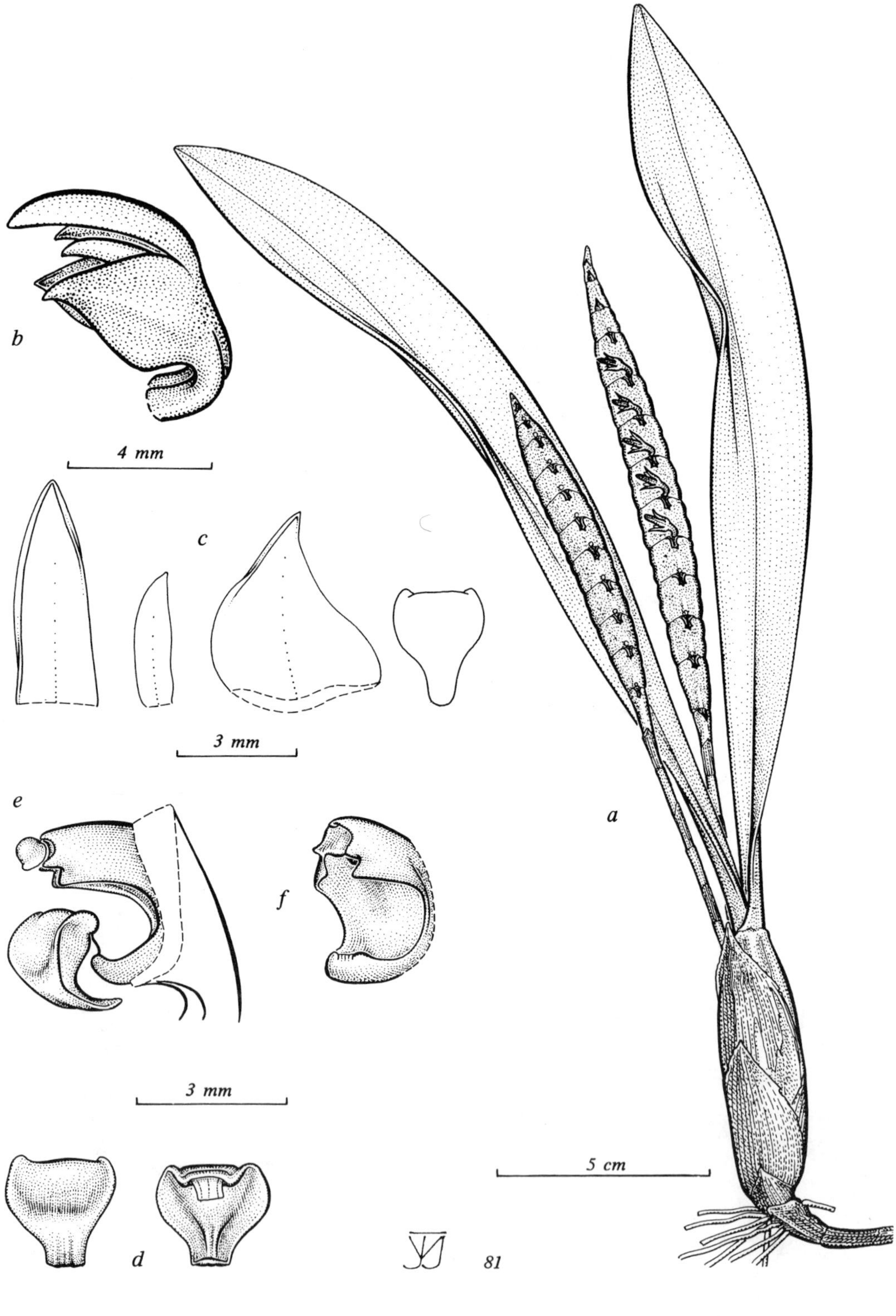

Fig. 69. *Bulbophyllum scaberulum* (Rolfe) Bolus var. *fuerstenbergianum* (De Wild.) J.J. Vermeulen.

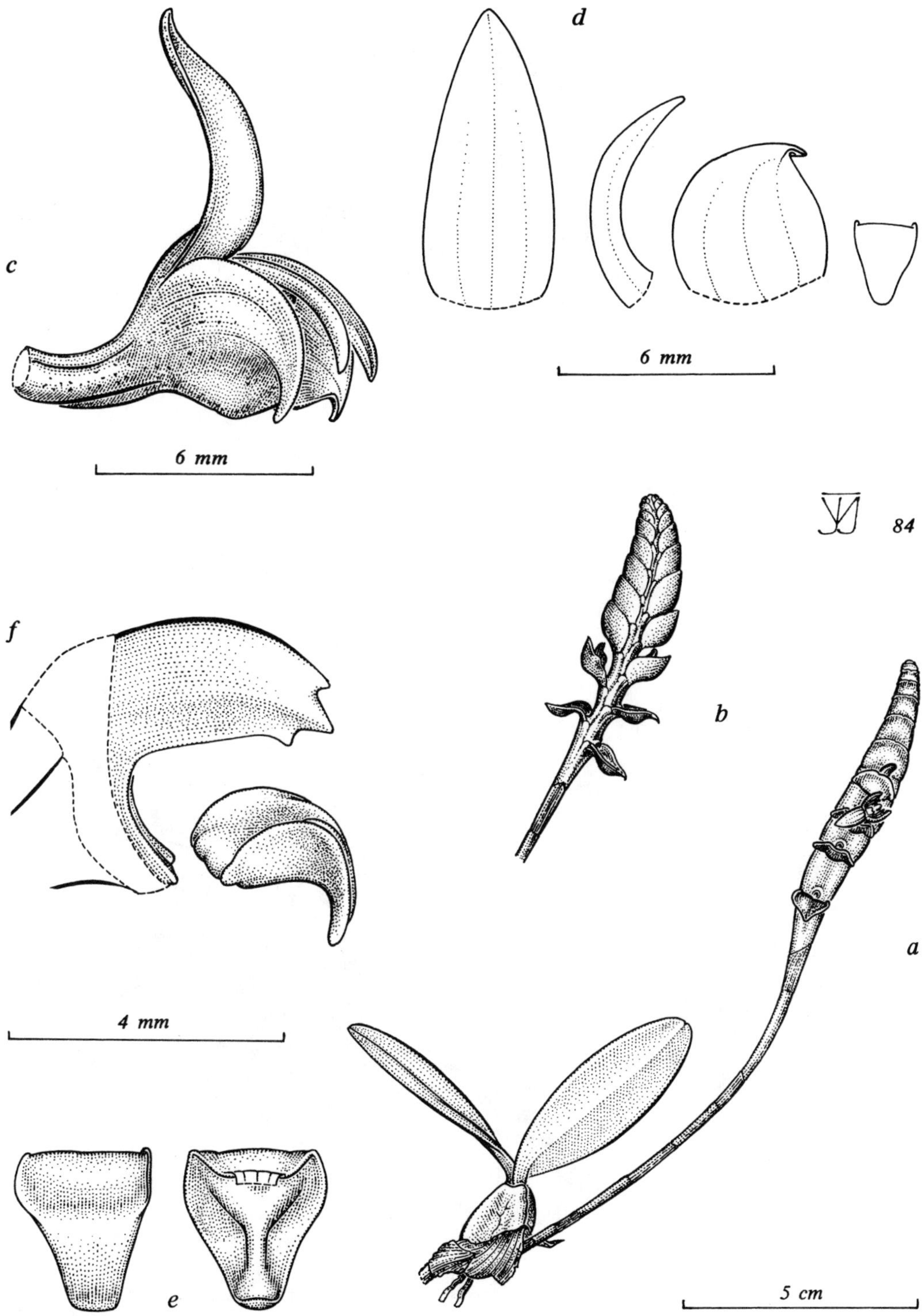

Fig. 70. *Bulbophyllum scaberulum* (Rolfe) Bolus var. *crotalicaudatum* J.J. Vermeulen.

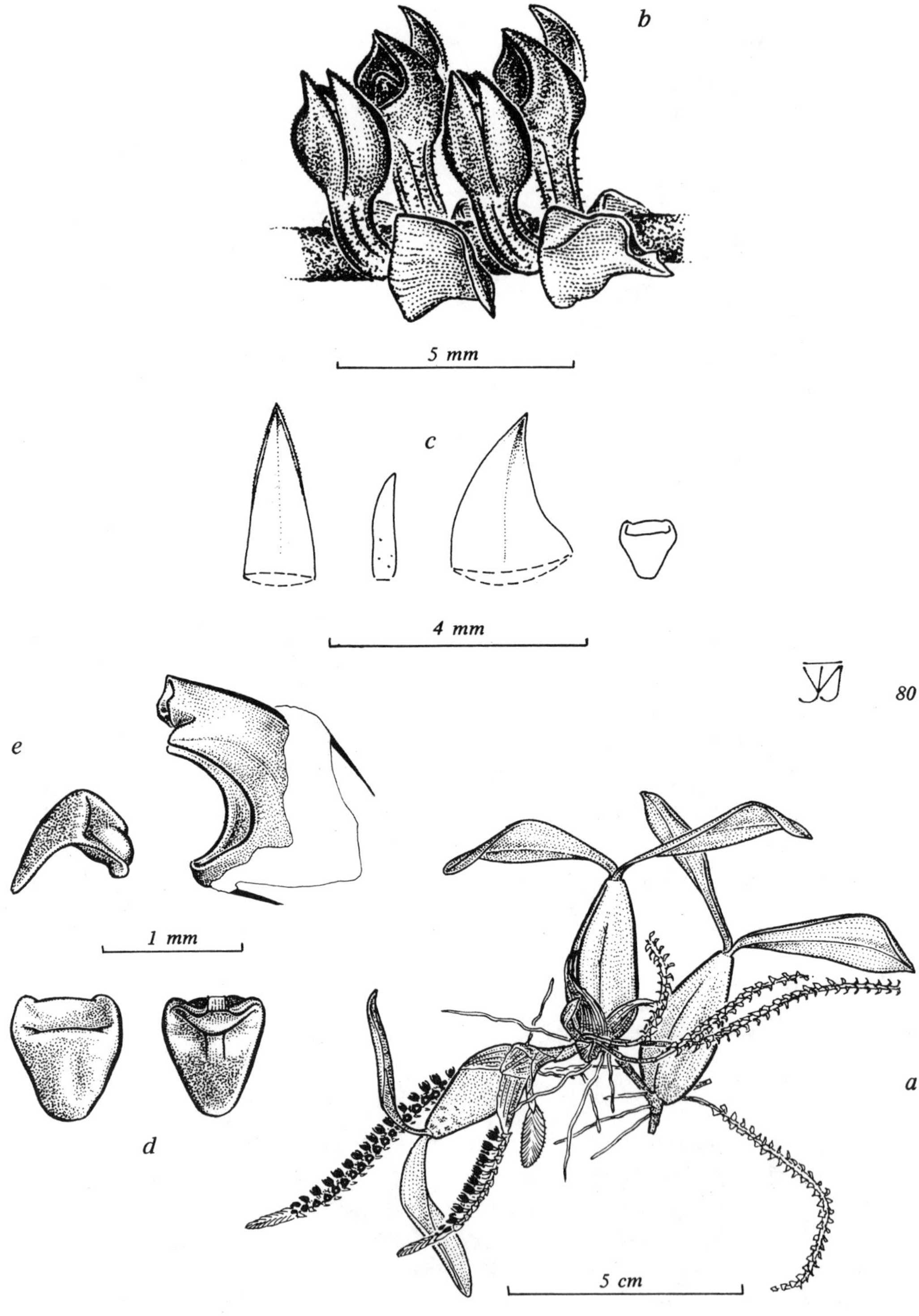

Fig. 71. *Bulbophyllum resupinatum* Ridl. var. *resupinatum*.

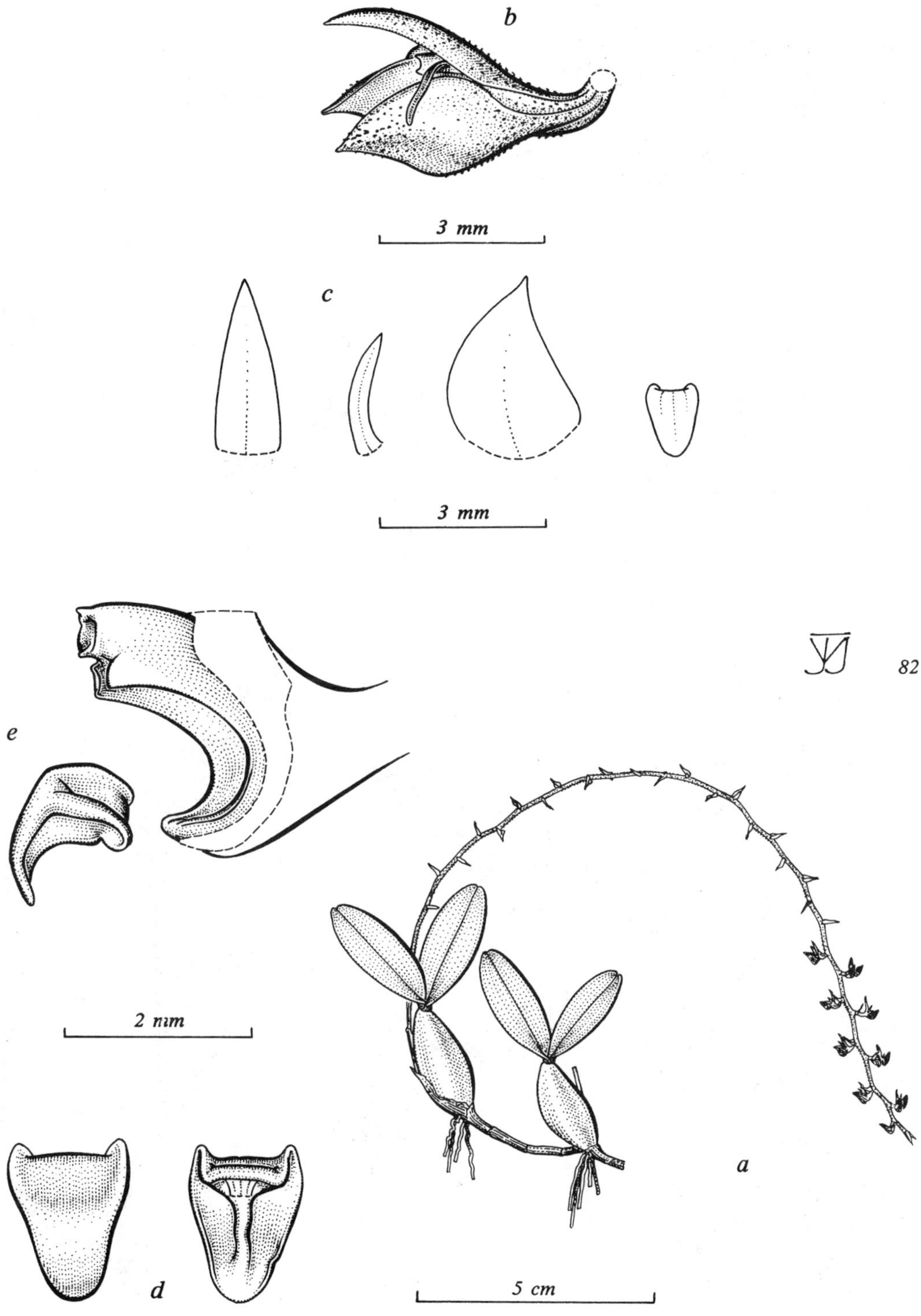

Fig. 72. *Bulbophyllum resupinatum* Ridl. var. *filiforme* (Krzl.) J.J. Vermeulen, entity a.

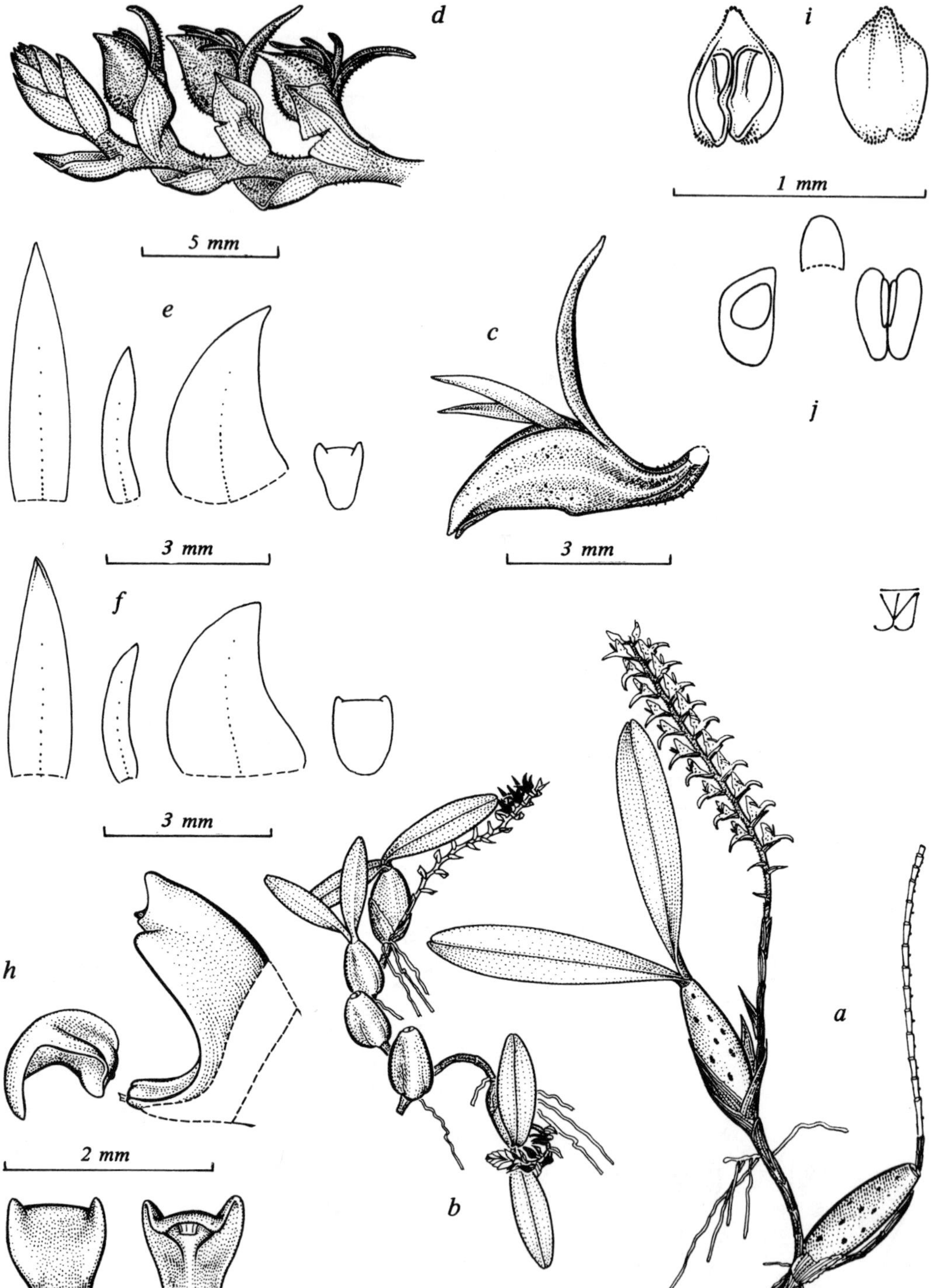

Fig. 73. *Bulbophyllum resupinatum* Ridl. var. *filiforme* (Krzl.) J.J. Vermeulen, entity b.

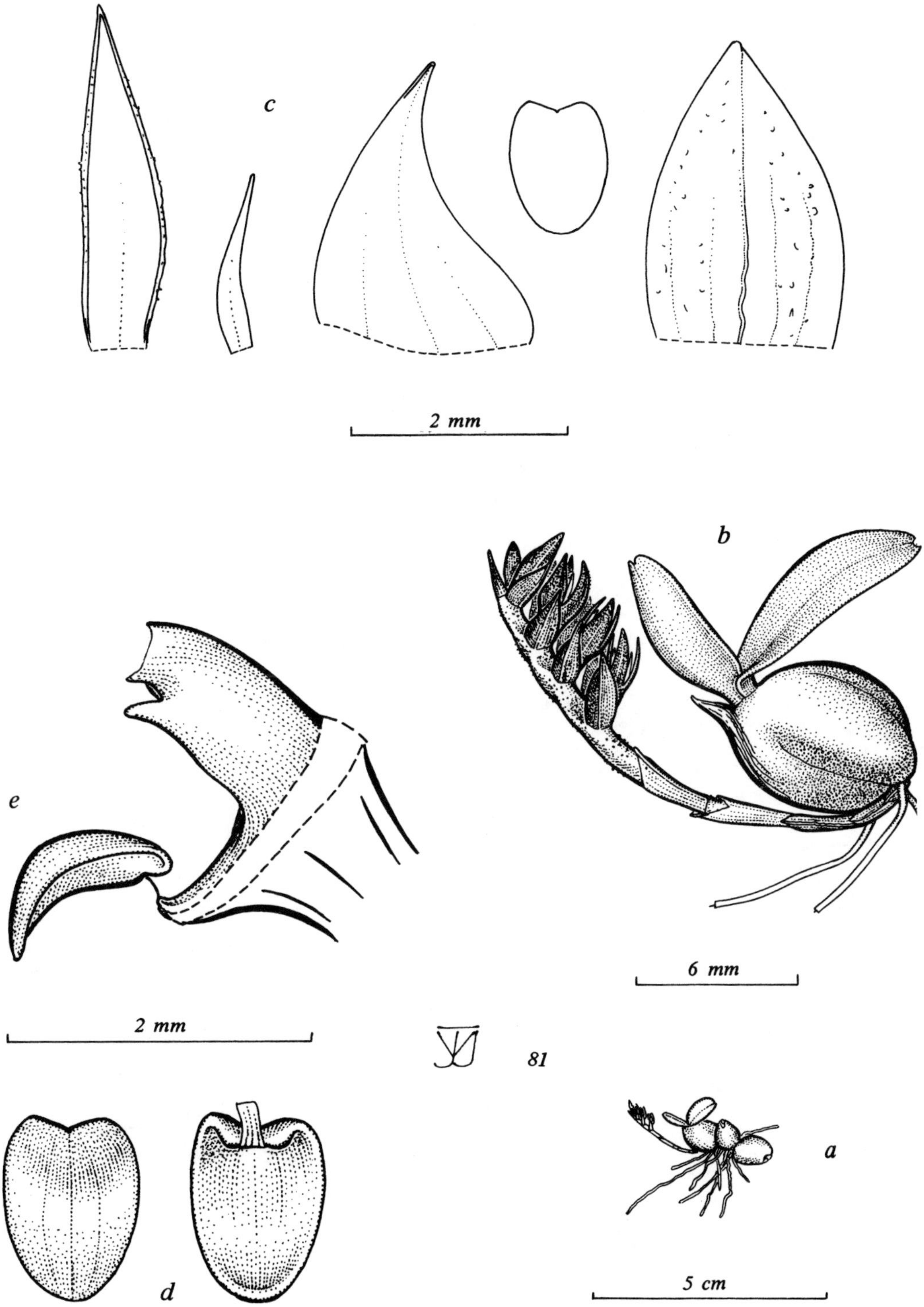

Fig. 74. *Bulbophyllum parvum* Summerh.

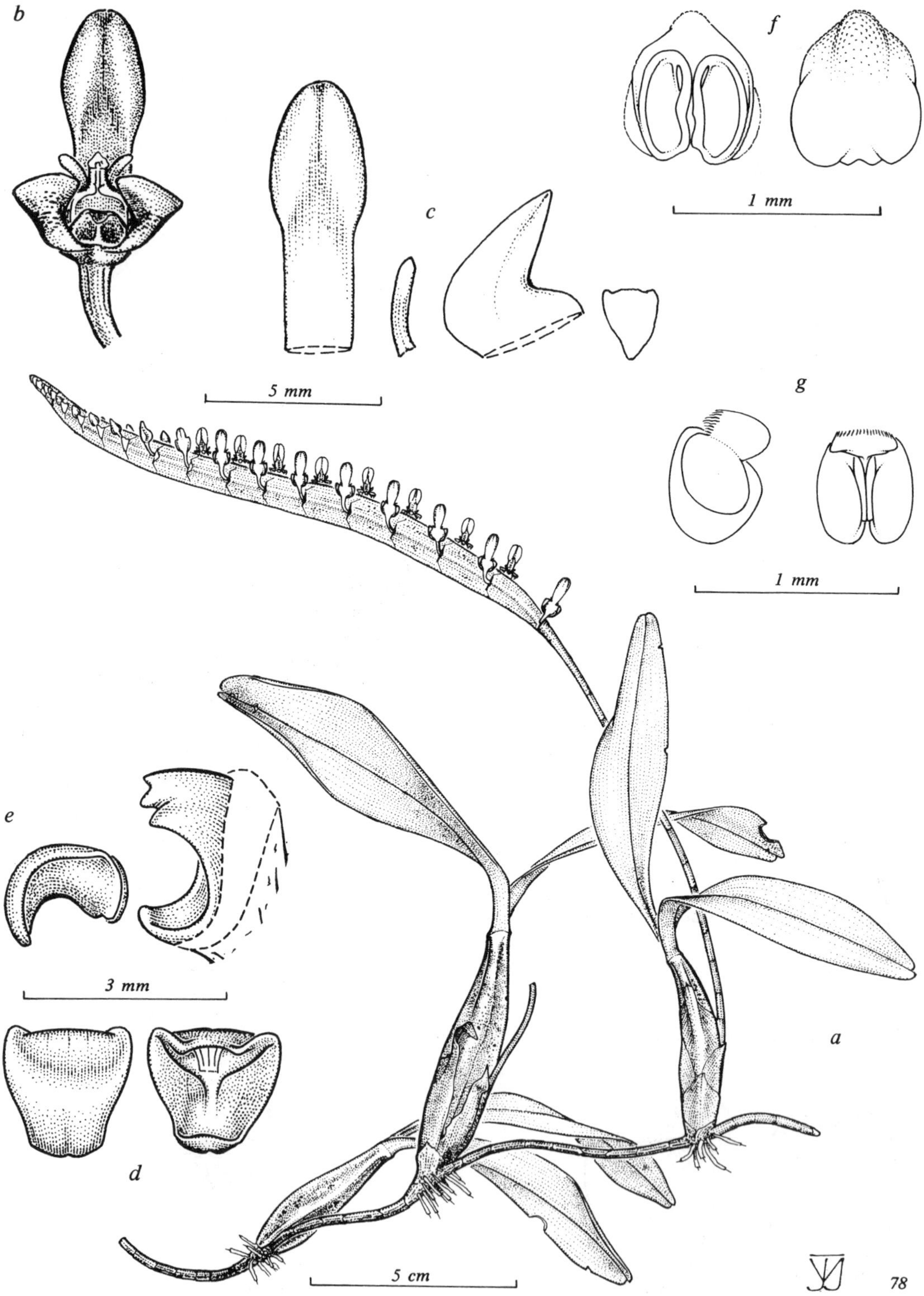

Fig. 75. *Bulbophyllum falcatum* (Lindl.) Rchb. f. var. *falcatum*.

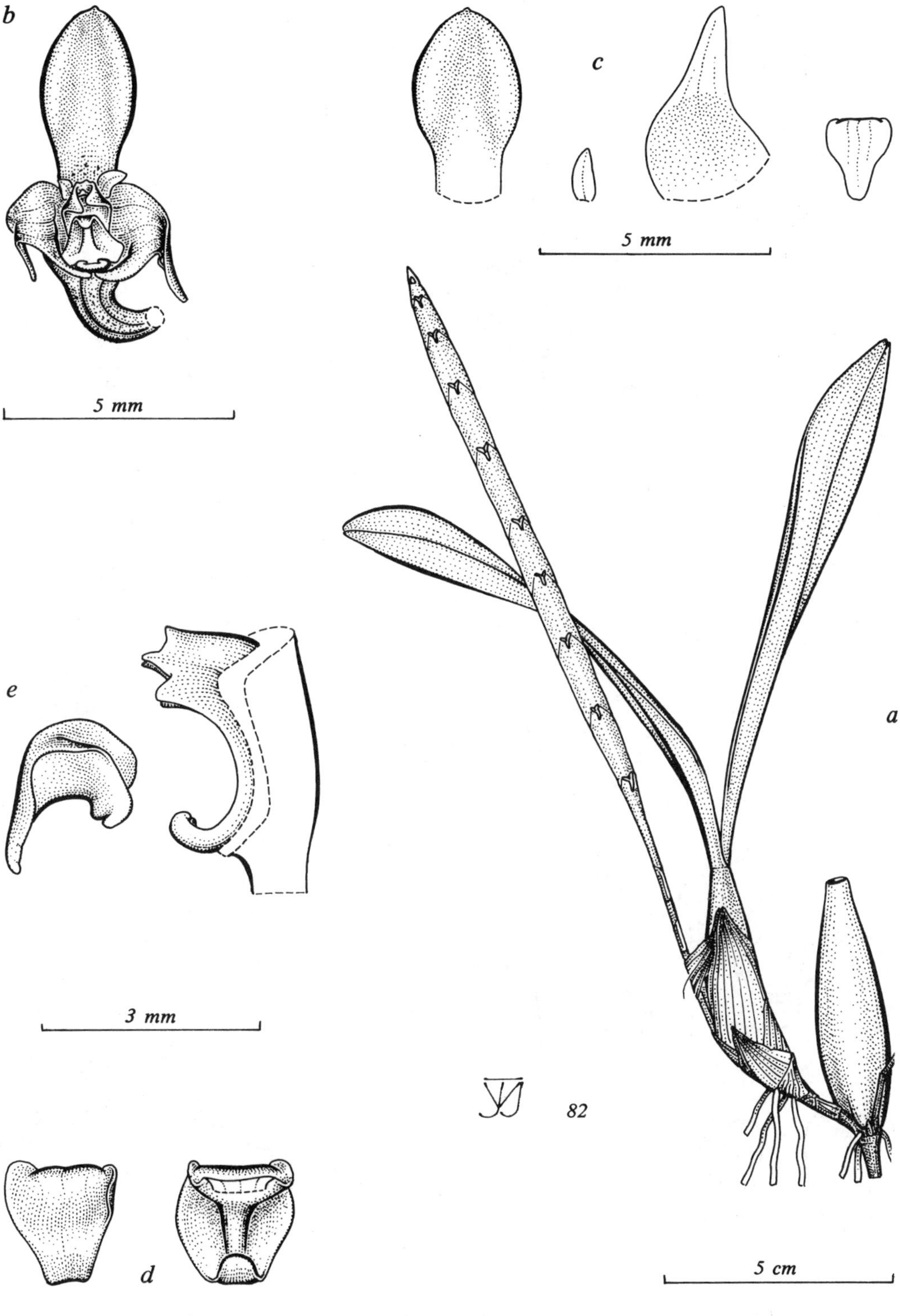

Fig. 76. *Bulbophyllum falcatum* (Lindl.) Rchb. f. var. *bufo* (Lindl.) J.J. Vermeulen.

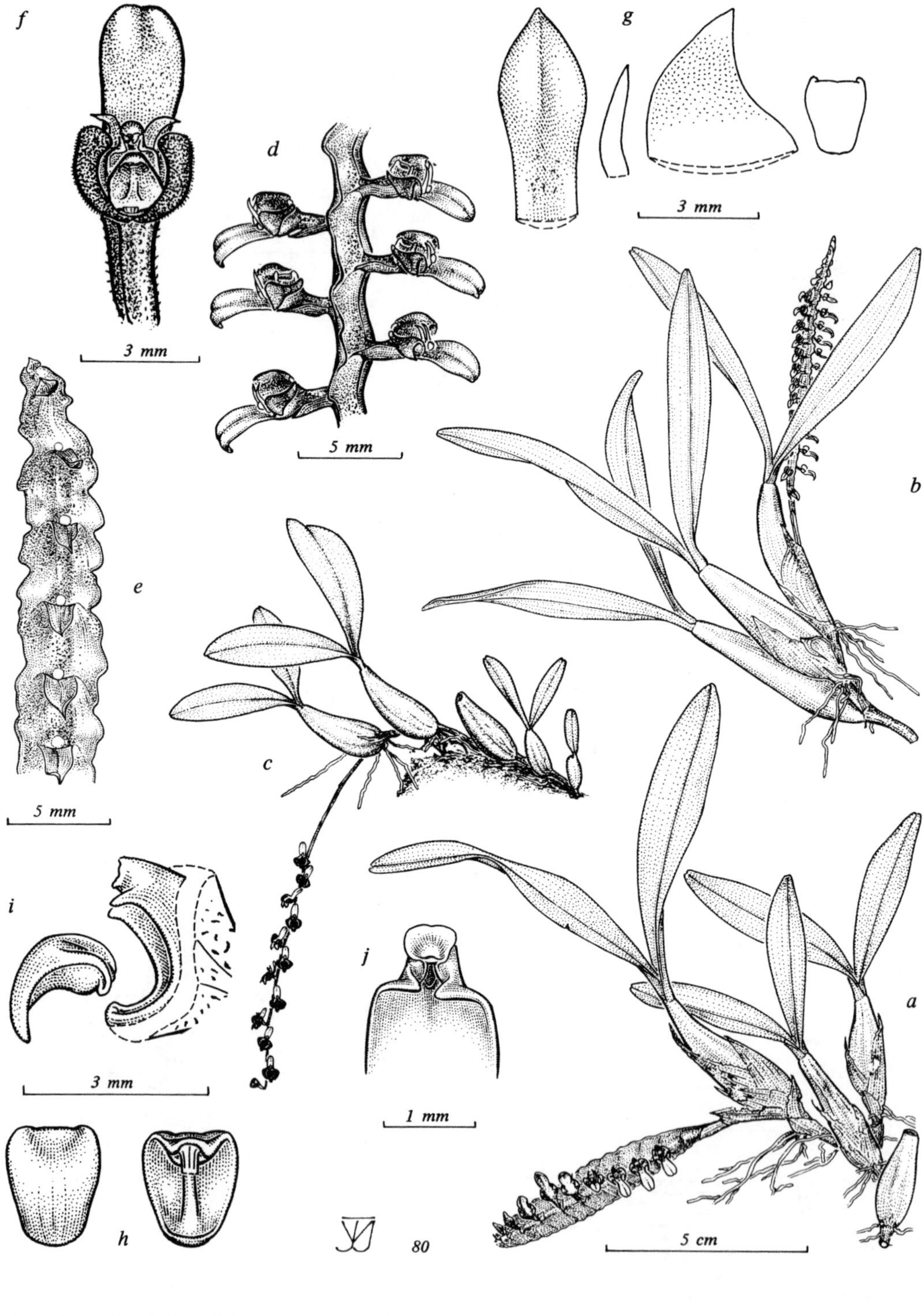

Fig. 77. *Bulbophyllum falcatum* (Lindl.) Rchb. f. var. *velutinum* (Lindl.) J.J. Vermeulen, entity a. (a, c, f–j), entity b. (b, d, e).

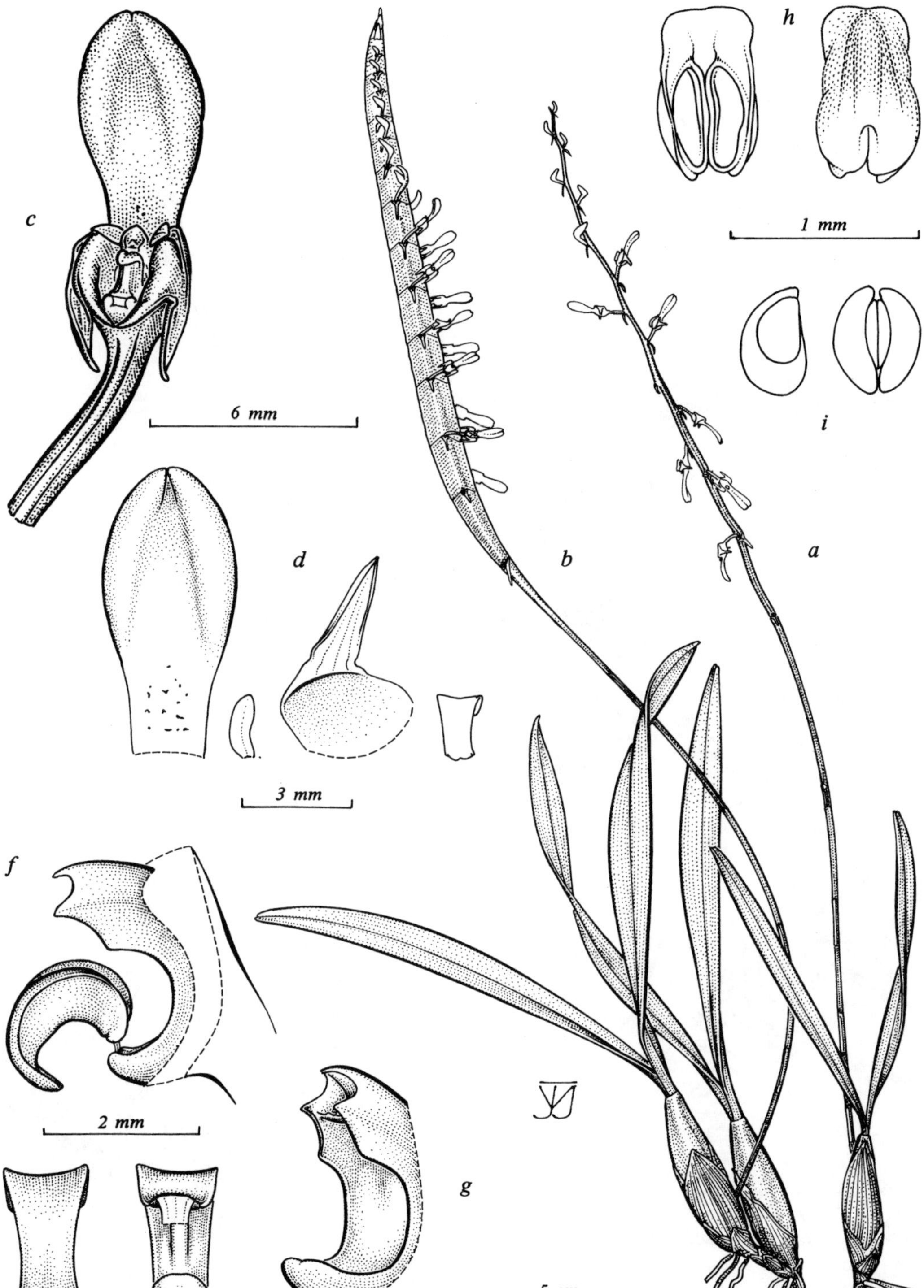

Fig. 78. *Bulbophyllum calyptratum* Krzl. *calyptratum*.

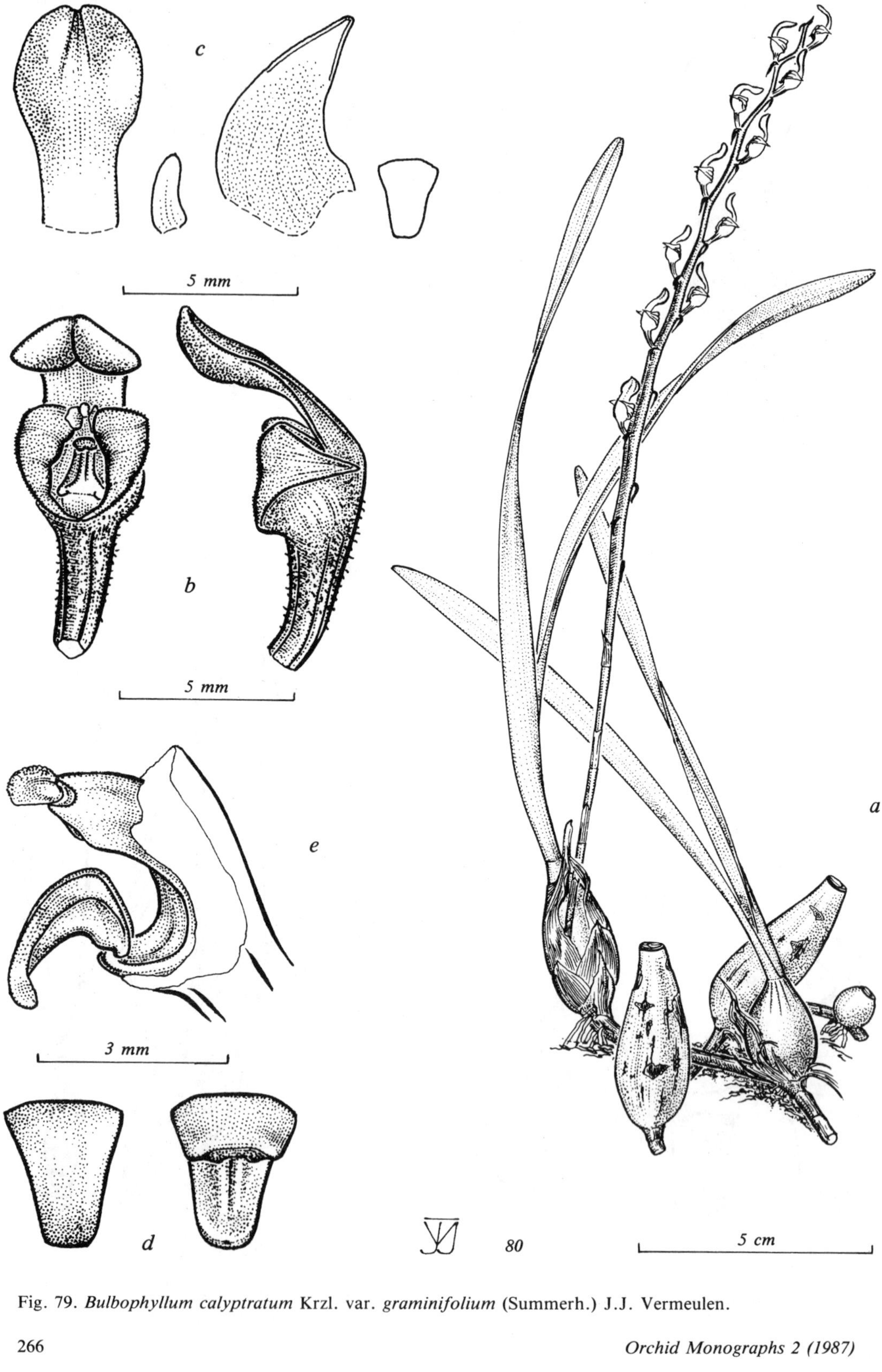

Fig. 79. *Bulbophyllum calyptratum* Krzl. var. *graminifolium* (Summerh.) J.J. Vermeulen.

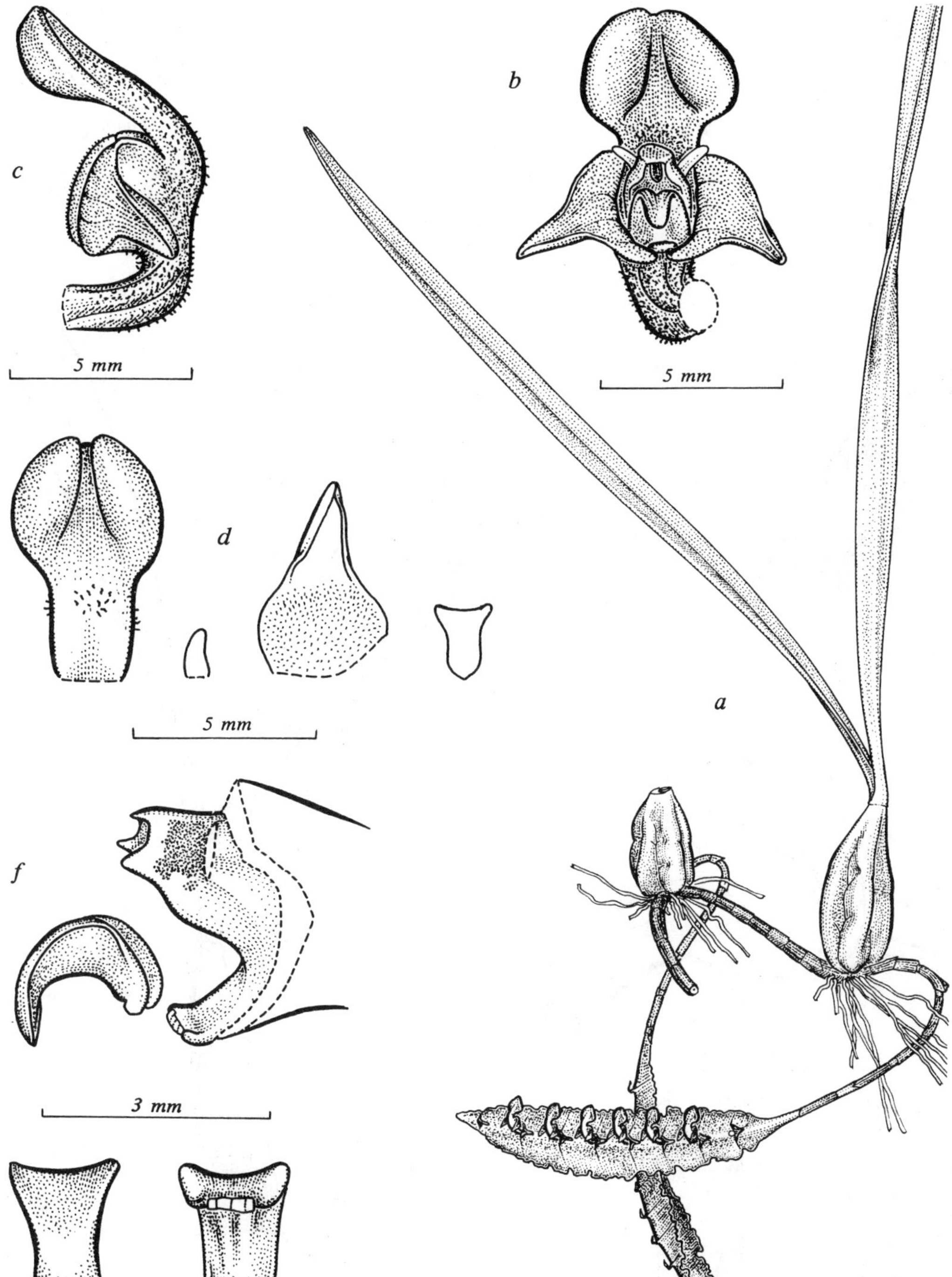

Fig. 80. *Bulbophyllum calyptratum* Krzl. var. *lucifugum* (Summerh.) J.J. Vermeulen.

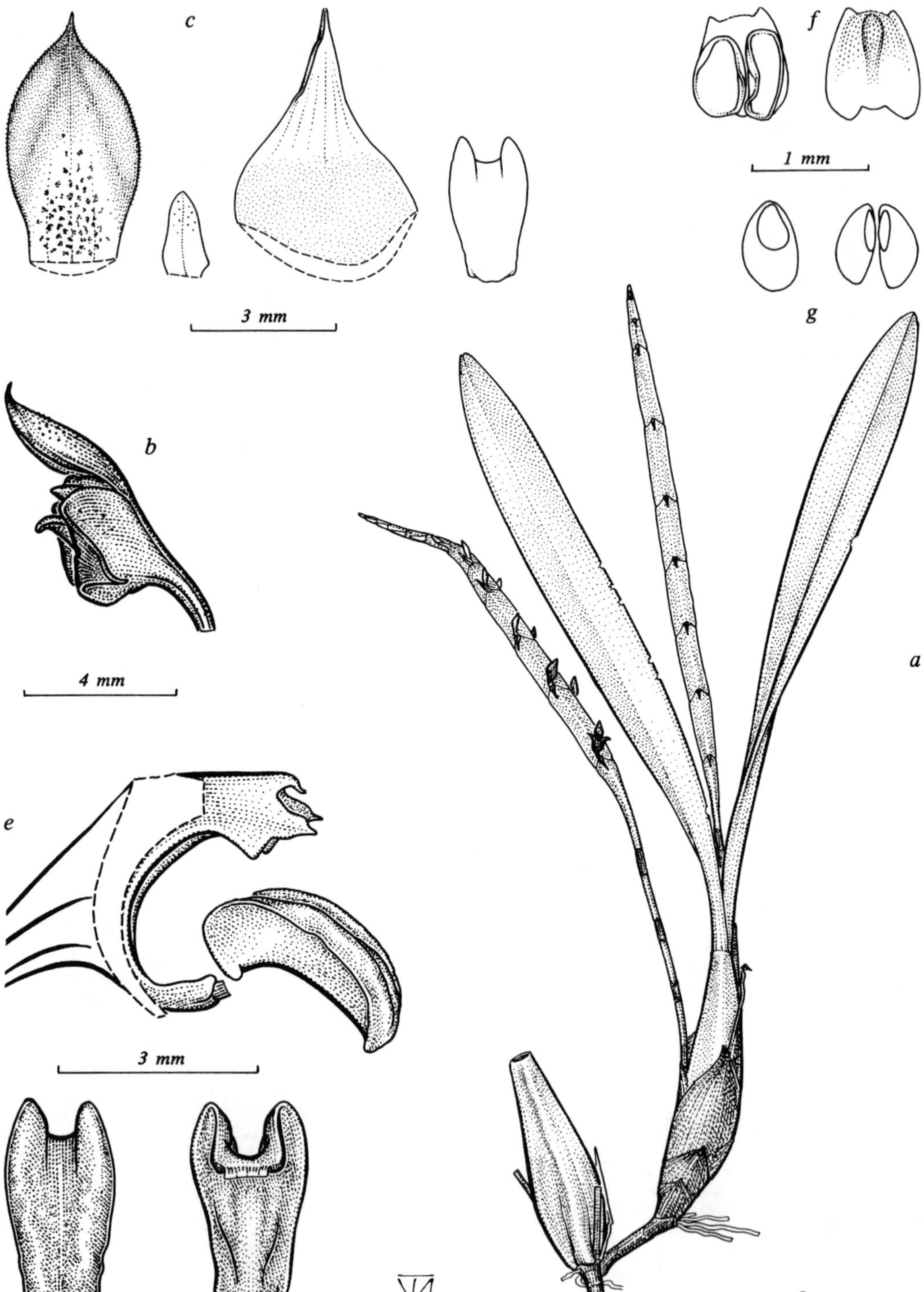

Fig. 81. *Bulbophyllum carnosilabium* Summerh.

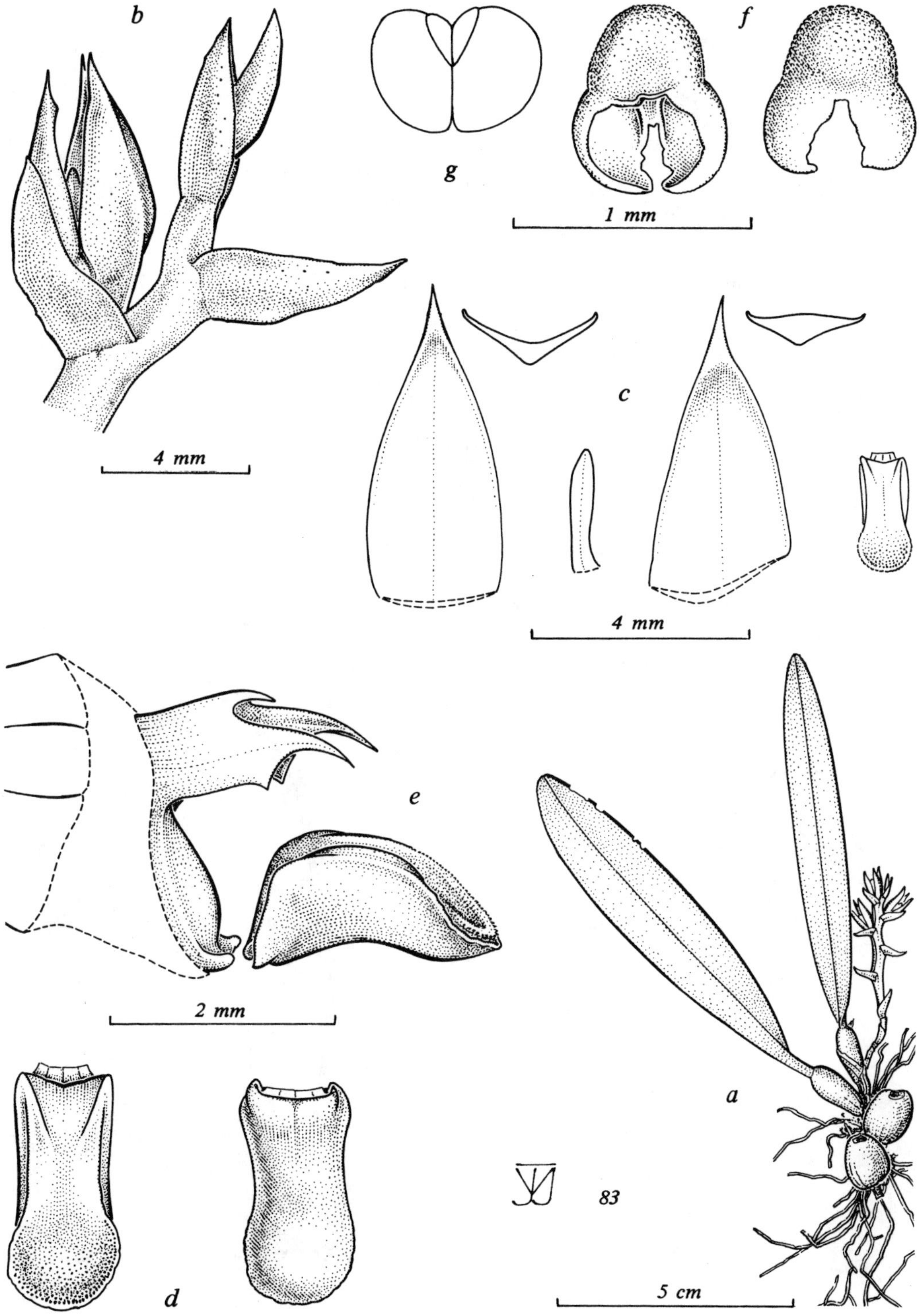

Fig. 82. *Bulbophyllum carnosisepalum* J.J. Vermeulen.

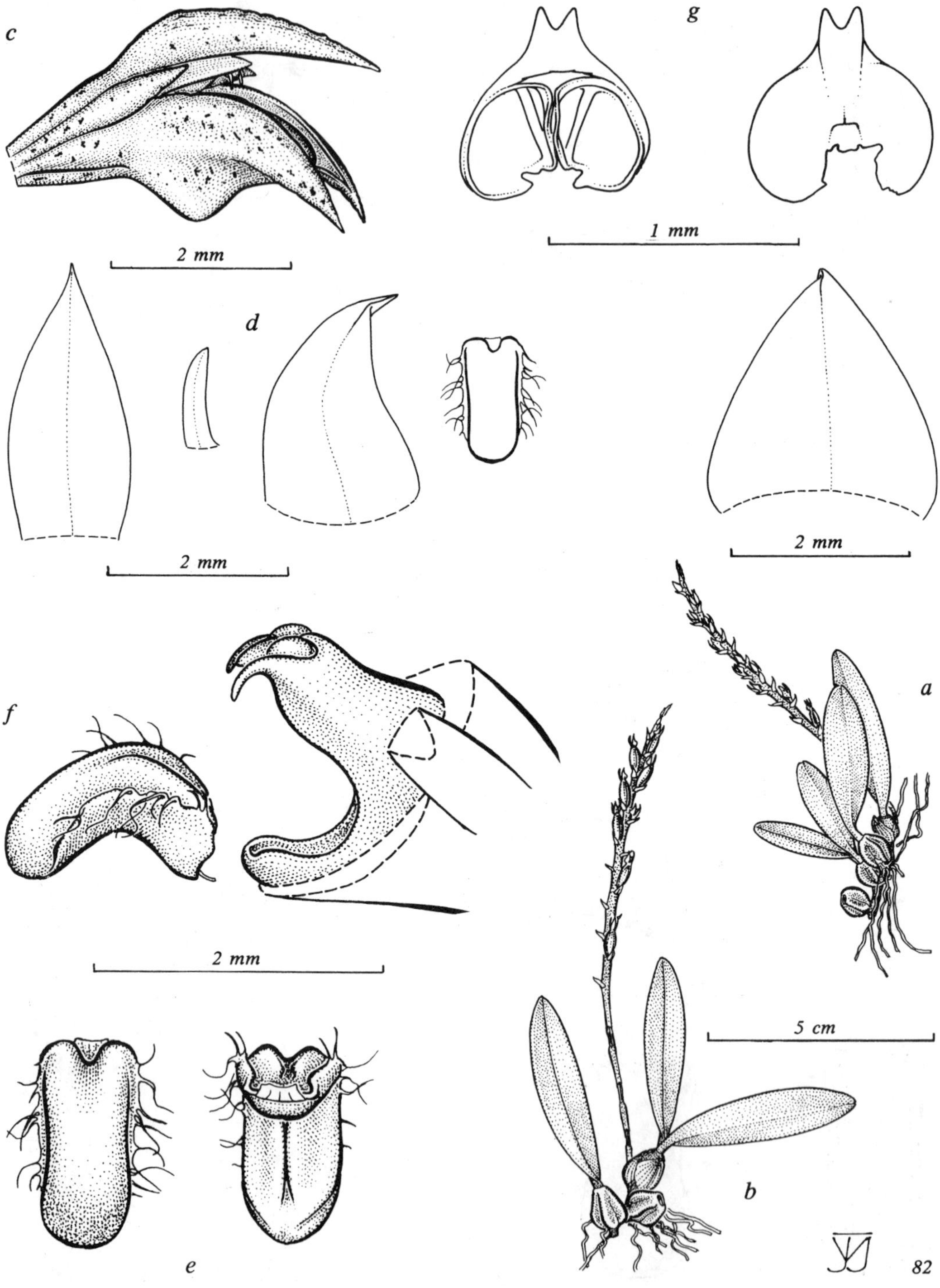

Fig. 83. *Bulbophyllum ballii* Cribb.

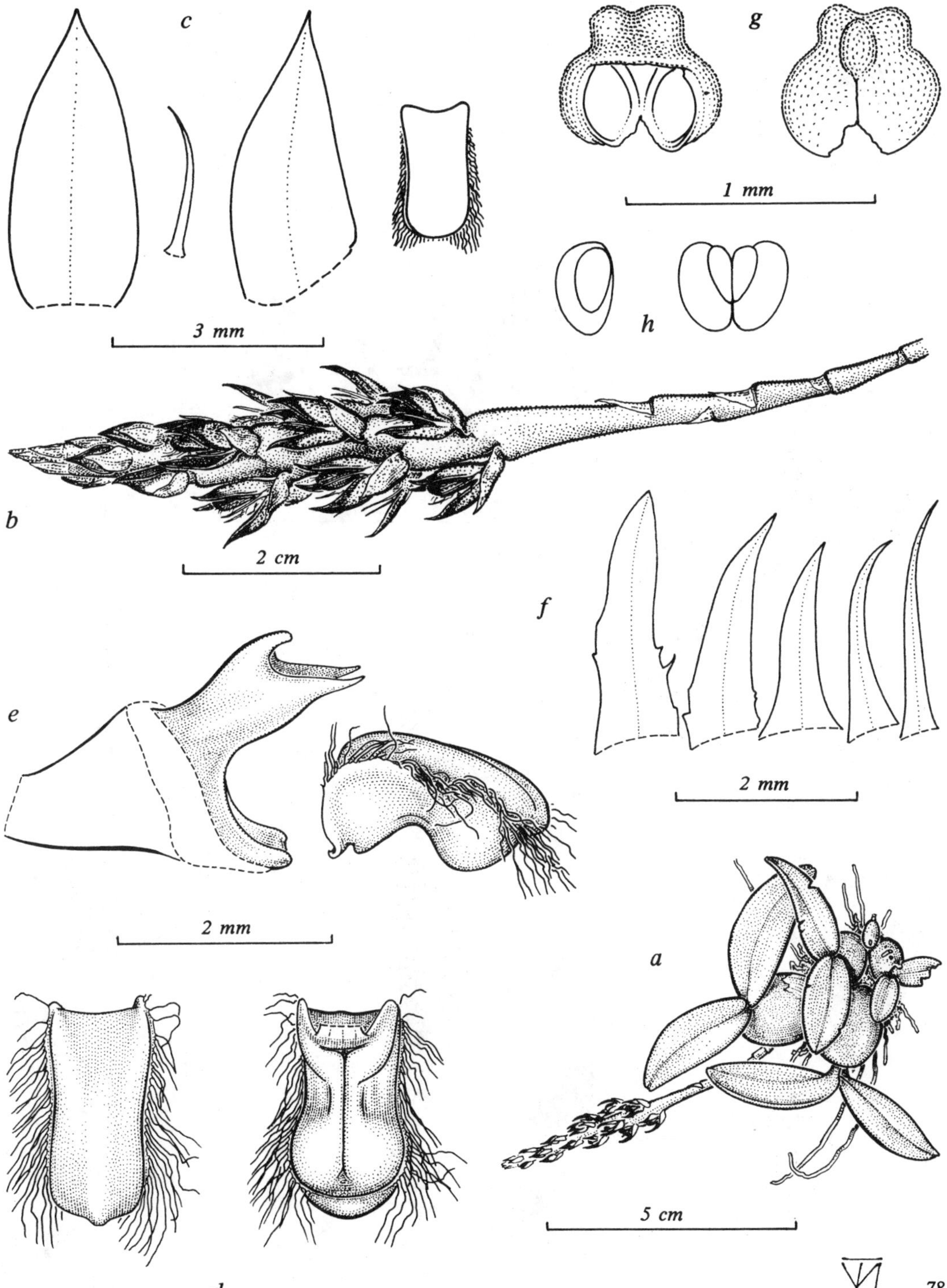

Fig. 84. *Bulbophyllum elliotii* Rolfe.

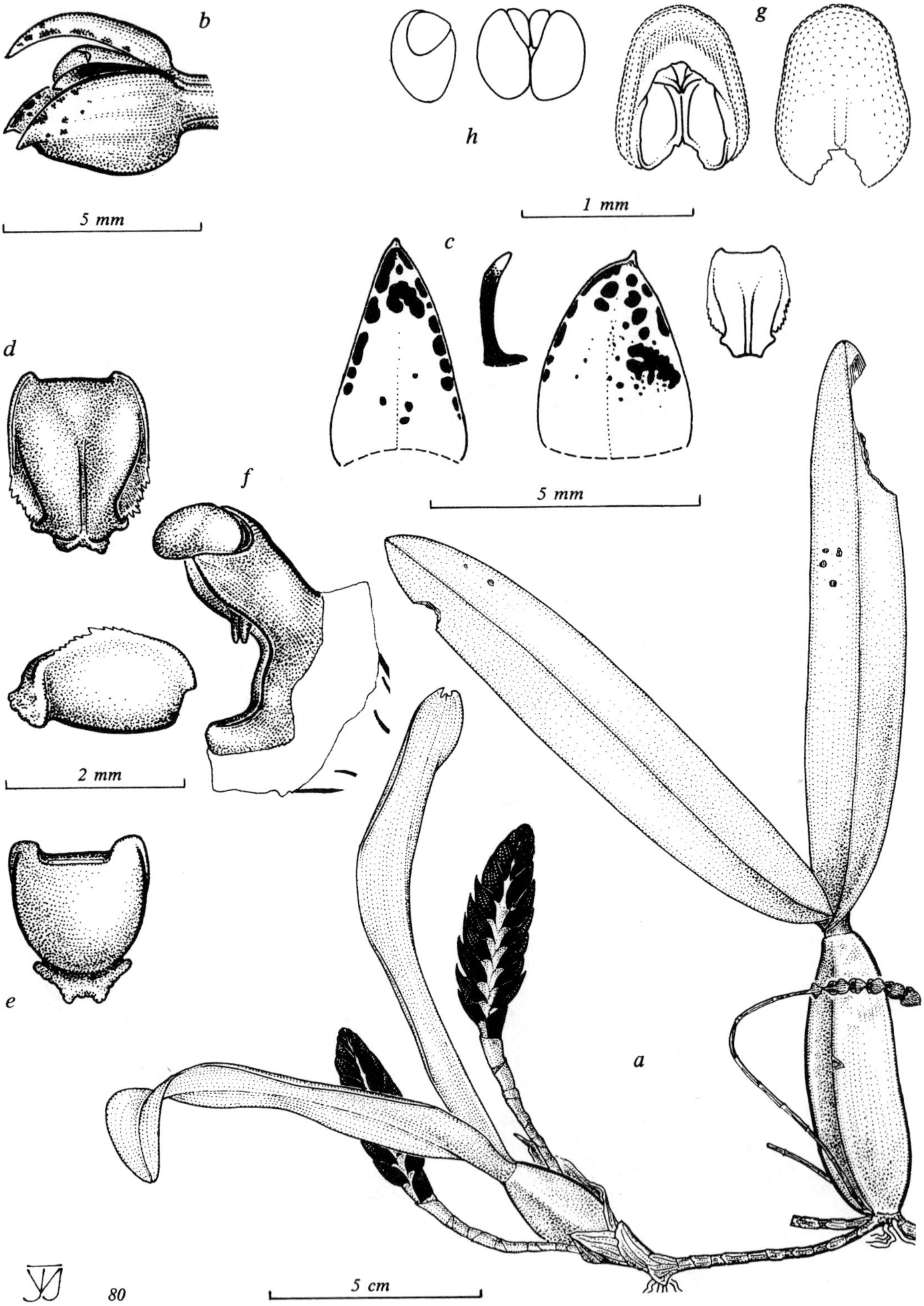

Fig. 85. *Bulbophyllum porphyrostachys* Summerh.

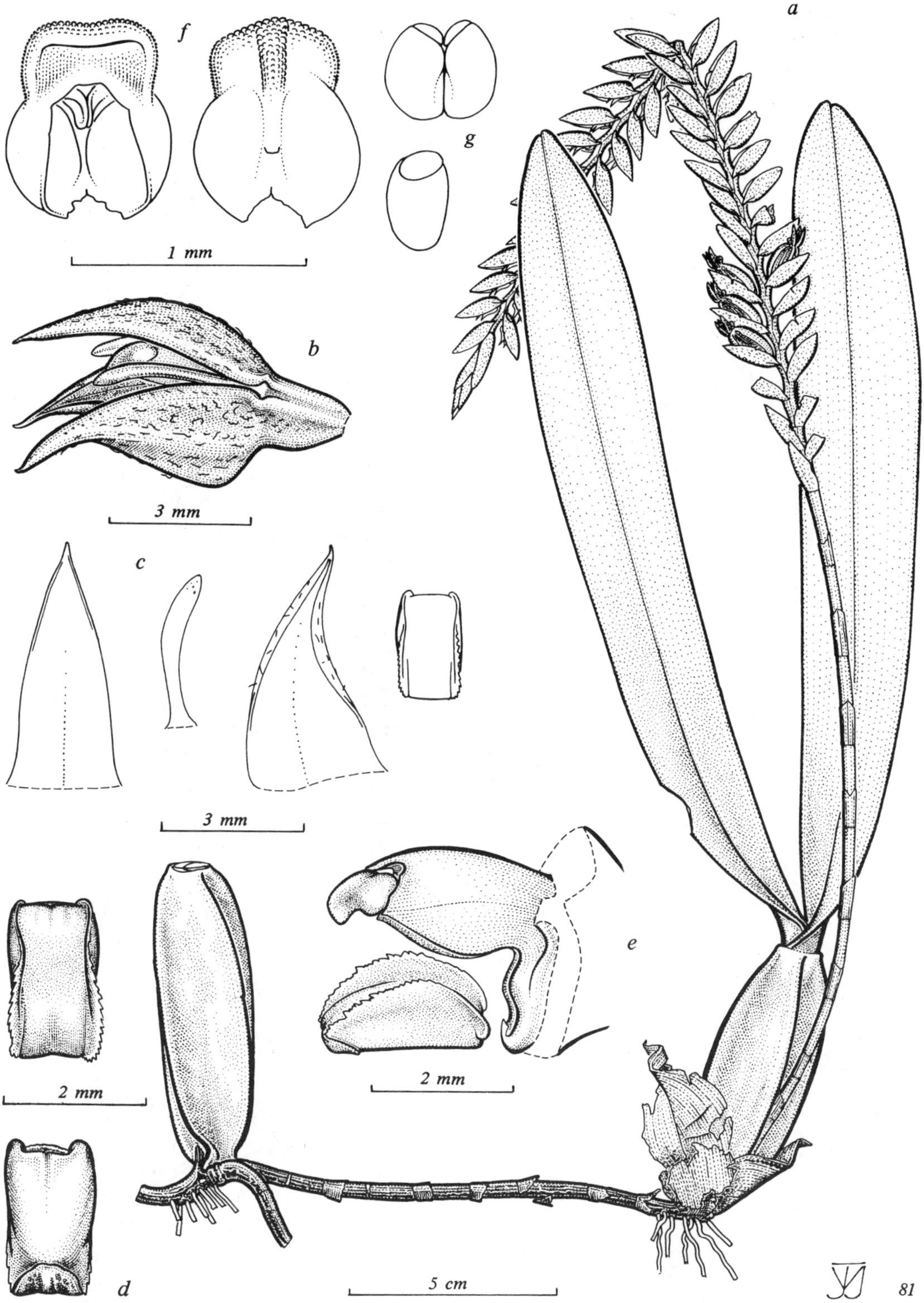

Fig. 86. *Bulbophyllum lupulinum* Lindl.

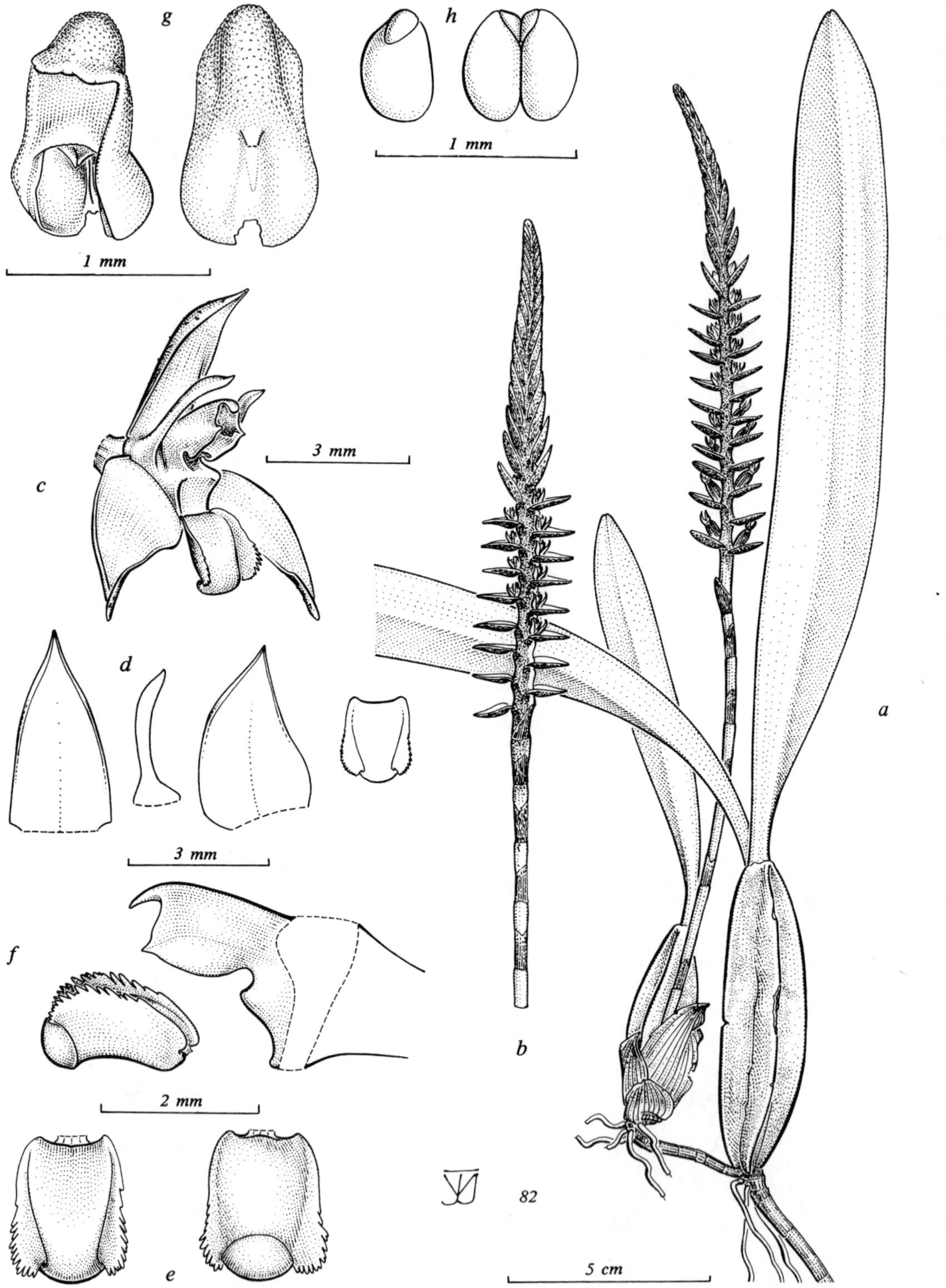

Fig. 87. *Bulbophyllum tetragonum* Lindl.

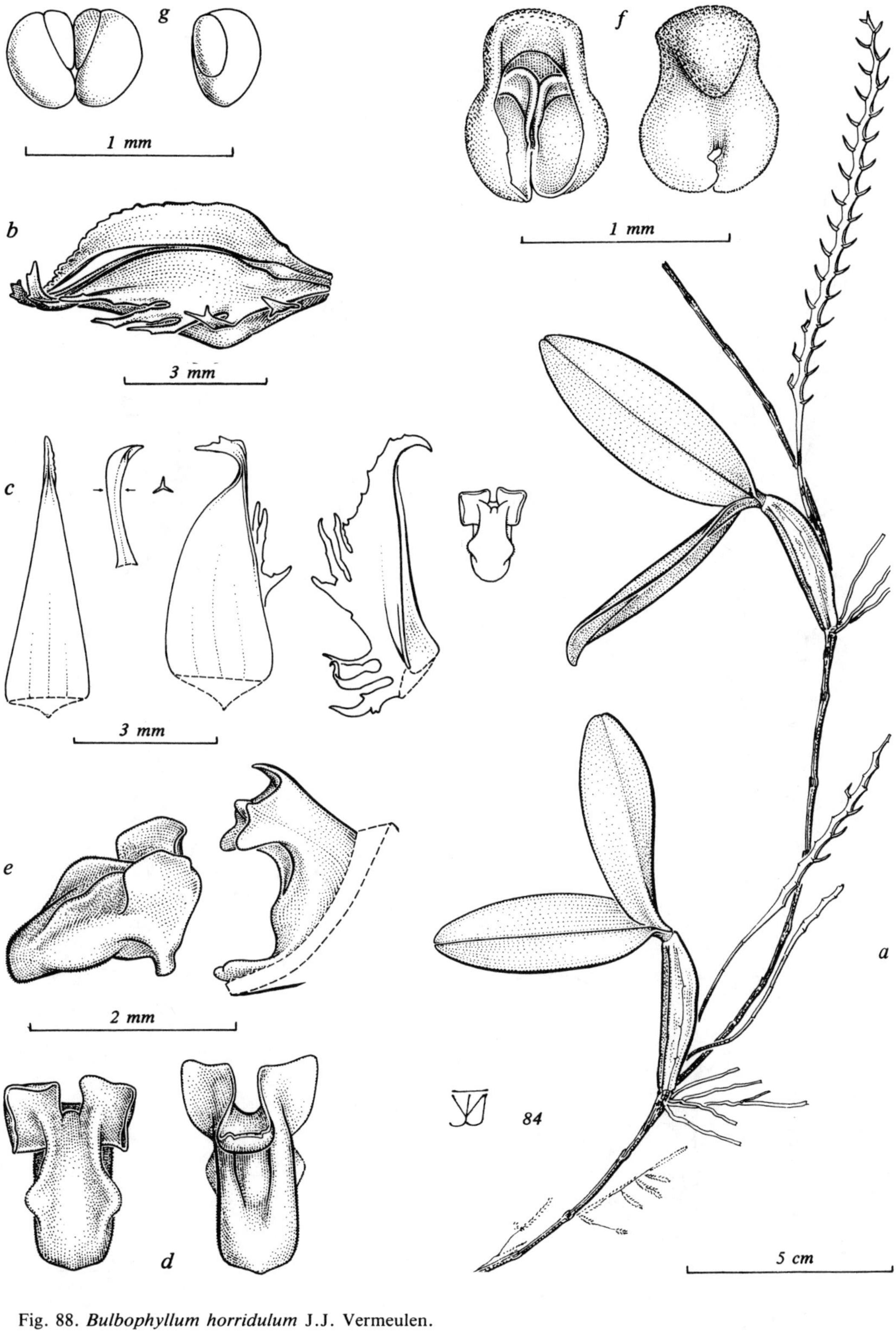

Fig. 88. *Bulbophyllum horridulum* J.J. Vermeulen.

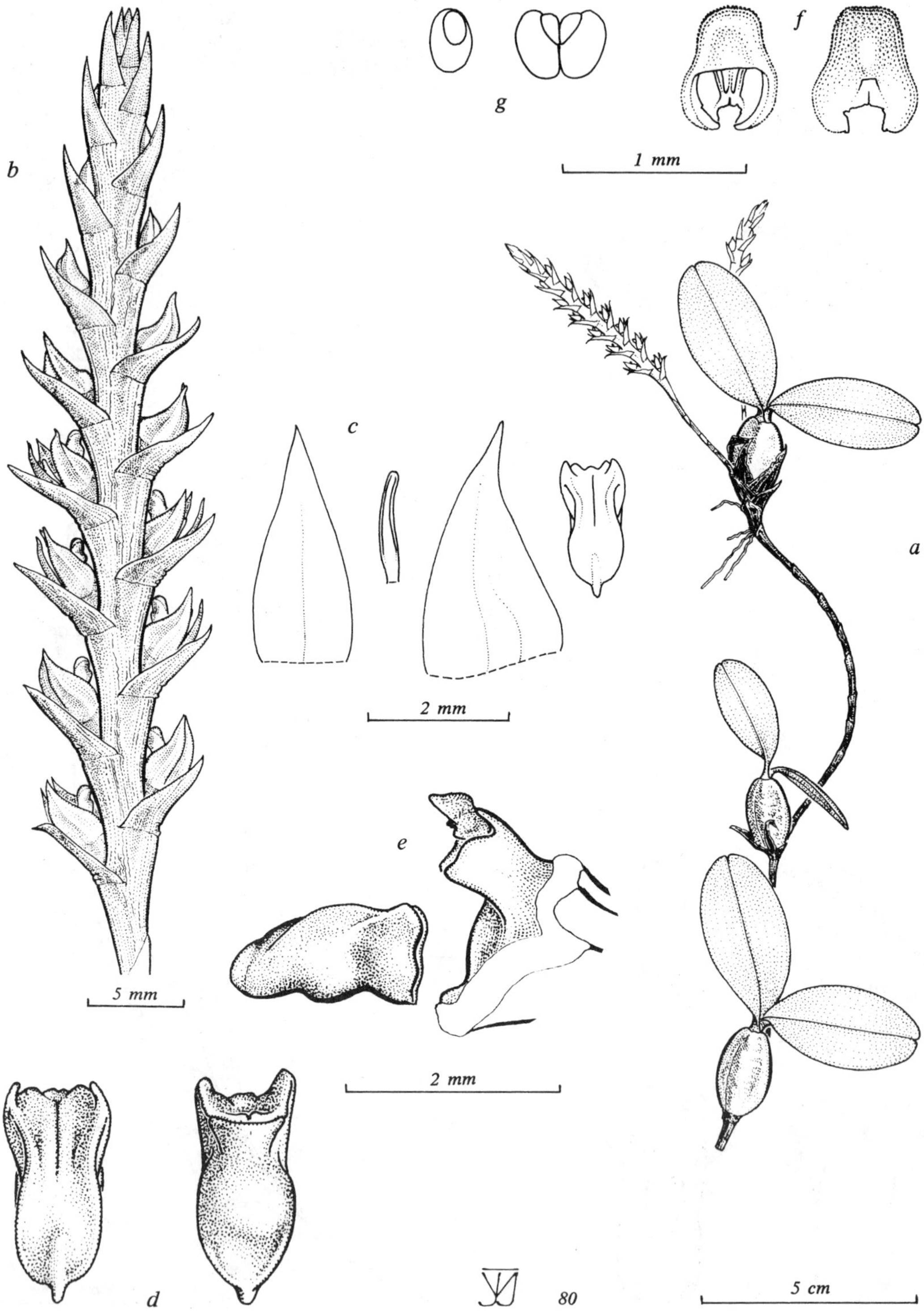

Fig. 89. *Bulbophyllum acutebracteatum* De Wild. var. *acutebracteatum*.

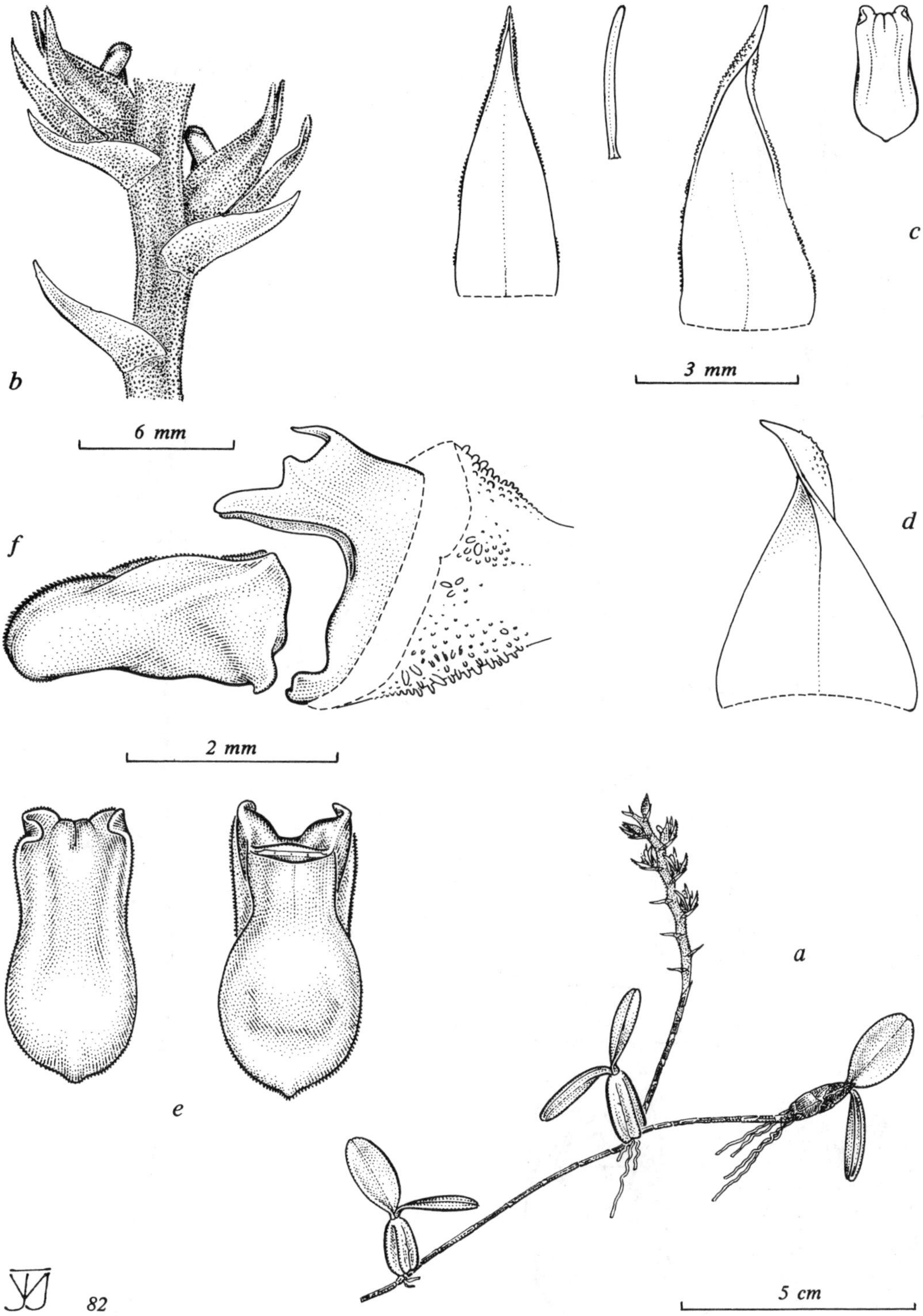

Fig. 90. *Bulbophyllum acutebracteatum* De Wild. var. *rubrobrunneopapillosum* (De Wild.) J.J. Vermeulen.

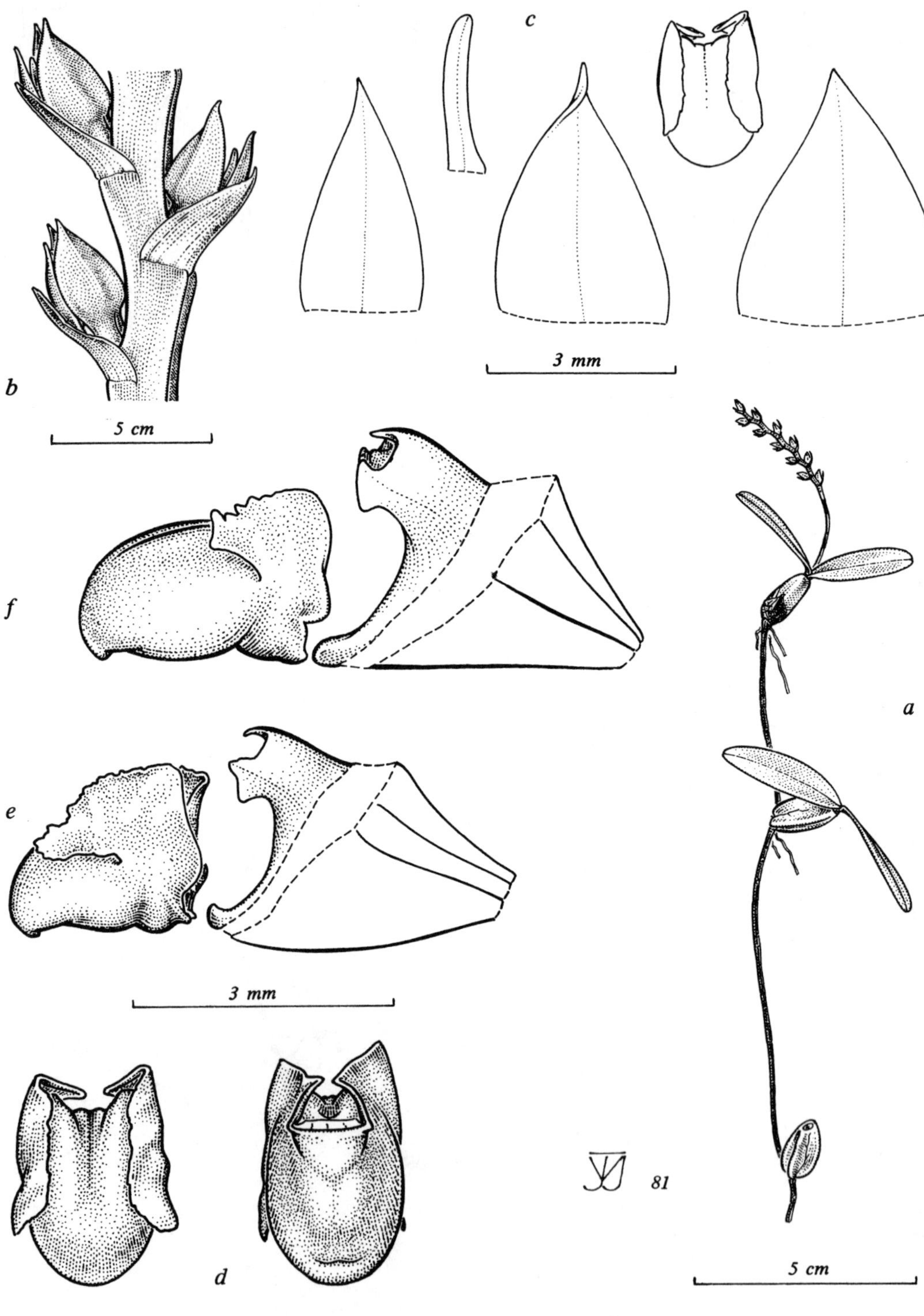

Fig. 91. *Bulbophyllum fuscum* Lindl. var. *fuscum*.

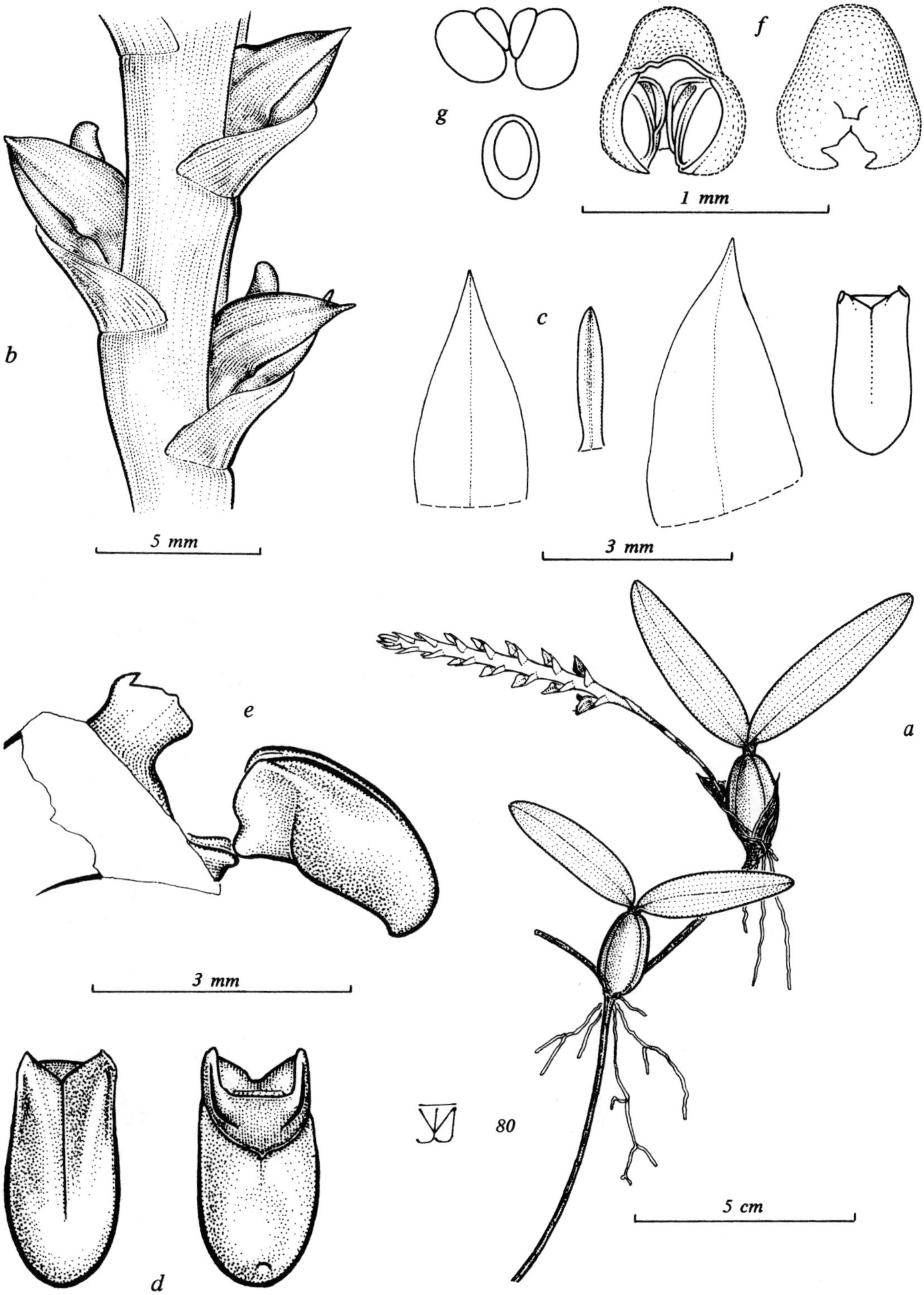

Fig. 92. *Bulbophyllum fuscum* Lindl. var. *melinostachyum* (Schltr) J.J. Vermeulen.

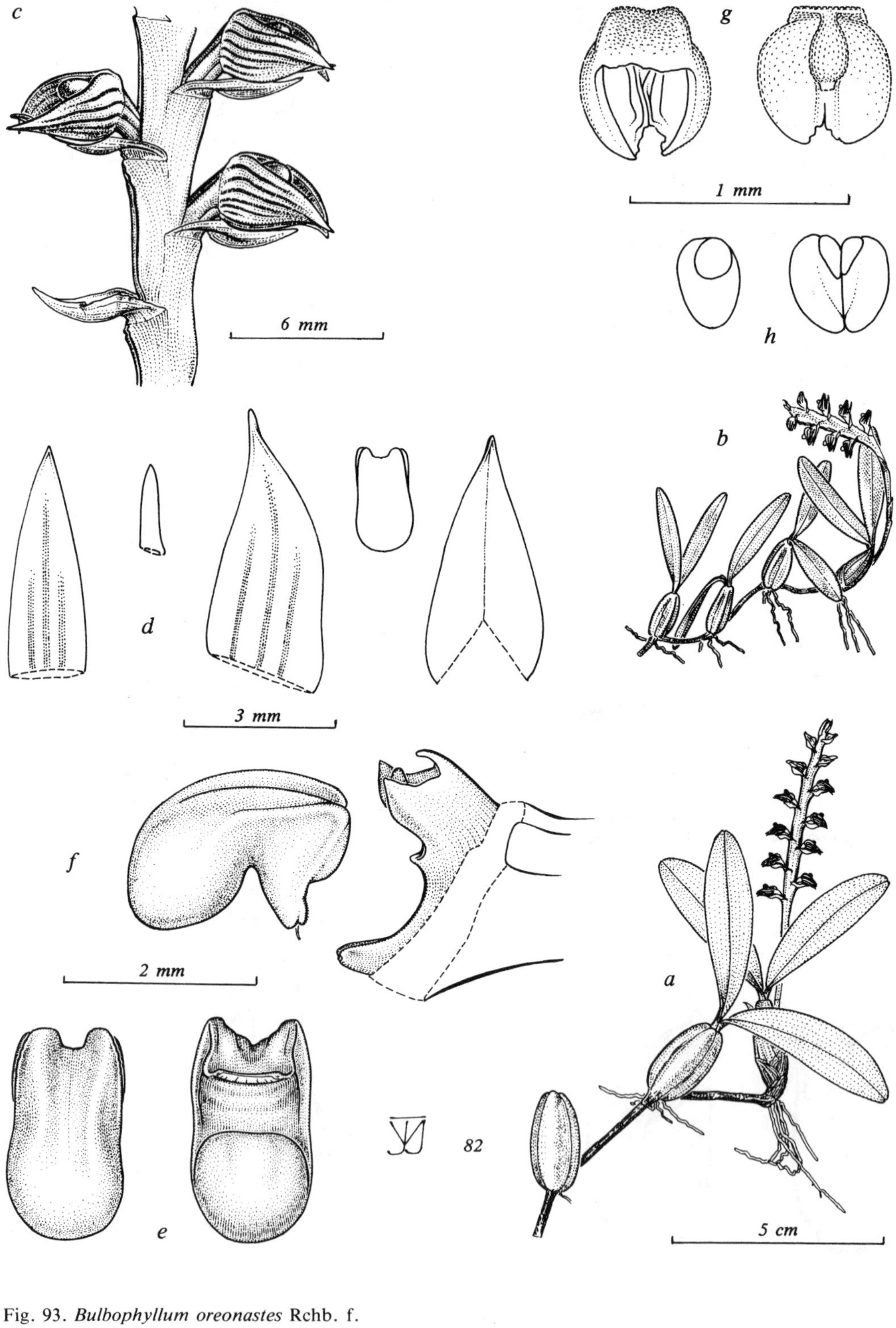

Fig. 93. *Bulbophyllum oreonastes* Rchb. f.

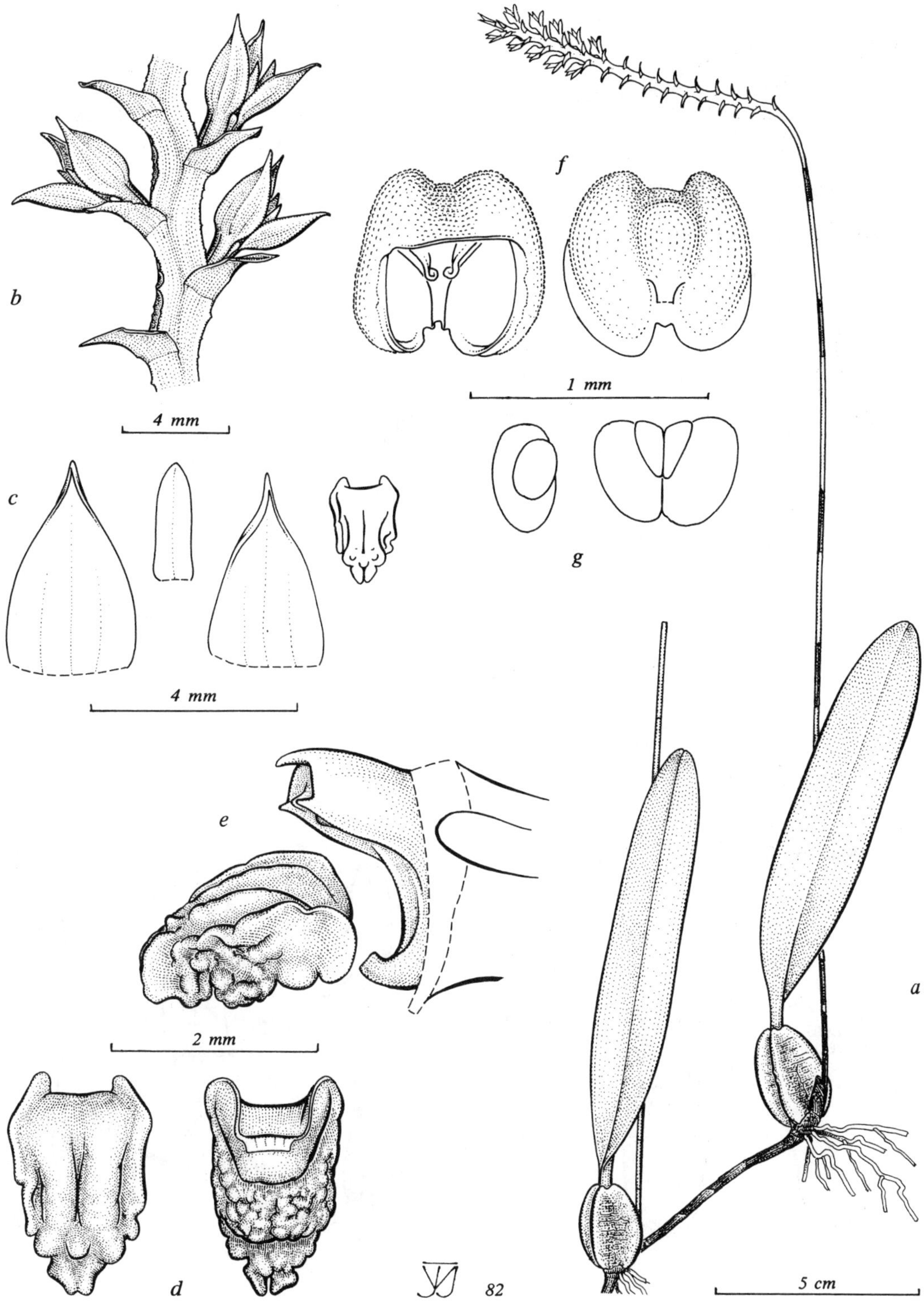

Fig. 94 *Bulbophyllum encephalodes* Summerh.

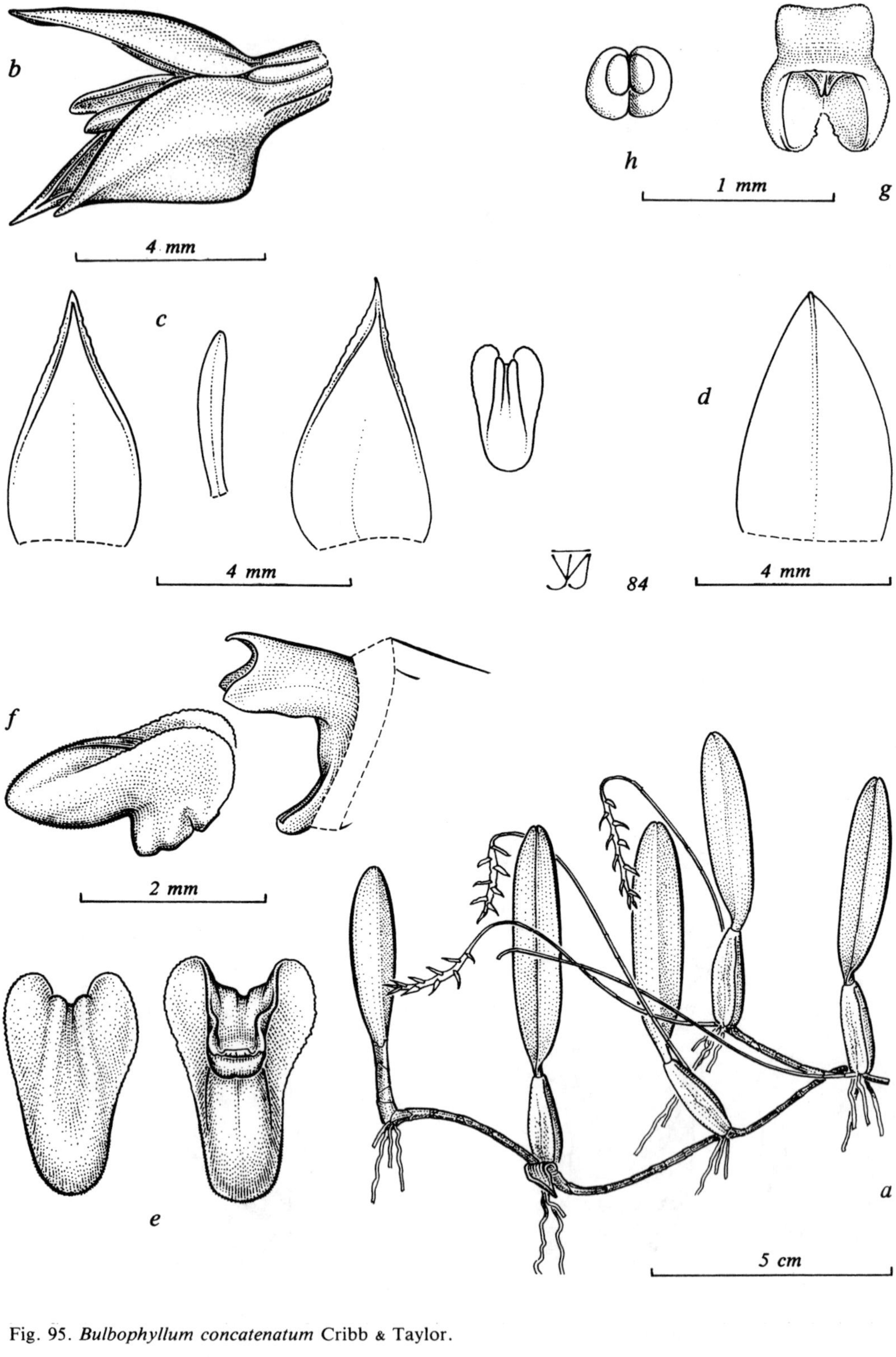

Fig. 95. *Bulbophyllum concatenatum* Cribb & Taylor.

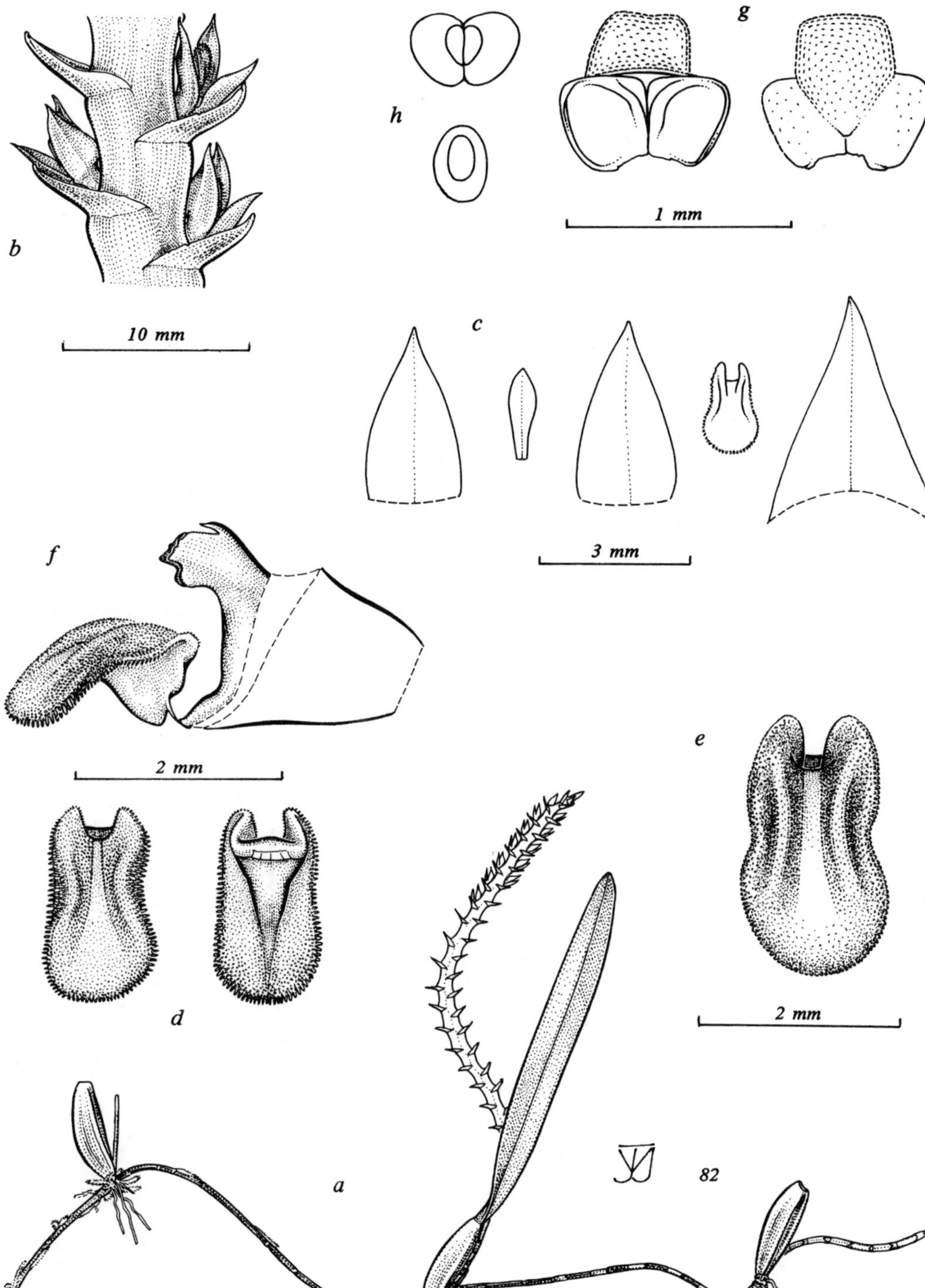

Fig. 96. *Bulbophyllum unifoliatum* De Wild. ssp. *unifoliatum*.

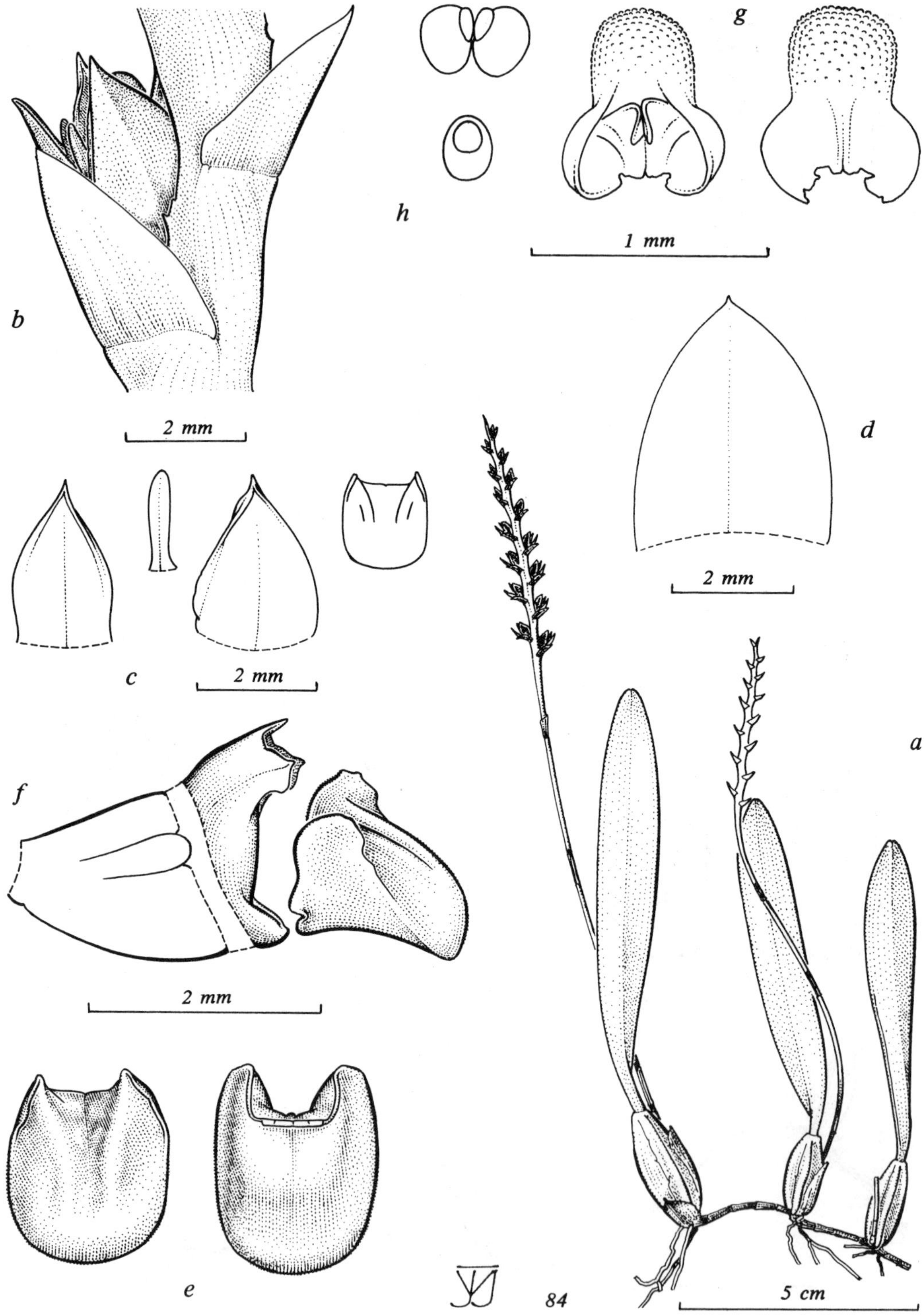

Fig. 97. *Bulbophyllum unifoliatum* De Wild. ssp. *infracarinatum* (Williamson) J.J. Vermeulen.

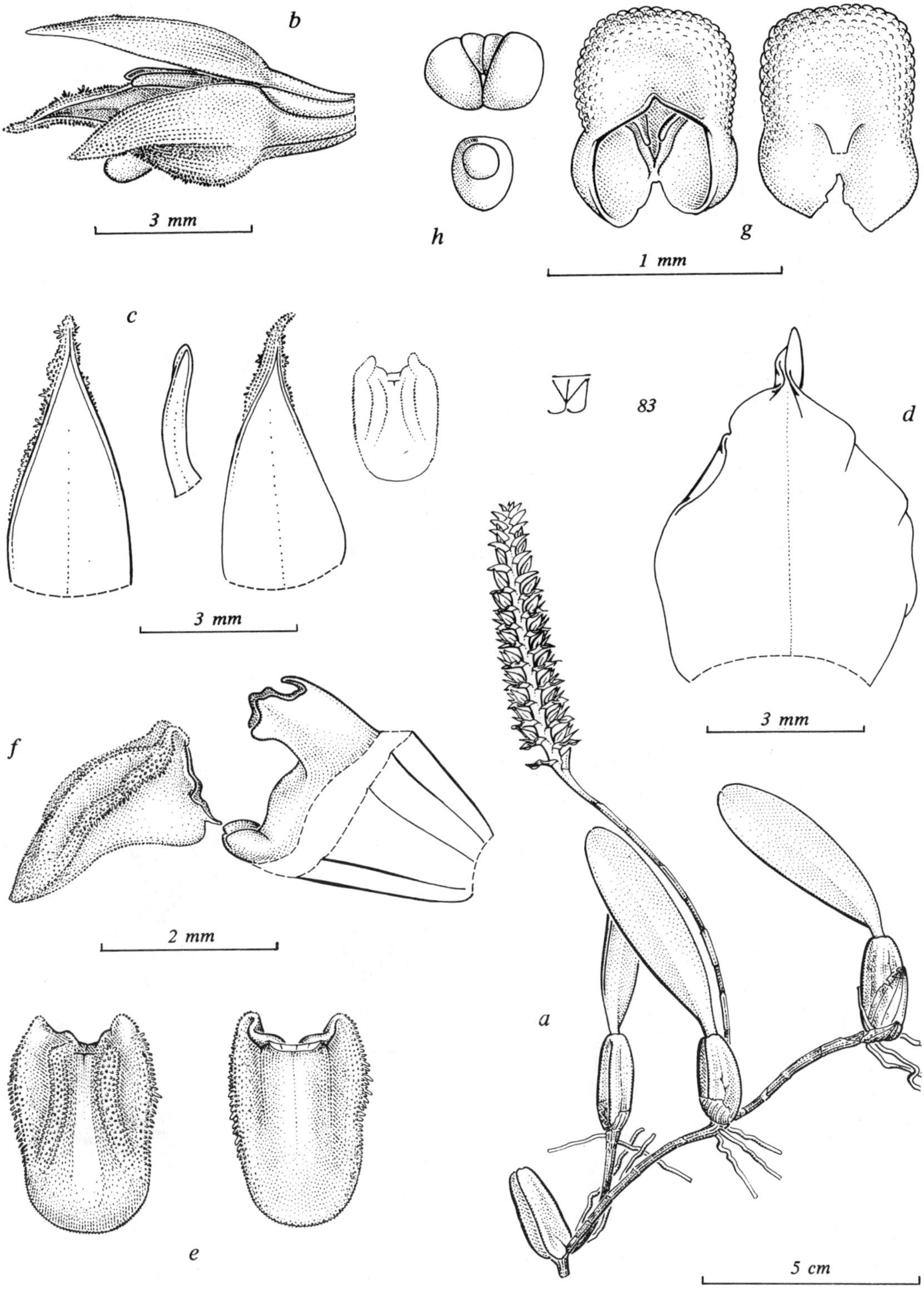

Fig. 98. *Bulbophyllum unifoliatum* De Wild. ssp. *flectens* (Cribb & Taylor) J.J. Vermeulen.

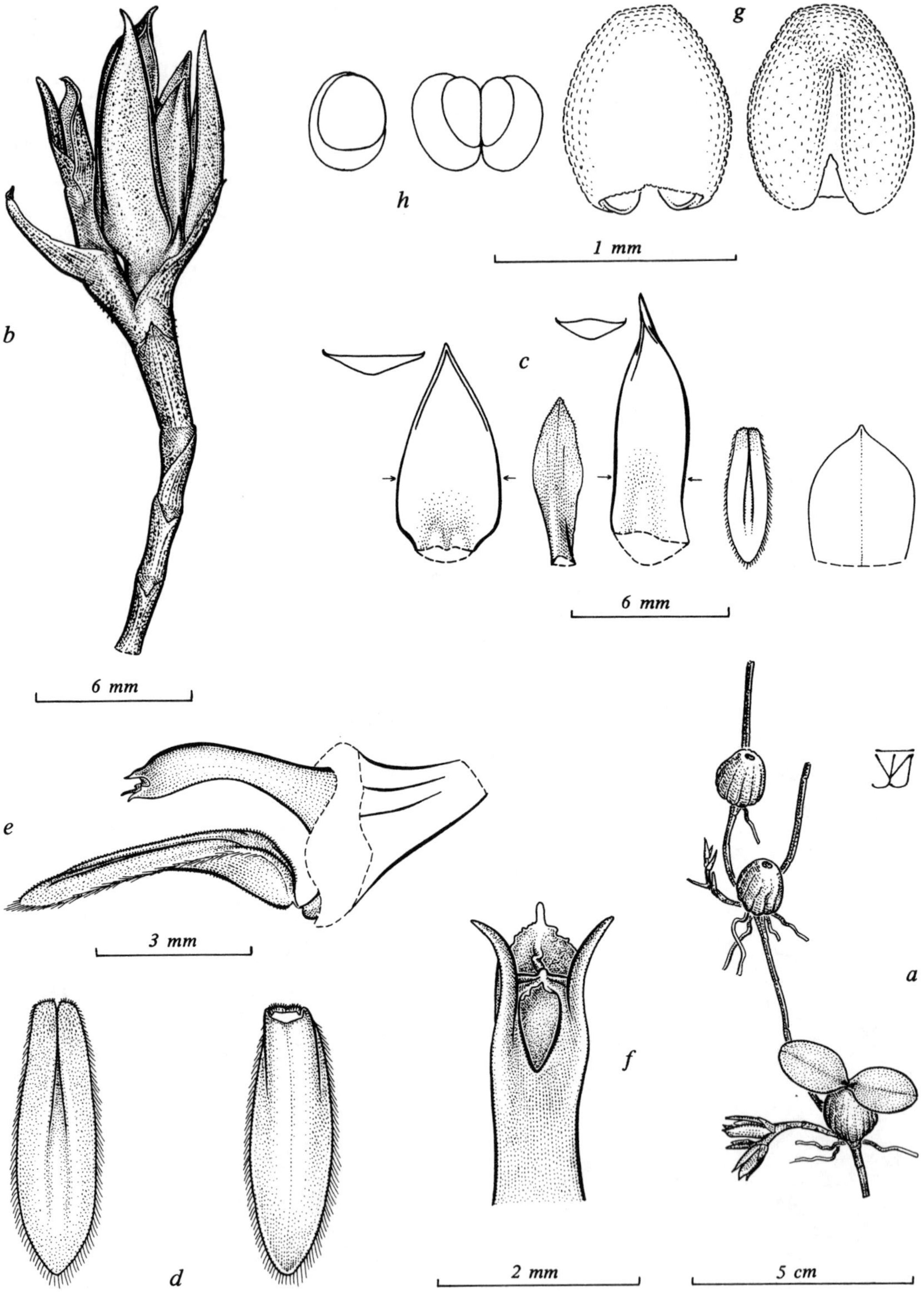

Fig. 99. *Bulbophyllum gilgianum* Krzl.

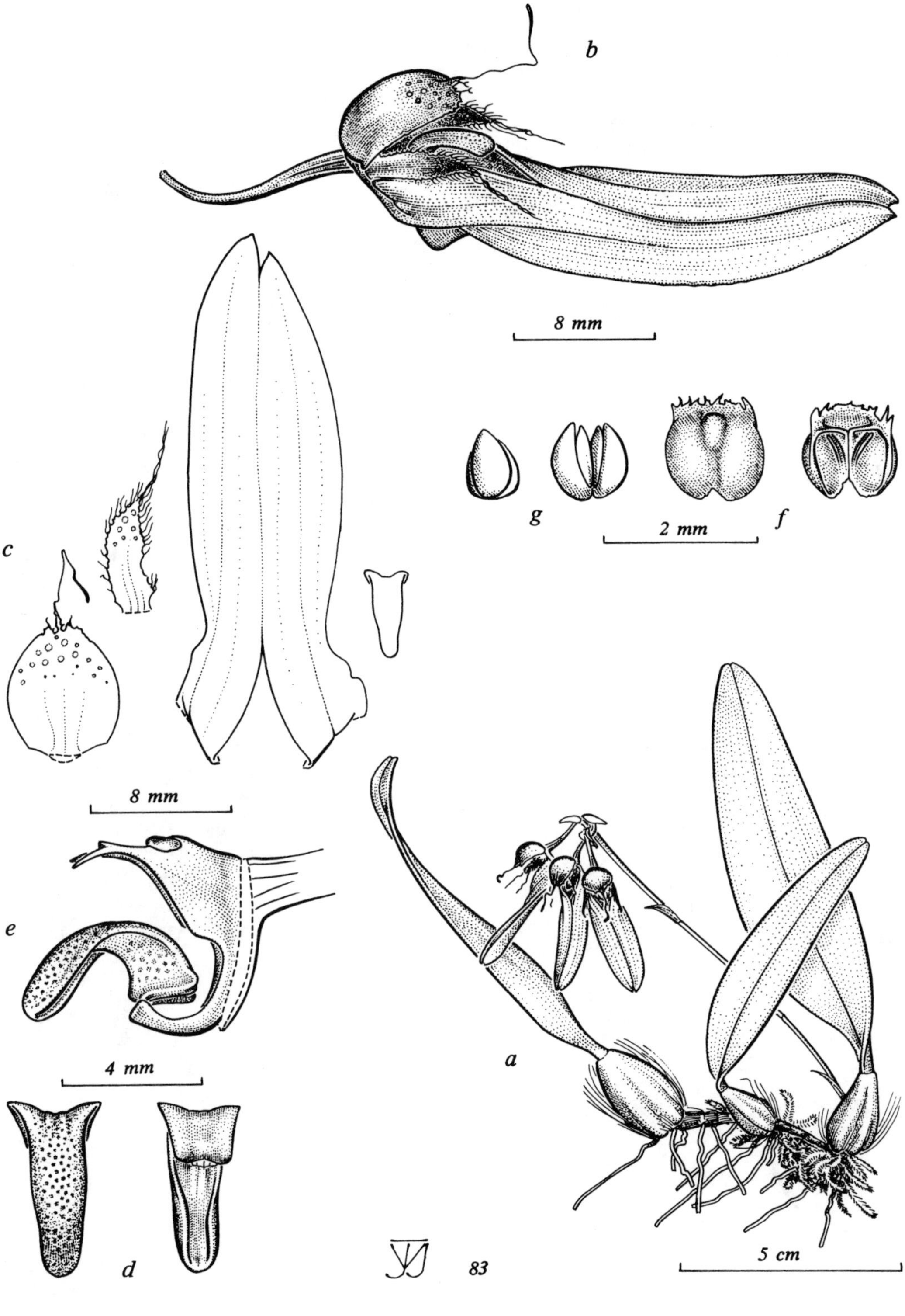

Fig. 100. *Bulbophyllum longiflorum* Thouars

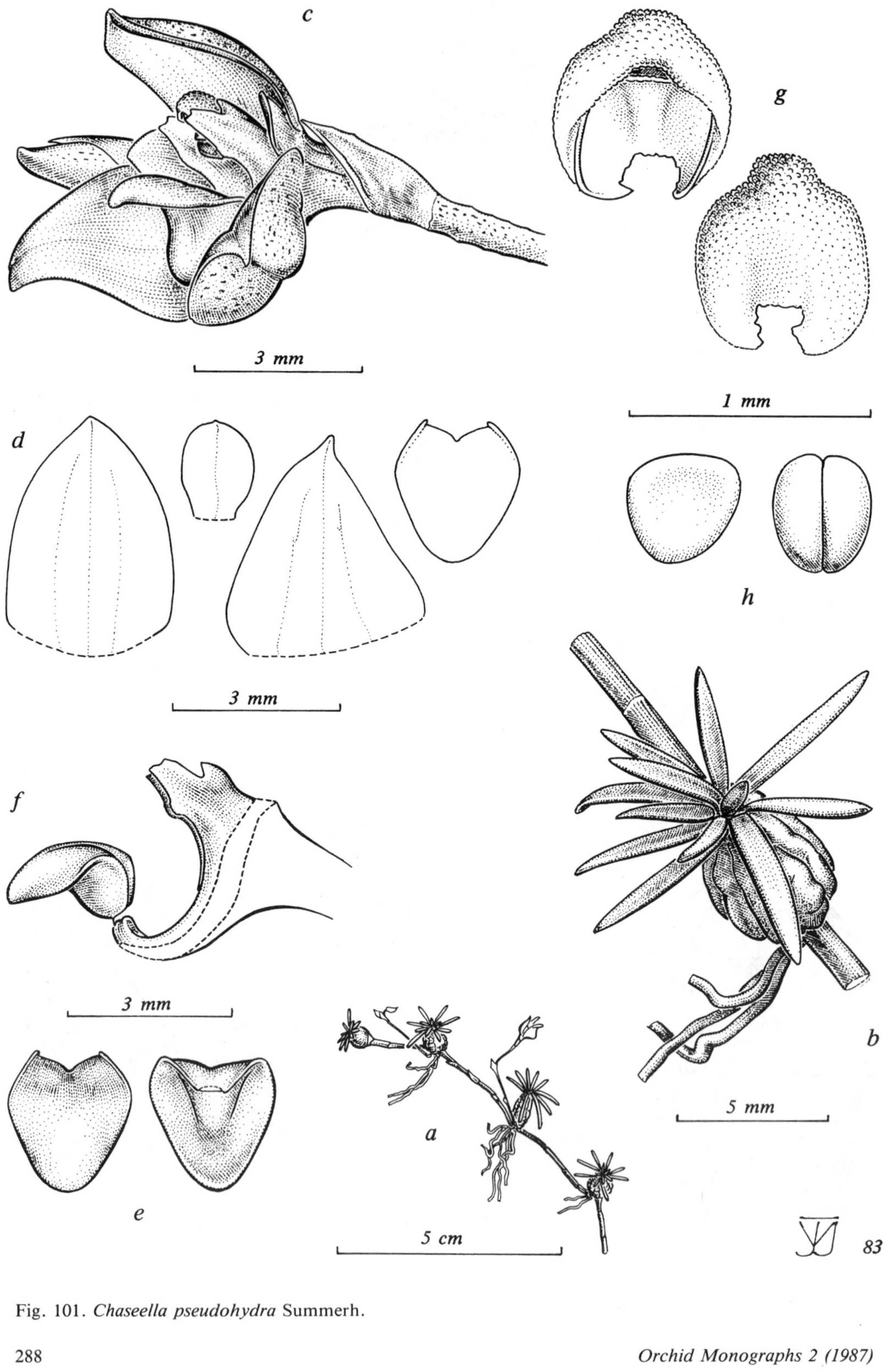

Fig. 101. *Chaseella pseudohydra* Summerh.

INDEX

The hyphen separates the names of taxa from page numbers and numbers of figures and plates. Printed in bold face are the names of accepted taxa and the page number on which the taxon is described. Fig. refers to a line-drawing, Plate to a coloured plate.

PLATES

LEGENDS OF THE PLATES

Plate 1

a. *B. schinzianum* Krzl. var. *schinzianum*
Jongejan, cult. 915, from Ivory Coast. Photograph Jongejan M-6-26
b. *B. schinzianum* Krzl. var. *phaeopogon* (Schltr) J.J. Vermeulen
Jongejan, cult. 877, from Zaire. Photograph Jongejan M-6-22
c. *B. subligaculiferum* J.J. Vermeulen
Wageningen, cult. Hort (de Wilde et al. [1983] 183) 83PTGA256, from Gabon. Photograph H.C.D. de Wit
d. *B. saltatorium* Lindl. var. *saltatorium*
Jongejan, cult. 878, from Africa, country not known. Photograph Jongejan M-6-10

Plate 2

a. *B. saltatorium* Lindl. var. *albociliatum* (Finet) J.J. Vermeulen
Leiden, cult. Hort. (v.d. Laan 1027) 23602, from Gabon. Photograph B.N. Kieft (collection Rijksherbarium, Leiden)
b. *B. saltatorium* Lindl. var. *albociliatum* (Finet) J.J. Vermeulen
Jongejan, cult. s.n., from Africa, country not known. Photograph Jongejan C-I-36 (collection Rijksherbarium, Leiden)
c. *B. saltatorium* Lindl. var. *calamarium* (Finet) J.J. Vermeulen
Jongejan, cult. s.n., from Liberia. Photograph Jongejan C-II-7 (collection Rijksherbarium, Leiden)
d. *B. barbigerum* Lindl.
Jongejan, cult. 48, from Africa, country not known. Photograph Jongejan M-6-16

Plate 3

a. *B. vulcanicum* Krzl.
Jongejan, cult. 60, from Rwanda. Photograph Jongejan O-4-11
b. *B. cochleatum* Lindl. var. *cochleatum*
Jongejan, cult. 23, from Africa, country not known. Photograph Jongejan J-4-20
c. *B. cochleatum* Lindl. var. *bequaertii* (De Wild.) J.J. Vermeulen
Jongejan, cult. s.n., from Africa, country not known. Photograph Jongejan C-1-19 (collection Rijksherbarium, Leiden)
d. *B. pipio* Rchb. f.
Jongejan, cult. 27, from Liberia. Photograph Jongejan A-II-29 (collection Rijksherbarium, Leiden)

Plate 4

a. *B. pumilum* (Sw.) Lindl.
Jongejan, cult. 26, from Africa, country not known. Photograph Jongejan J-4-8
b. *B. pumilum* (Sw.) Lindl.
Jongejan, cult. 874, from Zaire. Photograph Jongejan M-4-24
c. *B. pumilum* (Sw.) Lindl.
Jongejan, cult. s.n., from Ivory Coast. Photograph Jongejan B-VI-26 (collection Rijksherbarium, Leiden)
d. *B. pumilum* (Sw.) Lindl.
Jongejan, cult. s.n., from Africa, country not known. Photograph Jongejan C-IV-26 (collection Rijksherbarium, Leiden)

Plate 5

a. *B. pumilum* (Sw.) Lindl.
Jongejan, cult. 51, from Africa, country not known. Photograph Jongejan M-4-32
b. *B. josephii* (Kuntze) Summerh. var. *josephii*
From Malawi. Photograph E.A.S. la Croix
c. *B. expallidum* J.J. Vermeulen
From Malawi. Photograph E.A.S. la Croix
d. *B. schimperanum* Krzl.
Jongejan, cult. 54, from Zaire. Photograph Jongejan I-2-25

Plate 6

a. *B. oxychilum* Schltr.
Jongejan, cult. s.n., from Ivory Coast. Photograph Jongejan A-II-15 (collection Rijksherbarium, Leiden)
b. *B. comatum* Lindl. var. *inflatum* (Rolfe) J.J. Vermeulen
Jongejan, cult. 55, from Liberia. Photograph Jongejan J-4-13
c. *B. imbricatum* Lindl.
Jongejan, cult. 914, from Zaire. Photograph Jongejan M-4-21
d. *B. imbricatum* Lindl.
Kew, cult. Hort. Photograph D. Menzies (collection Rijksherbarium, Leiden)

Plate 7

a. *B. maximum* (Lindl.) Rchb. f.
Jongejan, cult. 19, from Kenya. Photograph Jongejan s.n. (collection Rijksherbarium, Leiden)
b. *B. magnibracteatum* Summerh.
Leiden, cult. Hort. (v.d. Laan 953) 25065, from Gabon. Photograph B.N. Kieft (collection Rijksherbarium, Leiden)
c. *B. sandersonii* (Hook. f.) Rchb. f. ssp. *sandersonii*
Jongejan, cult. 922, from Africa, country not known. Photograph Jongejan O-6-28
d. *B. sandersonii* (Hook. f.) Rchb. f. ssp. *stenopetalum* (Krzl.) J.J. Vermeulen
Jongejan, cult. 897, from Africa, country not known. Photograph Jongejan M-4-17

Plate 8

a. *B. lizae* J.J. Vermeulen
Wageningen, cult. Hort. (de Wilde et al. [1980] 129) 80PTST061, from Sao Tomé. Photograph Jongejan J-4-28
b. *B. purpureorhachis* (De Wild.) Schltr
Jongejan, cult. 875, from Zaire. Photograph Jongejan M-4-3
c. *B. purpureorhachis* (De Wild.) Schltr
Jongejan, cult. 875, from Zaire. Photograph Jongejan M-4-5
d. *B. resupinatum* Ridl. var. *resupinatum*
Jongejan, cult. 99, from Ivory Coast. Photograph Jongejan J-6-17

Plate 9

a. *B. falcatum* (Lindl.) Rchb. f. var. *falcatum*
Jongejan, cult. 76, from Zaire. Photograph Jongejan I-2-19

b. *B. falcatum* (Lindl.) Rchb. f. var. *bufo* (Lindl.) J.J. Vermeulen
Paris, cult. Hort. s.n., from Africa, country not known. Photograph B.N. Kieft (collection Rijksherbarium, Leiden)

c. *B. falcatum* (Lindl.) Rchb. f. var. *velutinum* (Lindl.) J.J. Vermeulen
Jongejan, cult. 30, from Africa, country not known. Photograph Jongejan A-I-11 (collection Rijksherbarium, Leiden)

d. *B. falcatum* (Lindl.) Rchb. f. var. *velutinum* (Lindl.) J.J. Vermeulen
Jongejan, cult. 930, from Africa, country not known. Photograph Jongejan A-I-31

Plate 10

a. *B. calyptratum* Krzl. *calyptratum*
Leiden, cult. Hort. (de Wilde et al. [1983] 938) 23633, from Gabon. Photograph B.N. Kieft (collection Rijksherbarium, Leiden)

b. *B. calyptratum* Krzl. var. *graminifolium* (Summerh.) J.J. Vermeulen
Jongejan, cult. s.n., from Liberia. Photograph Jongejan C-I-6 (collection Rijksherbarium, Leiden)

c. *B. porphyrostachys* Summerh.
Jongejan, cult. 15, from Africa, country not known. Photograph Jongejan s.n. (collection Rijksherbarium, Leiden)

d. *B. porphyrostachys* Summerh.
Jongejan, cult. 15, from Africa, country not known. Photograph Jongejan s.n. (collection Rijksherbarium, Leiden)

Plate 11

a. *B. lupulinum* Lindl.
Jongejan, cult. 448, from Cameroun. Photograph Jongejan O-4-30

b. *B. acutebracteatum* De Wild. var. *acutebracteatum*
Jongejan, cult. 6, from Liberia. Photograph Jongejan A-II-33 (collection Rijksherbarium, Leiden)

c. *B. fuscum* Lindl. var. *fuscum*
Jongejan, cult. 405, from Africa, country not known. Photograph Jongejan J-4-25 (collection Rijksherbarium, Leiden)

d. *B. longiflorum* Thouars
From Malawi. Photograph E.A.S. la Croix

a. *B. schinzianum* Krzl.
var. *schinzianum*

b. *B. schinzianum* Krzl.
var. *phaeopogon* (Schltr) J.J. Vermeulen

c. *B. subligaculiferum* J.J. Vermeulen

d. *B. saltatorium* Lindl.
var. *saltatorium*

Plate 2

a. *B. saltatorium* Lindl.
var. *albociliatum* (Finet) J.J. Vermeulen

b. *B. saltatorium* Lindl.
var. *albociliatum* (Finet) J.J. Vermeulen

c. *B. saltatorium* Lindl.
var. *calamarium* (Finet) J.J. Vermeulen

d. *B. barbigerum* Lindl.

a. *B. vulcanicum* Krzl.

b. *B. cochleatum* Lindl.
var. *cochleatum*

c. *B. cochleatum* Lindl.
var. *bequaertii* (De Wild.) J.J. Vermeulen.

d. *B. pipio* Rchb. f.

Plate 4

a. *B. pumilum* (Sw.) Lindl.

b. *B. pumilum* (Sw.) Lindl.

c. *B. pumilum* (Sw.) Lindl.

d. *B. pumilum* (Sw.) Lindl.

a. *B. pumilum* (Sw.) Lindl.

b. *B. josephii* (Kuntze) Summerh. var. *josephii*

c. *B. expallidum* J.J. Vermeulen

d. *B. schimperanum* Krzl.

Plate 6

a. *B. oxychilum* Schltr.

b. *B. comatum* Lindl.
var. *inflatum* (Rolfe) J.J. Vermeulen

c. *B. imbricatum* Lindl.

d. *B. imbricatum* Lindl.

a. *B. maximum* (Lindl.) Rchb. f.

b. *B. magnibracteatum* Summerh.

c. *B. sandersonii* (Hook. f.) Rchb. f.
ssp. *sandersonii*

d. *B. sandersonii* (Hook. f.) Rchb. f.
ssp. *stenopetalum* (Krzl.) J.J. Vermeulen

Plate 8

a. *B. lizae* J.J. Vermeulen

b. *B. purpureorhachis* (De Wild.) Schltr

c. *B. purpureorhachis* (De Wild.) Schltr

d. *B. resupinatum* Ridl.
var. *resupinatum*

a. *B. falcatum* (Lindl.) Rchb. f.
var. *falcatum*

b. *B. falcatum* (Lindl.) Rchb. f.
var. *bufo* (Lindl.) J.J. Vermeulen

c. *B. falcatum* (Lindl.) Rchb. f.
var. *velutinum* (Lindl.) J.J. Vermeulen

d. 77 *B. falcatum* (Lindl.) Rchb. f.
var. *velutinum* (Lindl.) J.J. Vermeulen

Plate 10

a. *B. calyptratum* Krzl.
var. *calyptratum*

b. *B. calyptratum* Krzl.
var. *graminifolium* (Summerh.)
J.J. Vermeulen

c. *B. porphyrostachys* Summerh.

d. *B. porphyrostachys* Summerh.